physica status solidi

AKADEMIE-VERLAG · BERLIN

VOLUME 19 · NUMBER 2 · 1967

Classification Scheme

1. Structure of Solids
   1.1 Alloys. Metallurgy
   1.2 Solid-State Phase Transformations
   1.3 Surfaces
   1.4 Films

2. Non-Crystalline State

3. Crystallography
   3.1 Crystal Growth
   3.2 Interatomic Forces

4. Microstructure of Solids

5. Perfectly Periodic Structures

6. Lattice Mechanics. Phonons
   6.1 Mössbauer Investigations

7. Acoustic Properties of Solids

8. Thermal Properties of Solids

9. Diffusion in Solids

10. Defect Properties of Solids (Irradiation Defects see 11)
    10.1 Defect Properties of Metals
    10.2 Photochemical Reactions. Colour Centres

11. Irradiation Effects in Solids

12. Mechanical Properties of Solids (Plastic Deformations see 10)
    12.1 Mechanical Properties of Metals (Plastic Deformations see 10.1)

13. Electron States in Solids
    13.1 Band Structure. Fermi Surfaces
    13.2 Excitons
    13.3 Surface States
    13.4 Impurity and Defect States

14. Electrical Properties of Solids. Transport Phenomena
    14.1 Metals. Conductors
    14.2 Superconductivity. Superconducting Materials and Devices
    14.3 Semiconductors
        14.3.1 Semiconducting Films
        14.3.2 Semiconducting Devices. Junctions (Contact Problems see 14.4.1)
    14.4 Dielectrics
        14.4.1 High Field Phenomena, Space Charge Effects, Inhomogeneities, Injected Carriers (Electroluminescence see 20.3; Junctions see 14.3.2)
        14.4.2 Ferroelectric Materials and Phenomena

15. Thermoelectric and Thermomagnetic Properties of Solids

16. Photoconductivity. Photovoltaic Effects

17. Emission of Electrons and Ions from Solids

18. Magnetic Properties of Solids
    18.1 Paramagnetic Properties
    18.2 Ferromagnetic Properties
    18.3 Ferrimagnetic Properties. Ferrites
    18.4 Antiferromagnetic Properties

(Continued on cover three)

physica status solidi

Volume 19 · Number 2 · Pages 453 to 874, K67 to K114, and A31 to A62

February 1, 1967

AKADEMIE-VERLAG · BERLIN

Subscriptions and orders for single copies should be addressed to

AKADEMIE-VERLAG GmbH, 108 Berlin, Leipziger Straße 3 – 4 or to Buchhandlung KUNST UND WISSEN, Erich Bieber, 7 Stuttgart 1, Wilhelmstr. 4 – 6 or to Deutsche Buch-Export und -Import GmbH, 701 Leipzig, Postschließfach 160

Editorial Note:

"physica status solidi" undertakes that an original paper *accepted for publication* before the 8th of any month will be published within 50 days of this date unless the author requests a postponement. In special cases there may be some delay between *receipt* and *acceptance* of a paper due to the review and, if necessary, revision of the paper.

Schriftleiter und verantwortlich für den Inhalt: Professor Dr. Dr. h. c. P. Görlich, 102 Berlin, Neue Schönhauser Str. 20 bzw. 69 Jena, Humboldtstr. 26. Redaktionskollegium: Dr. S. Oberländer, Dr. E. Gutsche, Dr. W. Borchardt. Anschrift der Schriftleitung: 102 Berlin, Neue Schönhauser Str. 20. Fernruf: 426788. Verlag: Akademie-Verlag GmbH, 108 Berlin, Leipziger Str. 3–4, Fernruf: 220441, Telex-Nr. 011773, Postscheckkonto: Berlin 35021. – Die Zeitschrift „physica status solidi" erscheint jeweils am 1. des Monats. Bezugspreis eines Bandes MDN 72,– (Sonderpreis für die DDR MDN 60,–). Bestellnummer dieses Bandes 1068/19. Jeder Band enthält zwei Hefte. Gesamtherstellung: VEB Druckerei „Thomas Müntzer" Bad Langensalza. – Veröffentlicht unter der Lizenznummer 1310 des Presseamtes beim Vorsitzenden des Ministerrates der Deutschen Demokratischen Republik.

Contents

Review Article

Y. P. VARSHNI Band-to-Band Radiative Recombination in Groups IV, VI, and III-V Semiconductors (I) 459

Original Papers

W. BARTH and B. FRITZ
Local Mode Absorption by U-Centres in Alkali-Halide Mixed Crystals 515

V. B. PARIISKII Effect of Stress State on the Shape of the Dislocation Rosette near an Indentation in Some Alkali Halide Crystals 525

S. N. KOMNIK, V. Z. BENGUS, and E. D. LYAK
On the Kinetics of Plastic Deformation of Some Alkali Halide Crystals 533

J. MAN, M. HOLZMANN, and B. VLACH
Microstrain Region and Transition to Macrostrain in 99.9% Polycrystalline Copper 543

G. DÖHLER On the Field Emission of Minority Carriers in Photoconductors . . . 555

O. E. FACEY and P. W. M. JACOBS
Photochemical Properties of SH^- Ions in KCl and KBr Crystals . . 565

A. R. BEATTIE and G. SMITH
Recombination in Semiconductors by a Light Hole Auger Transition 577

R. D. LEVINE and A. T. AMOS
On the Theory of Surface and Impurity States 587

W. FRANZ und P. MIOSGA
Verteilungsfunktion und Driftgeschwindigkeit warmer Elektronen in Germanium 597

J. TREUSCH Green's Function Method for Energy Band Calculations Including Spin-Orbit Coupling 603

E. I. GAVRILITSA and S. I. RADAUTSAN
Influence of Ordering on the Properties of Solid Solutions of the System $(HgSe)_{3x}$-$(In_2Se_3)_{1-x}$ 609

B. F. ROTHENSTEIN, A. POLICEC, C. ANGHEL, and M. LUPULESCU
Contributions to the Mechanism of the Procopiu and Inverse Wiedemann Effects 613

J. KOŁODZIEJCZAK and E. KIERZEK-PECOLD
Free Carrier Electro-Magneto-Optical Phenomena in Semiconductors 623

A. C. McLAREN, J. A. RETCHFORD, D. T. GRIGGS, and J. M. CHRISTIE
Transmission Electron Microscope Study of Brazil Twins and Dislocations Experimentally Produced in Natural Quartz 631

R. C. RAU and G. A. CHASE
Clustering of Defects in Neutron-Irradiated Beryllium Oxide . . . 645

Page

E. KÖSTER — Verformungs- und Temperaturabhängigkeit der Magnetisierungsprozesse in Nickeleinkristallen (II) 655

R. ENDERLEIN and R. KEIPER
On the Theory of Electro-Absorption and Electro-Reflectance in Semiconductors 673

J. SPYRIDELIS, P. DELAVIGNETTE, and S. AMELINCKX
On the Superstructures of Ta_2O_5 and Nb_2O_5 683

B. I. BOLTAKS and T. D. DZHAFAROV
The Effect of Applied Electric Field on Diffusion of Impurities in Gallium Arsenide 705

R. LABUSCH — Elastische Konstanten des Flußfadengitters in Supraleitern zweiter Art 715

A. AXMANN und W. GISSLER
Streuung langsamer Neutronen an polykristallinem Selen und Tellur 721

M. M. SHUKLA and B. DAYAL
The Lattice Vibrations of Nickel in Krebs's Model 729

K.-H. PFEFFER — Wechselwirkung zwischen Versetzungen und ebenen Blochwänden mit starrem Magnetisierungsverlauf 735

M. P. VERMA and B. DAYAL
Lattice Dynamics of MgO 751

R. FISCHER — Zur Theorie der Erzeugung der zweiten Harmonischen an der Oberfläche eines unmagnetischen, quadratischen, optisch einachsigen Kristalls 757

J. VILLAIN — Oscillations collectives de spins et ordre à courte distance. 767

P. T. LANDSBERG
The Condition for Negative Absorption in Semiconductors 777

R. STOCKMEYER and H. H. STILLER
Lattice Dynamics of Solid Adamantane 781

A. M. STONEHAM
Interference Effects in Spin Lattice Relaxation by Two-Phonon Processes 787

O. VÖHRINGER and E. MACHERAUCH
The Yield Point of α-Copper–Tin Alloys 793

B. JIMÉNEZ, J. MENDIOLA, and E. MAURER
Orientation Dependence of the Ferroelectricitiy of T.G.S. 805

G. A. JONES — Preferred Orientation and Domain Structure of Iron Films Deposited on (100) Copper Substrates 811

E. GUTSCHE and E. JAHNE
Spin-Orbit Splitting of the Valence Band of Wurtzite Type Crystals 823

M. HENZLER — Leitfähigkeit und Feldeffekt reiner Siliziumspaltflächen 833

R. H. WADE — Electron Diffraction from a Magnetic Phase Grating 847

Page

L. I. van Torne Electron Charge Distribution about Zinc Ions in Aluminum 855

G. Björkman and G. Grimvall
Polarization Vectors for Lattice Vibrations in Sodium 863

F. Brouers The Pair Distribution Function in the R.P.A. and in the Hubbard Approximation 867

Short Notes

E. Igras and T. Warmiński
Electron Mirror Observation of Lithium Precipitation at Scratch-Induced Dislocations in Silicon K67

K. Wetzig Investigations of Surfaces of the Binary System Ta–Fe by Means of an Emission Electron Microscope K71

L. Jahn Crystal Anisotropy and Saturation Magnetization of Hard Ferrites with a Composition near the M-Compound Prepared by Means of the Verneuil Method K75

M. Arndt and A. F. Rudolfh
Density Changes due to Dislocations in Silicon Single Crystals . . . K79

B. Sprušil and P. Vostrý
Resistivity Changes in Air-Quenched Gold Wires K83

R. E. Newnham and R. P. Santoro
Magnetic and Optical Properties of Dioptase K87

C. Cherki and R. Coelho
On Charge Storage in Anodic Tantal Oxide Layers K91

P. Guyot Camel-Humps Peierls Hills in Iron. K95

H. M. Otte, H. A. Lipsitt, P. F. Lindquist, M. L. Hammond, and R. H. Bragg
Further Comments on the Paper "On the Interpretation of Electron Diffraction Patterns from 'Amorphous' Boron" K99

R. Fischer Ein Beispiel für die Existenz eines „Brewster-Winkels“ in der nichtlinearen Optik anisotroper Kristalle K103

J. H. P van Weeren, R. Struikmans, and J. Blok
Test of the Read Model Concerning the Electron Mobility in Plastically Deformed n-Type Ge K107

M. A. Nizametdinova
The Reflection Spectrum of GaSe and GaS Single Crystals near the Fundamental Absorption Edge K111

Pre-printed Titles and Abstracts

of papers to be published in this or in the Soviet journal „Физика Твердого Тела“ (Fizika Tverdogo Tela) A31

Systematic List

Principal subject classification:	Corresponding papers begin on the following pages:
1	683, K99
1.2	609
1.3	K67, K71
1.4	811
6	721, 729, 751, 781, 863
9	705
10	525, 533, 631, 645, K79
10.1	543, 793, K95
10.2	515, 565
13	597, 787, 867, K87
13.1	577, 603, 823
13.3	587, 833
13.4	855
14.1	K83
14.2	715
14.3	K107
14.4.1	555, K91
14.4.2	805
18	767
18.2	613, 655, 735, 847
18.3	K75
20	623, 673, 757, K103
20.1	K111
20.2	777
20.3	459

The Author Index of Volume 19 Begins on Page 875

(It will be delivered together with Volume 20, Number 1.)

Review Article

phys. stat. sol. **19**, 459 (1967)

Subject classification: 20.3; 13.1; 13.2; 13.4; 20.1; 20.2; 22.1; 22.1.1; 22.1.2; 22.1.3; 22.2.1; 22.2.2; 22.2.3

Department of Physics, University of Ottawa

Band-to-Band Radiative Recombination in Groups IV, VI, and III–V Semiconductors (I) [1]

By

Y. P. Varshni

Contents

1. Introduction

2. Optical properties of semiconductors
- 2.1 Notation for optical constants
- 2.2 Absorption of photons
- 2.3 Empirical expressions for the absorption coefficient

3. Theory of radiative recombination rate
- 3.1 van Roosbroeck-Shockley theory
- 3.2 More refined calculations
- 3.3 Semiconductor masers
- 3.4 Quantum yield or quantum efficiency

4. Experimental methods of excitation and study of radiative recombination

5. General references

6. General remarks on results presented here

7. Results for elements
- 7.1 Diamond
- 7.2 Silicon
- 7.3 Germanium
- 7.4 Amorphous selenium
- 7.5 Hexagonal selenium
- 7.6 Tellurium

References (Part I)

List of Principal Symbols

Note: The subscripts "i" and "d" refer to indirect and direct transitions respectively. Sometimes, when it is clear from the context which transition is meant, these subscripts have been dropped.

B Probability for radiative recombination ($cm^3\ s^{-1}$)
c Velocity of light
e Electronic charge

[1]) Supported by the Defence Research Board under contract no. PW68-200027 with R. C. A. Victor Research Laboratories, Montreal.

E	Energy of photon
E_{ex}	Binding energy of exciton
E_g	Energy gap
E_{gd}	Energy gap, direct
E_{gi}	Energy gap, indirect
h	Planck's constant
$h\nu$	Energy of photon
k	Boltzmann constant
$\mathbf{k}$	Wave vector
m	Free electron mass
m_e	Effective mass in the conduction band
m_h	Effective mass in the valence band
m_r	$= m_e m_h/(m_h + m_e)$
n	Real part of the refractive index
n^*	Complex refractive index
n_i	Intrinsic carrier concentration
R	Radiative recombination rate ($cm^{-3} s^{-1}$)
T	Temperature
u	$= h\nu/kT$
α	Optical absorption coefficient
$\varkappa$	Absorption index
λ	Wavelength
ν	Frequency
σ	Recombination cross section (cm^2)
τ	Lifetime

1. Introduction

A number of mechanisms have been suggested to account for the recombination of electrons and holes in semiconductors.

General reviews of the various recombination processes have been given by Bemski [1], Smith [2], Hall [3], Bube [4], and Blakemore [5]. Accounts of selected topics in this field have been provided by Shockley [6], Vavilov [7, 8], and Guro [9]. These processes can be classified according to the answers to the following questions:

1. Is the recombination direct or through an imperfection?
2. How is the energy of the excited carriers released in the recombination?

The answer to the first question leads to three possibilities:

(i) The electron and hole combine through an electron dropping from a state in the conduction band to an empty state in the valence band in a single transition.

(ii) It is known that foreign atoms of a number of metals (e.g. Ni, Fe, Co, Cu, Au, etc.) form in semiconductors recombination and trapping centers for electrons and holes — such centers having levels in the forbidden band. The centers formed by these impurities are usually (depending on the temperature) both trapping centers (the center interacting only with one carrier type) and recombination centers (interaction takes place with both carrier types). Usually at a given temperature only one of the levels in such centers is effective. Quite often the deepest level is recombinative. Besides impurities, dislocations, and other mechanical imperfections of the crystalline lattice can act as recombination centers.

(iii) Surface recombination. This process is very similar to the second one, the traps being now at the surface of the semiconductor.

For the dissipation of the excess energy of these carriers, the following mechanisms have been considered:

a) Radiative recombination [10, 11]. Most of the energy is carried away by a photon.

b) Multiphonon recombination [12, 13, 6]. The energy given out in the recombination process is transformed into thermal energy of lattice vibrations.

c) Auger recombination, also called "impact" or three-body recombination [14 to 19]. Energy released is given to a free electron or hole to help satisfy the laws of energy and momentum conservation.

d) Plasma oscillations [20]. The energy can be given to the entire assembly of free carriers leading to the excitation of plasma oscillations.

Two or more of these processes may be involved simultaneously.

Thus in all we have 12 possibilities. Recombination radiation is usually classified in two broad categories: 1. "intrinsic recombination radiation", (i) above; 2. "extrinsic recombination radiation", (ii) and (iii) above.

In the period since 1952, the study of the recombination radiation spectra has given important information on the band structure, the defect and impurity energy levels, and the lattice vibrations of groups IV and III—V semiconductors, and has led to the development of semiconductor lasers.

The purpose of the present paper is to survey the field of intrinsic radiative recombination in semiconductors, in particular for diamond, Si, Ge, Se, Te, GaP, GaAs, GaSb, InP, InAs, and InSb. For all these semiconductors we have carried out calculations of the radiative recombination rate, intrinsic lifetime, and radiative recombination cross section and have compared them with experimental data where available. Such cases where the energy of the "extrinsic" recombination radiation lies very near the energy gap, have also been discussed.

A proper analysis of the observations on the recombination radiation of any semiconductor requires a knowledge, for some detailed and for some elementary, of its absorption coefficient, refractive index, phonon energies, band structure, and quantum efficiency.

For some of these properties we have attempted to summarize the experimental and theoretical results, while for others adequate references are given so that the interested reader may trace the sources for more detailed information.

2. Optical Properties of Semiconductors

The optical properties of semiconductors have been discussed in recent years by Fan [21], Moss [22], Smith [2], McLean [23], Parkinson [24], and Tauc [25]. In the following we give a brief resume of the essential features of the absorption in semiconductors, relevant to the later discussion of radiative recombination. More detailed discussion may be found in the above mentioned articles and books.

2.1 Notation for optical constants

It is customary to write the complex refractive index in the form

$$n^* = n\,(1 - i\,\varkappa) \tag{1}$$

or

$$n^* = n - i\,k\,. \tag{2}$$

$\varkappa$ is called the absorption index and k ($= n\,\varkappa$) the extinction coefficient.

For a plane electromagnetic wave of frequency ν and wavelength λ propagating through an absorbing medium, the absorption coefficient α is given by

$$\alpha = \frac{4\,\pi\,\varkappa}{\lambda} = \frac{4\,\pi\,n\,\varkappa}{\lambda_0}\,, \tag{3}$$

where λ_0 is the free space wavelength at frequency ν.

2.2 Absorption of photons

Photons having energy greater than E_g will be able to excite electrons from the valence band to the conduction band; the absorption coefficients for such photons usually are of the order of $10^4\ \text{cm}^{-1}$. On the other hand if the photon energy is much less than E_g, the residual absorption is due to the free carriers (electrons or holes) left in the semiconductor at the particular temperature at which the spectrum is measured. This absorption is only a few cm^{-1} in pure semiconductors, but continues to rise through to the microwave region of the semiconductor. Other factors which also contribute to the infrared absorption are the lattice bands and the impurity bands. Before discussing the types of transitions, we distinguish between the two types of semiconductors. The energy bands for the two are shown diagrammatically in Fig. 1a and b.

In the first type, the wave vector for the lowest energy state in the conduction band, $\boldsymbol{k}_{\text{min}}$, has the same value as for the highest energy state in the valence band, $\boldsymbol{k}_{\text{max}}$. Usually both are equal to zero. There is good evidence to show that InSb belongs to this class.

In the second type, shown in Fig. 1b, $\boldsymbol{k}_{\text{min}} \neq \boldsymbol{k}_{\text{max}}$.

For electrons in a periodic potential field, it can be shown that the matrix element involved in the transition of an electron from the valence band to the conduction band with the absorption of a photon vanishes except when the

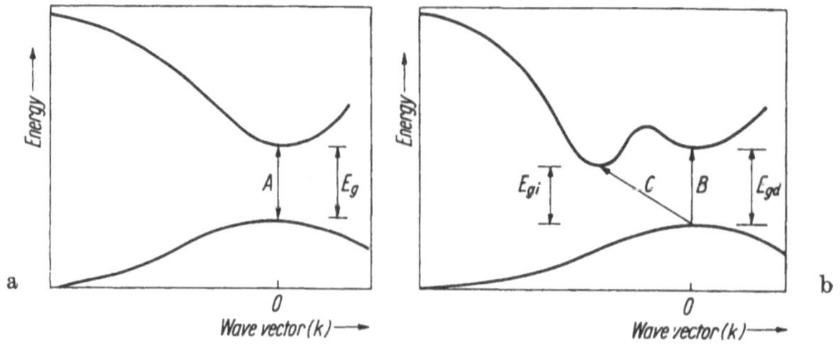

Fig. 1a and b. Optical transitions in semiconductors.
A and B represent direct transitions while C represents an indirect transition

condition for the wave number vectors

$$\boldsymbol{k}_i - \boldsymbol{k}_f = \pm \boldsymbol{k}_\lambda \tag{4}$$

is satisfied. Here $\boldsymbol{k}_i$ and $\boldsymbol{k}_f$ are the wave vectors of the electron in its initial and final state and $\boldsymbol{k}_\lambda$ is the wave vector of the radiation. This selection rule corresponds to the condition of conservation of momentum. The wave number vector $\boldsymbol{k}_\lambda$ of the radiation is small for the wavelengths of interest; therefore the selection rule is approximately

$$\boldsymbol{k}_i = \boldsymbol{k}_f . \tag{5}$$

Thus we see that in transitions from one band to another, only "vertical" transitions are allowed, and that there is a selection rule: $\Delta \boldsymbol{k} = 0$. Transitions A and B in Fig. 1a and b belong to this category.

Now we are in a position to discuss the types of transitions. There are two types of transitions:

1. Direct transitions. The electromagnetic radiation is absorbed by a semiconducting crystal due solely to the interaction between the radiation and the electrons in the crystal.
2. Indirect transitions. These involve the simultaneous interaction of the electrons with the electromagnetic radiation and lattice vibrations. The possibility of such transitions was first pointed out by Cheeseman [26].

We summarize the basic features of these two types of transitions.

2.2.1 Direct transitions

The absorption coefficient due to direct transitions in the case A (Fig. 1a) is determined primarily by the number of occupied states in the valence band and unoccupied states in the conduction band which are energetically within $h\nu$ of each other, and on the transition probability. Further, these transitions are divided in two categories: allowed and forbidden. Allowed transitions take place, for example, if the valence-band wave functions are derived from s-states of the individual atoms, and the conduction-band wave functions from p-states; if instead the conduction-band wave functions were derived from d-states the transition is forbidden.

Bardeen, Blatt, and Hall [27] have derived the following expressions for the absorption coefficients:

Allowed transitions:

$$\left.\begin{aligned} \alpha_d &= b\,(h\nu - E_g)^{1/2}; && h\nu > E_g, \\ \alpha_d &= 0; && h\nu \leqq E_g \end{aligned}\right\} \tag{6a}$$

in which

$$b = \frac{e^2\,(2\,m_r)^{3/2}}{n\,c\,m\,\hbar^2} f_{if}, \tag{6b}$$

where n is the real part of the refractive index, m_r is a reduced mass given by $m_e\,m_h/(m_h + m_e)$, f_{if} is the oscillator strength of the transition. It is of the order of unity. $\sum f_{if} = 1 + m/m_e + m/m_h$.

Here, following Smith [28] we have omitted an extra π occurring in the denominator of equation (6) of the paper by Bardeen et al. [27].

Also the m_h in the original equation of Bardeen et al. has been replaced by m_r in equation (6) to include the case when the conduction band is not flat [27].

Forbidden transitions:

$$\alpha_d = C\,(h\,\nu - E_g)^{3/2} \tag{7}$$

in which

$$C = \frac{2}{3}\,\frac{b}{h\nu}\,\frac{m_r}{m}\,.$$

When the lowest energy state in the conduction band dces not have the same value of k as the highest energy state in the valence band, direct transitions may still take place, but these are not the transitions corresponding to the lowest value of $h\,\nu$ for which transitions between the bands are possible. This is shown as transition B in Fig. 1 b.

The above expression for α_d was obtained on the assumption that the electron raised to the conduction band was free to move through-out the crystal uncorrelated with the motion of the hole. However, the experimental measurements of the absorption in germanium and silicon showed features which could be explained only if one takes into account the mutual attraction of the electron and hole produced in the absorption process. This attraction allows bound states of the electron and hole — excitons — to be formed in the crystal with energy less than that of the free pair of particles. In these states the relative motion of the electron and hole is localized and the pair of particles move through the crystal together.

Elliott [29] has shown that, for direct optical transitions between spherical non-degenerate conduction and valence bands with extrema at the centre of the Brillouin zone, one would expect a series of absorption lines at photon energies

$$h\,\nu = E_g - \frac{E_{ex}}{n_q^2}\,, \qquad n_q = 1, 2, 3 \ldots, \tag{8}$$

where E_g is the energy gap and E_{ex} the binding energy of an exciton of zero wave vector. These lines are followed by a region of continuous absorption beginning at $h\,\nu = E_g$ where the absorption coefficient

$$\alpha \propto \frac{e^x}{\sinh x} \qquad \text{in which} \qquad x = \pi\left[\frac{E_{ex}}{h\,\nu - E_g}\right]^{1/2}. \tag{9}$$

When $h\nu - E_g \gg \pi^2 E_{ex}$ this expression gives

$$\alpha \propto [(h\,\nu - E_g)^{1/2} + P]\,, \tag{10}$$

where P is a constant. The above expression is similar to equation (6). However, for smaller energies, the absorption deviates from the simple power law (6) due to the effects on the wave functions of Coulomb interaction between electron and hole and flattens off to a constant value close to the threshold E_g. A figure, showing the behaviour of α with E, with and without the electron–hole interaction, may be found in McLean [23].

2.2.2 *Indirect transitions*

We have seen that there is a selection rule for direct transitions: $\Delta\boldsymbol{k} = 0$, i.e. a transition indicated by a vertical line on a plot of E vs. k. Such vertical transitions are the only important transitions when both conduction- and valence-band extrema are located at the same value of k. However, in Ge and Si, and in many other semiconductors, the conduction band minimum occurs for different k-values from the valence band maximum. In such situations there is not only the possibility of direct transitions, but also of indirect transitions $\Delta \boldsymbol{k} \neq 0$, as indicated in Fig. 1b (transition C). Indirect transitions involve either the absorption of both a photon and a phonon or the absorption of a photon and the emission of a phonon. Thus the selection rule for conservation of momentum is satisfied by the phonon participation.

The theory of such transitions has been presented by Bardeen, Blatt, and Hall [27] who give a formula for the absorption coefficient. The transitions are considered as taking place through a virtual state which may be regarded as very short-lived. Energy is not conserved in the transition to the virtual state but momentum is conserved; energy is conserved only in the overall transition. The result obtained again depends on whether the transition at $\boldsymbol{k} = 0$ is allowed or not. If it is allowed they find

$$\alpha \propto (h\nu - E_g)^2 \tag{11}$$

and if the transition is not allowed

$$\alpha \propto (h\nu - E_g)^3 \,. \tag{12}$$

It may be shown [30, 2] that with absorption of a phonon the transition rate (and hence α) should be proportional to $(e^{\theta/T} - 1)^{-1}$, while for phonon emission it should be proportional to $(1 - e^{-\theta/T})^{-1}$, where $k\theta$ is the energy of the phonon involved. In equation (11), the energy of the phonon which is emitted or absorbed during the transition has not been taken into account. Macfarlane and Roberts [31] found, however, that to fit the details of the absorption curve it is essential to include the energy of the phonon. The final result [2] may thus be put as

$$\alpha_i = A\left\{\frac{(h\nu - E_g + k\theta)^2}{e^{\theta/T} - 1} + \frac{(h\nu - E_g - k\theta)^2}{1 - e^{-\theta/T}}\right\}; \qquad h\nu > E_g + k\theta\,, \tag{13a}$$

$$\alpha_i = A\,\frac{(h\nu - E_g + k\theta)^2}{e^{\theta/T} - 1}; \qquad E_g - k\theta < h\nu \leqq E_g + k\theta\,, \tag{13b}$$

$$\alpha_i = 0; \qquad h\nu \leqq E_g - k\theta\,. \tag{13c}$$

A and $k\theta$ are usually obtained from an analysis of the absorption coefficients, but θ = Debye temperature, is a reasonable approximation. For germanium, a slightly different set of equations was used by Macfarlane and Roberts [32]. In these the multiplying factor on the right-hand side is $A_2/(h\nu)^2$ instead of the A above. A strict analysis [23] shows that this multiplying factor should be of the form $A_1/(h\nu)$. Here A_2 and A_1 are constants. Thus

$$\alpha = \frac{A_1}{(h\nu)}\left\{\frac{(h\nu - E_g + k\theta)^2}{e^{\theta/T} - 1} + \frac{(h\nu - E_g - k\theta)^2}{1 - e^{-\theta/T}}\right\}; \qquad h\nu > E_g + k\theta\,, \tag{14a}$$

$$\alpha = \frac{A_1}{(h\nu)}\,\frac{(h\nu - E_g + k\theta)^2}{e^{\theta/T} - 1}; \qquad E_g - k\theta \leqq h\nu \leqq E_g + k\theta\,, \tag{14b}$$

$$\alpha = 0; \qquad h\nu \leqq E_g - k\theta\,. \tag{14c}$$

The above expressions have been obtained when exciton effects are neglected. The modification of these expressions by the inclusion of these effects has been discussed by McLean [23]. We quote here the modified expressions corresponding to equation (13b); the modification in the previous equation (13a) is analogous:

$$\alpha = \frac{\text{const}}{(e^{\theta/T} - 1)} \cdot \frac{E_{ex}^2}{h\nu} \int\limits_0^{\Delta} \frac{\sqrt{\Delta - z}}{1 - \exp\left(-\frac{2}{z^{1/2}}\right)} dz, \tag{15}$$

where

$$\Delta = \frac{1}{\pi^2 E_{ex}} (h\nu - E_g + k\theta)$$

and E_{ex} is the binding energy of an exciton.

For $h\nu \approx E_g$, equation (15) yields

$$\alpha = \frac{\text{const}}{(e^{\theta/T} - 1)} \cdot \frac{2 E_{ex}^{1/2}}{3\pi^3 h\nu} (h\nu - E_g + k\theta)^{3/2} \tag{16}$$

so that the component rises from its threshold as the three-halves power of the energy above this threshold. For $h\nu \gg E_g$, equations (15) is not valid physically, but it has some analytical significance as this corresponds to the region when the mutual interaction of the electron and hole becomes negligible. For this region, equation (15) gives

$$\alpha = \frac{\text{const}}{(e^{\theta/T} - 1)} \cdot \frac{1}{16\pi^3 h\nu} (h\nu - E_g + k\theta)^2, \tag{17}$$

which is similar to equation (14b).

For many approximate purposes, equations (13) may be used instead of the more accurate equation (15).

Hartman [33] has obtained formulas for the indirect absorption coefficient as a function of photon energy, for both degenerate and non-degenerate semiconductors, taking into account the dependence of the energy denominator on the energy of the intermediate states, and has carried out calculations for germanium.

2.3 *Empirical expressions for the absorption coefficient*

There have also been attempts to express the variation of α with $h\nu$ and T by empirical or semi-empirical expressions.

Urbach [34, 35] in 1953 found that for AgBr crystals, in the range of absorption coefficient 10^{-2} cm^{-1} to 10 cm^{-1}, $\log \alpha$ is linear with $h\nu$, also that the slope is given by $1/kT$ at corresponding temperatures. In more general terms we may write

$$\alpha = S \exp\left(\sigma \frac{h\nu}{kT}\right), \tag{18}$$

where S and σ are constants. This is now known as the "Urbach rule". Later work has shown that often σ is less than 1.

Fochs [36] found support for such a relation from measurements on the diffuse reflection spectra of several semiconductors. Urbach's rule has been found to hold for a number of substances: CdS [37, 38], TlCl [39], KI [40, 41], KCl [42], AgCl [43], SbI_3 [44], BiI_3 [45], etc. The constant S in equation (18) has been found to show an exponential dependence on $(-1/kT)$, e.g. Martienssen [46, 47] gives for KBr and KI

$$\alpha = \alpha_0 \exp\left[\frac{-\sigma(h\nu_0 - h\nu)}{kT}\right]; \qquad \sigma = 0.79. \tag{19}$$

Piper [48] found that for hexagonal ZnS $\log \alpha$ decreases linearly with λ. A similar behaviour in the case of AgCl was noticed by Aline [49].

No single and decisive explanation of the "Urbach rule" is recognized at present, though some possible causes have been suggested.

For example, assuming a strong *quadratic* coupling of the electronic transition to certain lattice displacements, Toyozawa [50, 51 52], has derived theoretically Urbach's rule for the exciton bands in alkali halides. A modification of the Urbach rule has been proposed by Mahr [53]. Rupprecht [54] has found for a number of substances like $BaTiO_3$, MgO, SiO_2, etc., that the absorption index decreases exponentially as a function of wave number, and has attempted to furnish an explanation for the same.

Values of the parameters, occurring in Urbach's rule, for a number of solids have been summarized by Knox [55].

It is known that for many of the substances the absorption coefficient flattens off at $\approx 10^5\ \mathrm{cm}^{-1}$. To incorporate such a behaviour with the Urbach rule, Moss [22] suggested the following equation:

$$\alpha = \frac{\alpha_\infty}{1 + \exp\left[\sigma\left(E^* - h\nu\right)\right]} . \tag{20}$$

If the absorption edge is defined by the condition that at that point $\mathrm{d}\alpha/\mathrm{d}(h\nu)$ is maximum, then

$$\frac{\mathrm{d}^2\alpha}{\mathrm{d}(h\nu)^2} = 0 \tag{21}$$

and equation (20) leads to

$$\exp\left[\sigma\left(E^* - h\nu\right)\right] = 1 . \tag{22}$$

Thus Moss identified E^* with the energy gap. Though this method of determining energy gap may be disputed, it is reasonable to assume that E^* is closely related to E_g though it may not be exactly equal to it.

3. Theory of Radiative Recombination Rate

3.1 van Roosbroeck-Shockley theory

A theoretical estimate of the rate of radiative recombination was first given by van Roosbroeck and Shockley [10] by an application of the principle of detailed balance. Their treatment is based on the assumption of unit quantum efficiency and is applicable only to non-degenerate materials.

Under conditions of thermodynamic equilibrium, the number of radiative recombinations in the frequency interval $\mathrm{d}\nu$ is equal to the number of electron-hole pairs generated by thermal radiation. The generation of pairs per unit time per unit volume is equal to $P(\nu)\,\varrho(\nu)\,\mathrm{d}\nu$, where $\varrho(\nu)\,\mathrm{d}\nu$ is the density of photons in the interval $\mathrm{d}\nu$ in the crystal and $P(\nu)$ is the probability per unit time of the absorption of a photon of frequency ν. The recombination rate in thermal equilibrium is equal to the generation rate for each interval $\mathrm{d}\nu$. We denote this by $R(\nu)\,\mathrm{d}\nu$. Thus

$$R(\nu)\,\mathrm{d}\nu = P(\nu)\,\varrho(\nu)\,\mathrm{d}\nu . \tag{23}$$

Let α be the absorption coefficient at frequency ν, the mean free path of a photon of frequency ν before producing a band-to-band transition is $\int\limits_0^\infty \mathrm{e}^{-\alpha x}\,dx =$

$= 1/\alpha$. Further let V_g be the group velocity of a photon; then the mean life time of such a photon will be given by $(\alpha\, V_g)^{-1}$ or

$$P(\nu) = \alpha\, V_g. \tag{24}$$

Also, we have

$$V_g = \frac{d\nu}{d\left(\frac{1}{\lambda}\right)} = \frac{c\, d\nu}{d(n\,\nu)}. \tag{25}$$

Thus

$$P(\nu) = \frac{\alpha\, c\, d\nu}{d(n\,\nu)} = \frac{4\,\pi\,\nu\, n\,\varkappa\, d\nu}{d(n\,\nu)} \tag{26}$$

on substituting $\alpha = 4\,\pi\,\nu\,n\,\varkappa/c$.

The density of photons $\varrho(\nu)\, d\nu$ can be obtained from the well-known Planck equation, according to which the number of photons $\varrho'(\lambda)\, d\lambda$ with wavelengths in the range λ to $\lambda + d\lambda$ is given by

$$\varrho'(\lambda)\, d\lambda = \frac{8\,\pi}{\lambda^4} \frac{d\lambda}{\exp\left(\frac{h\,\nu}{kT}\right) - 1}. \tag{27}$$

Also

$$\lambda = \frac{c}{n\,\nu}, \tag{28a}$$

hence

$$d\lambda = \frac{c}{n^2\,\nu^2} \frac{d(n\,\nu)}{d\nu} d\nu\,. \tag{28b}$$

Substituting (28a) and (28b) in (27), we get

$$\varrho(\nu)\, d\nu = \frac{8\,\pi\,\nu^2 \left[n^2\, \frac{d(n\,\nu)}{d\nu}\right] d\nu}{c^3 \left[\exp\left(\frac{h\,\nu}{kT}\right) - 1\right]}. \tag{29}$$

Finally we have

$$R(\nu)\, d\nu = \frac{32\,\pi^2\, n^3\,\nu^3\,\varkappa\, d\nu}{c^3 \left[\exp\left(\frac{h\,\nu}{kT}\right) - 1\right]}. \tag{30}$$

Integration over all frequencies gives the total number of recombination per unit volume per second. The rate of recombination R is

$$R = \frac{32\,\pi^2\, k^4\, T^4}{c^3\, h^4} \int_0^\infty \frac{\varkappa\, n^3\, u^3\, du}{e^u - 1} = \tag{31}$$

$$= 2.2094 \times 10^{12}\; T^4 \int_0^\infty \frac{\varkappa\, n^3\, u^3\, du}{e^u - 1}, \tag{32}$$

where $u = h\,\nu/kT$.

In terms of the absorption coefficient α, equation (31) may be put as

$$R = \frac{8\,\pi\, k^3\, T^3}{c^2\, h^3} \int_0^\infty \frac{\alpha\, n^2\, u^2}{e^u - 1}\, du\,. \tag{33}$$

We note from (31) that $P(\nu)\,\varrho(\nu)$ is a product of three factors, U, n^3, and $\varkappa$, where

$$U = 2.2094 \times 10^{12}\, T^4 \frac{u^3}{e^u - 1}\,. \tag{34}$$

If the dispersion is neglected [56] and n is assumed to be constant then (32) may be put as

$$R = 2.2094 \times 10^{12}\, T^4\, n^3 \int_0^\infty \frac{\varkappa\, u^3}{e^u - 1}\, du\,. \tag{35}$$

In a steady-state deviation from thermal equilibrium, the rate of radiative recombinations in a non-degenerate material is given by

$$R_c = \frac{n\,p}{n_i^2} R\,, \tag{36}$$

since the number of direct recombinations must be proportional to the product of the concentrations of electrons n and holes p, and must coincide with R when $n\,p = n_i^2$, as occurs under equilibrium conditions.

From the above expression, the decay time for a small disturbance in the carrier concentration may be calculated. Suppose the equilibrium carrier concentrations are n_0, p_0 and that these are increased slightly in some manner to $n = n_0 + \delta n$, $p = p_0 + \delta p$, then the recombination rate becomes

$$R + \delta R = R\,(n_0 + \delta n)\frac{p_0 + \delta p}{n_0\, p_0} \quad \text{or} \quad \frac{\delta R}{R} = \frac{\delta n}{n_0} + \frac{\delta p}{p_0}\,. \tag{37}$$

If $\delta n = \delta p$,

$$\frac{\delta R}{R} = \delta n \left(\frac{1}{n_0} + \frac{1}{p_0}\right). \tag{38}$$

The radiation lifetime τ is given by

$$\tau = \frac{\delta n}{\delta R} = \frac{n_0\, p_0}{R\,(n_0 + p_0)}\,. \tag{39}$$

For the intrinsic case $n_0 = p_0 = n_i$, and

$$\tau = \frac{n_i}{2\,R}\,. \tag{40}$$

The recombination probability, B, is given by [3]

$$B = \frac{R}{n_i^2} \tag{41}$$

and the recombination cross section σ by

$$\sigma = \frac{R}{n_0\, p_0\, v} = \frac{R}{n_i^2}\, v\,, \tag{42}$$

where v is a relative velocity of thermal motion and may be obtained from $\frac{1}{2}\, m\, v^2 = kT$.

A general review on the lifetime of excess carriers in semiconductors may be found in Many and Bray [57]. The effect of degeneracy on radiative recombinations has been considered by Landsberg [58].

3.1.1 *Intrinsic carrier concentration*

For some semiconductors the intrinsic carrier concentration has been determined over a range of temperatures by conductivity and Hall constant measurements. The results are usually expressed in the form

$$n_i^2 = \text{const}\ T^3 \exp\left(\frac{-E_{g0}}{kT}\right), \tag{43}$$

where E_{g0} is the extrapolated energy gap at 0 °K assuming a linear variation of E_g with T.

In other cases, n_i may be calculated theoretically from the well known expression

$$n_i^2 = N_c\, N_v \exp\left(\frac{-E_g}{kT}\right), \tag{44}$$

where

$$N_c = 2\, m_e^{3/2}\, M_c \left(\frac{2\pi kT}{h^2}\right)^{3/2},$$

and

$$N_v = 2\, m_h^{3/2}\, M_v \left(\frac{2\pi kT}{h^2}\right)^{3/2}, \tag{45}$$

in which M_c is the number of equivalent minima in the conduction band, M_v is the number of equivalent maxima in the valence band, m_e is the 'density of states' effective mass for electrons, and m_h is the effective hole mass obtained from

$$m_h^{3/2} = m_{h1}^{3/2} + m_{h2}^{3/2}, \tag{46}$$

m_{h1} and m_{h2} being the masses of 'heavy' and 'light' holes. When $M_v = 1$, equation (44) may be written as

$$n_i^2 = 4\left(\frac{2\pi kT}{h^2}\right)^3 M_c\, (m_e\, m_h)^{3/2} \exp\left(\frac{-E_g}{kT}\right) = \tag{47}$$

$$= 2.332\times 10^{31}\, M_c\, T^3 \left(\frac{m_e\, m_h}{m^2}\right)^{3/2} \exp\left(\frac{-E_g}{kT}\right). \tag{48}$$

3.1.2 *Analytical expressions*

Using the theoretical expressions for the absorption coefficient α, one can obtain analytical expressions for R and other associated quantities. These expressions are specially useful in those cases where detailed absorption-coefficient measurements are lacking but data on m_e, m_h, E_g, etc., are available.

3.1.2.1 *Direct transitions*

We shall assume

$$\alpha_d = b\,[(h\nu - E_g)^{1/2} + P], \tag{49}$$

where b is given by (6b) and P is a constant. Substituting (49) in (33) and assuming n to be constant we obtain

$$R_d = J\left\{(kT)^{7/2}\int_{u_0}^{\infty}\frac{(u-u_0)^{1/2}\,u^2}{e^u-1}\,du + (kT)^3\, P\int_{u_0}^{\infty}\frac{u^2}{e^u-1}\,du\right\}, \tag{50}$$

where

$$J = \frac{32\,\pi^3\, e^2\,(2\, m_r)^{3/2}\, n\, f}{c^3\, h^5\, m}$$

and $u_0 = E_g/kT$. Assuming that $u > 1$, to a good degree of approximation, we derive

$$R_d = J(kT)^3 e^{-u_0} \left\{ (kT)^{1/2} \left[\frac{\pi^{1/2}}{2} \left(\frac{15}{4} + 3 u_0 + u_0^2 \right) + \frac{e^{-u_0}}{4} \left(\frac{\pi}{2} \right)^{1/2} \left(\frac{15}{16} + \frac{3}{2} u_0 + u_0^2 \right) \right] + P \left[(2 + 2 u_0 + u_0^2) + \frac{e^{-u_0}}{2} \left(\frac{1}{2} + u_0 + u_0^2 \right) \right] \right\}. \quad (51)$$

If $u_0 \gg 1$, the above expression may be approximated by

$$R_d = J(kT)^3 e^{-u_0} u_0^2 \left\{ \frac{(kT)^{1/2} \pi^{1/2}}{2} + P \right\}. \quad (52)$$

Further if the electron–hole interaction is neglected, (52) may be simplified to

$$R_d = \frac{16 \pi^3 e^2 (2 m_r)^{3/2} n (kT)^{7/2} \pi^{1/2} e^{-u_0} u_0^2}{c^3 h^5 m} f. \quad (53)$$

For most of the semiconductors at ordinary temperatures, u_0 is much greater than 1, e.g. for Ge at 300 °K, $u_0 \approx 27$. In such cases expression (53) is quite adequate. But for semiconductors with low E_g and at high temperatures, the full expression (51) should be employed.

Expressions (51), (52), or (53) can be employed to obtain the recombination probability B and the recombination cross sections σ. Here we give expressions for B and τ corresponding to (53):

$$B_d = \frac{R_d}{n_i^2} = \frac{(2\pi)^{1/2} h e^2 n}{M_c c^3 m^{5/2} (kT)^{3/2}} \left(\frac{m}{m_e + m_h} \right)^{3/2} \left(1 + \frac{m}{m_e} + \frac{m}{m_h} \right) E_{gd}^2 \exp\left(\frac{E_{gi} - E_{gd}}{kT} \right), \quad (54)$$

when $E_{gi} < E_{gd}$; if the case is otherwise, the exponential term drops out. Expression (54) is similar to the one given by Hall [3].

The lifetime τ may be expressed in either of the following forms:

$$\tau_d = \frac{n_i}{2R} = \frac{c^3 h^2 m M_c^{1/2} (m_e m_h)^{3/4}}{8 \pi^2 e^2 n E_{gd}^2} \left(\frac{m_h + m_e}{m_e m_h} \right)^{3/2} \left(1 + \frac{m}{m_e} + \frac{m}{m_h} \right)^{-1} \exp\left(\frac{2 E_{gd} - E_{gi}}{2 kT} \right) \quad (55)$$

(for direct gap semiconductors, $E_{gd} < E_{gi}$, and the exponential factor in (55) becomes $\exp(E_{gd}/2kT)$)
or

$$\tau_d = \frac{n_i^2}{R(n_0 + p_0)} = \frac{c^3 m^{5/2} m_c (kT)^{3/2}}{(2\pi)^{1/2} h e^2 n E_{gd}^2} \left(\frac{m_e + m_h}{m} \right)^{3/2} \left(1 + \frac{m}{m_e} + \frac{m}{m_h} \right)^{-1} \frac{1}{n_0 + p_0} \exp\left(\frac{E_{gd} - E_{gi}}{kT} \right). \quad (56)$$

3.1.2.2 *Indirect transitions*

We can obtain an expression for R by using the analytical expressions for indirect transitions given in equation (13)

$$R = \frac{8\,\pi\,n^2\,k^3\,T^3}{c^2\,h^3} \int\limits_0^\infty \frac{u^2\,\alpha}{e^u - 1}\,du\,, \tag{57}$$

where $u = h\,\nu/kT$. The lower limit of the integral corresponds to the point where $\alpha = 0$.

On substituting equation (13) in the above expression and putting

$$\frac{E_g}{kT} - \frac{\theta}{T} = s_1\,, \qquad \frac{E_g}{kT} + \frac{\theta}{T} = s_2 \tag{58}$$

we obtain

$$R_i = \frac{8\,\pi\,n^2\,k^3\,T^3(kT)^2}{c^2\,h^3\,(e^{\theta/T} - 1)}\,A\,\{e^{-s_1}\,[2\,s_1^2 + 12\,s_1 + 24] + e^{\theta/T}\,e^{-s_2}\,[2\,s_2^2 + 12\,s_2 + 24]\}\,. \tag{59}$$

At ordinary temperatures we may assume $u_0 \gg 1$ and $u_0 \gg \theta/T$, then we have $s_1 = s_2 = u_0$ and

$$R_i = \frac{8\,\pi\,n^2(kT)^5}{c^2\,h^3\,(e^{\theta/T} - 1)}\,A\,\{2\,u_0^2\,e^{-u_0}\,(1 + e^{\theta/T})\}\,. \tag{60}$$

The corresponding expression for the recombination probability B is

$$B_{\text{indirect}} = \frac{h^3\,A\,E_{gi}^2\,n^2}{2\,\pi^2\,c^2\,M_c\,(m_e\,m_h)^{3/2}}\,\frac{e^{\theta/T} + 1}{e^{\theta/T} - 1}\,. \tag{61}$$

Equation (61) was obtained by Hall [3].

The lifetime τ may be put in either of the two forms

$$\tau_i = \frac{n_i}{2\,R} = \frac{c^2\,M_c^{1/2}}{2\,\pi\,n^2\,A\,E_{gi}^2}\left(\frac{\pi}{2\,kT}\right)^{3/2}(m_e\,m_h)^{3/4}\;e^{u_0/2}\,\frac{e^{\theta/T} - 1}{e^{\theta/T} + 1} \tag{62}$$

or

$$\tau_i = \frac{n_i^2}{R\,(n_0 + p_0)} = \frac{2\,\pi^2\,c^2\,(m_e\,m_h)^{3/2}\,M_c}{h^3\,A\,n^2\,E_{gi}^2}\,\frac{1}{n_0 + p_0}\,\frac{e^{\theta/T} - 1}{e^{\theta/T} + 1}\,. \tag{63}$$

We can also obtain analogous expressions corresponding to the theoretically sounder equations (14). These are

$$R_i = \frac{8\,\pi\,n^2(kT)^3\,(kT)\,A_1}{c^2\,h^3\,(e^{\theta/T} - 1)}\,\{e^{-s_1}\,[2\,s_1 + 6] + e^{\theta/T}\,e^{-s_2}\,[2\,s_2 + 6]\}\,, \tag{64}$$

$$R_i \approx \frac{8\,\pi\,n^2(kT)^3\,A_1}{c^3\,h^3\,(e^{\theta/T} - 1)}\,2\,E_{gi}\,e^{-u_0}\,(1 + e^{\theta/T})\,, \tag{65}$$

$$B_i = \frac{h^3\,A_1\,E_{gi}\,n^2}{2\pi^2\,c^2\,M_c\,(m_e\,m_h)^{3/2}}\;\coth\left(\frac{\Theta}{2\,T}\right), \tag{66}$$

$$\tau_i = \frac{c^2\,M_c^{1/2}}{2\,\pi\,n^2\,A_1\,E_{gi}}\left(\frac{\pi}{2\,kT}\right)^{3/2}(m_e\,m_h)^{3/4}\,e^{u_0/2}\;\coth\left(\frac{\theta}{2\,T}\right), \tag{67}$$

or

$$\tau_i = \frac{2\,\pi^2\,c^2(m_e m_h)^{3/2}\,M_c}{h^3\,A_1\,E_{gi}\,n^2}\,\frac{1}{n_0 + p_0}\;\coth\left(\frac{\theta}{2\,T}\right). \tag{68}$$

Baryshev [58a] has obtained the form of analytical expressions for σ by using other expressions for the absorption coefficients.

3.2 More refined calculations

Dumke [59] has derived expressions for the radiative-recombination lifetime based upon a microscopic analysis of both direct and indirect transitions. His results are as follows.

Indirect recombination:

$$\tau = \frac{1}{4\pi}\frac{c^2}{\hbar^3}\,\frac{\sum_v m_{\mathrm{v}}^{+3/2}\sum_c m_{\mathrm{c}}^{+3/2}}{A\,E_{\mathrm{g\,i}}^2\,n^2}\,\frac{1}{n_0+p_0}\,\frac{\mathrm{e}^{\theta/T}-1}{\mathrm{e}^{\theta/T}+1}\,. \tag{69}$$

Here the sums over v and c are over the extrema in the valence and conduction bands, and E_{gi} is the indirect energy gap. A is the factor involved in the MacFarlane-Roberts expression (13) and is obtained from an analysis of the absorption data. It may be noted that equation (63) is almost identical with the above expression. For intrinsic Ge at room temperature, Dumke calculated $\tau_{\mathrm{indirect}} = 1.98$ s.

Direct recombination:

$$\tau = \frac{1}{2(2\pi)^{3/2}}\left(\frac{m^{7/2}\,c^3\,k^{3/2}}{\hbar\,e^2}\right)\left(\frac{T^{3/2}}{E_{\mathrm{g\,d}}^2\,n^3\,|P_{\mathrm{if}}|^2}\right)\times$$
$$\times\exp\left(\frac{E_{\mathrm{g\,d}}-E_{\mathrm{g\,i}}}{kT}\right)\frac{\sum_c \alpha_{\mathrm{c}}^{-3/2}\sum_v \alpha_{\mathrm{v}}^{-3/2}}{\sum_v(\alpha_{\mathrm{c\,0}}+\alpha_{\mathrm{v}})^{-3/2}}\left(\frac{1}{n_0+p_0}\right). \tag{70}$$

Here m/α_{v} and m/α_{c} are the valence- and conduction-band effective masses, m/α_{c0} is the effective mass of conduction-band electrons at $\boldsymbol{k} = 0$, E_{gd} is the direct energy gap, P_{if} is the optical matrix element, the value of which can be estimated. From this expression, Dumke found the lifetime in intrinsic Ge at 300 °K to be $\tau = 0.29$ s. Equation (70) may be compared with equation (56).

Recently Lasher and Stern [11] have calculated the spectral line shapes of the radiation produced by band-to-band recombination under the assumption that the momentum matrix element is the same for all initial and final states, i.e. that there is no momentum selection rule. The peak of the stimulated radiation falls at a lower photon energy than does the peak of the spontaneous radiation, except when $T = 0$ °K. The limitations of the van Roosbroeck-Shockley approach have been pointed out. These authors have also considered the effect of a tail in the density of states on the spontaneous line shape.

3.3 Semiconductor masers

The possibility of achieving maser operation in semiconductors was considered theoretically by several workers [60, 61, 62]. The first successful observations on a semiconductor laser were made by Hall et al. [63] and by Nathan et al. [64] in 1962. Since that time, this field has developed tremendously. Recent reviews may be found in Lax [65], Hilsum [66], Geusic and Scovil [67], Birnbaum [68], and Rediker [69].

When photons of an appropriate energy fall on a semiconductor, they are absorbed with excitation of electrons from the valence band to the conduction band. This process is one we are familiar with; but there is a small but finite possibility of a second process occurring. Here one of the incident photons would interact with an electron in the conduction band and induce it to combine with a hole in the valence band. The recombination energy of the electron would be emitted as a photon, which would have to be identical to the photon which initiated the recombination. By this process, stimulated emission can only occur with high probability when there are present in the semiconductor a large number of free carriers and many holes. In thermal equilibrium these conditions are not satisfied and absorption is then more probable than emission, but when a p–n junction is operated at extremely high-current densities a large number of holes and electrons are injected into the transition region and the conditions become more favourable for emission than for absorption. A photon travelling into this region is more likely to stimulate the emission of another photon — in other words the first photon will be amplified. The stimulated photon is identical in every way with the original photon and is emitted so that it travels in the same direction. The recombination region near the junction is only a few μm thick, but if the junction is very flat and the original photon travelled exactly in the junction plane the stimulated photon will also be in this plane, and it too may be amplified. An incident light wave can thus grow as it travels across the crystal. Initiation of amplification by light incident from outside the crystal is not necessary. The spontaneous recombination radiation can itself be amplified, provided it is travelling in the junction plane. At low current densities the junction will emit radiation isotropically but as the current increases the light in the junction plane increases rapidly and the polar diagram of the emission becomes anisotropic.

Possible radiative transitions associated with laser action in semiconductors are illustrated schematically in Fig. 2. The four mechanisms are: 1 conduction to valence band, 2 donor to valence band, 3 conduction band to acceptor, and 4 donor to acceptor. Transitions between acceptor levels and the conduction band were studied by Eagles [69a]. Possibility of maser action from band-to-band transitions was considered by Dumke [62]. Lasher and Stern [11], on the basis of a simple model, have calculated the spectral line shape of the spontaneous and stimulated radiation emitted in band-to-band recombination.

In the case of isolated atoms it is usually easy to determine the modes in which generation of stimulated radiation occurs (the "singular" modes). For interband transitions, as a result of the discontinuity in the energy spectrum, the determination of the singular modes represents a more difficult problem. An attempt to solve this problem by determining the parameters of singular modes has been made by Vinetskii et al. [70].

Callaway [71] has calculated the transition probabilities and absorption constants for the band-to-acceptor and donor-to-acceptor transitions. Recently Dumke [72] has derived expressions for the optical absorption and radiative

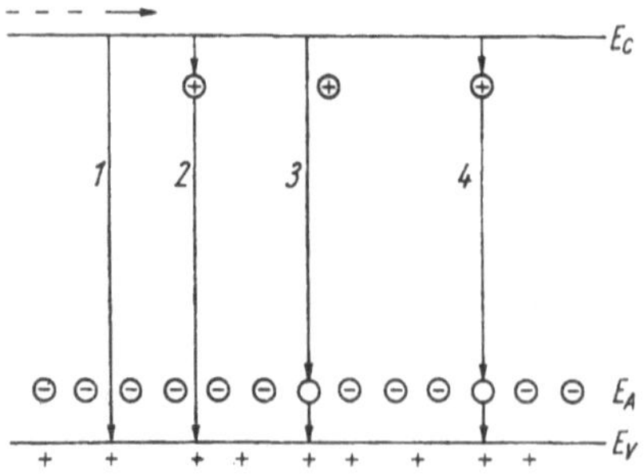

Fig. 2. Four mechanisms for radiative transitions:
1 — conduction to valence band;
2 — donor to valence band;
3 — conduction band to acceptor;
4 — donor to acceptor (after [69])

lifetime for transitions between donor levels and the valence band or between acceptor levels and the conduction band.

The available evidence appears to indicate that in some cases the radiation is due to band-to-band transitions while in others "band-to-acceptor" transition (mechanism 3) is operative. Thus the mechanism of the radiative transition is not unique; indeed it is thought that even in the same laser diode, it may be possible, by changing the junction current, to change the recombination mechanism.

3.4 Quantum yield or quantum efficiency

The quantum yield is defined as the number of electron–hole pairs produced per incident quantum. The van Roosbroeck-Shockley theory assumes unit quantum efficiency. However, if energy of the incident photon is high enough, it is no longer true. Investigations bearing on this topic are summarized below.

Goucher's [73] photoconductivity measurements, made on a single crystal of n-type germanium indicated a yield of one electron-hole pair per photon absorbed, for wavelengths in the vicinity of, but less than, the long-wave limit. This finding was further supported by the experiments of Goucher et al. [74] on the photo-response of a germanium p–n junction.

If the photon has a sufficient excess energy above threshold, additional electrons can be liberated in the crystal as a result of impact ionization. The possibility of a quantum yield of photo luminescence in excess of unity, upon excitation by quanta with energies more than double the energy of the luminescence quantum was indicated by Vavilov [75]. This phenomenon was observed experimentally by Butaeva and Fabrikant [76, 77].

To measure the quantum yield, either the electrons or the holes are collected at a p–n junction. Such measurements have been made on germanium by Koc [78] and by Vavilov and Britsyn [79], on silicon by Vavilov and Britsyn [80, 81], and on InSb by Tauc and Abraham [82].

In Fig. 3, the quantum yield curve obtained for germanium by Vavilov and Britsyn [79] is shown. It will be seen that below a certain threshold energy of the incident quanta, the quantum yield is unity. Above this threshold the quantum yield rises steadily with increasing energy. Similar behaviour is shown by Si and InSb.

Britsyn and Vavilov [83] have also examined the temperature dependence of the quantum yield in silicon. Their results are shown in Fig. 4. It is seen that

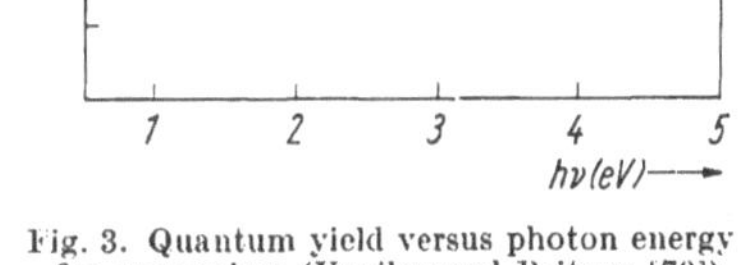

Fig. 3. Quantum yield versus photon energy for germanium (Vavilov and Britsyn [79])

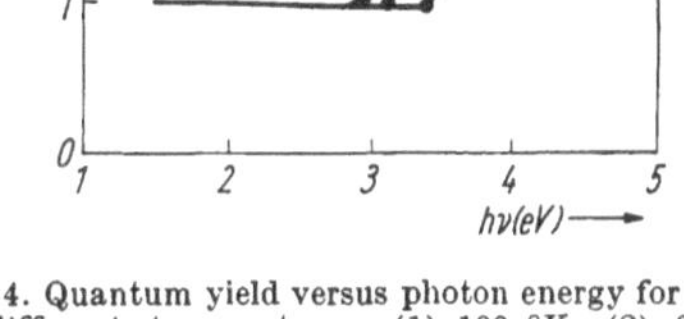

Fig. 4. Quantum yield versus photon energy for silicon at different temperatures: (1) 100 °K, (2) 300 °K, (3) 400 °K (Britsyn and Vavilov [83])

the increase in the quantum yield (>1) begins at different photon energies, depending on the temperature; also the shape of the curve for the region where $Q > 1$ is different at different temperatures.

Recently Tuzzolino [84] has carried out a measurement of the quantum efficiency of silicon over the photon energy range from 4.9 to 21 eV by measuring the photo-response of silicon surface-barrier photo-diodes. The quantum efficiency increases from 2.0 at $h\nu = 4.9$ eV to approximately 3 at $h\nu = 6$ eV; between $h\nu = 6$ eV and $h\nu \approx 10$ eV the quantum efficiency is approximately constant.

The theory of impact ionization has been developed by Chuenko [85], Antončik [86, 87], Shockley [88], and Hodgkinson [89, 90].

All explanations of quantum yield curves have been founded on the following basic mechanism. The excess energy $(h\nu - E_g)$ of a quantum of energy $h\nu$ falling on a semiconductor of lowest energy gap E_g is divided in some way between the electron and hole produced. If one of these has sufficient energy, it can create another electron–hole pair by collision ionization, and thus enhance the quantum yield. Antončik assumes that these processes occur in limited regions of phase space which an electron or hole must occupy if it is to take part in an ionizing collision. Those which do not occupy such a position are supposed not to ionize. Antončik's theory does not predict a temperature dependence of the quantum yield.

Shockley [88] uses an explanation in which some of the more energetic carriers lose their energy in phonon collisions before they take part in an ionizing collision.

Both of these treatments were based on specific assumptions concerning the way in which the excess energy of the absorbed quantum was divided between the electron and hole produced. Antončik assumed that all of the excess energy went to the electrons which were then distributed uniformly in $\boldsymbol{k}$-space around the conduction band minimum. Shockley assumed that the excess energy was divided equally between the electron and hole produced.

Recently Hodgkinson [89] has applied optical absorption and photoemission data to analyse quantum yield in silicon and has shown that most of the electrons and holes with energy above a certain threshold will take part in an ionizing collision before losing their energy. This is taken to mean that they are scattered into a region of phase space in which they can cause ionization, before their energy falls below the threshold. In a succeeding paper, Hodgkinson [90] has investigated the conditions governing impact ionization in semiconductors and has applied them to determine the position in energy and $\boldsymbol{k}$-space of the various impact ionization thresholds in silicon and germanium. Combining these results with the band structure and with the known mechanisms of optical absorption, he has succeeded in explaining the enhanced quantum yield thresholds in silicon and germanium.

The quantum yield curve for InSb has been interpreted by Beattie [91].

4. Experimental Methods of Excitation and Study of Radiative Recombination

In studies of the spectra of recombination radiation, the following methods of generating the necessary excess concentration of carriers have been used:

a) Excitation by photons having energy greater than the energy gap (photoluminescence) [92]. Laser has also been used recently by Basov et al. [93].

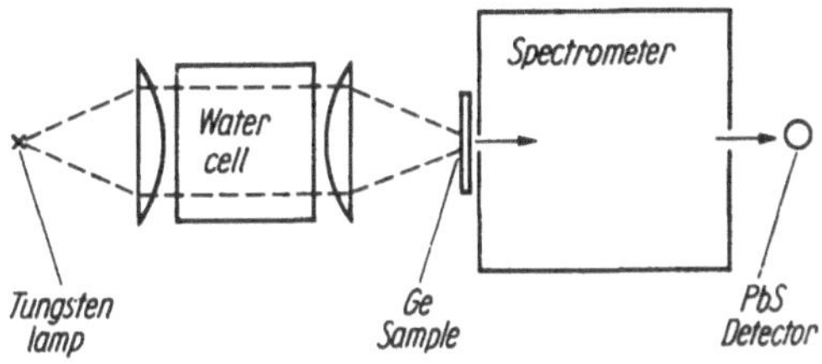

Fig. 5. Schematic arrangement used by Haynes [92] to measure recombination radiation in germanium

b) Injection of carriers by passing a current through a n–p junction (electroluminescence) [94, 95].

c) Excitation with a beam of fast electrons (cathodoluminescence) [96].

Kopylovskii et al. [97] have described electronic equipment which is used for generating and studying recombination radiation in semiconductors. A generator circuit is presented for high-power pulses having a duration of 1 to 5 μs with a continuous current adjustment from 0.5 to 400 A. A registration system for the radiation is described that makes it possible to observe on an oscillograph and to record on an automatic recording instrument the spectra of recombination radiation.

The essential principles behind the experiments may be understood from the arrangement used for germanium by Haynes [92] which we describe. The technique is shown schematically in Fig. 5. The light from a ribbon filament tungsten lamp was passed through a water absorption cell 10 cm long and focused on a thin slice of germanium (1.2×10^{-2} cm thick). The radiation from the opposite side of the slice was analysed with a spectrometer and detected by a PbS cell. The water cell transmits less than 10^{-10} of the incident radiation at wavelengths longer than 1.4 μm and the germanium transmits less than 10^{-10} of the radiation shorter than 1.4 μm; thus no measurable light from the tungsten source will enter the spectrometer. Any detectable radiation must arise from the recombination of electrons and holes produced by light. The results obtained by this arrangement are discussed in Section 7.3.

We may note that the spectral distribution of the emergent radiation from the experimental sample may be different from the originally produced one due to self absorption in the material and other effects. Corrections for these have been discussed by Haynes [92] and Moss [98].

When injection across a p–n junction is used, an important problem is that of bringing about the emergence of the radiation from the specimen. In fact, for the simplest experimental geometry (Fig. 6a) a considerable part of the light flux undergoes total internal reflection and does not reach the entrance slit of the spectrometer. This is due to the very high index of refraction of groups IV and III–V semiconductors. Aigrain and Benoit à la Guillaume [99, 100] overcame this difficulty by taking the specimen in the form of a Weierstrass sphere (Fig. 6b) and having injecting contact at the Weierstrass point of the sphere. In this case the radiative volume is approximately a hemisphere with a radius of the order of magnitude of the diffusion distance of the minority carriers. However, such a specimen geometry is best only for the study of extrinsic radiation [7]; the spectrum of the intrinsic radiation is strongly distorted because of absorption. With a complicated geometry this distortion is difficult to take into account.

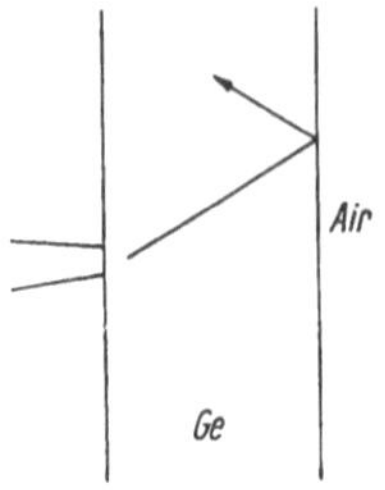

a

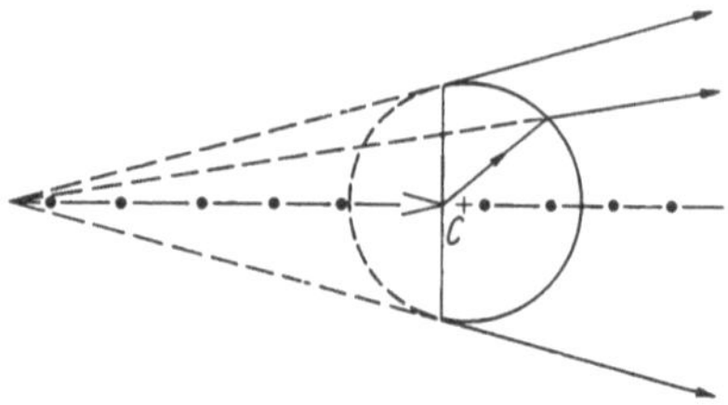

b

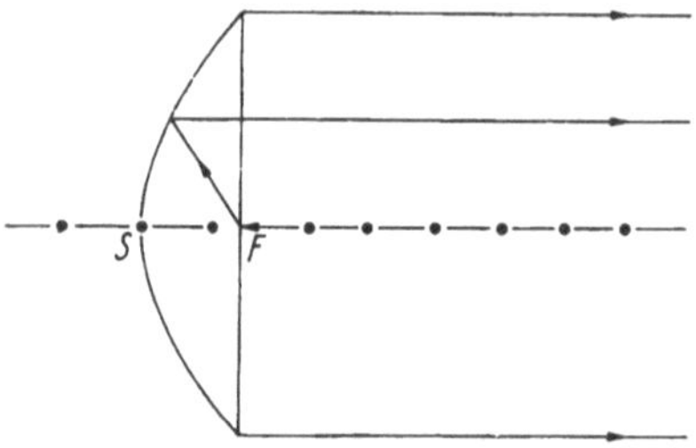

c

Fig. 6a. Total reflection at the surface of germanium

Fig. 6b. Weierstrass sphere, principle

Fig. 6c. Plane paraboloid combination (Aigrain [99])

Aigrain [99] has also suggested another shape which can be used but is much more critical, namely the plane-paraboloid combination shown in Fig. 6c in which light is first totally reflected in the paraboloid face, then passes normally through the plane face.

A good review of the various means of measuring the lifetime of carriers in semiconductors has been given by Suryan and Susila [101].

It has been pointed out by Dumke [59] that even in those semiconductors where radiative recombination is the primary mode of recombination, the observed lifetime may or may not be equal to that theoretically calculated. If E_R denotes the energy of the radiation peak we can distinguish between two cases. If E_R lies in the region where the absorption coefficient is fairly low ($\approx 10\ \mathrm{cm}^{-1}$), an emitted photon has got a chance of escaping the crystal before being reabsorbed with the resulting production of an electron–hole pair. But if E_R lies in a region where the absorption coefficient is very high ($\approx 10^3\ \mathrm{cm}^{-1}$) there would be heavy reabsorption and consequently no net decrease in the number of hole–electron pairs due to radiative recombination observed. The observed lifetime of these carriers due to radiative recombination would be very large. However, in very thin films, the reabsorption may be avoided.

5. General References

The investigations on the properties of III–V compounds up to 1960 have been summarized in the book by Hilsum and Rose-Innes [102]. Recently an excellent and up to date monograph on the physics of III–V compounds by Madelung [103] has come out.

Under each semiconductor we have given a brief resume of its optical properties and other necessary data. There are a number of papers and review articles in which theory or experimental data pertaining to three or more of the semiconductors considered here is presented. It is convenient to summarize them here.

Reviews of optical properties of semiconductors are to be found in the books of Moss [104] and Smith [2]. A brief survey of the optical properties of III–V semiconductors has been given by Stern [105]. In the summary given with each substance, we have confined ourselves to giving references to those works which are relevant to radiative recombination calculations and those which have been publishing during the last few years.

The infra-red lattice reflection spectra of AlSb, GaAs, GaSb, InP, InAs, and InSb were measured at liquid helium temperature by Hass and Henvis [106]. By comparing the observed spectrum with that calculated using a single classical dispersion oscillator, the values of TO and LO phonon frequencies for long-wavelength vibrations are obtained. Mitra and Marshall [107, 108] have noted a number of correlations among the characteristic phonon frequencies of III–V and II–VI compounds.

Johnson and Loudon [109] have presented a critical point analysis of the phonon spectra of diamond, silicon, and germanium. In a recent excellent review, Johnson [110] summarises the results and presents the interpretations of the infra-red lattice-band spectra for diamond, silicon, and germanium and also for AlSb, GaP, GaAs, and InSb.

The fundamental reflectivity spectrum of a number of semiconductors with zinc-blende structure has been discussed by Cardona [111] and by Lukeš and Schmidt [112].

Philipp and Ehrenreich [113, 114] have presented reflectance data for Si, Ge, GaP, GaAs, InAs, and InSb in the range of photon energies between 1.5 and 25 eV. The real and imaginary parts of the dielectric constant are deduced from the Kramers-Kronig relations.

The absorption spectrum of thin films of Ge and several III–V semiconductors has been studied by Cardona and Harbeke [115]. Besides the fundamental absorption edge and its spin–orbit splitting, additional structure at higher energies was found. The subject has been reviewed by Cardona [116].

Reviews on the band structure of groups IV and III–V semiconductors have been given by Paul [117], Long [118], Hilsum [119], and Herman et al. [120].

Ehrenreich [121] has discussed the band structure of III–V compounds with reference to their transport properties. Progress in the band theory of semiconductors has been reviewed by Pincherle [122]. A simple equation for the variation of the energy gap with temperature has been proposed by Varshni [123].

6. General Remarks on Results Presented here

Calculations of the radiative recombination rate and other related quantities were carried out within the framework of the van Roosbroeck-Shockley theory for the following substances: diamond, Si, Ge, Se, Te, GaP, GaAs, GaSb, InP, InAs, and InSb, in some cases at two or more temperatures. The values of absorption coefficients are usually given by investigators in figures. These figures for various semiconductors were enlarged by a projector and traced on a large graph, wherefrom the values could be read with reasonable accuracy. In two cases, Ge [124] and Si [125], the original large size graphs were kindly supplied by Mr. Roberts and the values were read directly from there. We consider each semiconductor individually and display the results in figures. In most of the figures we show the radiative recombination rate $P(\nu)\,\varrho(\nu)$, the absorption index $\varkappa$, and the quantity U defined in equation (34). In all figures,

Table 1a

Data for m_e, m_{h1}, m_{h2}, and m_h. — Most of the values are from the compilation of Cardona [127]. Values followed by a "T" are theoretical, estimated by Cardona. All m_h values, except that of diamond, were calculated from equation (46)

Substance	m_e/m	Ref.	m_{h1}/m	Ref.	m_{h2}/m	Ref.	m_h/m
Diamond	0.2	[144]					0.4 [148]*)
Si	0.13 T	[127]	0.52	[127]	0.16	[127]	0.578
Ge	0.041	[127]	0.35	[127]	0.045	[127]	0.361
GaP	0.13 T	[127]	0.56 T	[127]	0.13 T	[127]	0.601
GaAs	0.07	[127]	0.68	[127]	0.12	[127]	0.713
GaSb	0.047	[127]	0.3	[127]	0.06	[127]	0.318
InP	0.073	[127]	0.4	[127]	0.086 T	[127]	0.426
InAs	0.026	[127]	0.41	[127]	0.025	[127]	0.414
InSb	0.0155	[127]	0.25	[127]	0.012	[127]	0.252

*) Clegg and Mitchell [445] (see Part II of this article, to be published in Volume 20 of phys. stat. sol.) have used the band constants deduced from the results of cyclotron resonance measurements to calculate the hole concentrations. The calculated hole concentrations may be expressed in terms of a density-of-states mass $m_h/m = 3.6$ to 4.8. Reasons for this discrepancy have been discussed by these authors.

Table 1b

Data for refractive indices and energy gaps

Substance	n	T (°K)	E_{gi} (eV)	Reference	E_{gd} (eV)	Reference
Diamond	2.75	295	5.4	[144]		
Si	3.53	20	1.137	[31]	≈ 3	[23]
		90	1.132			
		195	1.110			
		249	1.098			
		290	1.086			
		333	1.074			
Ge	4.15	77	0.74	[124]	0.883	[212]
		300	0.665		0.802	
GaP	2.91	300	2.205	text		
GaAs	3.63	21			1.521	[336]*)
		90			1.511	
		185			1.479	
		294			1.435	
GaSb	3.72	80			0.80	[382]*)
		300			0.72	
InP	3.37	77			1.4135	[390]*)
		298			1.3511	
InAs	3.54	77			0.416	[404]*)
		298			0.355	
InSb	4.01	77			0.23	[103]
		295			0.18	

*) See Part II of this article.

Table 1c

Data for M_c and constants A and θ (occurring in equation (13)). For diamond, GaAs, GaSb, InP, InAs, and InSb, θ has been given a value equal to the Debye temperature.

Substance	M_c	A (eV^{-2} cm^{-1})	Source	θ (°K)
Diamond	6 ?	≈ 16000	estimated	2240
Si	6	2682	see text	600
Ge	4	2231 (77 °K) 2722 (300 °K)	[32]	260
GaP	6 ?	3760	see text	302
GaAs	1	≈ 3500	estimated	344
GaSb	1	≈ 3500	estimated	266
InP	1	≈ 3500	estimated	301
InAs	1	≈ 3500	estimated	248
InSb	1	≈ 3500	estimated	202

Table 2

Results for R, n_i, B, τ, and σ. Values of R were obtained by numerical integration and those of n_i by methods described under individual semiconductors. Figures in parentheses give powers of 10 by which the figure is multipled. Thus 4.0(−66) stands for 4.0×10^{-66}

Substance	T (°K)	R (cm^{-3} s^{-1})	n_i (cm^{-3})	B (cm^3 s^{-1})	τ (s)	σ (cm^2)
Diamond	295	4.0 (−66)	6.68(−28)	8.96(−12)	8.35(+37)	9.48(−19)
Si	20	4.8 (−267)	≈ 1(−126)	≈ 1(−15)		
	90	5.9 (−42)	6.81(−14)	1.27(−15)	5.77(+27)	2.44(−22)
	195	3.4 (−6)	4.88(+4)	1.43(−15)	7.18(+9)	1.86(−22)
	249	2.3 (+1)	1.2 (+8)	1.6 (−15)	2.61(+6)	1.84(−22)
	290	9.2 (+4)	7.16(+9)	1.79(−15)	3.89(+4)	1.91(−22)
	333	6.1 (+7)	1.8 (+11)	1.88(−15)	1.48(+3)	1.87(−22)
Ge	77	2.4 (−26)	2.42(−7)	4.1 (−13)	5.04(+18)	8.48(−20)
	300	2.85(+13)	2.33(+13)	5.25(−14)	4.09(−1)	5.5 (−21)
Se (hex.)	298	≈7 (−4)				
Te	100	1.6 (+8)	1.24(+9)	1.04(−10)	3.87(0)	1.9 (−17)
	300	3.0 (+20)	5.93(+15)	8.53(−12)	9.88(−6)	8.95(−19)
GaP	300	4.0 (−13)	2.73(0)	5.37(−14)	3.41(+12)	5.63(−21)
GaAs	21	1.1 (−336)				
	90	8.0 (−58)	2.14(−25)	1.75(−8)	1.34(+32)	3.34(−15)
	185	1.6 (−13)	9.15(−3)	1.91(−9)	2.86(+10)	2.55(−16)
	294	1.2 (+3)	1.29(+6)	7.21(−10)	5.37(+2)	7.64(−17)
GaSb	80	2.4 (−24)	9.3 (−9)	2.77(−8)	1.94(+15)	5.64(−15)
	300	2.2 (+14)	9.6 (+11)	2.39(−10)	2.18(−3)	2.50(−17)
InP	77	2.2 (−65)	1.32(−20)	1.26(−7)	3.0 (+35)	2.61(−14)
	298	6.0 (+4)	6.9 (+6)	1.26(−9)	5.75(+1)	1.33(−16)
InAs	77	≈1 (−1)	2.65(+3)	1.42(−8)	1.32(+4)	2.95(−15)
	298	5.8 (+19)	8.26(+14)	8.5 (−11)	7.12(−6)	8.94(−18)
InSb	77	4.5 (+9)	1.36(+9)	2.43(−9)	1.51(−1)	5.04(−16)
	295	1.03(+22)	1.50(+16)	4.58(−11)	7.28(−7)	4.84(−18)

Table 3

Calculated values of R, B, and τ for indirect and direct radiative recombination obtained from equations (53), (54), (55), (60), (61), and (62). The "indirect" values for GaAs, GaSb, InP, InAs, and InSb were obtained on the assumption that the observed energy gap is "indirect". Figures in parentheses give powers of 10 by which the figure is multiplied. Thus 2.5(−67) stands for 2.5×10^{-67}

Sub-stance	Temp. (°K)	Indirect R_i (cm^{-3} s^{-1})	Indirect B_i (cm^3 s^{-1})	Indirect τ_i (s)	Direct R_d (cm^{-3} s^{-1})	Direct B_d (cm^3 s^{-1})	Direct τ_d (s)
Diamond	295	2.5(−67)	5.6(−13)	1.3(+39)			
Si	20		7.5(−15)				
	90	6.4(−42)	7.5(−15)	2.3(+27)			
	195	3.5(−6)	7.9(−15)	3.0(+9)	1.9(−50)	4.3(−59)	1.1(+54)
	249	2.2(+1)	8.5(−15)	1.1(+6)	1.8(−33)	6.7(−49)	2.9(+40)
	290	8.4(+4)	8.9(−15)	1.8(+4)	8.6(−25)	9.2(−44)	3.5(+33)
	333	5.6(+7)	9.4(−15)	6.9(+2)	5.7(−18)	9.7(−40)	1.3(+28)
Ge	77	1.9(−27)	6.8(−14)	4.4(+19)	1.5(−32)	5.2(−19)	1.1(+25)
	300	4.7(+12)	1.5(−13)	5.9(−1)	1.9(+13)	6.4(−13)	2.8(−1)
GaP	300	4.1(−13)	5.5(−14)	3.3(+12)			
GaAs	21		2.1(−13)			1.9(−8)	
	90	9.9(−63)	2.1(−13)	1.1(+37)	9.7(−59)	2.1(−9)	2.2(+33)
	185	2.3(−17)	2.7(−13)	2.0(+14)	5.7(−14)	6.8(−10)	1.6(+11)
	294	5.8(−1)	3.5(−13)	1.1(+6)	5.3(+2)	3.2(−10)	2.4(+3)
GaSb	80	3.4(−29)	4.0(−13)	1.4(+20)	3.0(−25)	3.4(−9)	3.1(+16)
	300	6.6(+11)	7.2(−13)	7.3(−1)	3.5(+14)	3.8(−10)	2.7(−3)
InP	77	5.8(−71)	3.3(−13)	1.1(+41)	7.7(−67)	4.3(−9)	1.7(+37)
	298	2.9(+1)	6.2(−13)	1.2(+5)	2.5(+4)	5.2(−10)	2.8(+2)
InAs	77	1.1(−6)	1.6(−13)	1.2(+9)	8.2(−3)	1.2(−9)	3.2(+5)
	298	1.9(+17)	2.7(−13)	2.2(−3)	7.6(+19)	1.1(−10)	1.1(−5)
InSb	77	7.0(+5)	3.1(−13)	1.1(+3)	3.2(+9)	1.4(−9)	4.7(−1)
	295	6.0(+19)	4.9(−13)	9.2(−5)	1.4(+22)	1.2(−10)	7.8(−7)

$P(\nu)\,\varrho(\nu)$ and U are in cm^{-3} s^{-1}. The energy shown on the abscissa is in eV. The total radiative recombination rate R was obtained by numerically integrating the $P(\nu)\,\varrho(\nu)$ curve over $u = h\nu/kT$.

In many cases, for calculating the intrinsic carrier density n_i, data for m_e and m_h is required. In some cases experimental values of m_e and m_h are available. Braunstein and Kane [126] and Cardona [127] have given methods of calculating the conduction- and valence-band effective masses from the $\boldsymbol{k} \cdot \boldsymbol{p}$ theory. Where experimental values of these masses were not available, Cardona's [127] theoretical values were used. The data on m_e, m_{h1}, m_{h2} are summarized in Table 1 . The final results for R, n_i, τ, B, and σ are recorded in Table 2.

The quantities R, τ, and B were also obtained from the analytical expressions (53), (54), (55), (60), (61), and (62). The necessary data employed in these calculations are shown in Tables 1a, b, and c and the results in Table 3. GaAs, GaSb, InP, InAs, and InSb are usually considered as direct gap semiconductors. However, it was of some interest to see what difference in numerical magnitudes of R, B, and τ is introduced if the transitions are assumed to be indirect. The constant A in equation (13) was assumed to be ≈ 3500 eV^{-2} cm^{-1} for these five semiconductors; and θ was put equal to the Debye temperature [128].

7. Results for Elements

7.1 Diamond

Diamond, in its pure state, is an insulator having a resistivity at room temperature of the order of $10^8\ \Omega$ cm.

Experiments show that there are two types of diamonds, type I and type II [129]. Both types of crystal have absorption bands near 3, 4.0, and 4.8 μm and the Raman shift is observed in both types and is 1332 cm^{-1} (7.51 μm). But type I possesses an absorption band at 8 μm while type II does not.

The type of diamond most studied as a semiconductor is that known as type IIb; this type was shown by Custers [130, 131] to behave as a semiconductor, having a quite low value for its resistivity (10 to 100 Ωcm at room temperature).

7.1.1 Optical properties

Optical properties of diamonds have been reviewed by Moss [132, 104]. In recent years the infrared spectra of diamonds has been investigated by Charette [133, 134, 135] and by Smith and co-workers [136, 137]. The dispersion curves along [100] and [111] directions have been deduced by Yarnell et al. [138, 139] from inelastic scattering of slow neutrons.

Clark, Ditchburn, and Dyer [140] measured the absorption spectra of 52 natural diamonds at 80 and 290 °K. Type I diamonds were shown to have two kinds of absorption centres which are not detected in type II diamonds. Differences in the infrared absorption spectra between type IIa and type IIb were reported. They defined the fundamental absorption-edge characteristic of type II diamonds as that energy E_f at which there is first observed a change of slope of the absorption spectrum due to fundamental absorption. E_f was found to be 5.40 ± 0.03 eV. Champion and Humphreys [141] have also reported spectra for a large number of diamonds. Absorption coefficients up to about 700 cm^{-1} obtained from transmission experiments have been reported by Custers and Raal [142].

The changes in the optical absorption of a relatively pure type II diamond (called D100 in the paper) were measured as a function of temperature between 80 and 522 °K by Clark [143]. The experimental curves, after correction for impurity absorption, could be fitted to two absorption contributions written in the form

$$\alpha \propto \left(\frac{1}{e^{\theta/T}-1}\right)\{(h\nu - E_i)^{1/2} + 0.4\,(h\nu - E_i - 0.075)^{1/2} + c\,(h\nu - E_i - 0.102)^2\} \tag{71}$$

with $\theta = 1050$ and 2900 °K. E_i is the threshold photon energy of the absorption in the branch and c is a constant.

The above results are attributed to indirect transitions into conducting and exciton states of the crystal. Theoretically if the band extrema are at different values of k, indirect transitions involving the absorption or creation of phonons occur giving absorption proportional to $(h\nu - E_i)^{1/2}$ for each exciton band, and $(h\nu - E_i)^2$ for the continuum. Thus the first term in $(h\nu - E_i)$ in the above expression represents transitions into the ground state of the exciton, the next term transitions into the first excited exciton band. The last term represents transitions into the continuum.

The above work was extended by Clark, Dean, and Harris [144]. They studied the optical transmission spectra of several diamonds over the range 5.0 to

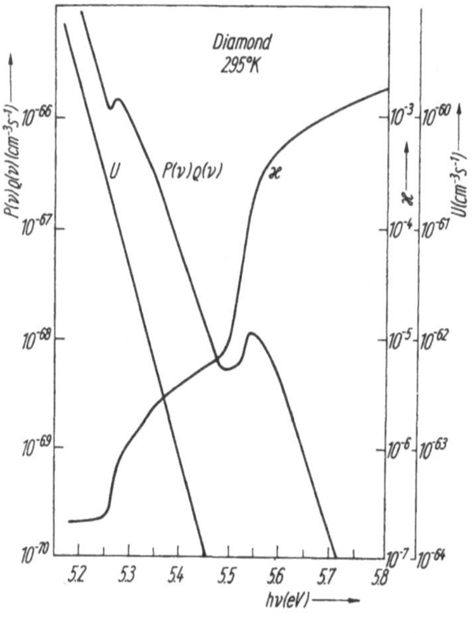

Fig. 7. The absorption index ϰ, the radiative recombination rate $P(\nu)\,\varrho(\nu)$, and the associated quantity U (defined by equation (34)) versus photon energy for diamond at 295 °K.

6.0 eV and at many temperatures between 90 and 600 °K. The intrinsic features of the absorption-edge spectrum could be interpreted in terms of allowed indirect electronic transitions into exciton and free-carrier states of the crystal. Analysis of the absorption data yielded six phonon energies. Three of these appear to represent combinations of two or more phonons.

The reflection spectrum of diamond has been investigated by Philipp and Taft [145] and by Clark et al. [144]. Walker and Osantowski [146] have measured the absolute reflection spectrum of type IIa diamond at room temperature from 4 to 30 eV and analysed it by dispersion techniques to obtain the curves for the optical parameters. These curves disagree substantially in the energy range 5.5 to 12 eV with the previous results of Philipp and Taft [145]. This discrepancy has been discussed by Philipp and Taft [147] who conclude that the results of Walker and Osantowski are not entirely correct. The analysis of [146] yields large negative values of the absorption coefficient in the region where diamond is transparent [142, 143]. This non-physical situation affects the computed values of the optical constants at higher energy and causes the disagreement with the previous work [145].

Data adopted here:

Absorption coefficient: At 295 °K, cf. Fig. 3 of Clark et al. [144].
Refractive index: $n = 2.75$ [146].
Intrinsic carrier concentration: Obtained from equation (48) using $m_c = 0.2\,m$ [144] and $m_h = 0.4\,m$ [148, 149, 150].

Calculated results are given in Fig. 7 and Table 2 (see Section 6).

7.1.2 Discussion of radiative transitions

The $P(\nu)\,\varrho(\nu)$ curve in Fig. 7 shows two maxima at 5.27 and 5.54 eV respectively. Below 5.25 eV, the $P(\nu)\,\varrho(\nu)$ curve shows an increasing behaviour instead of decreasing. This appears to indicate that there is still some impurity absorption left over. The absorption coefficient in this region is very small (≈ 0.3 cm^{-1}) and accurate subtraction of the impurity absorption is difficult.

Intrinsic recombination radiation in diamond was first observed by Male and Prior [151]. Their intensity curve is reproduced in Fig. 8. The radiation emitted has a maximum energy of about 5.6 eV, and intensity maxima occur at energies 5.278 and 5.127 eV. The 5.278 eV maximum is in satisfactory accord with the expectations from Fig. 7. Dean and Male [152] have interpreted this peak as

Fig. 8. Spectrum of intrinsic recombination radiation from diamond (Male and Prior [151])

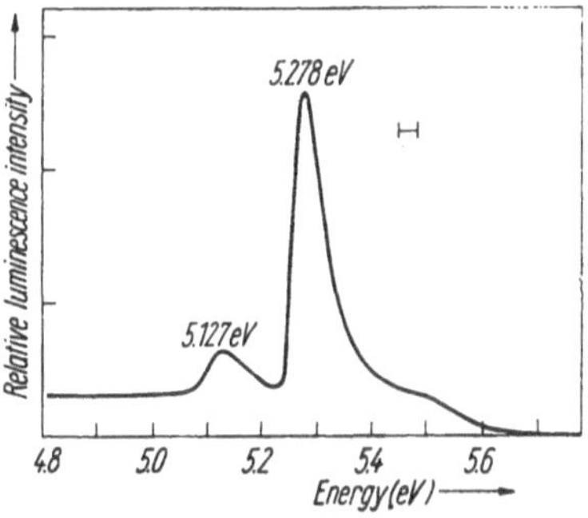

due to recombination occuring from exciton states having a Maxwell-Boltzmann kinetic energy distribution, since the low-energy threshold of the peak falls near the main exciton–phonon absorption threshold of the absorption spectrum [144]. The low-energy dispersion in $\boldsymbol{k}$-space of the associated optical phonons means that the shape of the emission band should be governed mainly by the kinetic energy distribution of the excitons.

Dean and Male initially calculated the points for a single Maxwell-Boltzmann distribution at a temperature of 350 °K (since it was anticipated that the crystal was subject to some Joule heating by the injected current). This distribution was corrected for the finite energy resolution of the spectrometer and was compared with the experimental curve. The experimental curve was observed to be steeper than the calculated one on the low-energy side, although the initial part of the high-energy side fitted reasonably well. The most likely explanation for this discrepancy is that there are, in fact, two unresolved components in the experimental peak. This is reasonable since two thresholds are found to occur in the absorption spectrum near 5.259 and 5.271 eV at 290 °K. The solid dots in the Fig. 9 are calculated on the assumption that each component follows a Maxwell-Boltzmann distribution for $T = 350$ °K. The energy separation and relative intensity of the two components were given the values observed for the corresponding absorption components, but the position of the pair of thresholds was adjusted to give the best fit to the experimental spectrum. Fig. 9 shows that the agreement obtained in this way is quite good.

More recently, Dean and Jones [153] have measured the recombination radiation spectrum over the photon energy range 4.9 to 5.5 eV at 90, 160, 207, and 320 °K. The optical emission was stimulated by bombardment with an electron

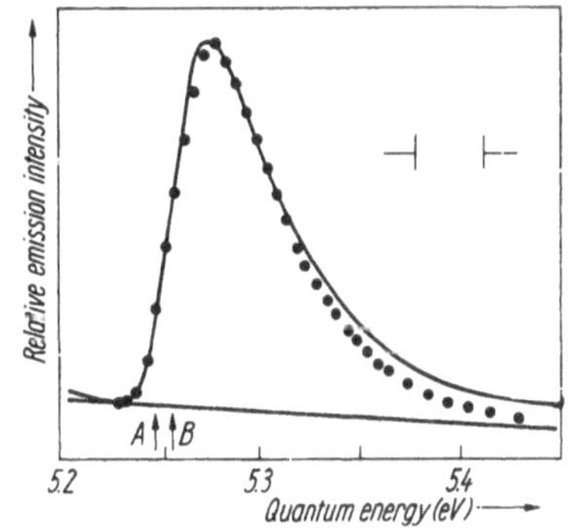

Fig. 9. The dominant peak of the intrinsic recombination radiation spectrum of diamond. Solid dots are points calculated on the assumption that recombination occurs from exciton states having a Maxwell-Boltzmann kinetic energy distribution corresponding to 350 °K. Points A and B indicate the low-energy thresholds of the two assumed components (unresolved experimentally) obtained by fitting the calculated points to the experimental curve (Dean and Male [152])

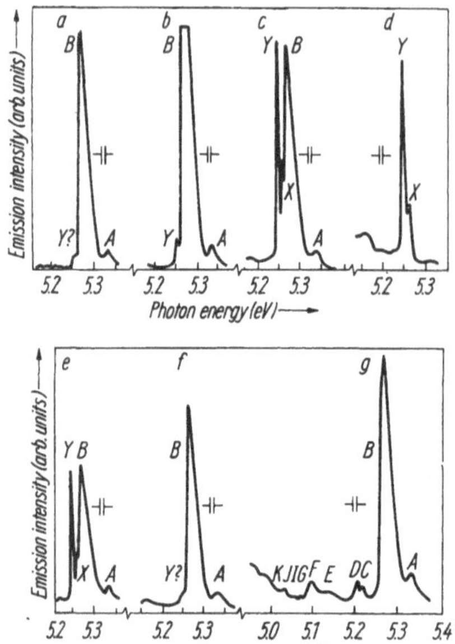

Fig. 10. Recombination radiation spectra from a selection of natural diamonds

a) Specimen Co72 (type IIa); 70 keV, 220 μA electron beam, 0.0063 eV resolution.
b) Specimen K80 (type IIa); 78 keV, 240 μA electron beam, 0.0063 eV resolution.
c) Specimen X60 (intermediate type); 80 keV, 250 μA electron beam, 0.0063 eV resolution.
d) Specimen WXR1 (type I); 80 keV, 250 μA electron beam, 0.0063 eV resolution.
e) Specimen C″116 (intermediate type); 80 keV, 220 μA electron beam, 0.0063 eV resolution.
f) Specimen D10 (type IIa); 72 keV, 240 μA electron beam, 0.0063 eV resolution.
g) Specimen E2 (type IIb); 30 keV, 220 μA electron beam, 0.0063 eV resolution (Dean and Jones [153])

beam of energy between 10 and 80 keV. A few of the emission spectra obtained at 90 °K are shown in Fig. 10. It is seen that the relatively broad bands A and B are the most persistent features of the spectra. The sharp bands C and D are present only in the spectrum from the semiconducting, type IIb, specimen (Fig. 10g), which also contains a number of other sharp and broad bands at lower energies (bands E to K). The effect of temperature on the recombination radiation spectra is shown in Fig. 11; the band B is seen to broaden appreciably with the increase in temperature.

Band B corresponds to the main feature in the spectrum recorded by Male and Prior [151]. Bands A and B are interpretable in terms of exciton annihilation with the emission into the lattice of one or more phonons.

7.2 Silicon

Silicon is a simple monatomic semiconductor and its properties have been intensively investigated. Reviews of its properties may be found in Fan [154] Moss [22], and Smith [2].

7.2.1 Optical properties

The optical properties of silicon have been extensively investigated.

Collins and Fan [155] and Johnson [156] determined the absorption spectra of silicon by means of optical transmission studies in the infrared region. Johnson in his study correlated the lattice absorption bands with multiple phonon

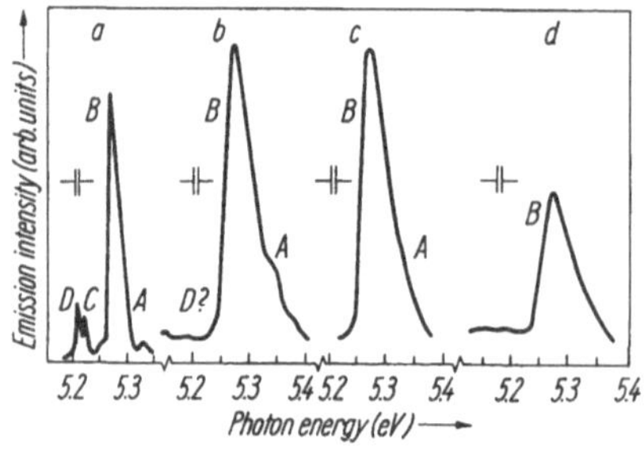

Fig. 11. Recombination radiation spectra from diamond at various specimen temperatures (type IIb, specimen A 100)

a) 90 °K; 80 keV, 220 μA electron beam, resolution 0.0063 eV.
b) 160 °K; 80 keV, 250 μA electron beam, resolution 0.0063 eV.
c) 207 °K; 80 keV, 250 μA electron beam, resolution 0.0063 eV.
d) 320 °K; 80 keV, 250 μA electron beam, resolution 0.0063 eV (Dean and Jones [153])

processes up to and including three phonons and covered the spectral region from 6.5 to 30 μm. His analysis of the lattice vibration spectra is in good agreement with that as determined by Brockhouse [157] from neutron scattering experiments on Si. Dolling [158] has reported more recent results from neutron scattering experiments. Two phonon bands in the infrared absorption spectra have been observed by Balkanski and Nusimovici [159]. Spectral emissivity measurements at wavelengths between 3 and 15 μm have been made on n- and p-type silicon by Stierwalt and Potter [160]. Infrared absorption of silicon at 7.5 °K has been measured by Aronson et al. [161]. Reflectance studies have been carried out by Vavilov et al. [162], Robin-Kandare and co-workers [163, 164], Philipp and Taft [165], Sasaki and Eshiguro [166] (10 to 19.2 eV), Philipp and Ehrenreich [114] (1.5 to 25 eV), and by Subashiev et al. [167] and in some cases the values of n and $\varkappa$ have also been derived.

Very precise measurements of the ultra-violet reflectance of silicon by Lukeš and Schmidt [112] have shown a hyperfine splitting of the M_1 absorption edge. Phillips [168] has pointed out that this structure is exactly what one must expect when multiphonon transitions are taken into account.

Investigations in the wavelength range from 2 to 7 μm have shown that the observed coefficient in n-type is greater than is predicted by theory. This anomalous behaviour has been discussed by Yakovlev [169]. The transmission of silicon between 40 and 100 μm has been measured by Walles [170].

The first detailed measurements on the absorption coefficient of Si in the vicinity of the absorption edge were those of Fan and co-workers [171, 30] who reported results at a number of temperatures between 90 and 602 °K. Soon after, Dash and Newman [172] measured the absorption coefficients of high-purity single crystal silicon at 77 and 300 °K in the energy range 1 to 3.4 eV.

The most exhaustive work on absorption at low values of the absorption coefficient is that of Macfarlane and collaborators. In their early work, Macfarlane and Roberts [31] measured the absorption of a polycrystalline specimen of silicon in the neighbourhood of the absorption edge at temperatures from 20 to 333 °K. It was found that in the region $\alpha < 100\ \text{cm}^{-1}$ the results can be well represented by an equation of the form (13). θ was found to be 600 °K.

This was followed by measurements of the absorption spectrum under higher resolution by Macfarlane, McLean, Quarrington, and Roberts [125, 173] near the main absorption edge at various temperatures between 4.2 and 415 °K. Fine structure was found in the absorption on the long-wavelength side of the edge. The structure was analysed and interpreted in terms of indirect transitions involving, in general, phonons with energies corresponding to temperatures of 212, 670, 1050, and 1420 °K. The original interpretation of the two components corresponding to the temperatures of 1050 and 1420 °K is now considered to be incorrect and instead these are now believed to be due to two-phonon processes [23].

Braunstein, Moore, and Herman [174] have reported measurements on the absorption coefficients at 5, 78, 197, 234, and 296 °K. At low energies they confirm the findings of Macfarlane and Roberts that the variation of α with E can be represented by equation (13).

Other measurements on the absorption coefficient are those of Huldt and Staflin [175], Dubrovskii et al. [176], and of Robin-Kandare and Robin [177].

The effect of heavy doping on the intrinsic absorption edge of silicon has been investigated by Dubrovskii and Subashiev [178, 179]. Dubrovskii [180] has calculated the energy dependence of the absorption spectrum in the vicinity of the intrinsic absorption edge in heavily doped silicon.

The variation of absorption coefficient with temperature for a fixed energy has been examined. Kessler and Schnell [181] and Ukhanov [182] found that for wavelengths between 1.5 and 13 μm, log α is linear with $1/T$. This dependence can be explained by the interaction of photons with free charge carriers [182]. We may note that at low temperatures deviations from linearity are found [181].

The dependence of the refractive index on temperature has been studied by Cardona, Paul, and Brooks [183] in the wavelength range 1 to 5 μm at 100, 194, and 297 °K. Another such investigation in the wavelength range 1.1 to 5.5 μm and in the temperature range 109 to 750 °K has been reported by Lukeš [184, 185], who finds that his experimental values of $\mathrm{d}n/\mathrm{d}t$ agree with the values predicted by the classical theory.

Data adopted here:

Absorption coefficient: Calculations were made using directly the experimental values of α as also by utilizing equation (13).

a) Experimental values at 291 °K, cf. Macfarlane et al. [173 and 31].

b) Equation (13) was used for 20, 90, 195, 249, 290, and 333 °K; $\theta = 600$ °K and E_g values given below.

The constant A was obtained by fitting equations (16) to the 290 °K curve of Macfarlane and Roberts [31] at 5 points. The mean value thus obtained was $A = 2682\ \mathrm{cm^{-1}\,eV^{-2}}$. We may note that Smith [2] gives $A = A'/E_g^2 = 5000/1.086^2 = 4239$ and Hall [3] gives $A = 3600$. We have been unable to locate the origin of these two high values.

Refractive index: The data given by Philipp and Taft [165] for the energy region 0.6 to 2.4 eV was used to obtain the following relation between n and E:

$$n^2 - 1 = \frac{100}{9.527 - 0.20175\,E - 0.49125\,E^2}\,. \tag{72}$$

The above relation gives values which are indistinguishable from those plotted by Philipp and Taft.

Intrinsic carrier concentration: Morin and Maita [186] give

$$n_i^2 = 1.5\times 10^{33}\,T^3 \exp\left(-\frac{1.21}{kT}\right), \tag{73}$$

where kT is in eV.

Such an expression assumes a linear variation of E_g with T. As the E_g values of Macfarlane and Roberts [31], specially at low temperatures, show strong departure from such a behaviour, it was felt that it would be better to express n_i in terms of E_g. Assuming

$$n_i^2 = D\,T^3 \exp\left(-\frac{E_g}{kT}\right), \tag{74}$$

values calculated from (73) were equated to (74) at two temperatures (249 and 333 °K), and an average value of D was determined. Thus

$$n_i^2 = 1.575\times 10^{31}\,T^3 \exp\left(-\frac{E_g}{kT}\right). \tag{75}$$

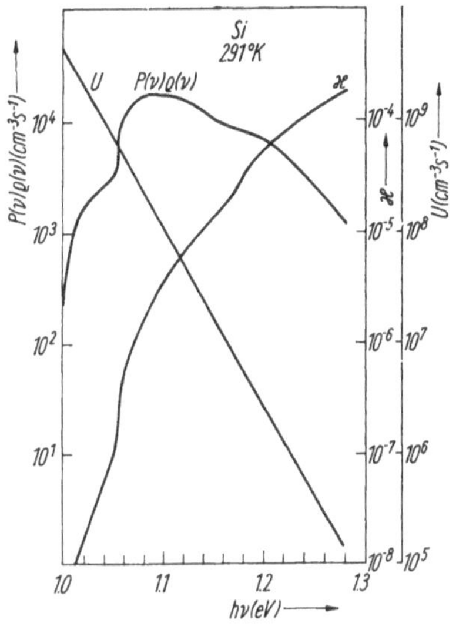

Fig. 12. $\varkappa$, $P(\nu)\,\varrho(\nu)$, and U for silicon at 291 °K

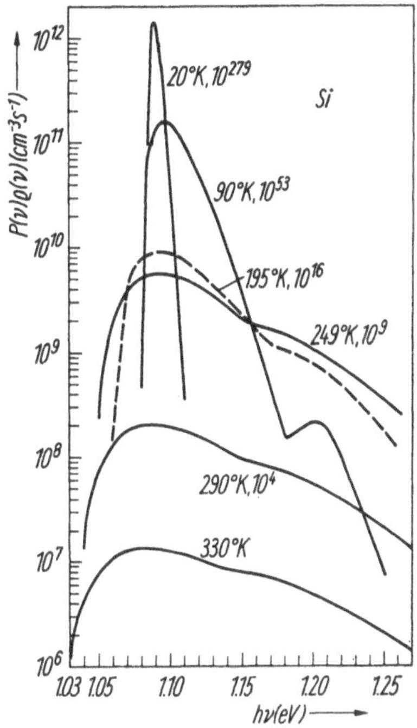

Fig. 13. Radiative recombination rate for silicon at several temperatures (For plotting purposes, $P(\nu)\,\varrho(\nu)$ has been multiplied by the factors shown)

We may note that at high temperatures (300 to 1000 °K) Kessler and Schnell's [181] results show

$$n_i = 5.71 \times 10^{16}\, T^{3/2} \exp\left(-\frac{1.16}{2\,kT}\right), \tag{76}$$

where kT is in eV.

Energy gaps: The following values were read from Fig. 2 of [31]:

T (°K)	E_g (eV)
20	1.137
90	1.132
106	1.129
195	1.110
249	1.098
290	1.086
333	1.074

Calculated values of $P(\nu)\,\varrho(\nu)$, $\varkappa$, and U at 291 °K using experimental values of absorption coefficients are shown in Fig. 12. Results for $P(\nu)\,\varrho(\nu)$ at a series of temperatures obtained from equation (13) are shown in Fig. 13. Calculated values of R, n_i, B, τ, and σ are recorded in Table 2 (see Section 6).

7.2.2 *Discussion of radiative transitions*

The maximum in the $P(\nu)\,\varrho(\nu)$ curve in Fig. 12 lies at ≈ 1.09 eV and is in reasonable accord with the observations of Haynes and Westphal [187] who found that at room temperature the peak of the recombination radiation occurs at 1.088 eV.

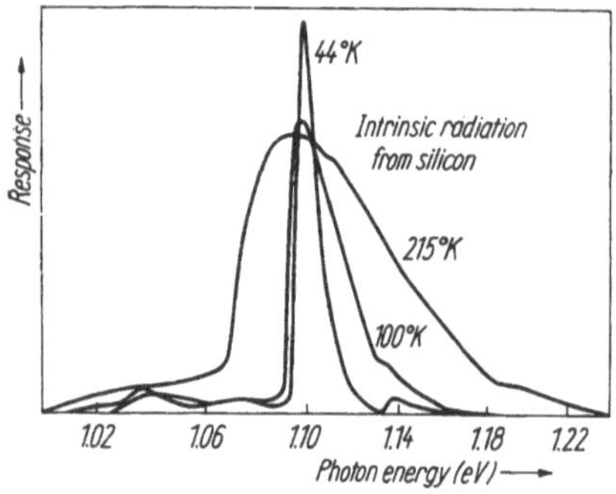

Fig. 14. Intrinsic recombination radiation from silicon obtained at three different temperatures (Haynes, Lax, and Flood [188])

The curves in Fig. 13 show that as the temperature is decreased the halfwidth decreases i.e. the radiation becomes sharper. It is interesting to compare Fig. 13 with some experimental results of Haynes, Lax, and Flood [188] reproduced in Fig. 14.

Silicon is a good illustration of the difficulties of accurate interpretation of experimental observations in this field.

The recombination radiation from silicon has been investigated by Haynes and collaborators over a number of years. The first observations were made by Haynes and Briggs [94] who found that a maximum intensity occurred at about 1.1 μm.

Haynes and Westphal [187] produced recombination radiation by applying current pulses to grown silicon p–n junctions with current flow from p- to n-type. Observations were made on a silicon sample containing boron and arsenic as added impurities at room temperature and at 77 °K. Three maxima were found to occur at 77 °K: a very sharp peak at 1.100 eV and two smaller maxima at 1.072 and 1.038 eV. The photondistribution of the 1.10 eV was found to be independent of the nature of the added impurities and was consistent with the optical absorption constants and the theory of van Roosbroeck and Shockley [10]. Thus it was interpreted as due to indirect transitions of electrons from (100) minimum to the top of the valence band involving phonon cooperation (indirect intrinsic radiation). The maximum at 1.039 eV was believed to be due to the recombination of electrons with holes on unionized boron atoms. Haynes and Westphal attempted to interpret the 1.072 eV maximum as due to the recombination of injected holes with electrons on unionized arsenic atoms, but they found that the ionization energy of As thus deduced was in serious disagreement with that obtained from thermal and from optical measurements.

Further investigations were reported by Haynes, Lax, and Flood [188, 189] at the Rochester and Prague conferences. Here we shall discuss the latter results.

The production of intrinsic recombination radiation in silicon involves a change in crystal momentum since for silicon the valence-band maximum is at the center of the Brillouin zone while the conduction-band minima are at a crystal momentum of k_0 which is near the zone edge. Recombination of electrons at the conduction-band minima with holes at the valence-band maximum therefore involves a change in crystal momentum from k_0 to $k = 0$ and this crystal momentum must be conserved. This can be accomplished by the electron either emitting or absorbing a phonon having a crystal-momentum wave vector of k_0. The phonon spectrum of silicon has four branches in the direction of the conduction-band minima and hence there are four phonons of different energies all having the required crystal momentum of k_0. Since energy as well as crystal momentum must be conserved in the transition we may expect the intrinsic radiation from silicon to consist of eight lines: four with energies less than the energy gap due to the emission of four phonons of four different energies, and four lines with energies higher than the energy gap produced by absorption of these four different phonons.

In the experiments of Haynes et al. [189] excess minority carriers were produced in silicon single crystals by illuminating them with light from a high-intensity mercury arc. These carriers recombine within a diffusion length and in thermal equilibrium with the crystal, giving recombination radiation. This radiation was analyzed with a spectrometer, detected by a lead sulphide cell and automatically recorded on a chart.

In Fig. 15 is shown a reproduction of a recorder trace of recombination radiation from a silicon crystal at 18 °K. The horizontal axis is the photon energy in eV. The vertical response is nearly proportional to the number of photons per unit energy interval. The crystal contained 2×10^{14} cm^{-3} phosphorus atoms and the small spike at C is due to them. The other four lines are all due to phonon emission since at this low temperature there are not enough of any of the required phonons to permit appreciable absorption. The phonon spectrum of silicon in the [001] direction has been measured by Brockhouse [157] by neutron scattering; these data enable an assignment of the particular phonon emitted in the transitions. The line at E is produced by the emission of the transverse acoustical (TA) phonon, the principal line at D is produced by the emission of the transverse optical phonon, and the lines at B and A are produced by the emission of sums of two phonons. The cause of the absence of transitions involving a single LO phonon is not clear [190].

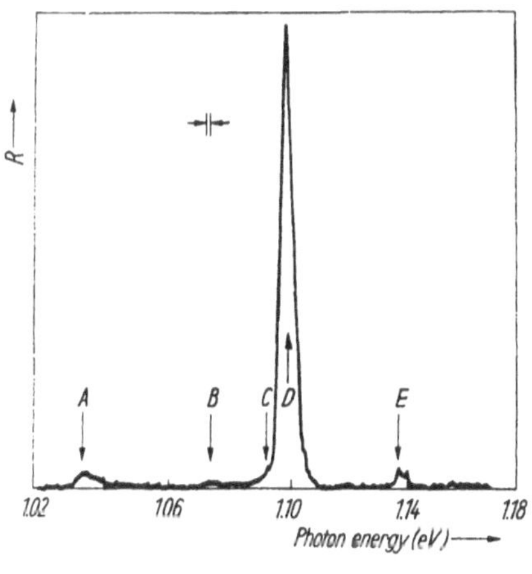

Fig. 15. Spectrometer response (nearly proportional to the number of photons/unit energy interval) as a function of the energy of the photons for a silicon crystal containing 2×10^{14} phosphorous atoms per cm^3 at 18 °K (Haynes, Lax, and Flood [189])

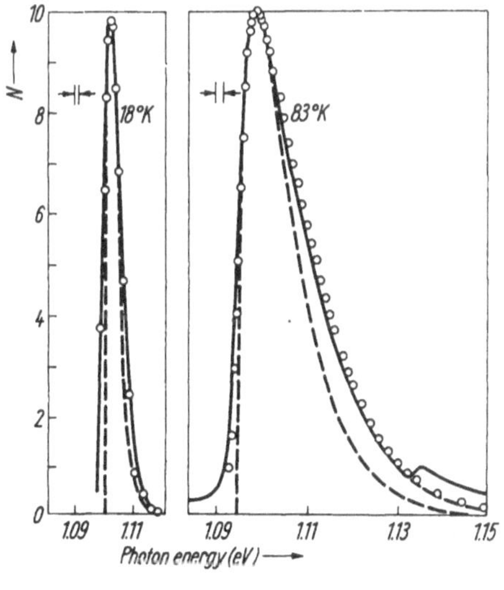

Fig. 16. Relative number of photons, N, as a function of photon energy at two temperatures. The solid lines are corrected spectrometer traces. The circles are calculated from optical absorption data. The dashed lines are Boltzmann energy distributions (Haynes, Lax, and Flood [189])

Using the value of the TO phonon and the van Roosbroeck-Shockley theory, the optical absorption data of Macfarlane et al. [125] can be brought into accord with the above results. This agreement can only be obtained, however, when it is realized that photon absorption at low temperatures is not the inverse process of photon emission. This is because both involve phonon emission; there are too few phonons of the required energy to permit of appreciable absorption. It is therefore essential that the optical absorption produced by the emission of different phonons be treated separately. When this is done the principle of detailed balance gives the radiation which occurs in thermal equilibrium due to phonon absorption. The measured radiation which occurs with phonon emission is obtained by shifting the calculated radiation to lower energy by twice the phonon energy, and the amount of radiation is obtained by multiplying by exp $(\hbar\,\omega/kT)$, where $\hbar\,\omega$ is the phonon energy.

The result obtained in this way is shown in Fig. 16, which is a plot of the number of photons per unit energy, N, as a function of photon energy in eV for two different temperatures. The solid lines are recorder traces of the radiation produced by emission of the TO phonon corrected for overall spectrometer and detector response. The circles are the values calculated from the optical absorption data. It will be seen that at 18 °K the accord is very good. At 83 °K the low energy fit is probably as good since a small part of the amplitude of the recorder trace is due to two-phonon emission. At higher energies the calculated values are a little too large. This probably reflects the error involved in separating the absorption involving TO from that involving TA phonons at the higher temperature.

Dumke [191] has calculated the ratio of the integrated intensity for a two-phonon process to the integrated intensity for the corresponding single-phonon process.

Recently Uchida [192] has observed the recombination radiation from forward biased p–n junctions of silicon using an infra-red image converter tube. The lifetime estimated from the decay time of the pulsed radiation was found to agree fairly well with that measured by the photoconductivity decay method. Smirnov, Vavilov, and Gerasimenko [193] have examined the feasibility of investigating the kinetics of the band-to-band recombination radiation of silicon using photomultipliers.

Newman [194] found that the passage of a reverse bias breakdown-current across a silicon p–n junction is accompanied by the emission of visible light. Subsequently, Chynoweth and McKay [195] made a detailed study of this light emission phenomenon, particularly its spectral distribution, and concluded that the visible light originated from recombinations occurring between energetic electrons and holes in the avalanche breakdown region. The spectrum of the radiation emitted from avalanche breakdown regions along the edge of a diffused silicon junction at 300 and 77 °K has been measured by Davies and Storm [196]. The results indicate no difference in the spectrum at 77 and 300 °K, within experimental error, in the range of photon energies 1.0 to 1.4 eV.

Davies and Storm [197, 196] have also measured the intensity distribution in the band-to-band recombination radiation from silicon under hot electron conditions. The results indicate that intrinsic recombination radiation in silicon, at least at temperatures below 77 °K, arises predominantly from the decay of excitons rather than from direct recombination of electrons and holes.

7.3 Germanium

Because of its technological importance, a considerable amount of research work has been carried out on the properties of germanium.

7.3.1 Optical properties

Earlier investigations on the optical properties of germanium have been discussed by Moss [132, 22].

Infra-red absorption due to lattice vibrations has been measured for pure Ge by Lax and Burstein [198]. Collins and Fan [155] have investigated the effect of impurities on these bands. Aronson, McLinden, and Gielisse [161] have presented high-resolution low temperature (7.5 °K) far-infrared data on intrinsic germanium.

More recently, Fray et al. [199] have remeasured the infra-red lattice bands of germanium and interpreted them as multiple phonon bands involving four characteristic phonon energies.

Stierwalt and Potter [200] have made infra-red spectral emittance studies, and neutron scattering experiments have been carried out by Brockhouse and Iyengar [201].

The optical constants of single-crystal Ge were obtained by Archer [202] from the ellipticity of reflected polarized light. The data were consistent with the Kramers-Kronig relations.

Computations of the optical constants n and k using reflectivity measurements and Kramers-Kronig dispersion relations have been carried out by Philipp and Taft [203] and by Rimmer and Dexter [204] over the range 0 to 10 eV and have been extended to 18 eV by Rustgi, Nodvik, and Weissler [205]. Nodvik [206] has pointed out a check on the over-all consistency of these calculated values by using these in a dispersion relation to evaluate the free-electron plasma frequency ω_p given by

$$\omega_p^2 = \frac{4\,\pi\,N\,e^2}{m},$$

where N is the free electron density. The results obtained for ω_p are consistent with the assumption that four valence electrons contribute to the electron density N.

Other recent reflectance measurements are those of Kiseleva and Pribytkova [207] and Philipp and Ehrenreich [113, 114].

Careful measurements [208 to 211, 112] on the reflectivity spectrum in the energy range 3 to 5 eV have revealed the existence of a fine structure. This fine structure is believed to be due to direct interband transitions.

The effect of temperature and doping on the reflectivity of germanium in the fundamental absorption region has been investigated by Cardona and Sommers [210].

The absorption coefficients of Ge have been measured at different temperatures. The studies of Dash and Newman [172] encompassed a range of absorption coefficient from 0.1 to 10^5 cm^{-1} at 77 and 300 °K. They succeeded in showing that the absorption spectrum is separated into two regions, one in which the absorption process is dominated by direct transitions, and another in which indirect transitions are dominant.

Accurate measurements at low values of the absorption coefficient were carried out by Macfarlane and Roberts [32] who found that at several temperatures α can be well represented by an equation similar to (16). These results were later refined by Macfarlane, McLean, Quarrington, and Roberts [124]. The measurements of these authors made with high resolution near the main absorption edge at various temperatures between 4.2 and 291 °K revealed a fine structure in the absorption on the long-wavelength side of this edge. This structure was analysed and was interpreted in terms of indirect transitions involving phonons with energies corresponding to temperatures of 90 and 320 °K. These authors have also investigated [212] the absorption around levels of about 4000 cm^{-1} where the absorption is thought to be due to direct optical transitions, and have confirmed exciton effects in absorption predicted by Elliott [29].

Moss and Hawkins [213] have extended the values of absorption coefficient near the band edge to much lower levels ($\approx 10^{-5}$ cm^{-1}) by measurements of photoconductivity of suitable specimens.

Hobden [214] has measured the optical absorption coefficient in the region of 1.1 eV and has found an absorption edge which he attributes to direct transitions from the split-off valence band to the conduction band.

Other absorption measurements are those of Braunstein et al. [174], Robin-Kandare and Robin [177], and Harbeke [215].

McLean and Paige [216] have discussed the mechanisms which can account for the different magnitude and temperature dependence of the broadening of the three optical absorption edges.

The dependence of the absorption coefficient on temperature has been investigated by Kornfeld [217] and Ukhanov [182]. Both authors find that log α decreases linearly with $1/T$. Kornfeld gives the following equation:

$$\alpha = S_2 \exp\left(-\frac{E_0 - E}{kT}\right), \tag{77}$$

where $S_2 = 2\times 10^4$ cm^{-1} and $E_0 = 0.89$ eV.

The variation of refractive index with wavelength has been studied by a number of workers: 1.8 to 2.6 μm [218], 2 to 16 μm [219, 220], 1.8 to 5.5 μm [203], 1.5 to 2.2 μm [221].

Investigations on the dependence of refractive index on the energy as well as on the temperature have been made by Cardona, Paul, and Brooks [183] and by Lukeš and Schmidt [222, 223, 224].

The theory of the temperature dependence of the refractive index has been discussed by Antončik [225] and Lukeš [226]. Lukeš [222, 226, 224] finds that the theoretical results agree with the experimental results quantitatively to a satisfactory degree.

Optical properties of thin films and layers of germanium have been investigated by several workers [227 to 233, 115]. We may note that results on the optical properties obtained by measurements made on evaporated films are known to differ not only from bulk material but also from film to film, depending upon method of preparation and history.

Data adopted here:

Absorption coefficient: At 77 °K; region 0.704 to 0.846 eV read from the figure given in Macfarlane et al. [124], region 0.840 to 0.94 eV read from Fig. 2 of Dash and Newman [172]. At 300 °K, cf. Fig. 2 and 3 of Dash and Newman [172].

Refractive index: Obtained from the following equation given by Kornfeld [221]:

$$n^2 = 12.7 + \frac{6}{1.8225 - E^2}. \tag{78}$$

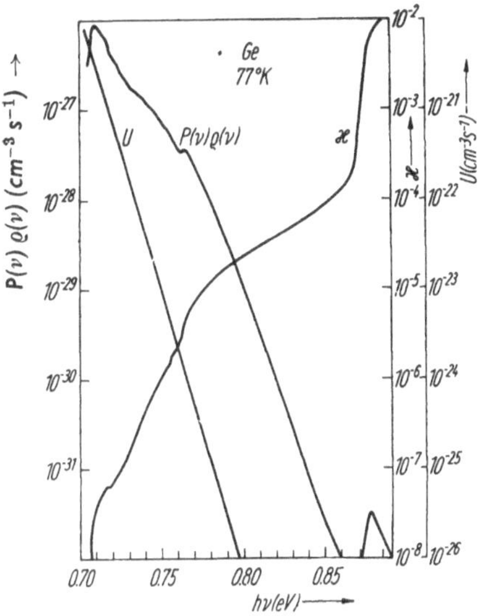

Fig. 17. ϰ, $P(\nu)\,\varrho(\nu)$, and U for germanium at 77 °K

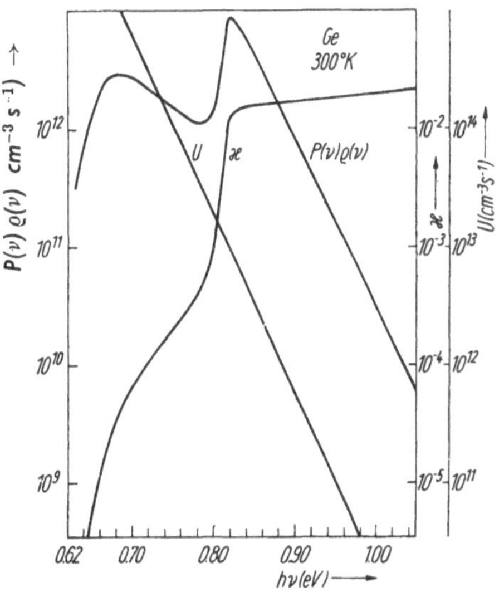

Fig. 18. ϰ, $P(\nu)\,\varrho(\nu)$, and U for germanium at 300 °K

Intrinsic carrier concentration: Calculated from the following equation of Morin and Maita [234, 186]:

$$n_i^2 = 3.10 \times 10^{32}\, T^3 \exp\left(-\frac{0.785}{kT}\right), \tag{79}$$

in which kT is in eV.

Results are shown in Fig. 17, 18, and Table 2 (see Section 6).

7.3.2 Discussion of radiative transitions

As compared to other substances, germanium has been relatively well investigated. The recombination radiation, radiative recombination rate R and cross section σ, all three have been measured. We discuss them separately.

7.3.2.1 Radiation

The first observations of the intrinsic radiation from germanium due to recombination of injected holes and electrons were made by Haynes and Briggs [94] and by Newman [95]. These investigations showed that the maximum of the energy radiated from the specimens used occurs at a wavelength of about 1.8 μm.

We first discuss here some early results obtained by Haynes [92]. Though these results have been superceded by later more accurate work, these illustrate the effect of absorption in the sample very well.

The absorption constant varies rapidly in the range of wavelengths considered and the emergent radiation has to be multiplied by a correction factor.

The spectrum shown by solid line in Fig. 19 was obtained by using a p–n junction with forward bias. The germanium sample was 1.3×10^{-3} cm thick. The diffusion lengths of the excess carriers were long compared to the sample thickness and the surface was treated to insure a low surface-recombination

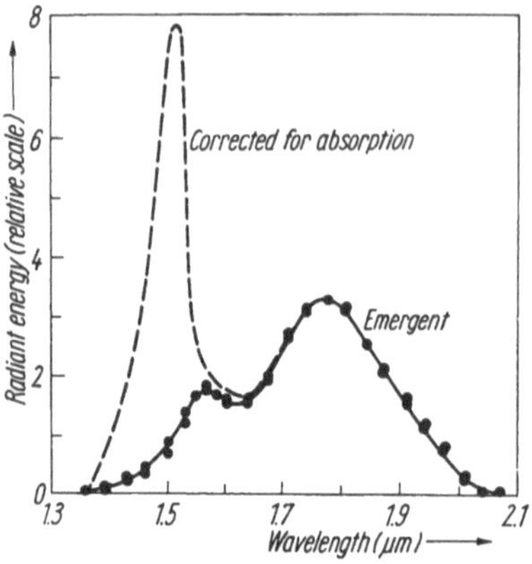

Fig. 19. Plot of experimental data obtained using a germanium sample 1.3×10^{-3} cm thick showing measured emergent radiation as a function of wavelength. The dotted curve was obtained from the solid line by correcting for absorption in the sample (Data taken by using a p–n junction with forward bias) (Haynes [92])

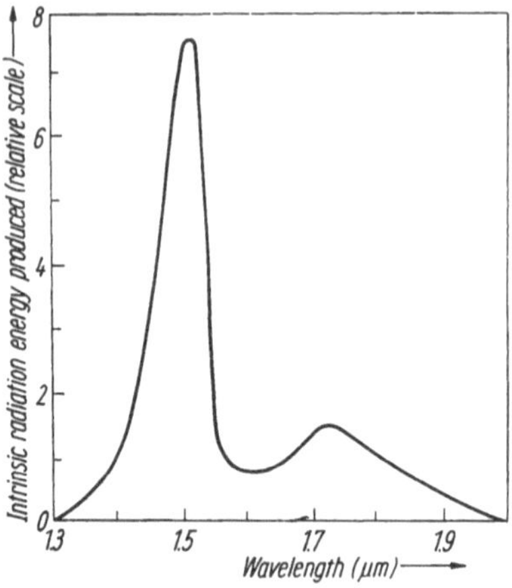

Fig. 20. Shape of radiant energy produced by the recombination of electrons and holes in germanium as a function of wavelength calculated using the optical absorption data of Dash, Newman, and Taft and the theory of van Roosbroeck and Shockley (Haynes [92])

velocity. In these circumstances it is reasonable to assume that both the minority-carrier density and volume recombination take place uniformly through-out the sample thickness. The recombination radiation energy produced in germanium, J_0, is then given by

$$J_0 = \frac{J \alpha d}{1 - e^{-\alpha d}}, \quad \text{for } \alpha d > 1, \tag{80}$$

where J is the measured emergent energy at a given wavelength, α is the optical absorption coefficient, and d is the sample thickness.

When this correction is applied to the experimental data, the resultant curve shows a large second peak at about 1.5 μm, shown by the dotted curve in Fig. 19. The resultant curve may be compared with the theoretical curve, calculated from the absorption data of Dash, Newman, and Taft [235] shown in Fig. 20. The agreement between the two curves is seen to be satisfactory.

Aigrain and Benoit à la Guillaume [100] observed the recombination radiation at 90 and 330 °K with high efficiency by using germanium in the form of a Weierstrass sphere. Further investigations were reported by Benoit à la Guillaume [236] and Newman [236a].

The role of phonon participation was recognized in the later work of Haynes, Lax, and Flood [188].

A very recent study of the radiative recombination in germanium in which both the indirect and direct transitions have been observed, has been presented by Haynes and Nilsson [237, 238]. Because of the very different absorption properties of photons emitted in direct and indirect transitions, it is possible to use the relative spectral intensity to measure the diffusion length of carriers in the material [238]. In the following we summarize the results of Haynes and Nilsson.

At low temperatures, for Ge, $\exp(h\nu/kT) \gg 1$ and the expression of van Roosbroeck and Shockley may be written as

$$R(h\nu)\, d(h\nu) = \frac{8 \pi k^3 T^3}{c^2 h^3} n^2 \left(\frac{h\nu}{kT}\right)^2 \alpha\, e^{-h\nu/kT}\, d(h\nu)\,. \tag{81}$$

This expression is valid for direct transitions. It is also valid for indirect transitions in the temperature range where α is known for both phonon emission and absorption. However, this is not true for germanium at low temperature since both measurable optical absorption and radiation occur by phonon emission. In this case it is essential that the optical absorption produced by the emission of different phonons be treated separately. When this is done equation (81) gives the thermal equilibrium radiation due to phonon absorption. The radiation which actually occurs with phonon emission is shifted to lower energy by twice the phonon energy and the radiation rate is multiplied by

$$B = \left(\frac{n_e h \nu_e}{n_a h \nu_a}\right)^2 \left(\frac{\Delta\varepsilon - \hbar\omega}{\Delta\varepsilon + \hbar\omega}\right)^2 e^{\hbar\omega/kT}, \tag{82}$$

where the subscripts "e" and "a" refer to phonon emission and absorption, $\hbar\omega$ is the phonon energy, and $\Delta\varepsilon$ is the difference in energy between the initial and intermediate states of the recombination transition. The first expression in parenthesis corrects for the change in density of photon states between the photon energy at phonon emission and that at phonon absorption. The second factor was explained by Macfarlane et al. [173] and the exponential term gives the usual difference between emission and absorption when its value is $\gg 1$.

The radiative recombination is bi-molecular, and the rate involves a product of the hole and electron concentrations. Thus the efficiency of the radiative process may be defined [237] by dividing equation (81) by the $n\,p$ product at equilibrium; this product is given by

$$n\,p = N_c N_v \exp\left(-\frac{E_g}{kT}\right), \tag{83}$$

where n and p are the concentrations of electrons and holes respectively, and N_c and N_v are the densities of states in the conduction and valence bands. Thus the efficiency $E(h\nu)$ is given by

$$E(h\nu) = \frac{R(h\nu)}{n\,p} = \frac{8\pi k^3 T^3}{c^2 h^3} \frac{n^2 \left(\frac{h\nu}{kT}\right)^2 \alpha\, e^{-\Delta h\nu/kT}}{N_c N_v}, \tag{84}$$

where $\Delta h\nu = h\nu - E_g$. Equation (84) is valid for the radiative efficiency of direct transitions when N_c is equal to the density of states at $k = 0$ and $\Delta h\nu$ is measured from the direct energy gap. For the indirect transitions both N_c and E_g refer to the multivalleyed band edge. Proper account must be taken of phonon emission and absorption in this case.

Using equation (84) and the optical absorption data for germanium [173, 124, 212, 23], the expected recombination radiation in Ge at 77 °K was computed by Haynes and Nilsson [237] and is shown in Fig. 21. The left-hand side of the figure is for indirect transitions which involve the emission of LA phonons. The right-hand side of the figure is for direct transitions at $k = 0$. The plotted points are the calculated values of equation (84). Experimental results for the indirect transitions at 70 °K are shown for comparison. The displacement of the experimental curve by 1.25 meV corresponds to the decrease of the energy gap between 70 and 77 °K.

It may be noted that the vertical scales for the two sides of Fig. 21 are very different, also that the efficiencies are referred in each case to a $n\,p$ product in which the n is that for the band in question. The peak efficiency for the direct recombination is approximately 3.2×10^6 times as large as that for the indirect

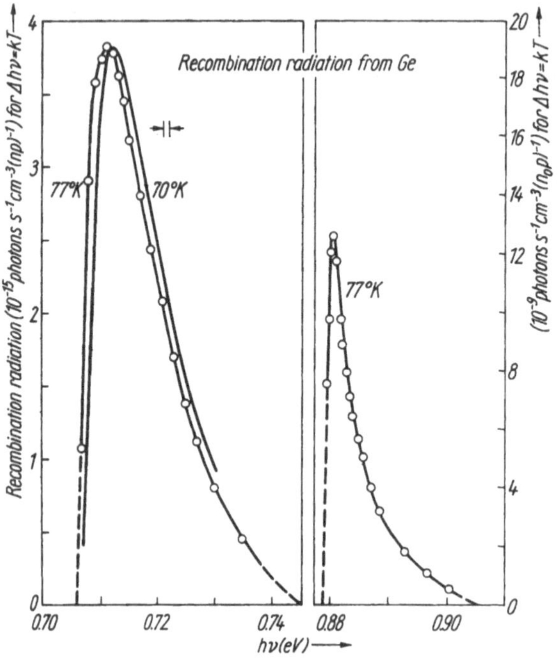

Fig. 21. Recombination radiation versus photon energy for germanium. The plotted points are calculated from equation (84). The curve labelled "70 °K" is the experimental one
(Haynes and Nilsson [237])

peak. Integrating over the lines, the efficiencies for the two processes are in the ratio of approximately 1.4×10^6. That is, for an electron in the $k = 0$ conduction-band minimum, the rate of radiative recombination is more than 10^6 times larger than for an electron in the many-valleyed minima in which radiation occurs with phonon assistance. In spite of the extremely high relative efficiency of the direct recombination process, it cannot be observed experimentally at these low temperatures because the probability of an electron being in the conduction band at $k = 0$ (approximately 0.15 eV above the indirect band minimum) is down by a factor of 10^{11} in comparison with the indirect band. At higher temperatures this is no longer true.

The shapes of the two lines in Fig. 21 are quite different. The indirect line at 77 °K is primarily due to recombination through excitons. Its width is determined by the Boltzmann energy distribution of these excitons and by exciton scattering with the lattice. The $k = 0$ line is made up of recombination via excitons at the low-energy end and by recombination of free holes and electrons at energies above about 0.8825 eV.

Basov, Lisitsyn, and Osipov [93] have used laser excitation for producing recombination radiation in germanium. The source of optical excitation was a ruby laser with 6934 Å waves. Basov and Bogdankevich [96] have detected the recombination radiation of germanium on irradiation with a beam of electrons of ≈ 0.6 MeV energy obtained from a linear accelerator.

The influence of degeneracy on recombination radiation in germanium was investigated by Pankove [239].

When the free carrier concentration is large, electron–hole interaction can cause screening for the carriers, which results in a decrease in their potential energy. In terms of the energy band concept, this is equivalent to a decrease in width of the forbidden energy gap of the semiconductor for large current carrier concentrations. This narrowing should lead to a displacement toward smaller energies of the long-wavelength edge of the intrinsic recombination radiation spectrum. Asnin and Rogachev [240] have carried out experiments on germanium under high-injection conditions which confirm the above expectation. The excess carriers were produced by applying forward voltage pulses to a germanium diode. It was observed that as the current density increased, the long-wavelength edge of the recombination radiation was displaced toward lower energies.

Nelson and Irwin [241] have observed visible light from a germanium reverse biased p–n junction. The emission is believed to arise from avalanche breakdown. Detailed investigation on the spectral properties of the emitted radiation have been carried out by Chynoweth and Gummel [242] and a theory to explain the spectrum has been developed by Wolff [243].

7.3.2.2 Radiative recombination rate

From equation (37) we find that the net rate of radiation per unit volume, δR, resulting from the recombination is given by

$$\delta R = R \frac{(n_0 + p_0)\,\delta p + (\delta p)^2}{n_i^2}, \tag{85}$$

where n_0 and p_0 are the equilibrium concentrations of electrons and holes respectively and δp is the excess concentration of holes or electrons. Equation (85) has been used by Brill and Schwarz [244, 245] to obtain R for germanium.

Light from a microscope lamp was chopped at 750 Hz, passed through a water filter, and focused on one face of an etched Ge sample. The recombination radiation, which peaked at 0.70 eV, was detected by a PbS cell whose absolute sensitivity was known. Simultaneous measurements of the change in resistance of the sample with light level were made. Both n- and p-type Ge ranging in resistivity from 0.2 to 12 Ωcm were used. The thickness of the illuminated regions varied from 0.2 to 0.4 mm.

If t is the thickness and A is the radiation area, we get from equation (85)

$$\delta R = \frac{R}{n_i^2} (n_0 + p_0)\, A\, t \left[\delta p + \frac{(\delta p)^2}{n_0 + p_0}\right]. \tag{86}$$

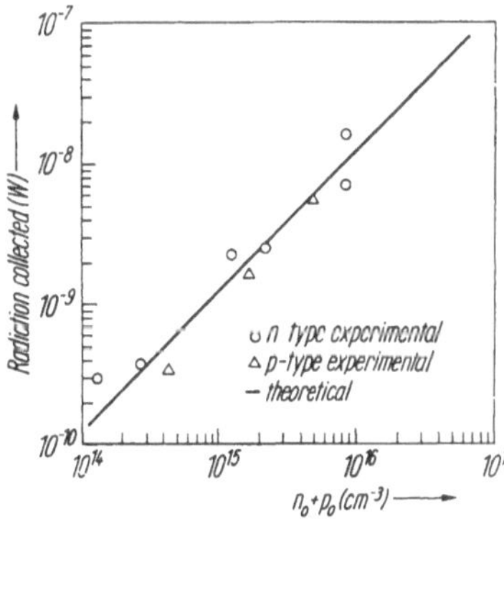

Fig. 22. Recombination radiation from various samples of germanium as a function of equilibrium carrier concentration for $t\,[\delta p + \delta p^2/(n_0 + p_0)] = 10^{12}\ cm^{-2}$. Radiating area = 0.228 cm² (Brill and Schwartz [244])

Koehler et al. [253] could be fitted in the following equation:

$$n^2 = 4.5 + \frac{9.714}{6.734 - E^2} . \tag{89}$$

The equation was assumed to hold up to 2.4 eV. — Calculated results are shown in Fig. 24.

7.4.2 *Discussion of radiative transitions*

It would be noticed from Fig. 24 that there is no maximum, the intensity continues to drop right through the absorption edge. The absorption coefficient increases exponentially at the absorption edge (see Fig. 26) and the above result was at first surprising. On closer analysis it was found that this absence of maximum is to be attributed to the fact that the absorption edge is not steep enough. For Se, for the energy interval 1.7 to 2.15 eV at room temperature,

$$\alpha = \text{const } e^{\sigma u} . \tag{90}$$

In this energy range $u \approx 70$ and the shape of the $P(\nu)\,\varrho(\nu)$ curve (equation (33)) is primarily governed by the product of the absorption coefficient and e^{-u}, i.e.

$$P(\nu)\,\varrho(\nu) \propto e^{\sigma u}\, e^{-u} \propto e^{(\sigma - 1)u} . \tag{91}$$

If $\sigma > 1$ we can expect a maximum around the transition point of the exponential rise to the flatter portion in the (α, E) curve. On the other hand, if $\sigma < 1$, no maximum can occur and no recombination radiation peak can be observed. In the case of Se, $\sigma \approx 1/3$ and no recombination radiation is possible. We may remark here that for substances like GaAs and InAs, $\sigma \approx 3$, at the absorption edge.

Recently Lanyon [262] has interpreted the exponential behaviour of the absorption coefficient of vitreous selenium at the absorption edge in terms of a distribution of states decreasing rapidly from the edge of the valence band. Lanyon's model is shown in Fig. 25. Absorption of light takes place by the excitation of an electron from the valence band to the conduction band. In the case of vitreous selenium there will also be absorption due to the exponential distribution of states in the forbidden gap. The exponential distribution of states to be considered is

$$N(\varepsilon) = N_0\, e^{-\varepsilon/\Delta} , \tag{92}$$

where N_0 is the density of states at the valence-band edge and Δ characterizes the rate at which the states fall off with energy. The demarcation line between the empty and the filled states in this distribution is the Fermi level, ζ_0. Absorption will take place by excitation from the filled states below the Fermi level to the conduction band. Excitation from the valence band into the empty

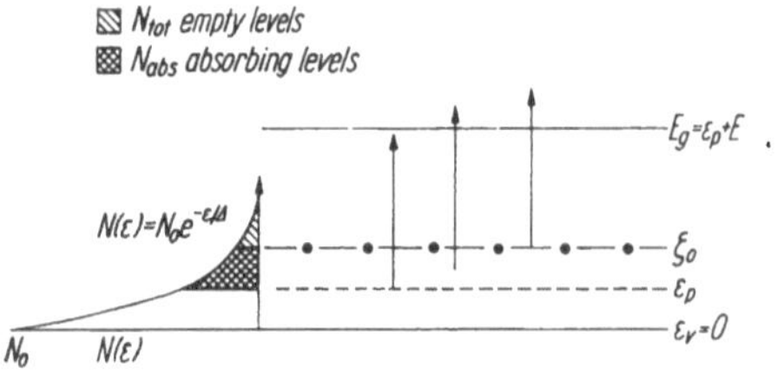

Fig. 25. Absorption scheme for vitreous selenium. The shaded levels above the Fermi level, being empty of electrons, cannot contribute to the absorption. Only those filled levels (cross-hatched) closer to the conduction band than the photon energy, E, contribute to the absorption (Lanyon [262])

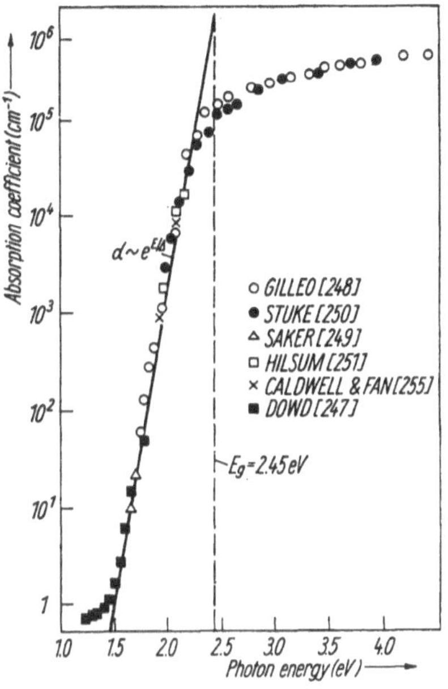

Fig. 26. Comparison of Lanyon's theoretical absorption edge for an exponential distribution of states in vitreous selenium [262] with the experimental results

states and from full states into the empty states in the distribution is negligible in the range of photon energies considered.

The absorption coefficient of a medium is given by [21]

$$\alpha = \left(\frac{\pi e^2}{n m}\right) N_{abs} f , \tag{93}$$

where e and m are the electronic charge and mass and N_{abs} is the density of states contributing to the absorption. f is the oscillator strength, which is of the order of unity for an allowed transition.

If photons of energy E are absorbed, all those filled states which are closer to the conduction band than E will contribute to the absorption as shown in Fig. 25. If the photon energy corresponds to excitation from an energy ε_p in the exponential distribution (which is measured from the valence band edge so that $\varepsilon_p + E = E_g$) all those states between the Fermi level and ε_p will contribute to the absorption. These states have been cross-hatched in Fig. 25. Provided that $\zeta_0 - \varepsilon_p \gg \Delta$, one may neglect the empty states above the Fermi level and the density of absorbing states is given by

$$N_{abs} = \int_{\varepsilon_p}^{\zeta_0} N_0 \, e^{-\varepsilon/\Delta} \, d\varepsilon \approx \Delta N_0 \, e^{-\varepsilon_p/\Delta} \approx \Delta N_0 \, e^{-(E_g - E)/\Delta} . \tag{94}$$

Thus

$$\alpha \propto e^{E/\Delta} . \tag{95}$$

In this interpretation $\Delta = 2.8\,kT$ and $E_g = 2.45$ eV. The agreement with observed values is shown in Fig. 26.

But it may be emphasized that too much physical significance should not be attached to the exponential nature of the distribution. Other distributions can also lead to the same density of states in the energy range considered. As a matter of fact, Lanyon finds not too different results for the absorption coefficient if a Gaussian distribution of states is assumed.

Gashimzade and Gubanov [263], using the band theory of amorphous semiconductors developed by Gubanov [264 to 268], have obtained theoretical expressions for the distribution function.

Moorjani and Feldman [269] have presented a review of the band theory of amorphous semiconductors.

7.5 Hexagonal selenium

7.5.1 Optical properties

There have been only a few investigations on the optical properties of hexagonal selenium.

Choyke and Patrick [270] used a photovoltaic method to measure the relative

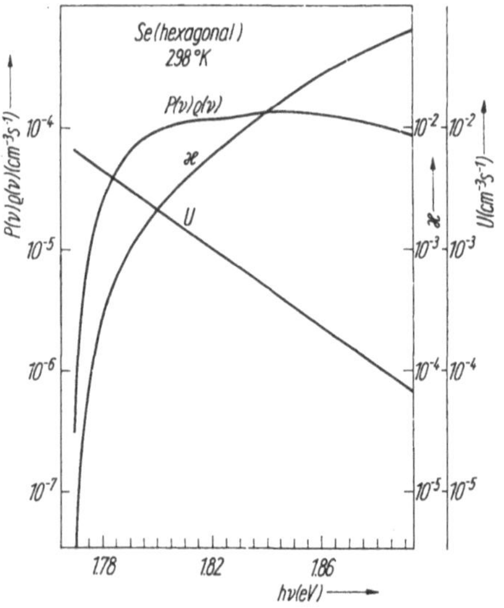

Fig. 27. $\varkappa$, $P(\nu)\,\varrho(\nu)$, and U for hexagonal selenium at 298 °K

absorption in polycrystalline hexagonal Se of photons of energy 1.6 to 2.0 eV. A quantity β proportional to the absorption coefficient was obtained. They attempted to interpret the results in terms of indirect transitions and β was represented by equation (13). (See, however, remarks by Moss ([104], p. 161).)

The optical constants of single crystals in the region of the intrinsic absorption edge in polarized light were investigated by Prosser [271, 272]. Eckart and Henrion [273, 273a, 273b] have measured the absorption coefficient in the region 1.6 to 1.8 eV, along the axis ($E \parallel c$) and perpendicular to the axis ($E \perp c$) They interpret their observations in terms of indirect transitions.

The refractive index has been measured by Prosser [272], Gobrecht and Tausend [257] and Eckart and Henrion [273].

Data adopted here:

Absorption coefficient: If we try to normalize the 298 °K data of Choyke and Patrick [270] with the data of Eckart and Henrion (at ≈ 1.80 eV) and of Prosser [271] (at ≈ 1.82 eV), we find that β has to be multiplied by a factor lying between 50 and 70. We have assumed this factor to be 60 and the absorption coefficient was obtained from equation (13) with $A = 1.8 \times 10^6$, $E_g = 1.794$ eV, and $\theta = 290.1$°K ($k\,\theta = 0.025$ eV).

Refractive index: Eckart and Henrion [273] give

$$n_{\parallel} = 3.2 \quad \text{and} \quad n_{\perp} = 2.5\,,$$

thus

$$n = \frac{1}{3}\,(2\,n_{\perp} + n_{\parallel}) = 2.73\,.$$

Intrinsic carrier concentration: No data to make a satisfactory estimate of this quantity is available.

The results are shown in Fig. 27 and Table 2 (see Section 6).

7.5.2 Discussion of radiative transitions

The $P(\nu)\,\varrho(\nu)$ curve is seen to have a very broad peak which is unlike any other substance considered here. Further investigations on the absorption coefficient of hexagonal selenium would be desirable.

7.6 Tellurium

The crystal structure of tellurium consists of spiral chains, each spiral having three atoms per turn. The chains are stacked parallel to each other with corresponding atoms in each chain forming plane hexagonal nets. A figure giving a perspective view of the atomic arrangement in the tellurium lattice may be found in Blakemore et al. [274]. Tellurium crystals are highly anisotropic, e.g., a) the compressibility coefficients have opposite signs parallel and perpendicular to the main crystal axis, b) the conductivity differs by a factor of two in directions parallel to and perpendicular to the crystal axis.

The properties of tellurium have been reviewed by Moss [132, 22] and more recently by Blakemore et al. [274].

7.6.1 Optical properties

As regards optical properties, tellurium is an optically uniaxial positive crystal. In such crystals, the refractive indices corresponding to the ordinary and extraordinary rays are usually expressed in terms of the direction of the E-vector for incident polarized light with respect to the optic axis. If c is the direction of the optic axis, the situations are described by the symbols $E \perp c$ and $E \parallel c$. In the case of tellurium the c-axis is the axis of the chains.

Transmission measurements of Loferski [275] in the neighbourhood of the absorption edge appeared to indicate that the actual edge energy might be different from the two polarizations (0.32 eV for $E \perp c$ but 0.37 eV for $E \parallel c$). Recently Blakemore and Nomura [276, 274, 277] have carried out accurate measurements of intrinsic optical absorption at 100 and 300 °K by transmission techniques; the resulting curves show a common energy origin.

Korsunskii and Lisitsa [278] have measured the transmissivity and reflectivity of Te monocrystals for natural and polarized radiation; from these they have calculated the absorption curves for polarized radiation in the region 0.06 to 0.34 eV. Their results for energies below 0.3 eV are very much higher than those of Blakemore and Nomura. The absorption in the infra-red region has been studied by Caldwell and Fan [255].

Hodgson [279] has determined the optical properties of polycrystalline tellurium by several methods. Absorption spectra in the ultraviolet (4000 to 110 Å) has been studied by Robin-Kandare and Robin [256, 177]. Optical properties of Te films have been measured by a number of workers [280 to 283].

The refractive index measurements of Hartig and Loferski [284] made on single crystals using polarized radiation, showed that the indices were different, depending on the orientation of the plane of polarization relative to the c-axis. They suggested that the weighted mean of their values, namely

$$n = \frac{1}{3}(2n_\perp + n_\parallel), \tag{96}$$

should give the average index observed in films if these are assumed to consist of randomly oriented microcrystals. More recently, measurements of refractive indices for 4 to 14 μm have been reported by Caldwell and Fan [255].

Data adopted here:

Absorption coefficient: As the absorption coefficients along the c-axis and perpendicular to it are different, a weighted absorption coefficient defined by the following equation was employed:

$$\alpha = \frac{1}{3}(2\alpha_\perp + \alpha_\parallel). \tag{97}$$

Data for both 100 and 300 °K:

0.26 to 0.36 eV, cf. Blakemore and Nomura [277];

0.36 to 0.40 eV, the curves given for α in [277] were extrapolated.

Blakemore and Nomura [277] have also attempted to represent their results by equations of the type (20). Their equations are satisfactory up to $\alpha \approx 3400\ \mathrm{cm}^{-1}$ for $E \perp c$ and to $\alpha \approx 10\ \mathrm{cm}^{-1}$ for $E \parallel c$. To examine whether such equations can be employed without much error, calculations were also carried out using the following equations:

$$\alpha_{100\,^\circ\mathrm{K}} = \frac{1}{3}\left\{\frac{13600}{1+\exp[302\,(0.351 - h\nu)]} + \frac{20}{1+\exp[225\,(0.351 - h\nu)]}\right\}, \tag{98}$$

$$\alpha_{300\,^\circ\mathrm{K}} = \frac{1}{3}\left\{\frac{13600}{1+\exp[267\,(0.346 - h\nu)]} + \frac{20}{1+\exp[197\,(0.346 - h\nu)]}\right\}, \tag{99}$$

in which $h\nu$ is in eV, and α in cm^{-1}.

Refractive index: The values of $n_\perp$ and $n_\parallel$ for the region 4 to 14 μm given by Caldwell and Fan [255] were substituted in equation (96) to obtain an average n. These values of n were employed to obtain the following empirical relation:

$$n = 5.2795 - 0.30357\,E + 2.3214\,E^2\,. \tag{100}$$

The above relationship was assumed to hold also for wavelengths a little shorter than 4 μm.

Intrinsic carrier concentration: Using a number of approximations, Blakemore, Long, Nomura, and Nussbaum [274] have obtained

$$n_i \approx 6\times 10^{15}\ \mathrm{cm}^{-3} \text{ at } 300\ ^\circ\mathrm{K}\,,$$

which in conjunction with $E_g = 0.34$ eV leads to

$$n_i = 8.2\times 10^{14}\, T^{3/2}\, e^{-E_g/2kT}\,. \tag{101}$$

These authors state that the constant "8.2×10^{14}" should be considered only as representative of the order of magnitude because of the assumptions involved in its derivation. A somewhat different value is given in Blakemore's book [5].

Energy gap: $E_{g\,100\,^\circ\mathrm{K}} = 0.35$ eV, $E_{g\,300\,^\circ\mathrm{K}} = 0.34$ eV.

Results are shown in Fig. 28 and 29 and Table 2 (see Section 6).

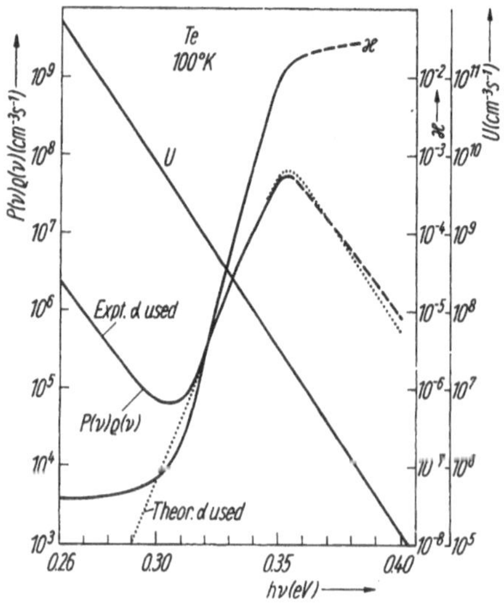

Fig. 28. ϰ and $P(\nu)\,\varrho(\nu)$ for tellurium at 100 °K

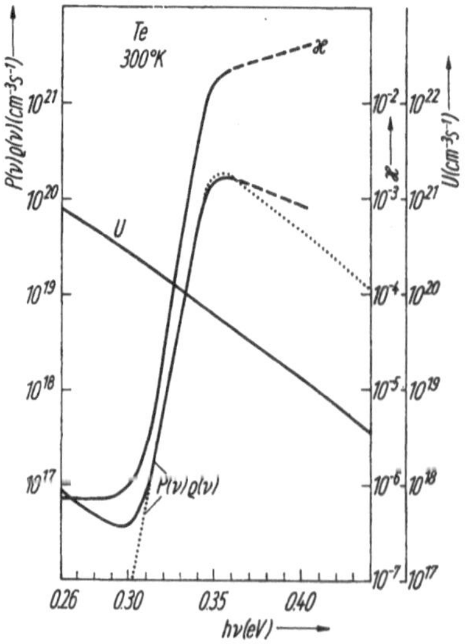

Fig. 29. ϰ and $P(\nu)\,\varrho(\nu)$ for tellurium at 300 °K. (Read at the dotted line: Theor. α used)

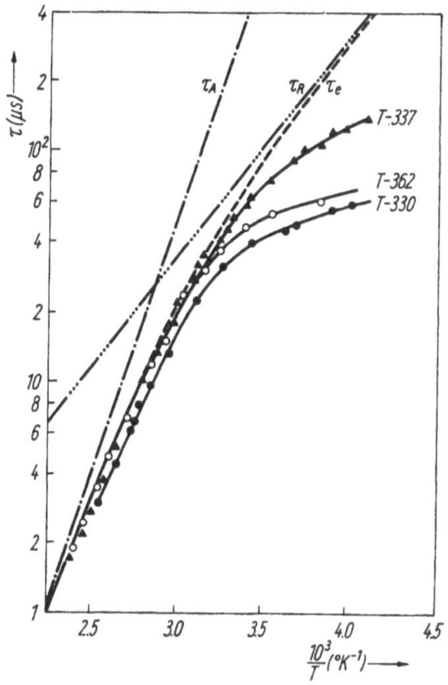

Fig. 30. Lifetime versus $1/T$ for three very pure and structurally perfect samples, compared with the theoretical expectations of radiative (τ_R) and Auger (τ_A) recombination. Effective lifetime $\tau_e = (1/\tau_A + 1/\tau_R)^{-1}$ (Blakemore [286])

7.6.2 *Discussion of radiative transitions*

Fig. 28 and 29 show that, except at very low energies, there is fair agreement between the $P(\nu)\,\varrho(\nu)$ curves obtained from the experimental data and those from the analytical expressions for the absorption coefficient.

The recombination processes in tellurium were investigated by Redfield [285] who also attempted to observe the recombination radiation but was unsuccessful, possibly because of experimental limitations.

Blakemore [286] has presented carrier lifetime data for tellurium samples with varying degrees of purity. In extensively purified monocrystals the lifetime behaviour above 300 °K suggests control by band-to-band transitions. Calculations were made of the radiative and Auger recombination rates, and experimental data between 300 and 420 °K fit the sum of these two rates. Blakemore's results are shown in Fig. 30. A curve in this figure marked τ_e shows the calculated lifetime limited by the two band-to-band mechanisms.

Recently Benoit à la Guillaume and Debever [287] have observed the recombination radiation from tellurium at 4 and 20 °K. The excitation was made by a 15 keV electron beam. The emission is polarized $E \perp c$. Stimulated emission has been obtained at current above 200 μA at 4 °K.

References

Note: The following abbreviations have been used:

Prague Conf. = Proceedings of the International Conference on Semiconductor Physics, Prague 1960, Academic Press, New York 1961.

Exeter Conf. = Report of the International Conference on the Physics of Semiconductors held at Exeter, July 1962, The Institute of Physics and the Physical Society, London 1962.

Paris Conf. = Comptes Rendus du 7e Congrès International Physique des Semiconducteurs, Paris 1964, Dunod, Paris 1964.

Paris Symp. = Symposium on Radiative Recombination, Paris 1964, Dunod, Paris 1964.

[1] G. Bemski, Proc. IRE **46**, 990 (1958).
[2] R. A. Smith, Semiconductors, Cambridge University Press, 1959.
[3] R. N. Hall, Proc. IEE **B 106**, Suppl. **17**, 923 (1959)
[4] R. H. Bube, Photoconductivity of Solids, Wiley, 1960.
[5] J. S. Blakemore, Semiconductor Statistics, Pergamon Press, New York 1962.
[6] W. Shockley, Proc. IRE **46**, 973 (1958).
[7] V. S. Vavilov, Uspekhi fiz. Nauk **68**, 247 (1959); Soviet Phys. – Uspekhi **2**, 455 (1959).

[8] V. S. Vavilov, Effects of Radiation on Semiconductors, Consultants Bureau, New York 1965.
[9] G. M. Guro, Uspekhi fiz. Nauk **72**, 711 (1960); Soviet Phys. – Uspekhi **3**, 895 (1961).
[10] W. van Roosbroeck and W. Shockley, Phys. Rev. **94**, 1558 (1954).
[11] G. Lasher and F. Stern, Phys. Rev. **133**, A 553 (1964).
[12] W. Shockley and W. T. Read, Phys. Rev. **87**, 835 (1952).
[13] R. N. Hall, Phys. Rev. **87**, 387 (1952).
[14] L. Pincherle, Proc. Phys. Soc. **B 67**, 319 (1955).
[15] H. Gummel and M. Lax, Phys. Rev. **97**, 1469 (1955).
[16] N. Sclar and E. Burstein, Phys. Rev. **98**, 1757 (1955).
[17] A. R. Beattie and P. T. Landsberg, Proc. Roy. Soc. **A 249**, 16 (1958).
[18] A. R. Beattie and P. T. Landsberg, Proc. Roy. Soc. **A 258**, 486 (1960).
[19] P. T. Landsberg, C. Rhys-Roberts, and P. Lal, Proc. Phys. Soc. **84**, 915 (1964).
[20] V. L. Bonch-Bruevich, Izv. Akad. Nauk SSSR, Ser. fiz. **21**, 87 (1957).
[21] H. Y. Fan, Rep. Progr. Phys. **21**, 107 (1956).
[22] T. S. Moss, Optical Properties of Semiconductors, Butterworths Publ. Ltd., London 1959.
[23] T. P. McLean, Progr. Semicond. **5**, 53 (1960).
[24] D. H. Parkinson, Contemporary Phys. **3**, 58 (1961).
[25] J. Tauc, Progr. Semicond. **9**, 87 (1965).
[26] I. C. Cheeseman, Proc. Phys. Soc. **A 65**, 25 (1952).
[27] J. Bardeen, F. J. Blatt, and L. H. Hall, Photoconductivity Conference 1954, Wiley, New York 1956 (p. 146).
[28] R. A. Smith, Wave Mechanics of Crystalline Solids, Wiley, New York 1961 (p. 409).
[29] R. J. Elliott, Phys. Rev. **108**, 1384 (1957).
[30] H. Y. Fan, M. L. Shepherd, and W. Spitzer, Photoconductivity Conference 1954, Wiley, New York 1956 (p. 184).
[31] G. G. Macfarlane and V. Roberts, Phys. Rev. **98**, 1865 (1955).
[32] G. G. Macfarlane and V. Roberts, Phys. Rev. **97**, 1714 (1955).
[33] R. L. Hartman, Phys. Rev. **127**, 765 (1962).
[34] F. Urbach, Phys. Rev. **92**, 1324 (1953).
[35] F. Moser and F. Urbach, Phys. Rev. **102**, 1519 (1956).
[36] P. D. Fochs, Proc. Phys. Soc. **69**, 70 (1956).
[37] D. Dutton, Phys. Rev. **112**, 785 (1958).
[38] D. L. Dexter, J. Phys. Chem. Solids **8**, 473 (1959).
[39] S. Tutihasi, Phys. Rev. **105**, 882 (1957).
[40] U. Haupt, Z. Phys. **157**, 232 (1959).
[41] H. Mahr, Phys. Rev. **125**, 1510 (1962).
[42] K. Kobayashi and T. Tomiki, J. Phys. Soc. Japan **16**, 1417 (1961).
[43] P. V. Meiklyar, Fiz. tverd. Tela **4**, 148 (1962); Soviet Phys. – Solid State **4**, 105 (1962).
[44] B. L. Evans, Proc. Roy. Soc. **A 276**, 136 (1963).
[45] B. L. Evans, Proc. Roy. Soc. **A 289**, 275 (1966).
[46] W. Martienssen, J. Phys. Chem. Solids **2**, 257 (1957).
[47] W. Martienssen, J. Phys. Chem. Solids **8**, 294 (1959).
[48] W. W. Piper, Phys. Rev. **92**, 23 (1953).
[49] P. G. Aline, Phys. Rev. **105**, 406 (1957).
[50] Y. Toyozawa, Progr. theor. Phys. **20**, 53 (1958).
[51] Y. Toyozawa, Progr. theor. Phys. **22**, 455 (1959).
[52] Y. Toyozawa, Suppl. Progr. theor. Phys. **12**, 111 (1959).
[53] H. Mahr, Phys. Rev. **132**, 1880 (1963).
[54] G. Rupprecht, Phys. Rev. Letters **12**, 580 (1964).
[55] R. S. Knox, Theory of Excitons, Academic Press, New York 1963 (p. 155).
[56] P. T. Landsberg and T. S. Moss, Proc. Phys. Soc. **B 69**, 661 (1956).
[57] A. Many and R. Bray, Progr. Semicond. **3**, 117 (1958).

[58] P. T. LANDSBERG, Proc. Phys. Soc. **B 70**, 282 (1957).
[58a] N. S. BARYSHEV, Fiz. tverd. Tela **3**, 1428 (1961); Soviet Phys. – Solid State **3**, 1037 (1961).
[59] W. P. DUMKE, Phys. Rev. **105**, 139 (1957).
[60] H. J. ZEIGER, Quarterly Progress Report on Solid State Research, Lincoln Laboratory, M.I.T., October 1959, AD231991 (p. 41).
[61] M. G. A. BERNARD and G. DURAFFOURG, phys. stat. sol. **1**, 699 (1961).
[62] W. P. DUMKE, Phys. Rev. **127**, 1559 (1962).
[63] R. N. HALL, G. E. FENNER, J. D. KINGSLEY, T. J. SOLTYS, and R. O. CARLSON, Phys. Rev. Letters **9**, 366 (1962).
[64] M. I. NATHAN, W. P. DUMKE, G. BURNS, F. H. DILL JR., and G. LASHER, Appl. Phys. Letters **1**, 62 (1962).
[65] B. LAX, Science (Washington) **141**, 1247 (1963).
[66] C. HILSUM, Sci. Progr. **52**, 91 (1964).
[67] J. E. GEUSIC and H. E. D. SCOVIL, Rep. Progr. Phys. **27**, 241 (1964).
[68] G. BIRNBAUM, Optical Masers, Academic Press, New York 1964.
[69] R. H. REDIKER, Phys. today **18**, 42 (1965).
[69a] D. M. EAGLES, J. Phys. Chem. Solids **16**, 76 (1960).
[70] V. L. VINETSKII, V. S. MASHKEVICH, and P. M. TOMCHUK, Fiz. tverd. Tela **6**, 2036 (1964); Soviet Phys. – Solid State **6**, 1607 (1965).
[71] J. CALLAWAY, J. Phys. Chem. Solids **24**, 1063 (1963).
[72] W. P. DUMKE, Phys. Rev. **132**, 1998 (1963).
[73] F. S. GOUCHER, Phys. Rev. **78**, 816 (1950).
[74] F. S. GOUCHER, G. L. PEARSON, M. SPARKS, G. K. TEAL, and W. SHOCKLEY, Phys. Rev. 81, 637 (1951).
[75] S. I. VAVILOV, Luminescence and its Duration, 1947, in: Collected Works, Vol. 2, Soviet Academy of Sciences Press, 1952 (p. 293).
[76] F. A. BUTAEVA and V. A. FABRIKANT, Izv. Akad. Nauk SSSR, Ser. fiz. **18**, 2 (1954).
[77] F. A. BUTAEVA and V. A. FABRIKANT, Izv. Akad. Nauk SSSR, Ser. fiz. **21**, 541 (1957).
[78] S. KOC, Czech. J. Phys. **7**, 91 (1957).
[79] V. S. VAVILOV and K. I. BRITSYN, Zh. eksper. teor. Fiz. **34**, 521 (1958); Soviet Phys. – J. exp. theor. Phys. **7**, 359 (1958).
[80] V. S. VAVILOV and K. I. BRITSYN, Zh. eksper. teor. Fiz. **34**, 1354 (1958); Soviet Phys. – J. exp. theor. Phys. **7**, 935 (1958).
[81] V. S. VAVILOV and K. I. BRITSYN, Optika i Spektroskopiya **8**, 861 (1960); Soviet Phys. – Optics and Spectroscopy **8**, 452 (1960).
[82] J. TAUC and A. ABRAHÁM, Czech. J. Phys. **9**, 95 (1959).
[83] K. I. BRITSYN and V. S. VAVILOV, Optika i Spektroskopiya **8**, 861 (1960); Soviet Phys. – Optics and Spectroscopy **8**, 452 (1960).
[84] A. J. TUZZOLINO, Phys. Rev. **134**, A205 (1964).
[85] V. A. CHUENKO, Izv. Akad. Nauk SSSR, Ser. fiz. **20**, 1550 (1956).
[86] E. ANTONČIK, Czech. J. Phys. **7**, 674 (1957).
[87] E. ANTONČIK, Czech. J. Phys. **8**, 492 (1958).
[88] W. SHOCKLEY, Solid State Electronics **2**, 35 (1961).
[89] R. J. HODGKINSON, Proc. Phys. Soc. **82**, 58 (1963).
[90] R. J. HODGKINSON, Proc. Phys. Soc. **82**, 1010 (1963).
[91] A. R. BEATTIE, J. Phys. Chem. Solids **23**, 1049 (1962).
[92] J. R. HAYNES, Phys. Rev. **98**, 1866 (1955).
[93] N. G. BASOV, L. M. LISITSYN, and B. D. OSIPOV, Dokl. Akad. Nauk SSSR **149**, 561 (1963); Soviet Phys. – Doklady **8**, 290 (1963).
[94] J. R. HAYNES and H. B. BRIGGS, Phys. Rev. **86**, 647 (1952).
[95] R. NEWMAN, Phys. Rev. **91**, 1313 (1953).
[96] N. G. BASOV and O. V. BOGDANKEVICH, Zh. eksper. teor. Fiz. **44**, 1115 (1963); Soviet Phys. – J. exp. theor. Phys. **17**, 751 (1963).

[276] K. C. NOMURA and J. S. BLAKEMORE, Bull. Amer. Phys. Soc. **5**, 62 (1960).
[277] J. S. BLAKEMORE and K. C. NOMURA, Phys. Rev. **127**, 1024 (1962).
[278] V. M. KORSUNSKII and M. P. LISITSA, Fiz. tverd. Tela **2**, 1619 (1960); Soviet Phys. – Solid State **2**, 1466 (1961).
[279] J. N. HODGSON, J. Phys. Chem. Solids **23**, 1743 (1962).
[280] M. P. LISITSA and N. G. TSVELYKH, Optika i Spektroskopiya **4**, 373 (1958).
[281] O. P. RUSTGI, W. C. WALKER, and G. L. WEISSLER, J. Opt. Soc. Amer. **51**, 1357 (1961).
[282] H. MERDY, S. ROBIN-KANDARE, and J. ROBIN, C. R. Acad. Sci. (France) **257**, 1526 (1963).
[283] H. MERDY, S. ROBIN-KANDARE, and J. ROBIN, J. Phys. Radium **25**, 223 (1964).
[284] P. A. HARTIG and J. J. LOFERSKI, J. Opt. Soc. Amer. **44**, 17 (1954).
[285] D. REDFIELD, Phys. Rev. **100**, 1094 (1955).
[286] J. S. BLAKEMORE, Prague Conf. (p. 981).
[287] C. BENOIT À LA GUILLAUME and J. M. DEBEVER, Solid State Commun. **3**, 19 (1965).

(Received June 28, 1966)

Original Papers

phys. stat. sol. **19**, 515 (1967)
Subject classification: 10.2; 22.5.2

2. Physikalisches Institut der Technischen Hochschule Stuttgart

Local Mode Absorption by U-Centres in Alkali-Halide Mixed Crystals

By

W. BARTH[1]) and B. FRITZ

The local mode absorption bands of U-centres are affected by the presence of alkali or halogen foreign ions in the host lattice in the following ways: a) new bands appear, which are displaced to higher and lower frequencies b) the entire spectrum undergoes a concentration-dependent shift and broadening. These effects are studied for U-centres in $KCl:Rb^+$, $KCl:Na^+$, $KCl:Br^-$, $RbCl:K^+$, $KBr:Rb^+$ and $RbBr:K^+$ with doping concentrations between 10^{-3} and 10^{-1}. Groups of new lines can be attributed to U-centres having doping ions in the next three shells of neighbouring ions. The intensities of these lines are slightly greater than expected from the statistical probabilities for such configurations. The splitting observed in a characteristic pair of lines compares favourably with a simple calculation for the case of a U-centre perturbed by a foreign cation in a nearest neigbour (100) position. The relation between the lattice constant and local mode frequency obtained in mixed crystals may be used to discuss lattice relaxation near U-centres and "quasi-harmonic" temperature-dependent band shifts.

Durch Alkali- oder Halogenfremdionen in Alkalihalogenid-Einkristallen wird das Absorptionsspektrum der lokalisierten U-Zentren-Schwingung folgendermaßen verändert: a) Es treten neue Banden mit erhöhter und erniedrigter Frequenz auf. b) Alle Banden werden, in Abhängigkeit von der Konzentration der Fremdionen, verschoben und verbreitert. Diese Beobachtungen wurden an U-Zentren in den Kristallen $KCl:Rb^+$, $KCl:Na^+$, $KCl:Br^-$, $RbCl:K^+$, $KBr:Rb^+$ und $RbBr:K^+$ gemacht. Die Fremdionenkonzentrationen lagen zwischen 10^{-3} und 10^{-1}. Die neuen Linien können auf U-Zentren zurückgeführt werden, die von Fremdionen in den drei nächsten Nachbarschalen gestört werden. Die beobachteten Intensitäten sind etwas größer, als auf Grund der Wahrscheinlichkeit der entsprechenden Konfigurationen im Fall statistischen Einbaus zu erwarten wäre. Eine für den Fall der Störung durch ein Fremdion unter den nächsten Nachbarn durchgeführte Modellrechnung ergibt gute Übereinstimmung mit der beobachteten Aufspaltung des entsprechenden Linienpaars. Der gefundene Zusammenhang zwischen Frequenzverschiebung und Gitterkonstante in Mischkristallen gestattet eine Abschätzung temperaturabhängiger „quasiharmonischer" Frequenzverschiebungen und der Gitterrelaxation in der Nähe des U-Zentrums.

Substitutional H^- ions at anion sites in alkali halides introduce an infrared active local mode vibration, which is threefold degenerated in otherwise unperturbed crystals. In this paper we report on the perturbation of this local mode absorption by foreign cations and anions. Such effects were observed for the first time by Schäfer [1], who found sharp new peaks, displaced to higher and lower frequencies from the strong local mode resonance peak near 500 cm^{-1} in $KCl:Rb^+$.

---

[1]) Present address: Institut für Höchstfrequenztechnik, Technische Hochschule Stuttgart.

This observation leads to the question of the types of U-centre foreign-ion aggregates that may be formed in such crystals, and of the displacements from their original positions of the neighbouring ions of such aggregates. These displacements produce shift and splittings of the sharp local mode resonance line, similar to the effects which are seen using external uniaxial stress [2, 3]. One may expect some of these results to bear on discussions of residual line widths and splittings occuring in the spectra of crystals of reagent grade purity [4].

1. Experiments

The mixed crystal systems investigated by us are listed in Table 1. The crystals were air-grown from the melt by the Kyropoulos method. The range of doping percentage and the method of determination thereof are also given. U-centres were introduced into the crystals using conventional methods [5]. The concentration of U-centres was determined by measuring the well-known ultraviolet band [5]. Apart from a small shift of this band, the UV absorption above 200 nm appeared unchanged by the doping. The shift at room temperature is typically −0.05 eV in KCl:Rb^+ (8%) and is thus larger than predicted by the Ivey relation for this band in different crystals [6].

Table 1

Crystal	Admixture	Concentration range (mol%)	
KCl	Rb^+	0.8−8	from flame photometry
	Na^+	1−5	
	Br^-	1−50	in the melt
KBr	Rb^+	1−10	
RbCl, RbBr	K^+	0.05−ca. 1	(different commercial purity grades)

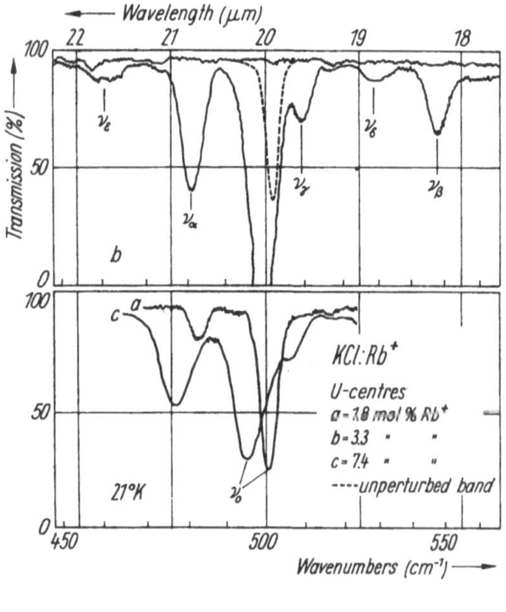

Fig. 1. Infrared absorption spectrum of U-centres in KCl:Rb^+, at $T = 21$ °K. ν_0 is a U-band of nearly unperturbed frequency. ν_a, etc. are new bands not seen in pure KCl

The infrared spectra were measured on a prism spectrometer allowing a maximum resolution of 2 cm^{-1}. The optical arrangement and the necessary corrections were described in a recent paper [4].

The experimental results obtained are as follows. Fig. 1 demonstrates the main features of U-centre spectra as found in KCl:Rb^+:

a) The strongest line denoted by ν_0 has nearly the same position as that in pure KCl with U-centres. This position is indicated by the dashed curve.

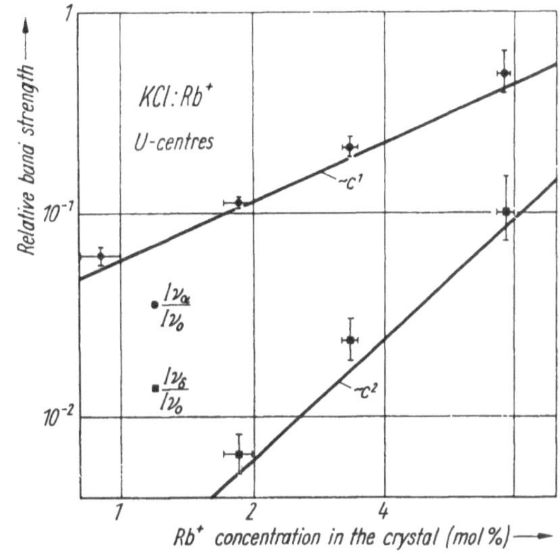

Fig. 2. Relative intensities versus concentration c. The straight lines approximate (small c) relative probabilities of configurations with one and two foreign cations among the nearest neighbours of a U-centre (random distribution of U-centres and foreign ions assumed!)

b) Five new bands, denoted by ν_α, ν_β, ν_γ, ν_δ, ν_ε are seen in the neighbourhood of ν_0 for Rb^+ dopings between 1 and 8%.

c) All bands, including that at ν_0, are shifted and broadened by like amounts, depending on the doping concentration (curves a, b, and c of Fig. 1).

Besides these sharp bands, the usual side band structure (near 560 cm^{-1}) is seen to be present in Fig. 1, in agreement with observations on undoped KCl:H^- crystals. This underlying side-band absorption was subtracted in order to determine relative intensities of the mixed crystal bands reported here.

The dependence on Rb^+ concentration c of the band at ν_α relative to that at ν_0 is given in Fig. 2. This relative intensity is found to be proportional to c. The bands at ν_β and ν_γ follow the same dependence, which is shown in particular by the linear relationship between the absorption at ν_α and ν_β (Fig. 3). The band at ν_β is weaker by 1:2.16 than ν_α (Fig. 3) and roughly equally intense as ν_γ. Quite a different behavior is exhibited by the bands ν_δ and ν_ε, whose relative intensities vary as c^2 (Fig. 2). These much weaker bands are again roughly equally strong.

In most systems listed in Table 1 groups of three lines are found, resembling those found in KCl:Rb^+ regarding relative intensities and amount of splitting.

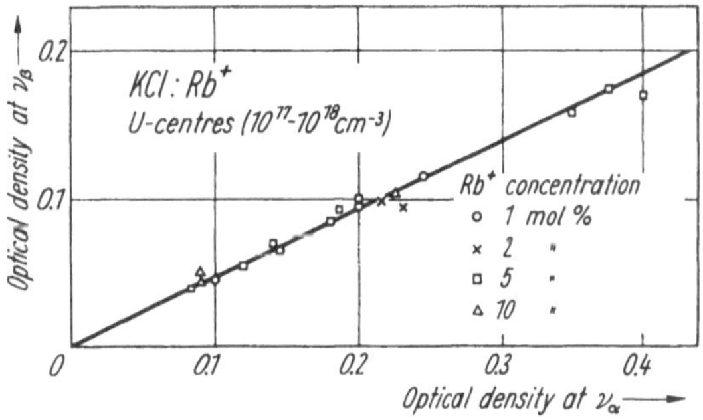

Fig. 3. Peak absorption of corresponding ν_α and ν_β bands. The intensity ratio $I(\nu_\alpha)/I(\nu_\beta) = 2.16$

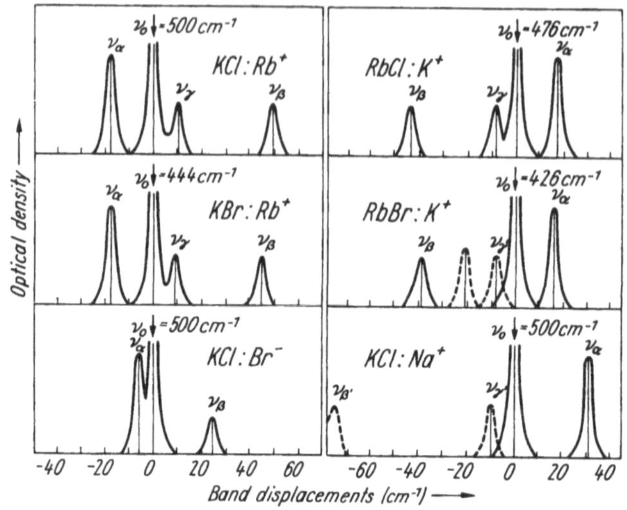

Fig. 4. Band positions in the different mixed crystal systems containing U-centres. $T \approx 20$ °K (schematic representation). Bands whose connection with the doping is not safely established are indicated by primes

The line with twofold intensity, ν_α, is seen on the low-frequecny side of ν_0 in KCl:Rb^+ and KCl:Br^-, whereas two other lines ν_β and ν_γ are on the high-frequency side of ν_0. In these cases an ion of the host lattice has been replaced by a larger doping ion (see Fig. 4). In the other systems, where the foreign doping ions have smaller ionic radii than the ions they substitute for, the strongest line ν_α is on the high-frequency side of ν_0 and the weaker ones on the low-frequency side, so that the splitting pattern is mirror symmetric to that above. In RbBr:K^+ and KCl:Na^+ the connection of the β- and γ-bands with the doping material could not be safely established, because of the lower doping concentration employed. These bands are denoted by primed letters.

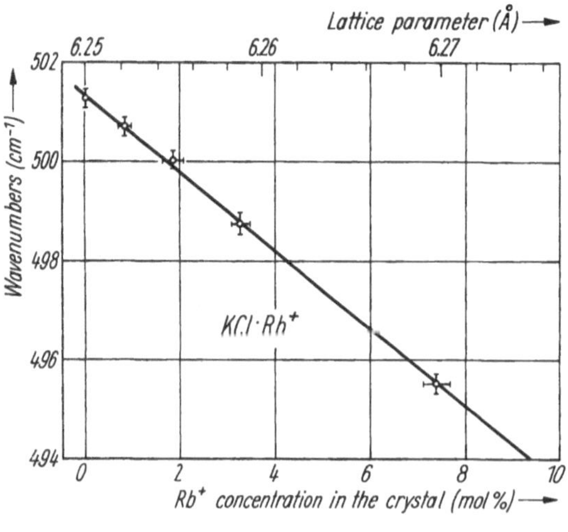

Fig. 5. Positions of ν_0 (KCl:Rb^+) versus doping concentration (21 °K). The lattice parameter is interpolated using a Vegard relation known to hold at room temperature. The bands $\nu_\alpha \ldots \nu_\gamma$ show the same dependence

Measurements on D^--U-centres in KCl:Rb^+ gave the result that the mixed crystal bands as described here undergo the same isotopic shift as the local mode frequency ν_0, namely

$$\frac{\nu_{H^-}}{\nu_{D^-}} = 1.39 \approx \sqrt{2}\,.$$

This behaviour is clearly different from that of the sidebands due to the two-phonon absorption [4].

The concentration-dependent shift and broadening of all lines including ν_0 were also investigated, and the results are given in Fig. 5 and 6. According to Fig. 5 the frequency of the band peak at ν_0, measured at 20 °K, is a linear function of the doping concentration in KCl:Rb^+. On top of this figure a lattice parameter scale is also given, which is based on the findings of Gnaedinger [7] who showed that the lattice constant in this system at room temperature follows closely a Vegard relation. We used this to extrapolate for the 20 °K lattice parameter values at a given Rb^+ concentration in the crystal. From these results we deduce the following relation between the frequency shift $\Delta\nu$ and the average lattice constant change Δa:

$$\frac{-\Delta\nu}{\nu_0} = 3.3\,\frac{\Delta a}{a_0}\,. \tag{1}$$

The halfwidths which turn out to be the same for ν_0, ν_α and ν_β bands, are plotted in Fig. 6 versus temperature with different concentrations of Rb^+ as an additional parameter. At low temperatures the bandwidths reach saturation values, which increase with the Rb^+ doping concentration. At higher temperatures the bands broaden according to a temperature dependence almost as strong as found in undoped KCl. This dependence for the undoped case, which is roughly T^2 above 100 °K and stronger below, is plotted for comparison.

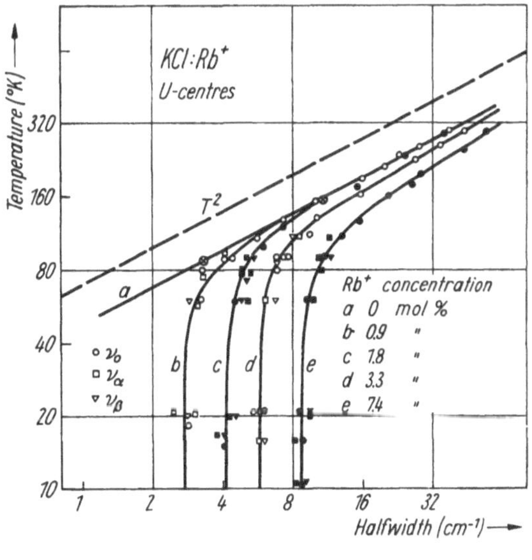

Fig. 6. Halfwidth versus doping concentration and temperature

2. Discussion

2.1 Additional lines

The additional lines appearing in mixed crystals may be interpreted by considering the influence of strain caused by nearby foreign ions on the frequency of the local mode of the U-centre.

This frequency is determined to a large extent by the repulsive overlap potential between the hydrogen ion and the surrounding ions; because of the O_h symmetry it is threefold degenerate in the unperturbed lattice. The strongest perturbation is expected from one foreign cation on a nearest neighbour site (100). The symmetry is then C_{4v} and leads to a splitting of the unperturbed local mode transition (ν_0) into two lines having an intensity ratio of 2:1. We interpret the ν_α and ν_β lines as arising from this configuration. The different positions of these lines, which are observed in the presence of larger and smaller foreign cations (for instance KCl:Rb^+ and KCl:Na^+) are in good agreement with this model (see discussion, Section 2.2). An equally strong perturbation may arise from two or more foreign cations in the nearest neighbour shell. The overall probability for such configurations is lower by a factor of 10 at $c = 4\%$, and varies as c^2 (see Fig. 2). This agrees with the behavior of ν_δ and ν_ε bands in KCl:Rb^+. These bands therefore may be interpreted as being due to U-centres with two Rb^+ ions adjacent to each other within the nearest neighbour shell. The symmetry is C_{2v}, and the splitting may be expected to be into three equally intense bands, one of them probably being hidden under the rest of the spectrum. The other possible configuration of two Rb^+ ions having D_{4h} symmetry occurs only in one out of five cases and is therefore unlikely to appear in the spectra. Less strong perturbations are to be expected from cations in the third shell. The probability is proportional to the doping concentration c (at low concentrations), and the intensity should be comparable to that of the ν_α and ν_β bands. The ν_γ band, which is seen very close to ν_0, is likely to be the most strongly shifted line of the two expected in this case (C_{3v} symmetry).

In the case of foreign anion doping (KCl:Br^-), the strongest effect should arise from a Br^- ion in a second neighbour position. In fact we find bands in intensities which are close to the expected values as estimated from the doping to the melt (ν_α and ν_β). Because of the difficulties in resolving the ν_α band from the ν_0 line (Fig. 4) this identification is not so well established as in case of cation doping.

2.2 Calculation of the splitting of the lines

The splitting of a series of lines associated with one of the configurations discussed above may be calculated from the displacements of lattice ions and the different repulsive interaction introduced by a foreign ion. In Fig. 7 we give results of a rough calculation for the ν_α and ν_β bands in KCl:Rb^+ and RbCl:K^+, wherein the local mode frequency was approximated by that of a H^- ion vibrating in the rigid potential well arising from nearest neighbour interactions with the H^- only.

According to Fig. 7, a foreign alkali ion is assumed to occupy one of the nearest neighbour sites 1 adjacent to the H^- ion in site 0. Such a substitutional ion generally causes its own nearest neighbours to shift their positions by about one half the difference between the radii of lattice and impurity ions, as shown by recent calculations [8, 9]. This should also be true in the present case where

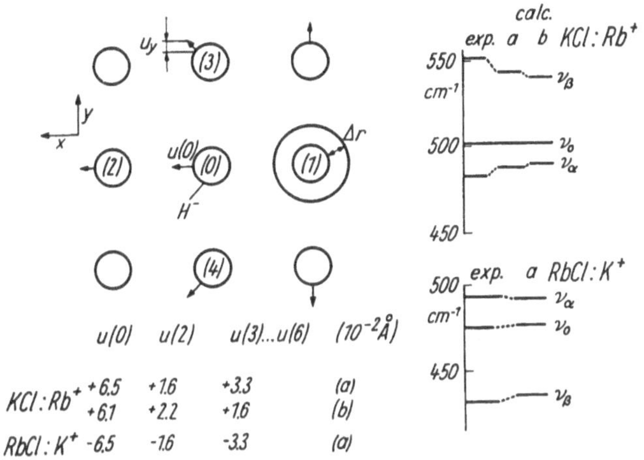

Fig. 7. Proposed model for a U-centre having local mode frequencies at ν_α and ν_β. The displacements occurring in the neighbourhood of a Rb^+ ion in KCl and a K^+ ion in RbCl are given in the table, according to different assumptions a) and b) (see text). These values are used to calculate the relative splitting from equations (2a) and (2b)

one of these neighbours is an H^- ion, if this H^- ion has a radius not too different from that of a halogen ion (e.g. Cl^-) and has a similar form for its repulsive potential. We make this assumption about nearest neighbour displacements in our calculation and let the other displacements follow a r^{-2} dependence around the foreign ion as the origin (continuum approximation [10].) The values of the displacements are given in Fig. 7, row a). The ionic radii of the Rb^+ and K^+ ion are assumed to differ by 0.130 Å according to calculations by Fumi and Tosi [11]. Under b) we quote for comparison displacements which have been calculated by Dick [9] for the case of KJ:Rb^+, to show the preferential strain in the [100] direction brought about by considering the lattice structure near the defect in detail.

The relative frequency shift may be calculated by fitting parameters of the overlap potential and the polarizability of the H^- ion to the observed frequency shift which results from a known variation of the lattice constant. We have noted an isotropic band shift to occur in mixed crystals of known average lattice parameter (equation (1)); we must assume, in the present application of this result, that da/a_0 at the defect is the same as the average change in the lattice. We use a potential of the Huggins-Mayer type

$$\Phi_{(0,i)} = c_{+-}\, b_i \exp\left(\frac{r_H + r_K - r_{0,i}}{\varrho}\right), \tag{2}$$

where $r_{0,i}$ is the distance between the H^- ion at site 0 and its nearest neighbour ion i, and r_H, r_K are the crystal radii of the H^- ion and the nearest neighbour cations, respectively. It has been shown (e.g. [13]) that taking the values of the parameters in equation (2) from alkali halide data and considering the ions as rigid leads to a local mode frequency which is 50% too high. One way of dealing with this situation is to consider a simple shell model for the vibrating H^- ion, which may be justified from the large polarizability of this ion. The local mode frequency ω_L is then given by

$$m_H \omega_L^2 = f\left(1 + \frac{f}{g_H}\right)^{-1}, \tag{3}$$

where the overlap force constant f may be derived from (2), whereas g_H is an internal shell–core spring constant introduced to take account of the deformation of the H^- in the (odd) local mode vibration (see also [12]). If we follow Fieschi et al. [13] in setting $\varrho = 0.34$ Å (equation (2)), which is the average alkali halide value, then we obtain $f/g_H = 0.37$ on our model (equation (3)), by fitting the result of equation (1).

The ratio of force constants $\bar{f}_\alpha$ changed by strain to unchanged ones f_α is obtained by taking the second derivatives of (2) and summing over six nearest neighbours at sites 1 to 6:

$$\frac{\bar{f}_x}{f_x} = \left(2 - \frac{4\varrho}{a_0}\right)^{-1}\left[\exp\frac{\Delta r_K - u(0)}{\varrho} + \exp\frac{u(0) - u(2)}{\varrho} - \frac{4\varrho}{a_0}\cdot\exp\frac{-u_y(3)}{\varrho}\right], \tag{2a}$$

$$\frac{\bar{f}_y}{f_y} = \left(2 - \frac{4\varrho}{a_0}\right)^{-1}\left[2\left(1 - \frac{\varrho}{a_0}\right)\cdot\exp\frac{-u_y(3)}{\varrho} - \right.$$
$$\left. - \frac{\varrho}{a_0}\cdot\exp\frac{\Delta r_K - u(0)}{\varrho} - \frac{\varrho}{a_0}\cdot\exp\frac{u(0) - u(2)}{\varrho}\right]. \tag{2b}$$

Δr_K is the difference of impurity and lattice cation radii, and the displacements are defined in Fig. 7. These expressions do not contain the parameters c_{+-} and b_i of equation (2), since these are assumed to be equal for H^-–K^+ and H^-–Rb^+ interactions and thus drop out. This assumption may be justified from the fact that ω_L differs by 5% only between KCl:H^- and RbCl:H^-. Equations (2a) and (2b) together with the value of g_H from above give the calculated frequencies ν_α and ν_β in Fig. 7, after inserting the sets of values a) and b) for the displacements as specified in the figure. The fair agreement with the observed amount of splitting strongly supports our explanation of the ν_α and ν_β bands and arising from the proposed configuration. The fact that a relatively larger displacement in site 2 than in sites 3 to 6 brought about by Dick's [9] theory diminishes the frequency shift for x-polarization probably is a hint that an H^- ion is more compressible than a halogen ion and hence does not transmit elastic forces to the same extent.

From a more accurate treatment of both elastic strain and vibrational properties of this defect we should expect predictions about band intensities as well. We have observed the ν_α and ν_β bands to be in a ratio 2.2 instead of the expected value 2.0 (Fig. 3); furthermore, their total relative intensity $(I_\alpha + I_\beta)/I_0$ exceeds the statistical expectation by 1.5 (Fig. 2). It cannot be decided at present whether the latter ratio indicates a slightly preferential formation of the defect type of Fig. 7 or is due to a change in local field and effective charge of the localised oscillator.

2.3 Dependence of the line position on the lattice constant

The empirical relation $-\Delta\nu/\nu_0 = 3.3\,\Delta a/a_0$ links the frequency change experienced by all bands on doping of the host crystal to the average lattice constant. This can only be understood if the lattice cells containing U-centres participate in a general expansion of the lattice, which is isotropic on the average. The numerical result above could also be obtained using external pressure on a crystal containing U-centres, e.g. uniaxial pressure. From an experiment with stress applied in the [100] direction the isotropic compression effect was calculated, starting from the band splitting observed with polarized light in the [100] and [010] directions [3]. The result was in satisfying agreement with our relation baove, giving a factor 3.6 instead of 3.3. The dependence on lattice constant of the local mode frequency is found to be somewhat weaker when different alkali halides are compared, e.g. the potassium salts with different anions (see Schäfer [1]). For the latter one finds $-\Delta\nu/\nu_0 = 2.4\ \Delta a/a_0$, a clear indication that an increasing relaxation of the lattice near the U-centre takes place on going from KCl to KBr or KJ. Assuming unperturbed positions of the nearest neighbours of a U-centre in KCl, one estimates that an inward displacement of these neighbours by 1.9% in KBr would be necessary to reconcile the observed frequency value and the prediction of equation (1). From lattice parameter measurements on $KBr:H^-$ by Hilsch and Pohl [14] one deduces a relaxation of 1.6% using Eshelby's theory [10], which may be compared with our estimate. — The temperature dependent shift of the infrared U-band [4] has been discussed in a recent paper [12]. It is assumed that the thermal lattice expansion contributes to the total observed shift a "quasi-harmonic" part which may also be described by equation (1).

2.4 Broadening of lines and remnant line width

The concept of an average lattice constant change produced by statistically distributed centres necessarily comprises a spectrum of local distortions, scattering around the average dimension of a lattice cell. These local deviations are sensed by U-centres and thus contribute to the broadening of their local mode absorption. The remnant line width in Fig. 6 which is concentration dependent, is thus explained.

We use the results concerning the line width in mixed crystals with known doping concentrations to extrapolate for the situation to be expected for crystals of standard as-grown impurity level. The concentration dependence of the remnant line width w_0 from Fig. 6 is

$$(w_0)_{\text{imp}} \approx A\ c^{1/2}$$

with

$$A = 3.1\ \text{cm}^{-1}\ \text{mol}\ \%^{-1/2}\ .$$

This gives $(w_0)_{\text{imp}} = 1\ \text{cm}^{-1}$ for $c = 0.1\%$ and $0.3\ \text{cm}^{-1}$ for $c = 0.01\%$. This may be compared with the major impurity concentration in alkali halides, which for reagent grade material is in the order of 0.01% (e.g. Na in KCl) and thus contributes a few tenths of a cm^{-1} to the low temperature line width. The very different temperature dependence of the line width of H^- and D^- centers in KCl and other crystals, reported and discussed in [4] and [12], shows, however, that phonon broadening is the predominant effect at least for D^- lines.

After completion of this work we learned about measurements by Mirlin and Reshina [15], who obtained similar results on U-centres in KCl doped with various anions and cations. These authors give a qualitative interpretation of the various mixed crystal bands in agreement with our discussion under 2.1.

Acknowledgement

We are indebted to Professor H. Pick for his interest and support of this work. One of us (W. B.) wishes to thank the Deutsche Forschungsgemeinschaft for financial support.

References

[1] G. Schäfer, J. Phys. Chem. Solids **12**, 233 (1960).
[2] W. Barth, B. Fritz, and U. Gross, Internat. Symp. Colour Centers in Alkali Halides, Urbana, 1965 (p. 12).
[3] W. Hayes, H. F. MacDonald, and R. J. Elliott, Phys. Rev. Letters **15**, 961 (1965).
[4] B. Fritz, U. Gross, and D. Bäuerle, phys. stat. sol. **11**, 231 (1965).
[5] H. Pick, Struktur von Störstellen in Alkalihalogenidkristallen, in: Tracts in Modern Physics, Vol. 38, Springer-Verlag, Berlin 1965.
[6] B. S. Gourary and F. J. Adrian, Solid State Phys. **10**, 128 (1960).
[7] R. Gnaedinger, J. chem. Phys. **21**, 323 (1953).
[8] Y. Fukai, J. Phys. Soc. Japan **18**, 1413 (1963).
[9] B. G. Dick, Phys. Rev. **145**, 609 (1966).
[10] D. Eshelby, Solid State Phys. **3**, 79 (1956).
[11] F. G. Fumi and M. P. Tosi, J. Phys. Chem. Solids **25**, 31 (1964).
[12] H. Bilz, B. Fritz, and D. Strauch, J. Phys. Radium, to be published.
[13] R. Fieschi, G. F. Nardelli, and N. Terzi, Phys. Rev. **138**, 203 (1965).
[14] R. Hilsch and R. W. Pohl, Trans. Faraday Soc. **34**, 883 (1938).
[15] D. N. Mirlin and N. N. Reshina, Fiz. tverd. Tela **8**, 152 (1966).

(Received September 9, 1966)

phys. stat. sol. **19**, 525 (1967)

Subject classification: 10; 22.5.2

Physico-Technical Institute of Low Temperatures,
Academy of Sciences of the Ukrainian SSR, Kharkov

Effect of Stress State on the Shape of the Dislocation Rosette near an Indentation in Some Alkali Halide Crystals

By

V. B. Pariiskii

The structure of a dislocation rosette near an indentation in KBr and KCl crystals is studied. The motion of dislocations out of the rays of the dislocation rosette due to external impulse stress is shown to depend on the amplitude and duration of the pulse. The effect of stress on the shape of the dislocation rosette near the indentation is shown for KBr, KCl, and NaCl single crystals. During indentation of the stressed crystal the uniaxial surface contraction elongates the rosette rays that adjoin the direction of contraction, whereas the uniaxial dilation elongates the rays that adjoin the direction normal to the direction of extension. The effect is enhanced by increasing the preliminary stresses on the surface to be indented. The observed phenomenon is explained qualitatively by considering the structure of the rosette near the indentation. This effect may be used to determine the magnitude and sign of the local residual elastic stresses at the surface.

In KBr und KCl Kristallen wird die Struktur der Versetzungsrosette in der Nähe des Eindrucks untersucht. Es wird gezeigt, daß die Bewegung der Versetzungen aus den Strahlen der Versetzungsrosette heraus von der Amplitude und Dauer der äußeren Impulsbeanspruchung abhängt. Der Effekt einer Spannung auf die Form der Versetzungsrossette in der Nähe des Eindrucks wird für KBr, KCl und NaCl Einkristalle angegeben. Während des Eindrückens verlängert eine einachsige Oberflächenkontraktion die Rosettenstrahlen, die an der Richtung der Kontraktion anschließen; eine einachsige Streckung verlängert die Strahlen, die an die Richtung senkrecht zur Streckrichtung anschließen. Der Effekt wächst mit wachsender Vorspannung an der Fläche, die eingedrückt wird. Die beobachtete Erscheinung wird qualitativ erklärt unter Berücksichtigung der Rosettenstruktur in der Nähe des Eindrucks. Dieser Effekt kann zur Bestimmung der Größe und des Vorzeichens der elastischen Restspannungen an der Oberfläche benutzt werden.

Studies of the dislocation structure of rosettes appearing on the surface of alkali halide single crystals under a localized force have been reported by many authors [1 to 9]. Distortions of this type are often present in these materials.

The influence of the stress state on the dislocation structure of the rosette near the indentation in KBr, KCl, and NaCl crystals is studied in this work.

1. Rosette Structure near an Indentation on an Unstressed Crystal

The dislocation structure of the rosette near an indentation on NaCl [2, 5] single crystals has been studied in detail. Fig. 1 and 2 give a schematical representation of the arrangement of dislocation loops around an indentation of a diamond indentor on a {100} plane; the direction of dislocation motion out of the rosette under external stress is also shown [5].

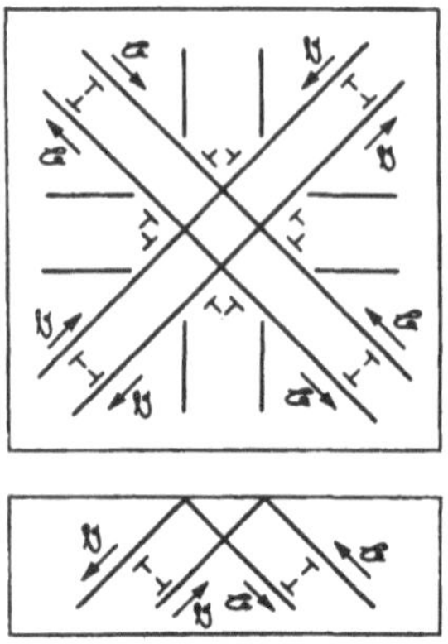

Fig. 1. Traces of {110} glide planes on {100} planes near the indentation and the corresponding Burgers vectors in edge planes. The intersection of the dislocation-loop edge component with the surface is marked by ⊥

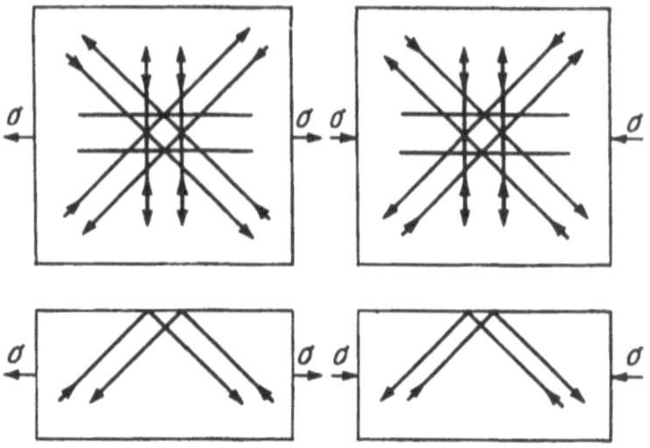

Fig. 2. Diagram of dislocation motion out of the rosette near the indentation under external stress. The external stress is marked by arrows.

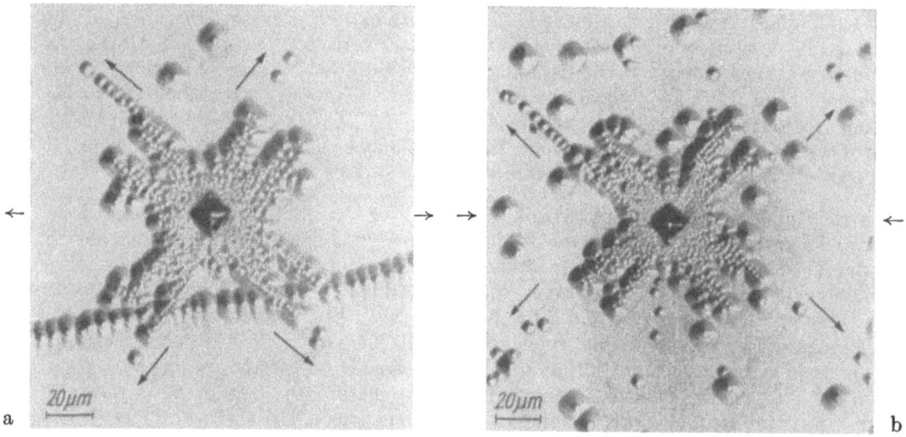

Fig. 3. Dislocation motion out of the rosette near the indentation under stress pulses of different sign in a KBr crystal. a) dilation, b) contraction. The external stress is shown by dashed arrows

The investigation of dislocation rosettes near indentations on KBr and KCl crystals by means of selective etching [10, 11] and the observation of the dislocation motion out of the rosette rays (Fig. 3) have shown that the arrangement of dislocation loops and their Burgers vector in principle agree with the above scheme.

The use of stress pulses of different sign, length and amplitude allowed to study the effect of amplitude and length on the character of the dislocation motion out of the rays of the rosette. Qualitatively the result is the following: When the external stress is low or a high-amplitude but short pulse is applied, only "correct"[1]) motion of dislocations out of the rosette is observed. This is in accord with Fig. 2. An increase of amplitude or time of stress application leads to a "mixed"[1]) motion when an elongation of all the rays of the disloca-

[1]) Here the terminology of the papers [2 and 8] is used.

tion rosette takes place. This result is illustrated for NaCl crystals. The experimental procedure was carried out in the following way: The indentations were made on the surface of a freshly cleaved crystal $5 \times 5 \times 20$ mm³ in size along its length at a distance of 250 μm from each other by means of PMT-3. Then the crystal was etched and subjected to pulses of bending stress (Fig. 4a) [11]. The stress pulses were measured by means of a tensiometer and photographed from the screen of oscillograph. The distribution of the shear stress acting on the edge dislocations in a thin surface layer calculated in the elastic approximation [13] is shown in Fig. 4b.

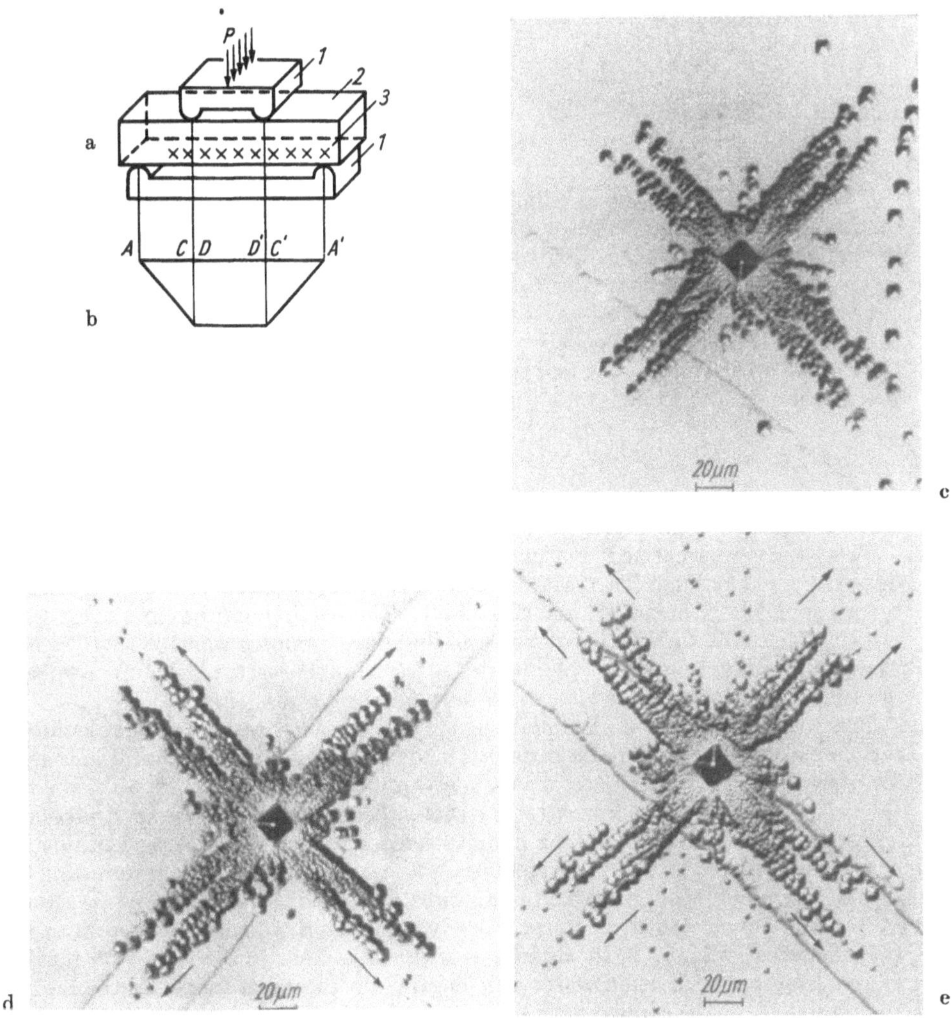

Fig. 4. NaCl crystal stressed by pulses of different length

a) Scheme of pulse application; 1 — knives, 2 — sample, 3 — indentations on the sample surface.
b) Epure of the distribution of resolved shear stress acting on edge dislocations in a thin surface layer.
c) No dislocation motion in the regions AC and A'C'.
d) Correct dislocation motion in the regions CD and C'D'.
e) Mixed motion of dislocations in the regions DD'.

When the amplitude of the pulse was high enough repeated etching revealed some areas differing by the character of dislocation motion out of the rosette near the indentation (Fig. 4b). In the area of low stress (AC and A'C') there was no dislocation motion out of the rosette near the indentation at all (Fig. 4c). As stress increased at first the "correct" (CD and C'D') (Fig. 4d) motion of dislocations began and then the "mixed" one (DD' area) (Fig. 4e). The stress at which the mixed motion of dislocations started decreased with increasing pulse length. In these experiments the pulse length varied approximately by a factor of 100 from 2.6×10^{-2} to 2.4 s. The relation between the stress τ', at which the mixed motion of dislocations begins, and the pulse length appears to be linear in semilogarithmic coordinates, and can be represented in the form

$$t = A \exp(-a\tau') , \tag{1}$$

where $A = 1150$ s and $a = 8.25\times10^{-7}$ dyn^{-1} cm^2. It should be noted that such a relation between the time of load action and the stress is typical of thermally activated processes [12].

The mixed motion of dislocations is due to the generation of dislocation loops of the opposite Burgers vector (opposite direction of by-pass [5]) in the rays of the rosette which, according to Fig. 2, should shorten for the chosen sign of stress. One of the mechanisms for the formation of these loops is double cross-slip, which was considered in connection with the motion of rosette rays in [5, 8]. The observed dependence of the character of dislocation motion out of the rosette on the amplitude and time of application of stress does not contradict the mechanism of double cross-slip.

2. Effect of Stress State on the Shape of the Dislocation Rosette near the Indentation

This study was carried out on KBr, KCl, and NaCl single crystals having different yield points. The samples were annealed for 50 to 100 h at temperatures about 50 °K below the melting point. The cooling and heating rates were 1 to 2°/min. The dislocation density in annealed samples was 10^3 to 10^4 cm^{-2}. The samples having a size of about $4\times4\times20$ mm^3 were placed in a special device where they were subjected to pure bending stress (Fig. 5a).

The device was fixed on the plate of PMT-3 in such a way as to enable indentations on stressed crystals at different loads on the diamond pyramid indentor. The indentations were made across the sample as shown in Fig. 5b, c on the side of stress gradient, the external stress being chosen so that no plastic deformation occurred before the indentation. Then the sample was unloaded and etched. A change of the symmetry of the rosette near the indentation on stressed samples was observed for all substances; schematically this is shown in Fig. 5b. In the contracted area the rays close to the direction of contraction are elongated, whereas in the dilated area the rays close to the direction normal to the direction of dilation increase in length. On the neutral axis all the rays of the rosette have the same length. Fig. 6a and b show the corresponding pictures for a KBr crystal. A similar change of the shape of the rosette near the indentation has been observed for KCl and NaCl crystals.

If the length difference of adjacent edge rays of a rosette is taken to be a measure of the effect the following is noticed:

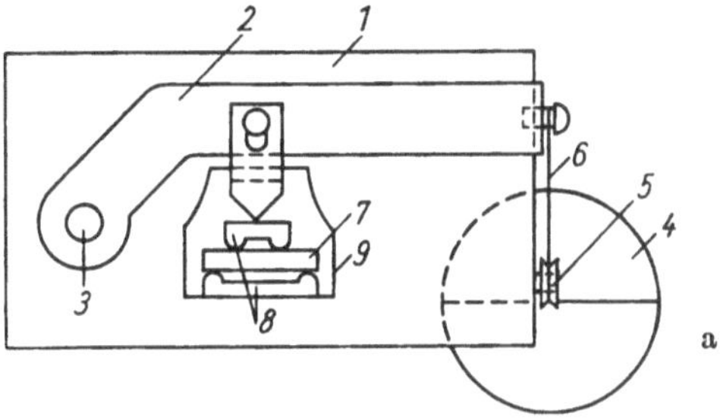

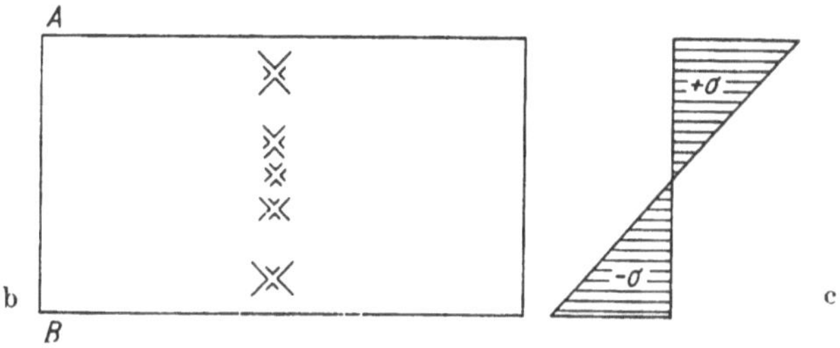

Fig. 5. a) Attachment to PMT-3 for four-point bend crystal loading; 1 – base, 2 – lever, 3 – bearing, 4 – scale for loads, 5 – block, 6 – string, 7 – crystal, 8 – knives, 9 – cuvette.

b) Scheme of intendation arrangement on the crystal. A and B: dilated and contracted surfaces, respectively.

c) Epure of normal stress distribution

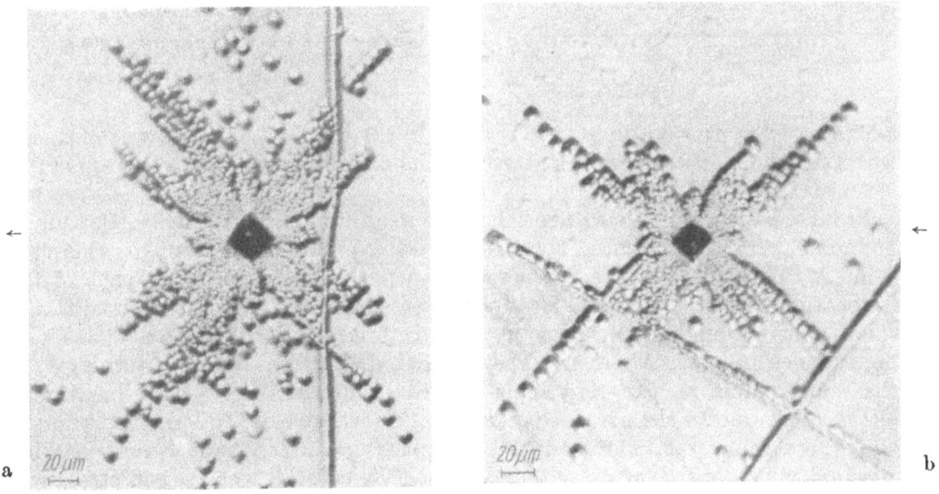

Fig. 6. Microphotographs of dislocation rosettes near the indentation on a stressed KBr crystal. a) Dilation, b) contraction. External stress is shown by arrows

1. The length difference is the larger the larger the magnitude of the stress acting in this region of the crystal.
2. At given stress the length difference of the rays decreases with increasing rigidity of the sample (increasing yield point).

Fig. 7 shows the dependence of the length difference Δ of dislocation-rosette rays, obtained for each rosette by averaging over all four rays, on the shear stress τ applied to sample surface before the indentation. This dependence is shown for a KCl single crystal with the critical shear stress $\tau_c = 270$ p/mm^2. Curve 1 was obtained for a dilation of the part adjoining the surface A and con-

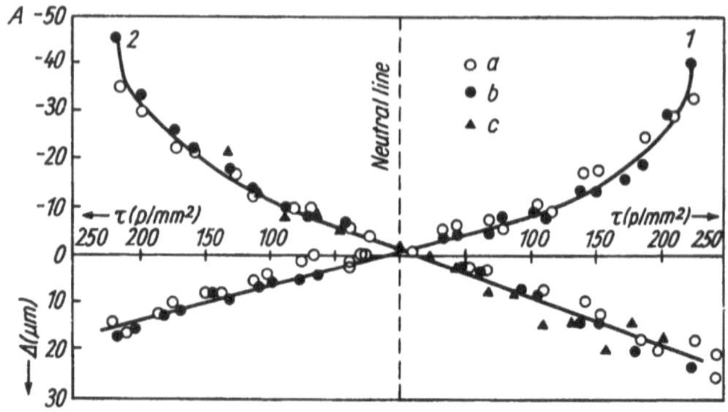

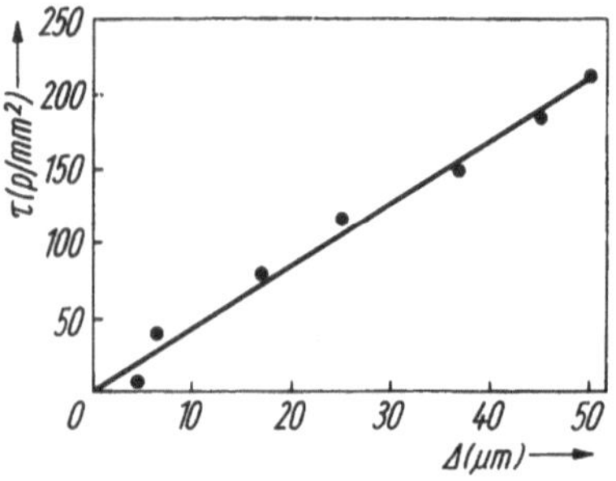

Fig. 7. Dependence of the length differences of the rays of a dislocation rosette (Δ) on the shear stress (τ) applied to sample surface before indentation for a KCl crystal. A and B: sample surfaces (see Fig. 5b); (1) dilation of the surface A, contraction of the surface B; (2) dilation of the surface B, contraction of the surface A

a, b, c: loads of 1, 2, and 5 p, respectively

Fig. 8. Dependence of the length differences of the rays of a dislocation rosette (Δ) on shear stress (τ) applied to the sample surface before indentation for a KBr crystal

traction of the part adjoining the surface B (see Fig. 5b). Curve 2 was obtained when the signs of stress were reversed. The experimental points correspond to indentor loads of 1, 2, and 5 p.

From the diagram it is evident that the magnitude of the indentor load does not affect this dependence in the limits of experimental error. In this case a linear relation is observed between the length difference of the rays of the dislocation rosette Δ and the local shear stress τ in the range of dilational and small contractional stress. In the range of larger contractional stress this dependence becomes stronger. The similarity of curves 1 and 2 obtained on the same sample for different signs of stress shows that the above mentioned effect is not due to the presence of residual elastic stress. To verify this condition the sample was examined in transmitted polarized light before loading. An analogous dependence was found for a KBr crystal, whose yield stress was determined from the appearance of the first slip-bands ($\tau_c = 250$ p/mm²) for the area of dilation (Fig. 8).

It should be mentioned that internal inhomogeneities in the sample cause some scatter of the experimental points, particularly in the presence of internal stress in the sample. Therefore, to get "calibration curves" similar to those of Fig. 7 and 8 carefully annealed undeformed crystals are to be used.

It is useful to analyse the forces which act on the head dislocations in the rays of a dislocation rosette. When the sample is indented elastic and plastic deformation occurs. Therefore the analysis can only be given in general outline. Two cases will be considered.

1. Diamond-pyramid indentor acting on an unstressed crystal. — In this case plastic deformation takes place forming a dislocation rosette near the indentation. At the same time an elastic stress field is built up around the pyramid indentor which hinders the further intrusion into the crystal. We can write the equilibrium condition for the head dislocation of the rosette ray under load

$$\tau^{\mathrm{I}} + \tau^{\mathrm{II}} = \tau_{\mathrm{c}} \,, \tag{2}$$

where τ^{I} is the stress due to plastic deformation, i.e. the superposition of the elastic fields of all dislocations formed, τ^{II} the elastic stress balanced by the external stress, and τ_{c} the shear strength.[2]) When the diamond pyramid is lifted the elastic stress τ^{II} disappears (elastic return of indentation). The stress τ^{I} is balanced by the lattice resistance that is always lower than or equal to τ_{c}. We write the equality (2) in the form $\tau^{\mathrm{I}} = \tau_{\mathrm{c}} - \tau^{\mathrm{II}}$.

Hence, in order to move the head dislocation out of the rosette after the indentor has been lifted, it is necessary to apply an external stress which exceeds or is equal to τ^{II}.

2. *Diamond-pyramid indentor acting on a stressed crystal.* — In this case the external stress tends to elongate some, and to shorten other rays of the dislocation rosette as shown in Fig. 1 and 2. The equilibrium condition for the head dislocation will be written as

$$\tau_1^{\mathrm{I}} + \tau_1^{\mathrm{II}} + \tau = \tau_{\mathrm{c}} \tag{3}$$

for an elongating ray and as

$$\tau_2^{\mathrm{I}} + \tau_2^{\mathrm{II}} - \tau = \tau_{\mathrm{c}} \tag{4}$$

for a shortening one. τ is the external shear stress. The stress τ^{II} decreases with increasing r, where r is the distance from the point of stress application to a given point in the crystal. From the relations (3) and (4) it is evident that $\tau_1^{\mathrm{II}} < \tau_2^{\mathrm{II}}$ (as $\tau_1^{\mathrm{I}} \approx \tau_2^{\mathrm{I}}$) and, consequently, $r_1 > r_2$, where r_1 is the distance from the centre of indentation to the head dislocation in the elongating and r_2 to that in the shortening rays. This is just the circumstance which leads to the change of the shape of dislocation rosettes on stressed crystals.

When $\tau = \tau_{\mathrm{c}}$ the condition of equilibrium (3) is not fulfilled and the rays pass through the whole crystal, which is just observed.

When the indentor is lifted and the crystal unloaded, τ^{II} and τ go to zero and a rearrangement of stress occurs in accordance with the equalities (3) and (4).

From the relations (3) and (4) it is seen that the equilibrium conditions for the head dislocation contain the value τ_{c} which in real crystals strongly depends on the quantitative, qualitative, and structural state of impurities. Hence it is evident that the length difference of dislocation-rosette rays near the indentation (Δ) depends not only on the value of external stress τ, but also on τ_{c}. Therefore the dependences similar to those shown in Fig. 7 and 8 are different for crystals of the same substance which have different yield stresses. In order to determine the values of internal stress in a crystal from the change of the form of the dislocation rosette it is necessary to obtain a calibration curve for the given substance similar to Fig. 7 and 8.

---

[2]) Here the term shear strength refers to the stress which is needed to displace the dislocations in the lattice in a definite fixed time of its influence. It is necessary to introduce this limitation because the process of dislocation movement is thermally activated, and the stress that leads to some displacement decreases to the definite value, (to athermal constituent) with increasing time of load application.

The sensitivity of the method depends on the rigidity of the crystal and is about 10% of the yield stress determined from the appearance of the first slip bands in the crystal. This method is rather sensitive for soft crystals and permits to determine internal stresses of some p/mm^2.

The advantages of this method lie in the fact that it is rather simple, rapid, and permits to measure the local stress in the sample. The effect observed is thought to be useful for the determination of the internal stress in single crystals where the stress-birefringence method cannot be used.

Acknowledgement

The author is greatly indebted to V. I. Startsev for his continuous attention to the work and S. V. Lubenets for helpful discussions.

References

[1] J. J. GILMAN and W. G. JOHNSTON, Dislocations and Mechanical Properties of Metals, Izd. innostr. Lit., Moscow 1960 (p. 82), (in Russian).
[2] A. A. PREDVODITELEV, V. N. ROZHANSKII, and V. M. STEPANOVA, Kristallografiya **7**, 418 (1962).
[3] M. P. SHASKOLSKAYA and G. F. DOBZHANSKII, Kristallografiya **7**, 103 (1962).
[4] A. A. SHPUNT, Kristallografiya **7**, 474 (1962).
[5] W. M. NADGORNYI and A. V. STEPANOV, Fiz. tverd. Tela **5**, 1006 (1963).
[6] A. A. URUSOVSKAYA, R. THYAGARAJAN, and M. V. KLASSEN-NEKLYUDOVA, Kristallografiya **8**, 625 (1963).
[7] A. A. URUSOVSKAYA and R. THYAGARAJAN, Kristallografiya **8**, 531 (1964).
[8] A. A. PREDVODITELEV, V. N. ROZHANSKII, V. M. STEPANOVA, and N. A. TUMANOVA, Kristallografiya **9**, 695 (1964).
[9] N. A. TOROPOV and YU. P. UDALOV, Izv. vuzov SSSR, Fiz. **9**, 187 (1966).
[10] S. V. LUBENETS and N. F. KOSTIN, Kristallografiya **7**, 328 (1962).
[11] V. B. PARIISKII, S. V. LUBENETS, and V. I. STARTSEV, Fiz. tverd. Tela **8**, 1227 (1966).
[12] H. G. VAN BUEREN, Defects in Crystals, Izd. innostr. Lit., Moscow 1962, (in Russian).
[13] S. TIMOSHENKO, Strength of Materials, Izd. Nauka, Moscow 1965, (in Russian).

(Received November 9, 1966)

phys. stat. sol. **19**, 533 (1967)

Subject classification: 10; 22.5.2

Physico-Technical Institute of Low Temperatures of the Ukrainian Academy of Sciences, Kharkov

On the Kinetics of Plastic Deformation of Some Alkali Halide Crystals[1])

By

S. N. Komnik, V. Z. Bengus, and E. D. Lyak

A study is made of the kinetics of plastic deformation in LiF, NaCl, KCl, and KBr crystals. This includes the non-uniformity of the plastic deformation, the strain rate dependence of this non-uniformity, the strain rate dependence of the flow stress, and the dislocation structure of the deformed crystals. In the above crystals a regular change of the kinetics is found for a decrease of the elastic anisotropy factor $A = 2\,C_{44}/(C_{11} - C_{12})$. The correlation of the macroscopic behavior of the crystals with their elastic anisotropy is assumed if other conditions are similar.

Die Kinetik der plastischen Deformation wird in LiF-, NaCl-, KCl- und KBr-Kristallen untersucht, wobei die Ungleichförmigkeit der plastischen Deformation, ihre Abhängigkeit von der Deformationsgeschwindigkeit, die Abhängigkeit der Flußspannung von der Deformationsgeschwindigkeit und die Versetzungsstruktur der deformierten Kristalle eingeschlossen werden. In den genannten Kristallen wird eine regelmäßige Änderung der Kinetik mit abnehmendem elastischen Anisotropiefaktor $A = 2\,C_{44}/(C_{11} - C_{12})$ gefunden. Zwischen makroskopischem Verhalten der Kristalle und ihrer elastischen Anisotropie wird eine Korrelation angenommen, wenn andere Bedingungen ähnlich sind.

1. Introduction

It is of great interest to predict the macroscopic behavior of a crystal during plastic deformation in terms of its basic physical constants (elastic, thermal, etc.). However, only properties of individual dislocations may probably be immediately correlated with physical constants of the crystal.

But in macroscopic plastic deformation of the crystal a great number of dislocations takes part simultaneously. The macroscopic behavior of a crystal is not so much determined by the properties of individual dislocations (for example, by start stress) as by the behavior of the great ensemble of dislocations. The macroplasticity of a crystal is largely determined by the interaction of dislocations and their multiplication in this ensemble with statistical regularities as well. Therefore one may expect some correlation between the macroscopic behavior of the crystal and the physical constants which essentially determine the interaction of dislocations. As was shown by Chou the elastic interaction of dislocations and the shape of the dislocation elastic stress field are determined by the elastic anisotropy of the crystal [1, 2].

It seems therefore expedient to look for a correlation between the macroscopic behavior of the crystal during plastic deformation and its elastic anisotropy.

---

[1]) The contents of this article was reported at the Seventh International Congress of Crystallography, Moscow, July 1966.

When studying the plastic behavior of different alkali halide crystals differing by the value of the anisotropy factor $A = 2\,C_{44}/(C_{11}-C_{12})$ we found a regular change in their behavior. This is the subject of the present paper.

2. Method

The investigations were made with LiF, NaCl, KCl, and KBr crystals having A-values of 1.82, 0.7, 0.375, and 0.35, respectively. Samples about $4\times4\times10$ mm^3 in size were obtained by cleavage and deformed by compression at room temperature, which implied nearly identical temperature (relative to the melting point) for all the crystals studied. The deformation was carried out in an apparatus of about 2×10^3 kp/mm rigidity, the time dependence $P(t)$ of the load P acting on a crystal at a given strain rate being recorded photographically [3]. The accuracy of the force measurement was about 10 p. The experiments were carried out on fairly pure (with respect to cation impurities) crystals grown from the melt in air by the Kyropoulus method. For example, the KCl crystals contained less than 30 ppm Mg^{2+}, less than 5 ppm Ca^{2+}, and much less than 2 ppm Ba^{2+}. The grown-in dislocation density was 3×10^4 to 10^6 cm^{-2} in different crystals.

The non-uniformity of plastic deformation, the strain rate dependence of this non-uniformity, the strain rate dependence of flow stress, and the dislocation structure of the deformed crystals were studied in this work.

3. Results

3.1 Non-uniformity of plastic deformation and the strain rate dependence of flow stress

The non-uniformity of plastic deformation during creep was studied by Klassen-Neklyudova in NaCl [4]. The non-uniformity of plastic deformation in KCl at constant strain rate manifests itself in sudden changes of load with time [5]. The original "microscopic" yield drops in the $P(t)$ curve appear simultaneously with the creation of the first slip bands (before the macroscopic yield stress) and accompany the whole process of plastic deformation of KCl crystals. Apparently each drop corresponds to the lateral broadening of existing slip bands or to the appearance of new ones. The drops in the $P(t)$ curve are found [5] to become more pronounced with decreasing strain rate, and the rise before the drop becomes often elastic. This elastic rise indicates the absence of a storage of moving dislocations in the crystal. It implies that the condition $\dot{n}/l_0 \ll N\,v$ is fulfilled, where $\dot{n}$ is the rate of multiplication of moving dislocations, N the density of moving dislocations, v their velocity, l_0 the sample length in the direction

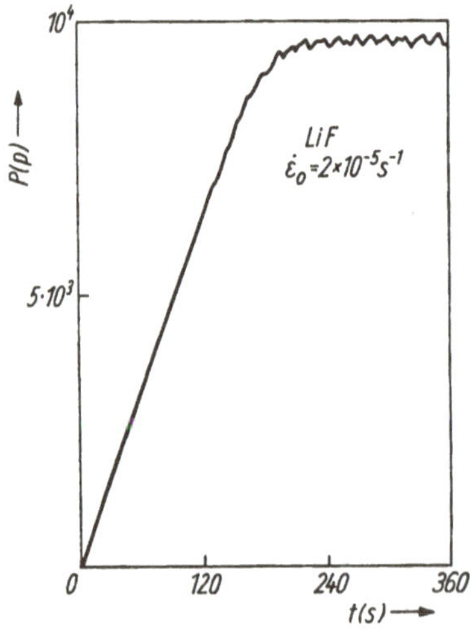

Fig. 1. Time dependence of load P near the yield point for LiF

of the normal to the slip plane ($l_0 = (l + a)/2$ here, l and a being the length and width of the sample). The appearance of moving dislocations at a certain time instant leads to an immediate elongation which ceases when the dislocations leave the crystal or are pinned in it. A similar behavior is characteristic of all alkali halide crystals. In particular for LiF crystals it is shown in Fig. 1.

Considering the shape of a certain drop in the $P(t)$ curve one can evaluate the number and average velocity of dislocations leading to the drop of load.

From the equation

$$v_0 t = \frac{\varepsilon l_0}{\sqrt{2}} + \frac{P}{K}, \tag{1}$$

where v_0 the velocity of the deforming rod, K the rigidity of the system "crystal + apparatus", ε the plastic shear strain, P the load, it is seen that if

$$v_0 t_c \ll \frac{\varepsilon l_0}{\sqrt{2}} \tag{2}$$

(where $t = t_c$ is the time of drop) the decrease of load ΔP is determined by the crystal elongation

$$\frac{\Delta P}{K} = \frac{\varepsilon l_0}{\sqrt{2}}. \tag{3}$$

When the velocity of the rod is large (when the condition (2) is not fulfilled) the recorded value of ΔP is no longer characteristic of the true value of deformation jump, which in this case should be determined from (1):

$$\frac{\varepsilon l_0}{\sqrt{2}} = v_0 t_c + \frac{\Delta P}{K} \equiv \frac{\Delta P^*}{K}. \tag{4}$$

Since the time of drop t_c is about 1 s at the strain rate of $1 \times 10^{-5}\ s^{-1}$ (see Fig. 2) the recorded value of ΔP (about 50 p) is no longer a measure of the true plastic deformation during the drop, as the $K v_0 t_c$-correction reaches 200 p. For the rate $6 \times 10^{-7}\ s^{-1}$ ΔP is 100 p and the $K v_0 t_c$-correction is about 10 p.

Supposing that the elongation is caused by a simultaneous exit of n dislocations one obtains $\Delta P^*/K = n b/\sqrt{2}$, from where

$$n = \frac{\Delta P^* \sqrt{2}}{K b}, \tag{5}$$

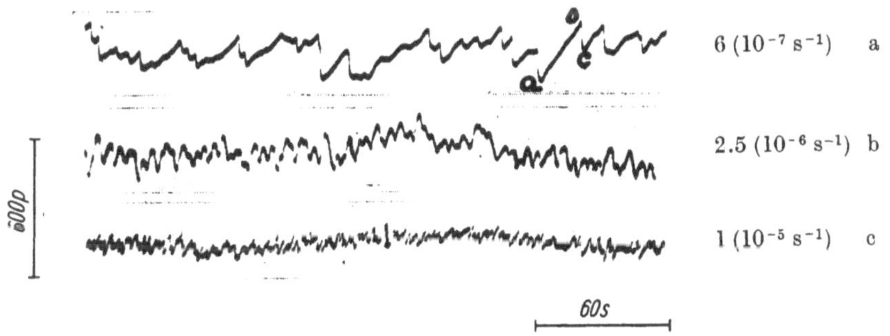

Fig. 2. Time dependence of load P in the stage of easy glide at different strain rates for KCl

where b is the Burgers vector. For a ΔP^* of about 100 p n is about 150 (at $K = 2\times 10^7$ p/cm, $b = 4.44\times 10^{-8}$ cm).

The average number of dislocations per drop increases with increasing strain rate (see Table 1). The values given in Table 1 are obtained from Fig. 2.

Table 1

v_0/l_0 (s^{-1})	Ω (min^{-1})	n	$n\,\Omega$	v (cm/s)
1×10^{-5}	27	360	10×10^3	1.0
2.5×10^{-6}	15	210	3×10^3	1.0
6×10^{-7}	6	160	1×10^3	1.0

The frequency of drops Ω also increases (as shown in Table 1) with increasing rate. If it is assumed that each drop corresponds to an act of multiplication of moving dislocations and the corresponding drop occurs each time when the definite value of resolved shear stress is reached, the frequency of drops will be determined only by the time which is needed to load the crystal to the required level (after the previous drop), provided the delay of dislocation multiplication is less than this time. Then an increase of strain rate must shorten the interval between drops, and this is just observed.

Let us evaluate the average velocity of the dislocations that cause some drop ΔP. Assuming that n dislocations move at the same velocity v and leave the crystal at the same time, v can be estimated from the equation

$$\cdot = \left(v_0 - \frac{1}{K}\,\frac{dP}{dt}\right)\frac{\sqrt{2}}{l_0} \tag{6}$$

considering $\dot{\varepsilon} = b\,n\,v/a\,l_0$, $n = \Delta P\,\sqrt{2}/b\,K$, where a is the average width of the crystal, $\dot{\varepsilon}$ is the plastic shear strain rate,

$$v = \frac{K\,v_0 - \dfrac{dP}{dt}}{\Delta P^*}\cdot\frac{a\,l}{l_0}\,. \tag{7}$$

The estimate of this velocity from (7) for KCl crystals shows that practically it does not vary with strain rate in the range studied (6×10^{-7} to 1×10^{-5} s^{-1}) and is about 1 cm/s. In this connection we may state that the increase of plastic shear strain rate in KCl crystals in the given range of strain rates is due not to the increase of dislocation velocity but mainly to the increase of the rate of their multiplication.

To check this statement a macroscopic deformation curve for a KCl crystal was measured for a step-like change of strain rate (Fig. 3). It is seen that there is practically no rate dependence of flow stress and hardening modulus (it is essential only before the macroscopic yield stress). This independence would not be easy to understand if we did not know the discrete nature of the process of deformation with time. It is apparent that, since the flow stress does not vary with strain rate, the dislocation velocity (as a function of stress) should be also constant. Thus the increase of plastic shear strain rate can be caused only by the increase of the number of moving dislocations.

The non-uniformity of the process of plastic deformation in different alkali halide crystals manifests itself in different ways (Fig. 4).

Fig. 3. Stress–strain curve for KCl for a step-like change of strain rate

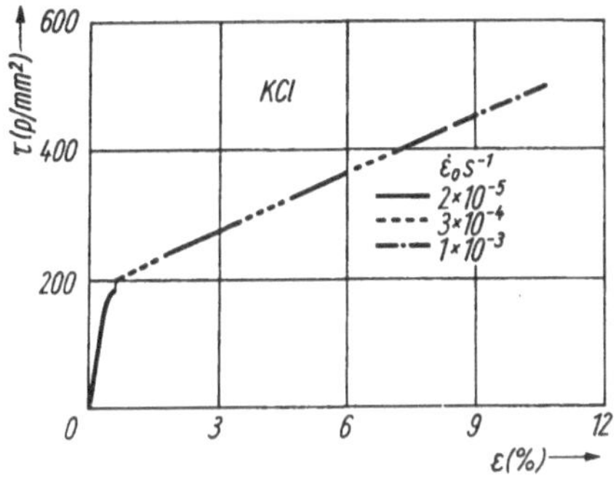

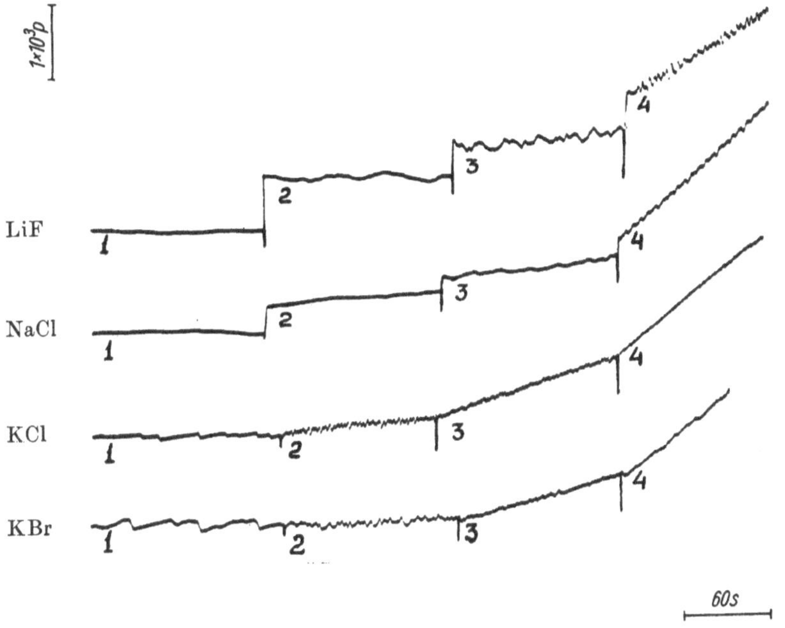

Fig. 4. Time dependence of load P for alkali halide crystals for a step-like change of strain rate (density of grown-in dislocations about 5×10^5 cm^{-2}). Strain rate: (1) 3×10^{-7} s^{-1}, (2) 5×10^{-6} s^{-1}, (3) 2×10^{-5} s^{-1}, (4) 8×10^{-5} s^{-1}

It is seen that for each crystal the non-uniformity is most clearly expressed in a definite range of strain rates. For KBr and KCl crystals these rates are about 10^{-6} s^{-1}; for NaCl and LiF they are about 10^{-4} s^{-1}. At small strain rates the plastic deformation of NaCl and LiF is mainly a uniform process, and the load drops (if they occur at all) are less drastic than for KCl. The oscillation of the plastic shear strain rate $\dot{\varepsilon}$ for LiF is much lower for than KCl. The load drops for LiF are very slow (strictly speaking, they cannot be called drops, see velocities 2 and 3), and this means that the density flow $N\,v$ of the dislocations leaving the crystal at the same time is not great. For these reasons one can assume that the multiplication of moving dislocations in LiF occurs in small portions. Thus at the rate 3×10^{-7} s^{-1} the non-uniformities correspond to portions of 10 to 30 dislocations, which do not give any measurable load drops, so that practically an ideal flow is observed. At the rate 5×10^{-6} s^{-1} only slow oscillations of load with periods of 10 to 50 s are observed. At great rates the behavior of LiF crystals

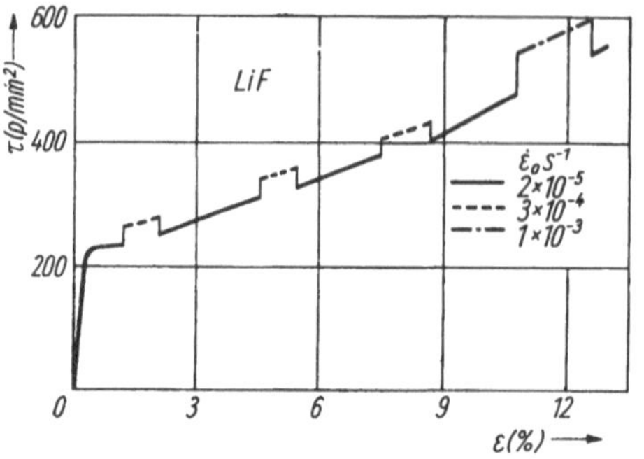

Fig. 5. Stress–strain curve for LiF for a step-like change of strain rate

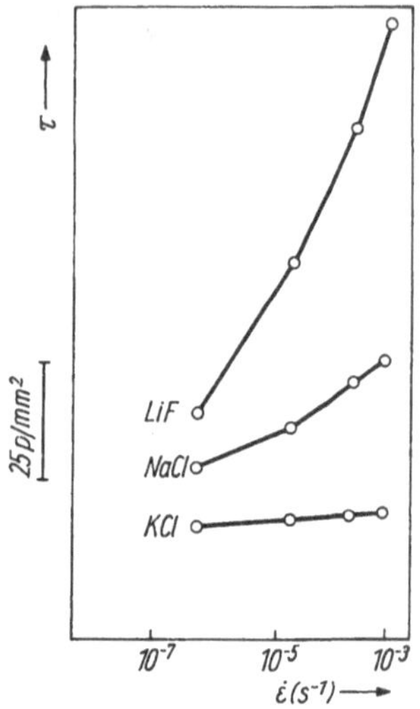

Fig. 6. Strain-rate dependence of flow stress for different alkali halide crystals (zero ordinate arbitrary)

resembles that of KCl: the $P(t)$ curve contains drops whose frequency increases with increasing rate, thus indicating an increase of the rate of dislocation multiplication.

The second characteristic difference between LiF and KCl is the appreciable rate dependence of flow stress (see Fig. 5 where the stress–strain curve obtained for a step-like change of strain rate is shown). The increase of flow stress with increasing strain rate should lead to an increase of the velocity of dislocations appearing in the crystal. An estimate of this velocity, which is similar to the above one, shows that the dislocation velocity in LiF increases with increasing strain rate. From here it follows that unlike KCl the increase of the plastic shear strain rate in the crystal is due both to the increase of the rate of dislocation multiplication and the speeding-up of their motion.

NaCl takes an intermediate position between KCl and LiF. The rate dependence of its flow stress is lower than for LiF. Fig. 6 shows the strain rate dependence of flow stress for KCl, NaCl, and LiF.

3.2 Dislocation structure of a crystal

The dislocation structure of deformed crystals consists of "trails" (or "debris") of the motion and multiplication of dislocations. The number of "trails" depends on the interaction of moving dislocations with one another and with other defects, and the type of interaction determines the kind of "trails" (dipoles and multipoles, dislocation loops of by-pass, individual dislocations, etc.).

Therefore, the stronger these interactions, the more "trails" are left in the crystal. Of course the "trails" are left in those crystal planes in which the dislocations were moving, i.e. in those crystal planes where they appeared as a result of multiplication. (The term "multiplication" refers to the creation of new moving dislocations in crystal planes adjacent to the initial one.)

Therefore the dislocation structure of a crystal depends on the localization of the process of multiplication: if the process of multiplication is strongly localized the crystal volume where the "trails" remain is small.

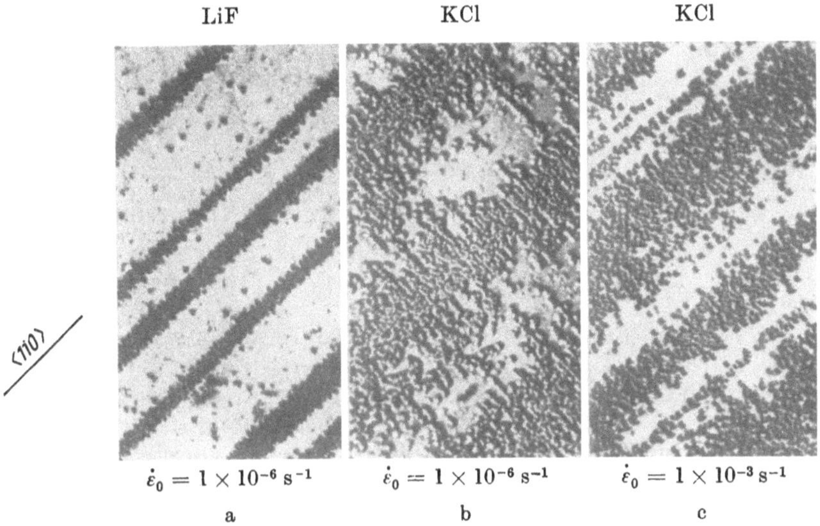

Fig. 7. Dislocation structure of LiF and KCl crystals in the initial stage of plastic deformation (180×)

The investigation of the dislocation structure of LiF, NaCl, KCl, and KBr crystals shows that the distribution of "trails" in the form of glide bands is typical of these crystals. But the appearance of glide bands for various crystals is different. This is well illustrated in Fig. 7 where the dislocation structures of KCl and LiF crystals are compared in the initial stage of deformation at the same strain rate (Fig. 7a and b). It is seen that in LiF the glide bands are sharply outlined, i.e. their boundaries are parallel to the traces of slip planes of the crystal. The dislocation density outside the glide bands is small and equal to the initial one (before the deformation). In KCl the glide bands are "smeared", there are short rows of dislocations perpendicular to the band adjacent to it. The dislocation density outside the bands exceeds the initial one and increases as the strain of the crystal increases. It can be concluded that, at the given strain rate, in LiF slip is more localized than in KCl.

As the strain rate increases the glide bands arising in KCl become clearer (see Fig. 7b). The dislocation structure of LiF does not change markedly when the rate is increased by a similar factor (by three orders of magnitude). With respect to dislocation structure, KBr is similar to KCl, and NaCl is more similar to LiF than to KCl.

4. Discussion of the Results

There have been several attempts to systematize the macroscopic behavior of pure alkali halide crystals. Comparisons between work hardening and ionic polarization and between microhardness and shear modulus were made by Alden [6] and Gilman [7], respectively. Nadeau [8] compared the macroscopic yield stress τ of different alkali halide crystals with their shear modulus G, and came to the conclusion that there is no sufficient reason, even for pure crystals, to suppose that the yield stress is proportional to the shear modulus. In fact, a calculation by means of the data [8] shows that the crystals of greater shear modulus (in the series LiF, NaCl, KCl, KBr) have lower relative yield stress:

Table 2

Material	τ (p/mm²)	$G = [C_{44}(C_{11} - C_{12})/2]^{1/2}$ (p/mm² × 10⁶)	$\frac{\tau}{G} \times 10^5$
LiF	75	4.67	1.61
NaCl	61	1.51	4.04
KCl	75	1.01	7.43
KBr	66	0.86	7.68

All the crystals studied in this work have identical lattice structure (rock salt type) and identical slip systems {110} ⟨110⟩ (at room temperature), hence the geometry of dislocation interaction must be the same. But the interaction of dislocations is essentially determined by the shape of the stress field near the individual dislocation, which depends on the elastic anisotropy of the crystal. For example, such a dependence for hexagonal crystals is shown in [1].

For alkali halide crystals the shape of stress rosettes can be calculated by the formulas of [2]. This shape changes regularly in the series LiF, NaCl, KCl, KBr (arranged in the order decreasing anisotropy factor A). The trend of the change is illustrated by Fig. 8, where the rosettes of shear stress are shown schematically in the plane and direction of slip, for screw and edge dislocations in LiF and KCl. It is seen that with decreasing anisotropy factor the lines of equal resolved shear stresses more and more extend into the regions near the glide plane of the dislocation. Thus the field of resolved shear stress of dislocations (especially of screw dislocation) becomes less localized near the glide plane of the dislocation considered. This results in a less localization of the internal residual elastic stresses which appear in the deformed part of the sample. The long-range action of these stresses in the direction normal to the glide plane of each dislocation (as compared with the long-range action in glide direction) increase with decreasing A. In this case the fluctuations of internal stresses in the deformed part of the sample must increase and the region occupied by each fluctuation must also become larger. The dimensions of the crystal regions where the overlap of the stress of individual dislocations increases the resolved shear stress and where, as we think, the multiplication of moving dislocations takes place (e.g., during the lateral expansion of the glide bands), are greater in crystals with smaller A (independent of supposition about the mechanism of the process: double cross slip, nucleation, Frank-Read mechanism). Therefore the non-uniformity of plastic deformation must depend regularly on A.

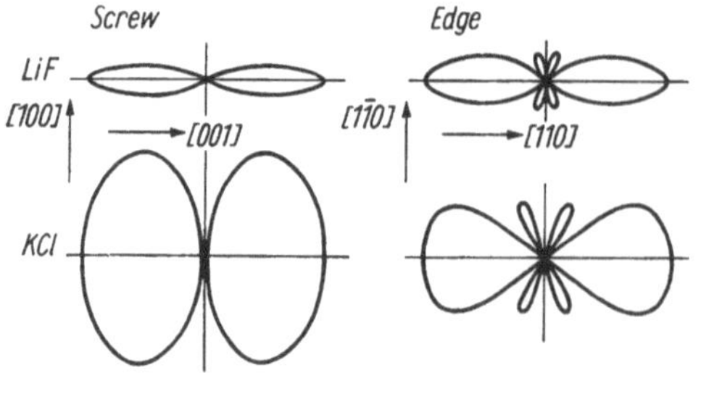

Fig. 8. Shape of the rosette of resolved shear stress in a slip plane for edge and screw dislocations in LiF (anisotropy factor $A = 2\,C_{44}/(C_{11} - C_{12}) = 1.82$) and KCl ($A = 0.375$)

The imperfect structure of the deformed part of the crystal must be rougher and more heterogeneous in crystals with smaller A. It should be mentioned that the increase of the heterogeneity of the deformed part of the crystal with decreasing A in the series LiF, NaCl, KCl, KBr is enhanced by the increase of the lattice parameters.

On this basis, considering the differences in the lattice parameters and elastic moduli of the crystals one can understand the change in the degree of localization of the glide trails in the series of crystals studied, and the change in the non-uniformity and heterogeneity of plastic deformation.

Acknowledgement

It is a pleasure to acknowledge many helpful and stimulating discussions with V. I. Startsev. Thanks are due also to S. Kalinchuk for carrying out the quantitative analysis of impurities in the crystals.

References

[1] Y. T. Chou, J. appl. Phys. **33**, 2747 (1962).
[2] Y. T. Chou, J. appl. Phys. **34**, 429 (1963).
[3] V. P. Regel and G. A. Dubov, Prib. i. Tekh. Eksper. No. 6, 102 (1953).
[4] M. V. Klassen-Neklyudova, Zh. Russk. Fisiko-Khim. Obshch. **59**, 509 (1927).
[5] V. Z. Bengus, S. N. Komnik, and O. B. Shtitelman, phys. stat. sol. **14**, 215 (1966).
[6] T. H. Alden, Trans. AIME **230**, 649 (1964).
[7] J. J. Gilman, Austral. J. Phys. **13**, 327 (1960).
[8] J. S. Nadeau, J. appl. Phys. **35**, 1248 (1964).

(Received November 9, 1966)

From experimental results achieved by etch-pit methods [10 to 12] and on foils [14] it seems that inhomogeneity is the characteristic feature of microstrain regions with the progressive formation of slip activity in more and more grains. The continuous transition into the macrostrain region is characterized by the fact that plastic strain is initiated nearly in all grains.

In this paper the influence of grain size and substructure on the microstrain region and on the transition to the macrostrain region is studied. The obtained results are compared with the conclusions of the Brown-Lukens theory to prove the deficiency of this theory. Finally, the applicability of the statistical theory to describe the stress–strain dependences in the microstrain region is shown.

Table 1

Grain diameter d (mm)	$d^{-1/2}$($mm^{-1/2}$)	Cold reduction (%)	Annealing temperature (°C)
0.59	1.3	30	850
0.077	3.6	30	650
0.051	4.4	30	500
0.018	7.5	65	500

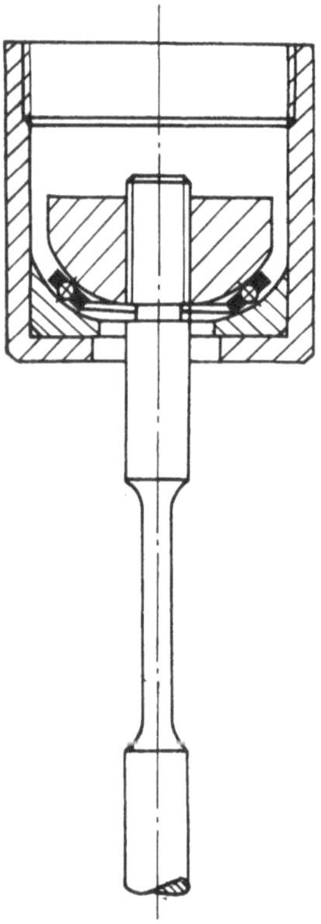

Fig. 1. Gripping device

Fig. 2. The dependence of σ on ε_p for recrystallized specimens plotted in logarithmic coordinates

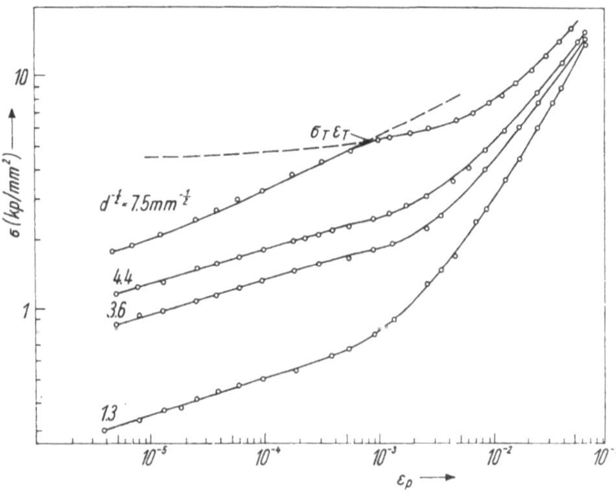

2. Material, Experimental Techniques

Two batches of specimens made from electrolytic copper (99.9%) were used for experimental examinations.

The first batch of specimens was produced from rods cold drawn to various reductions. By recrystallization annealing at different temperatures samples with different grain size were obtained. Transmission electron microscope technique revealed that the dislocation density of these samples was different and depended on the annealing temperature.

The cold reduction, annealing temperatures, and the resulting grain size are given in Table 1.

Specimens of the second batch were taken from as-received materials that had a considerable dislocation density (as observed by transmission electron microscopy). By annealing at temperatures of 500, 600, and 700 °C for 1 h this density was reduced so that specimens annealed at higher temperature had a lower dislocation density. The initial grain size $d^{-1/2} = 3.5\ \mathrm{mm}^{-1/2}$ did not change by the annealing process.

The specimens used were of cylindrical form (diameter 4 mm, length 55 mm) with threaded ends and the middle part of them was chemically polished.

The specimens were gripped in the testing machine by means of a special device (Fig. 1) in order to minimize the bending. Tests were carried out on the ZWICK electronic testing machine, the cross head motion was 1.2 mm/min. Strain measurements were made using inductive strain gauge (gauge length $l = 50$ mm) that enables the strain to be measured with maximum sensitivity 4×10^{-6}.

Microplastic strain was determined as deviation from the elastic stress–strain line. Some specimens in the microstrain region were also tested by step loading and unloading technique. Results obtained by these two techniques did not essentially differ from one another. Three samples of each sort were tested, the reproducibility was very good.

3. Experimental Results

The obtained dependences $\sigma(\varepsilon_p)$ (σ is the true stress, ε_p the true plastic strain) were plotted in the coordinates $\log \sigma$ vs. $\log \varepsilon_p$. These dependences for recrystallized samples are shown in Fig. 2. Two strain stages can be seen on the curves. The first stage, the microstrain, is the approximately straight-line dependence between $\log \sigma$ and $\log \varepsilon_p$, ending at strains of about 2×10^{-4}. After a short transition part the macrostrain region follows in which the dependence between $\log \sigma$ and $\log \varepsilon_p$ is given by a bent curve.

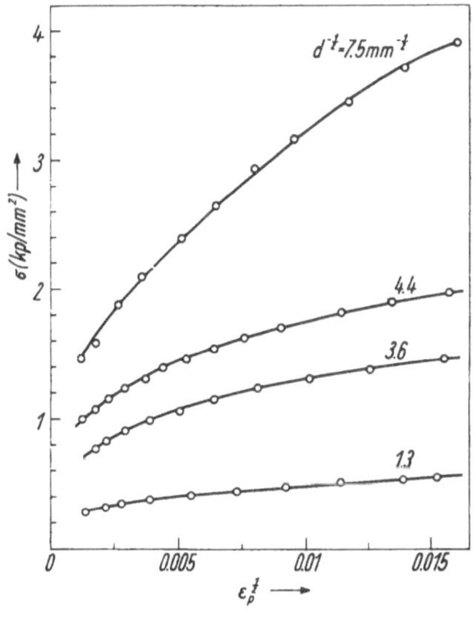

Fig. 3. The dependence of σ on $\varepsilon_p^{1/2}$ in the microstrain region for recrystallized specimens

The obtained results do not prove the validity of equation (1) in the microstrain region, because there is no linear dependence between σ and $\varepsilon_p^{1/2}$ (Fig. 3). An equation of a more general type,

$$\sigma = \sigma_0 + K\, \varepsilon_p^n\,, \tag{3}$$

was used therefore to describe the dependence of σ on ε_p. Equation (3) was used for the macrostrain region as well.

The method used for determining the constants σ_0, K, and n is given in [13]. Values of σ_0, K, and n for both batches of specimens are shown in Table 2. In this table values of σ_T and ε_T are also quoted that characterize the transition from the microstrain to the macrostrain region. These values were determined as the points of intersection of the curves in Fig. 2.

Values of the exponent n in the microstrain region were a little different for the variously treated specimens, but they did not show any dependence on grain size or on the annealing temperature. n in the Table 2 represents the mean value.

In Fig. 4 and 5 the dependences of σ on $\varepsilon_p^{0.25}$ are plotted for the microstrain region for both batches of specimens. The value σ_0 can be obtained by extrapolation to zero plastic strain, K is given by the slope of the straight line. From the plot and from Table 2 it can be seen that with recrystallized samples the stress–strain rate increases with decreasing grain size. At the same time the value of σ_0 changes only very slightly, but with increasing annealing temperature σ_0 decreases (Fig. 5).

The dependences of σ on ε_p for the macrostrain region plotted in the coordinate system σ vs. $\varepsilon_p^{0.76}$ form for various grain size a system of mutually displaced, approximately parallel straight lines. In Fig. 6 the dependences of σ on $\varepsilon_p^{0.76}$ for

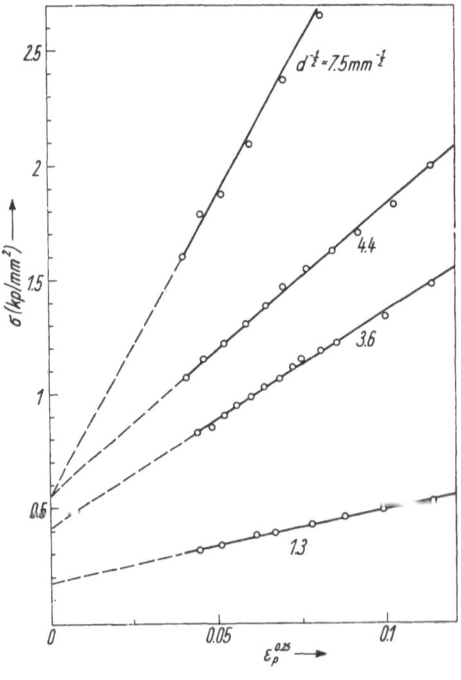

Fig. 4. The dependence of σ on $\varepsilon_p^{0.25}$ for recrystallized specimens in the microstrain region

Fig. 5. The dependence of σ on $\varepsilon_p^{0.25}$ for annealed specimens in the microstrain region

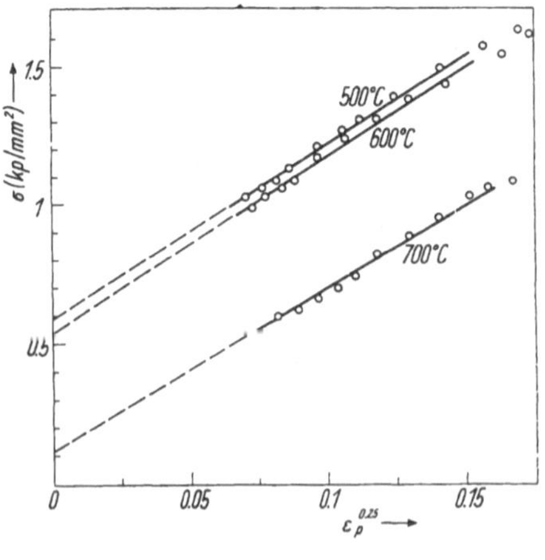

Table 2

	$d^{-1/2}$ (mm$^{-1/2}$)	Annealing temperature (°C)	Microstrain σ_0 (kp/mm^2)	Microstrain K (kp/mm^2)	Microstrain n	Transition to macrostrain σ_T (kp/mm^2)	Transition to macrostrain ε_T (10^{-4})	Macrostrain σ'_0 (kp/mm^2)	Macrostrain K' (kp/mm^2)	Macrostrain n'
Recrystallized specimens	1.3	850	0.18	3.1		0.59	3	0.45	86	
	3.6	650	0.41	9.2	0.25	1.60	4	1.30	94	
	4.4	500	0.55	12.5		2.20	4	1.90	93	0.76
	7.5	500	0.53	26.5		5.00	7	4.60	93.5	
Annealed specimens	3.5	500	0.58	10.8		1.80	2	1.60	93.5	
	3.5	600	0.52	10.5	0.25	1.70	2	1.50	93.5	0.76
	3.5	700	0.12	10.2		1.25	2	1.10	93	

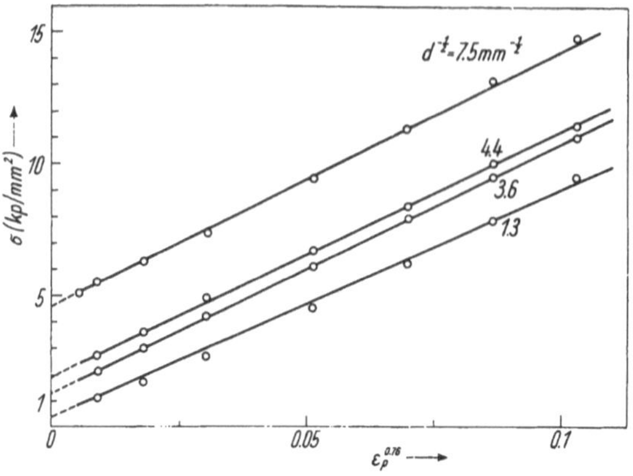

Fig. 6. The dependence of σ on $\varepsilon_p^{0.76}$ for recrystallized specimens for the macrostrain region

recrystallized samples are given. The magnitude of the displacement of the straight lines in direction to the stress axis equals the stress difference at the end of the first deformation stage. It follows therefore that, contrary to the first strain region, the increment of flow stress with strain in the second stage is not a function of grain size (at least in the investigated strain region).

This fact is also proved by Fig. 7 where σ vs. $d^{-1/2}$ is plotted for given strain. In the microstrain region the slope of the stress–grain size dependence increases with increasing strain, in the macrostrain region it does not change any more.

In the second batch the difference between flow stress for specimens annealed at various temperatures was the same in the whole strain range (up to 4×10^{-2}) and was equal to the difference of the stress σ_0 in the microstrain region.

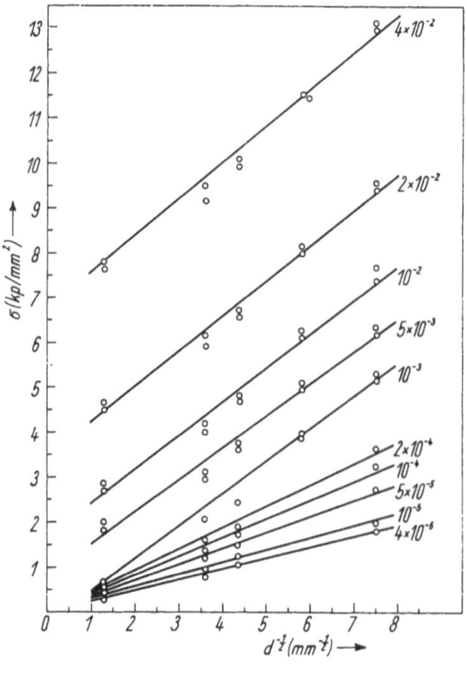

Fig. 7. Flow stress of recrystallized specimens at various plastic strain levels

4. Discussion

4.1 Microstrain as a distinct stage of strain taken from the mechanical point of view

From our results and from those mentioned in the introduction it may be thought that microstrain in polycrystalline material is a distinct process that differs from the macrostrain process by the following features:

1. From results obtained by the etching technique and those achieved from transmission electron microscopy it is evident that plastic deformation is heterogeneous and that elastically and plastically deformed grains exist in the microstrain region. The dependence of σ on ε_p in the macrostrain region is determined by the work-hardening process in individual grains. The dependence of σ on ε_p in the microstrain region is determined by a successive slip activation in the individual grains.

2. In the microstrain region an appreciable effect of grain size on the deformation becomes evident (Fig. 4, Table 2), whereas in the macrostrain region the grain size has no influence on the increase of flow stress with strain (up to $\approx 4 \times 10^{-2}$).

3. In a plot of $\log \sigma$ vs. $\log \varepsilon_p$ for Cu there is a change in slope that divides the micro-from the macrostrain stage. If we approximate the dependence of σ on ε_p in both regions by using the equation of the type $\sigma = \sigma_0 + K\,\varepsilon_p^n$, then the values of the three constants are different in the micro- and macrostrain regions.

It may be assumed therefore that the heterogeneity of plastic microstrain will be of decisive importance for the microstrain process, all other effects may result from this heterogeneity.

4.2 Comparison of the results with the Brown-Lukens theory, use of the statistical theory

The only theory describing quantitatively the dependence of σ on ε_p in the microstrain region is the Brown-Lukens (B-L) theory. Results of our measurements on polycrystalline Cu and those of other authors on various metals did not prove the general validity of the relation derived by B-L. There is generally no linear dependence between σ and $\varepsilon_p^{1/2}$; σ_0 is not constant, but varies with the change of substructure.

The cause of this disagreement with experimental results appears to be as follows:

a) B-L assume that dislocations emitted from a source pile-up at grain boundaries. As it was proved by different authors the dislocation distribution in the microstrain region did not show pile-ups [6, 10, 11, 12].

b) B-L consider the polycrystalline material as a group of grains that deform independently from each other whereby all grains are loaded with the same stress. In polycrystalline metals in which plastically and elastically deformed grains exist, the behaviour of these grains will be mutually influenced. It is reasonable to assume that the stress in the cross section will not be uniform. Therefore, the applied stress is the mean value taken from stresses resolving in elastically and plastically deformed grains.

c) The frequency distribution that determines the number of grains plastically deforming at a given stress is too simple according to B-L. Suits and Chalmers [10] using the etch-pit technique obtained quite different curves for fine-grained and coarse-grained material in Fe–Si.

To describe the microstrain we shall use the statistical method of calculating the stress–strain relation according to Afanasev [16].

The polycrystalline material is supposed to consist of grains having in their volume the same properties. The individual grains have different yield stress σ_y in the direction of the acting tensile force. It is assumed that such grains follow Hook's law before slip activation at the yield stress σ_y occurs. Let us assume further that in the microstrain region after the onset of yielding further slip in the grain takes place at the stress σ_y. It is assumed, therefore, that during the deformation in the microstrain region the stress fields acting against dislocation movement do not change. (This assumption might be thought as reasonable because any considerable increase in dislocation density was not observed.) The applied stress is considered as a mean value taken from the stresses resolving in elastically and plastically deformed grains.

Let us assume further that the deformation of all grains is the same and is given by the deformation of elastically deformed grains. The frequency distribution according to this theory is given by

$$y = f(\sigma_y) = -\frac{1}{E^2}\frac{d^2\sigma}{d\varepsilon^2}\,. \tag{4}$$

From this relation results that the frequency curve is the second derivative of the stress–strain curve.

By using equation (4) it is possible to determine the frequency distribution in the microstrain region.

We may obtain the frequency distribution for different grain sizes by graphic differentiation of the stress–strain curve. By this procedure we can get only a part of the frequency distribution (the dashed line in Fig. 8, 9), because the graphic differentiation can be realized only from the first measurable plastic strain.

We can also obtain the frequency distribution in another way, if we use equation (3) for description of the stress–strain relationship. The stress as a function of strain for $\sigma \leqq \sigma_0$ is given by

$$\varepsilon = \frac{\sigma}{E}\,,$$

the frequency distribution is $y = 0$.

For $\sigma \geqq \sigma_0$ we have

$$\varepsilon = \frac{\sigma}{E} + \left(\frac{\sigma - \sigma_0}{K}\right)^{1/n} \tag{5}$$

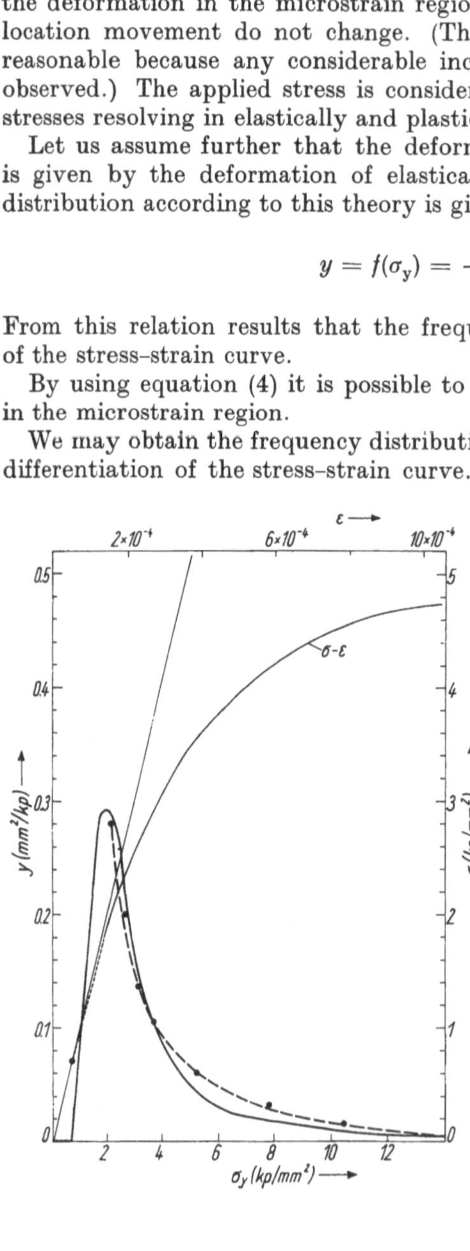

Fig. 8. Frequency distribution curve for fine-grained Cu ($d^{-1/2} = 7.5\ mm^{-1/2}$) graphically determined (dashed line) and calculated (solid line)

and the frequency distribution is

$$y = \frac{1}{E^2} \frac{\frac{1}{n}\left(\frac{1}{n} - 1\right)(\sigma - \sigma_0)^{\frac{1}{n} - 2}}{K^{1/n}\left(\frac{1}{E} + \frac{\frac{1}{n}(\sigma - \sigma_0)^{\frac{1}{n} - 1}}{K^{1/n}}\right)^3}. \tag{6}$$

Equation (6) gives the dependence of frequency on the applied stress for the known constants σ_0, K, and n. By calculating σ from (5) and substituting it into (6) it is possible to obtain the dependence $y = f(\varepsilon)$, or eventually $y = f(\sigma_y)$, as required by (4). The calculation of σ from (6) is rather difficult. Therefore, values of the function $y = f(\sigma_y)$ have been calculated from equations (5) and (6) by a simultaneous substitution of the given value σ. Curves thus calculated are plotted in Fig. 8 and 9 for fine-grained and coarse-grained structures ($d^{-1/2} = 7.5\ \text{mm}^{-1/2}$ and $1.3\ \text{mm}^{-1/2}$). By comparing both curves it follows that the frequency curve for coarse-grained material is narrower and has a much larger maximum value than the frequency curve for fine-grained material. (The scales in the figures are different.)

By analysing equation (6) we can see that the form and position of the frequency curve is given by the values of the constants σ_0, K, n. σ_0 gives the starting point of the frequency curve, n and K define the form of the frequency curve (for the same n a higher K value gives a wider frequency curve).

On the basis of this analysis we may try to attribute a physical significance to the constants σ_0, K, n in (3) for the microstrain region.

According to the accepted deformation model σ_0 is the stress at which slip activation occurs in the first grains (equation (6)). The significance of σ_0 is the same as that of σ_u^0 in the theory of B-L. σ_0 correlates to the resolved shear stress τ_0, which is necessary to bring up the yielding in the slip plane, $\sigma_0 = \overline{m}\,\tau_0$, where $\overline{m}$ is not given only by the orientation factor but is influenced also by other factors. From the experimental results and from the analysis mentioned above it is not possible to decide unambigously about the physical meaning of σ_0. On the basis of etch-pit studies it seems that τ_0 is associated with the activation of Frank-Read (F.-R.) sources, because a small increase of dislocations in the slip band may be observed. As it results from Fig. 5,

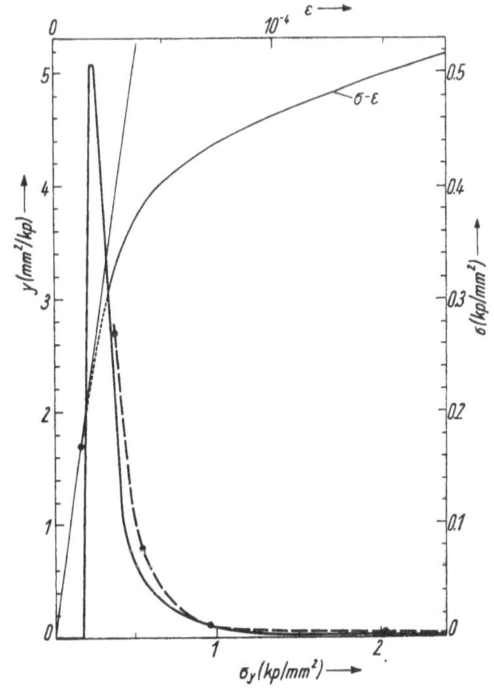

Fig. 9. Frequency distribution curve for coarse-grained Cu ($d^{-1/2} = 1.3\ \text{mm}^{-1/2}$) graphically determined (dashed line) and calculated (solid line)

the dislocation substructure certainly influences the value τ_0 (and thus also σ_0). At higher dislocation density the dislocation segments acting as F.-R. sources can be shorter, similarly the resistance against dislocation movement caused by the dislocation forest will depend on the dislocation substructure. Therefore, it can be expected that a higher dislocation density will result in a higher σ_0. The dislocation density in 99.9% Cu is then determined by the annealing temperature.

The constants n and K are given by the form of the frequency curve; this form is probably a function of the structure and of the material.

For copper n seems to be constant, consequently the type of the curve is the same for different grain sizes. For other materials the value of n can be different.

For a given value of n the constant K is associated with the mean value and the deviation of the frequency distribution. As it is shown in Fig. 4 and 5, K does not depend on substructure, it depends only on the grain size. Coarse-grained 99.9% Cu has a narrow frequency curve (Fig. 9), i.e. on reaching the stress σ_0 a much greater amount of grains start to deform plastically.

4.3 Transition into the macrostrain region

According to the accepted conception the number of plastically deformed grains in the microstrain region increases with increasing strain. The increment of external stress is caused by the stress increment inside elastically deformed grains. If the process can be described by equation (3), then this process is believed to take place before a deviation from the linear dependence $\sigma \sim \varepsilon_p^n$ appears. The deviation from relation (3) may be expected to appear when slip activation is initiated in nearly all the grains. From this movement the increment of the external stress must be balanced by the work-hardening associated with the increase of dislocation density and another strain stage sets in which according to the general conception can be considered as the macrostrain stage. Under the assumption that the density increases and the change of dislocation distribution in the macrostrain region does not depend on the grain size, then from this moment a plot of σ vs. $\varepsilon_p^{n'}$ should present for different grain size a system of parallel straight lines with the same values K' and n' (Fig. 6, Table 2). It may be believed that the change of the curves in the diagram $\log \sigma$ vs. $\log \varepsilon_p$ is associated with the beginning of the macrostrain region in polycrystalline 99.9% Cu.

Experimental results in Fig. 6 show that the difference in flow stresses in the macrostrain region for various grain sizes d is given by the stress difference during transition from the micro- to the macrostrain region. If we express the dependence of flow stress σ_f on the grain size d by the relation

$$\sigma_f = \sigma_{f0} + k_f \, d^{-1/2} , \tag{7}$$

then the magnitude of the flow stress component $k_f \, d^{-1/2}$ is mainly determined by processes taking place in the microstrain region which result in macroplastic strain with fine grained structure starting at higher stress (Fig. 4, Table 2).

These results show that it is necessary to alter the opinion about the meaning of the constant k_f (theory of the blocking effect [15]) in (7). and it appears that the work-hardening processes in polycrystalline 99.9% Cu are not so much influenced by the grain size at least in the investigated strain range (similarly to steel [13]).

5. Conclusions

1. The dependence of σ on ε_p in recrystallized and annealed polycrystalline 99.9% Cu in the micro- and macrostrain region can be described by the equation

$$\sigma = \sigma_0 + K\, \varepsilon_p^n \quad \text{or} \quad \sigma = \sigma_0' + K'\, \varepsilon_p^{n'} . \tag{8, 9}$$

In the microstrain region σ_0 depends on substructure (dislocation density), K depends on the grain size, and n depends neither on substructure nor on grain size.

In the macrostrain region σ_0' is determined by the stress at the end of the microstrain stage,

$$\sigma_0' \doteq \sigma_T = \sigma_0 + K\, \varepsilon_T^n \tag{10}$$

(ε_T strain at which the transition from micro- to macrostrain occurs) is thus a function of substructure and grain size, K', n' depend neither on the grain size nor on substructure (in the investigated strain range 4%).

2. The dependence of σ on ε_p in the microstrain region is determined by the frequency distribution that gives the frequency of plastically deformed grains as a function of the applied stress. With the change of substructure the starting point of the frequency curve varies, and with the change of grain size the form of the frequency distribution alters.

3. The difference between flow stresses for different grain sizes in the macrostrain region is not determined by the different influence of the blocking effect of the grain boundaries, but is given by the stress difference at the end of the microstrain stage.

References

[1] D. A. Thomas and B. L. Averbach, Acta metall. **7**, 69 (1959).
[2] N. Brown and K. F. Lukens, Acta metall. **9**, 106 (1961).
[3] W. Bonfield and C. H. Li, Acta metall. **11**, 585 (1963).
[4] W. Bonfield and C. H. Li, Acta metall. **12**, 577 (1964).
[5] W. Bonfield, Acta metall. **13**, 551 (1965).
[6] T. J. Koppenaal and P. R. Evans, J. Inst. Metals **92**, 238 (1964).
[7] R. F. Tinder and J. Washburn, Acta metall. **12**, 129 (1964).
[8] A. R. Rosenfield and B. L. Averbach, Acta metall. **8**, 624 (1960).
[9] R. D. Carnahan and J. E. White, Phil. Mag. **10**, 513 (1964).
[10] J. C. Suits and B. Chalmers, Acta metall. **9**, 854 (1961).
[11] P. J. Worhington and E. Smith, Acta metall. **12**, 1277 (1964).
[12] W. D. Brentnall and W. Rostoker, Acta metall. **13**, 187 (1965).
[13] M. Holzmann and J. Man, J. Iron Steel Inst. **204**, 230 (1966).
[14] W. Bonfield and C. H. Li, Acta metall. **13**, 317 (1965).
[15] R. Armstrong, I. Codd, R. M. Douthwaite, and N. J. Petch, Phil. Mag. **7**, 45 (1962).
[16] N. N. Afanasev, Statisticheskaya teoriya ustalostnoi prochnosti metallov, Izdatelstvo Akademii Nauk Ukrainskoi SSR,

(Received November 9, 1966)

1. The saturation of the photocurrent occurs when the enhanced field strength at the anode tends to a critical value E_K (p-conductor) [8, 9].

2. The field strength at the anode cannot exceed this critical value E_K so that the crystal turns out to be divided in two regions with different field strengths when the applied voltage lies in the saturation range [7, 8].

3. With increasing voltage the width of the high-field region will increase, while the range of the low-field region decreases so that a homogeneous electric field E_K in the whole crystal exists when the applied voltage takes the value $E_K L$ (L specimen length) [8].

4. When the voltage exceeds this value $E_K L$, the field strength will first increase at the cathode while the field strength at the anode retains the critical value E_K [7, 8].

5. Kiess [8] observed that the field strength at the anode immediately takes on the critical value E_K at the saturation voltage while Polke and Hasler [9] observed with increasing voltage an increase of the field strength in the high-field domain at the anode until the critical value E_K is reached.

6. Kiess [8] did not observe any instabilities while Polke and Hasler [9] did observe low-frequency current oscillations in Se.

7. For the occurrence of these oscillations a minimum of light intensity was necessary [9].

8. Oscillations with frequencies in the range of some 10^3 Hz in the high-field domain were reported by Böer and Wilhelm [1].

Further, this model must be able to explain the observed

9. temperature quenching [9], and

10. the quenching by additional infrared irradiation [9].

2. The Model

The model was first proposed by Kiess and Stöckmann [7] and is schematically represented for a p-conductor in Fig. 1. G_0 and G_1 represent the optical generation rates due to irradiation of the crystal: both of them must be taken into consideration because the electron traps are separated from the conduction band by a very small energy $\Delta\varepsilon$.

The electron traps which are neutral at the beginning of the experiment inside the crystal with a concentration H_0 will be partly occupied by electrons (concentration H^-). The trapped electrons can be emitted into the conduction band by generation (G_2) or return into the valence band by the transition R_3.

A lifetime τ_1 will be assumed relating to the band-band recombination R_1. Likewise, a lifetime τ_3 will be assumed concerning the "trap-valence band" recombination (R_3). Finally, it will be assumed that the concentration of traps

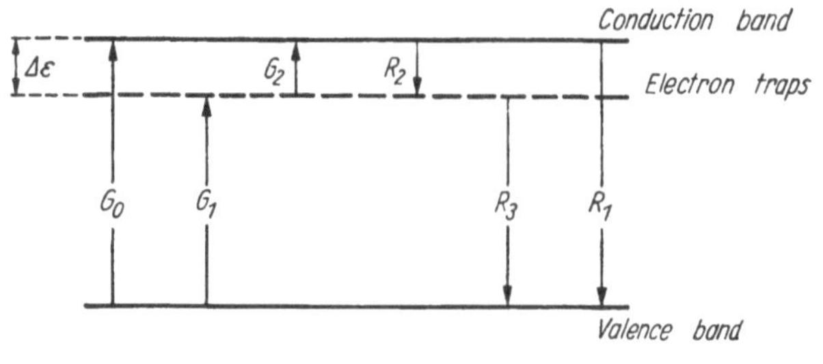

Fig. 1. Schematic representation of the model

H_0 is large as compared with the trapped electron concentration H^-:

$$R_1 = \frac{n}{\tau_1}, \qquad R_2 = r\,n\,(H_0 - H^-) \approx r\,n\,H_0 = \frac{n}{\tau_2}, \qquad R_3 = \frac{H^-}{\tau_3},$$

where n is the concentration of free electrons in the conduction band. Without irradiation the concentration n will be zero.

The rate of generation G_2 is proportional to the trapped electron concentration H^- and proportional to a field dependent emission probability:

$$G_2 = H^- \, W(E, T),$$

where T represents the absolute temperature. The field dependence of the probability can be enhanced by

a) collision ionization,

b) tunneling [10], which is important if the depth $\Delta\varepsilon$ of the traps is small:

$$W(E, T) = W(E) + W(T) = w_0 E \exp\left(\frac{-\gamma}{E}\right) + \alpha_0 \exp\left(\frac{-\Delta\varepsilon}{kT}\right). \tag{1}$$

This form of the emission probability given in [5] must be corrected since the energie $\Delta\varepsilon$ will decrease with increasing field. A trapped electron will move in the case of zero field in a potential $-e/4\,\pi\,\varepsilon\,\varepsilon_0$ below the conduction band. If an electric field E is applied along the z-axis, the relevant effective mass potential for the trapped electron will be

$$V = -\frac{e}{4\,\pi\,\varepsilon\,\varepsilon_0} - E\,z\,.$$

An estimation gives for the temperature dependent emission probability defined in (1)

$$\alpha_0 \exp\left(-\frac{\Delta\varepsilon - \Delta\psi}{kT}\right) \quad \text{with} \quad \Delta\psi = 2\,e\left(\frac{e\,E}{4\,\pi\,\varepsilon\,\varepsilon_0}\right)^{1/2}. \tag{2}$$

We will further use the following assumptions to simplify the transport equations:

The specimen is homogeneous and isothermic. The optical generation rates G_0 and G_1 are time- and space-independent. Surface effects are neglected.

Further, we will neglect the current due to free electrons: we assume the mobility of minority carriers to be small compared to the mobility of majority carriers. Furthermore we will see that the concentration n of free electrons is small compared to the hole concentration p.

The role of the acceptors responsible for the p-conduction of the specimen will be neglected in this work.

With all these conditions we obtain the following system of equations from the Poisson and transport equations for the one-dimensional case:

$$\frac{1}{e}\frac{\partial j}{\partial x} = -\frac{\partial p}{\partial t} + G - \frac{n}{\tau_1} - \frac{H^-}{\tau_3}; \qquad G = G_0 + G_1, \tag{3}$$

$$\frac{\partial n}{\partial t} = -\frac{\partial H^-}{\partial t} + G - \frac{n}{\tau_1} - \frac{H^-}{\tau_3}; \qquad G = G_0 + G_1, \tag{4}$$

$$\frac{\partial H^-}{\partial t} = \frac{n}{\iota_2} + G_1 - \frac{H^-}{\tau_3} - H^- \, W(E, T), \tag{5}$$

$$\frac{\partial E}{\partial x} = \frac{e\,(p - p_0 - n - H^-)}{\varepsilon\,\varepsilon_0} = \frac{\varrho}{\varepsilon\,\varepsilon_0}, \tag{6}$$

$$j = j_p = e\left(\mu\,p\,E - D\frac{\partial p}{\partial x}\right). \tag{7}$$

3. The Stationary Case

3.1 The effective lifetime

For the stationary case the concentrations n and H^- may be eliminated from the system of equations (3) to (7). Thus, we obtain from the continuity and the Poisson equations a non-linear system:

$$j = e\left(\mu p E - D\frac{\mathrm{d}p}{\mathrm{d}x}\right) = \text{const}\,, \tag{8}$$

$$\frac{\mathrm{d}E}{\mathrm{d}x} = \frac{e\,(p - p_0 - G\,\tau(E))}{\varepsilon\,\varepsilon_0}\,, \tag{9}$$

where we introduced an effective lifetime

$$\tau(E) = \tau_1 + \frac{\left(b + \frac{\tau_1}{\tau_2}\right)\left(1 - \frac{\tau_1}{\tau_3}\right)}{W(E, T) + \frac{\tau_1 + \tau_2}{\tau_2\,\tau_3}}\,. \tag{10}$$

The coefficient $b = G_1/G$ ($G = G_1 + G_0$) cannot exceed 1 and can be neglected if $\tau_2 \ll \tau_1$. Thus, the dependence of the effective lifetime on light intensity may be neglected when the trap concentration H_0 is high enough ($H_0 \sim 1/\tau_2$). Thus, without space charges Poisson's equation gives a field-dependent hole concentration

$$p = p_0 + G\,\tau(E)$$

so that a decreasing volume conductivity with increasing field strength may only be excepted if the predominating recombination of minority carriers takes place via recombination centres (R_1) and not via electron traps ($\tau_3 > \tau_1$).

From the relation $j = \sigma E$ between current density and field strength the current–voltage characteristic in the case of no space charges may be obtained. Thus, a voltage-controlled negative differential resistance may only be expected if the relation

$$1 + \frac{\mathrm{d}\ln\sigma}{\mathrm{d}\ln E} < 0 \tag{11}$$

is fulfilled in a range of field strengths. Yet, for zero electric field the effective lifetime takes the value τ_0; for very high fields it takes the value τ_1. Thus, the relation (11) will only hold if the inequality

$$\tau_0 \gg \tau_1 \tag{12}$$

is fulfilled. This inequality is necessary to describe the experimental results and will be assumed to be fulfilled in the following. It may be remarked that an effective lifetime of the order of some 10^{-4} s was measured in Se by Polke [11], whereas τ_1 usually may be of the magnitude of some 10^{-6} s.

3.2 Quenching effects

We saw that field emission can only be important if the relation (12) is satisfied. Thus, we obtain from equation (10)

$$1 - \frac{2\,\tau_1}{\tau_3} \gg \tau_2\,W(0, T) + \frac{\tau_2}{\tau_3}\,.$$

Suppose now, the relation $\tau_2\, W(0, T) \ll 1$ is true at low temperatures. Then, because of the exponential dependence of the probability $W(0, T)$ on temperature, a critical temperature T_0 can be defined. Therefore, for $T > T_0$ no field emission can be observed and thermal quenching dominated (point 9 of the introduction).

Suppose the critical temperature T_0 to be 300 °K. From the equation

$$\tau_2\, W(0, T) = \tau_2\, W(0, 300\ °\mathrm{K}) = 10^{-2}$$

we obtain with the values [5]

$$r = 10^7\ \mathrm{cm^3\, s^{-1}},\ H_0 = 10^{16}\ \mathrm{cm^{-3}},\ \alpha_0 = 10^{10}\ \mathrm{s^{-1}}$$

an activation energy of the electron traps of $\Delta\varepsilon \approx 0.2$ eV.

Suppose now, by additional infrared illumination that more electrons will be excited from the recombination centres into the conduction band (or into traps). Then the effective value of the lifetime τ_1 will be enlarged. Thus, even if $\tau_{1,\mathrm{eff}}$ becomes of the order of τ_3, the inequality (12) can no longer be fulfilled: the optical quenching effects is plausible (point 10 of the introduction).

3.3 Proposed approximation for the effective lifetime

The non-linear system of equations (8) and (9) is an autonomous system which was discussed by Böer using the method of characteristics. With some assumptions, Böer and Wilhelm [5, 12] defined a critical field strength E_c where field inhomogeneities and instabilities may occur (layer-like instabilities). But this method is not able to give a solution of the system considered here.

Thus, in the following we neglect the diffusion current: The voltage drop across a Debye length must be of the order of 1 V or higher. With a hole concentration of 10^{12} cm^{-3} and at 100 °K the Debye length will be about 10^{-4} cm so that for this case the field strength must be larger than 10^4 V/cm [12].

Now, with diffusion current neglected, the hole concentration p may be eliminated so that only the Poisson equation

$$\frac{\varepsilon\, \varepsilon_0}{e} \frac{\mathrm{d}E}{\mathrm{d}x} = \frac{j}{e\, \mu\, E} - p_0 - G\, \tau(E) \tag{13}$$

must be solved. This equation was numerically integrated with the results shown in Fig. 2.

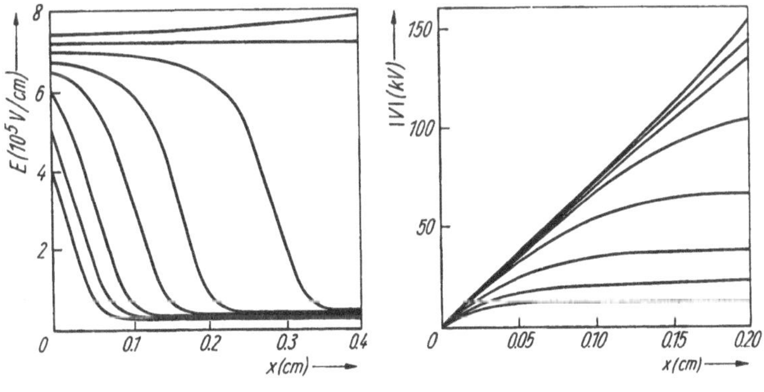

Fig. 2. Field and potential variations in the semiconductor. $p_a = 2 \times 10^{12}$ cm^{-3}, $p_0 = 5 \times 10^{11}$ cm^{-3}, $G = 3.7 \times 10^{17}$ cm^{-3} s^{-1}, $\tau_0 = 1.01 \times 10^{-4}$ s, $\tau_1 = 10^{-7}$ s, $\Delta\varepsilon = 0.25$ eV, $T_0 = 300$ °K (emission probability (1))

Contrary to Adirovich [13] who assumed a discontinous approximation function for the field dependent probability $W(E, T)$, we will use the following approximation for the effective lifetime [10]:

$$\tau(E) = \begin{cases} \tau_0; & E \leqq E_0 , \\ \tau_0 - \Gamma\left(1 - \frac{E_0}{E}\right); & E_0 \leqq E \leqq E_1 , \\ \tau_1; & E_1 \leqq E , \end{cases} \tag{14}$$

where the parameters Γ and E_0 can be calculated, for example, by equating the approximation function (and its derivative) to the real function at the value $\tau_0/2$. Thus, we obtain for Γ the magnitude $\Gamma \approx (5 \text{ to } 10)\ \tau_0$ [14].

With the approximation (14) for the effective lifetime, the form of equation (13) will not be altered in the three field regions:

$$\tau_r \mu E \frac{dE}{dx} = E_i - E , \tag{15}$$

where the distinction between these regions is made by the values of τ_r and E_i:

1. Region I ($E \leqq E_0$):

$$\tau_r = \frac{\varepsilon \varepsilon_0}{e \mu (p_0 + G \tau_0)} , \qquad E_i = \frac{j}{e \mu (p_0 + G \tau_0)} .$$

2. Region II ($E_0 \leqq E \leqq E_1$):

$$\tau_r = -\frac{\varepsilon \varepsilon_0}{e \mu (G \Gamma - p_0 - G \tau_0)} , \qquad E_i = E_0 \frac{1 - \dfrac{j}{e \mu G \Gamma E_0}}{1 - \dfrac{p_0 + G \tau_0}{G \Gamma}} .$$

3. Region III ($E_1 \leqq E$):

$$\tau_r = \frac{\varepsilon \varepsilon_0}{e \mu (p_0 + G \tau_1)} , \qquad E_i = \frac{j}{e \mu (p_0 + G \tau_1)} .$$

In the field regions I and III we obtain the effective dielectric relaxation time and the corresponding ohmic field strength in the case of no space charges. In region II we obtain a characteristic negative time ($G \Gamma > p_0 + G \tau_0$) which has the meaning of a dielectric relaxation time, if we start from the usual definition [15]

$$\frac{\partial j}{\partial x} = \frac{\partial \sigma E}{\partial x} = \frac{\varrho}{\tau_r} = -\frac{\partial \varrho}{\partial t}$$

and if we take in account the approximation (14) and the definition of the conductivity: $\sigma = e \mu (p_0 + G \tau(E))$. Without approximation (14) we remark that a relaxation time thus defined will have the schematic dependence shown in Fig. 3: It is infinity at the values E_{c1} and E_{c2}, whereas E_{c1} is just the critical value defined by Böer using the method of characteristics.

Remembering that the value E_i in field region II is just the ohmic field strength in the absence of space charges, we obtain in this region a negative differential resistance

$$r_2 = \frac{dE_i}{dj} = -\frac{1}{e \mu (G \Gamma - p_0 - G \tau_0)} .$$

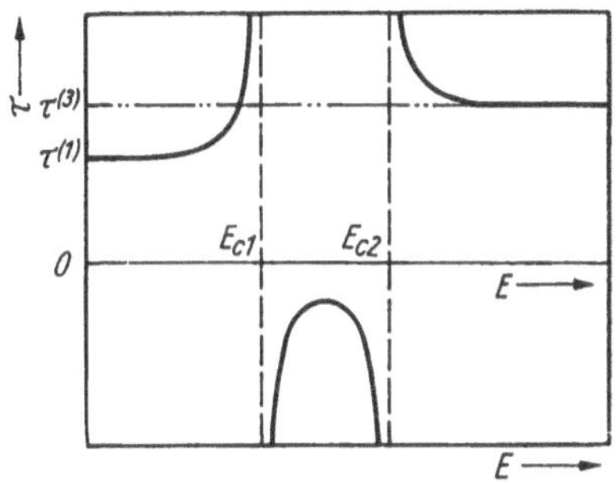

Fig. 3. Schematic variation of the dielectric relaxation time as a function of the electric field strength ($\tau^{(1)} = \varepsilon\,\varepsilon_0/e\,\mu\,(p_0 + G\,\tau_0)$, $\tau^{3} = \varepsilon\,\varepsilon_0/e\,\mu\,(p_0 + G\,\tau_1)$)

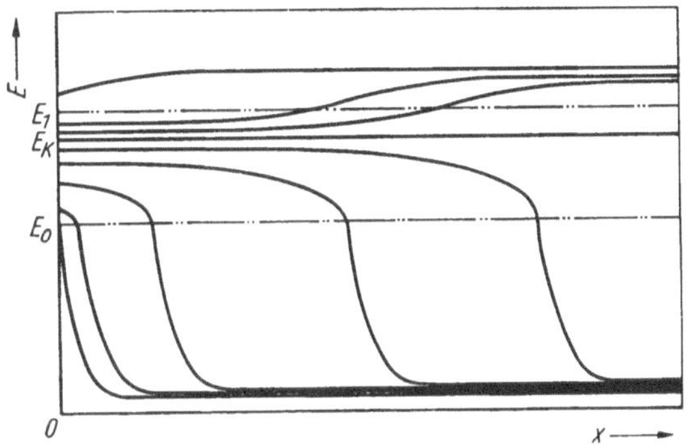

Fig. 4. Field distribution (qualitative behaviour) for different boundary values of E_a ($p_0 + G\,\tau_1 < p_a < p_0 + G\,\tau_0$)

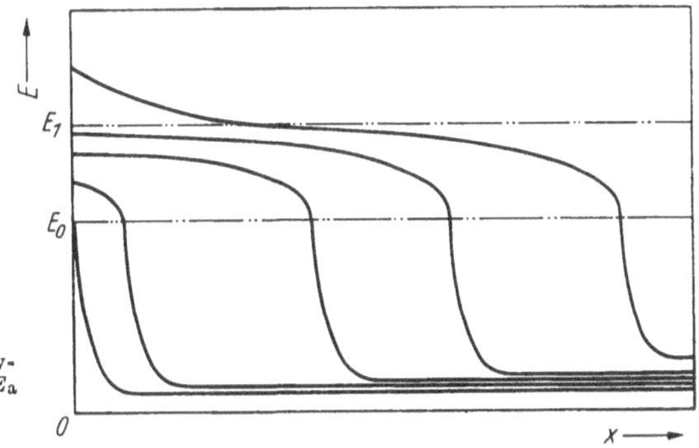

Fig. 5. Field distribution (qualitative behaviour) for different boundary values of E_a ($p_a < p_0 + G\,\tau_1 < p_0 + G\,\tau_0$)

3.4 Solution and discussion

The integration of equation (15) presents no substantial difficulties. The solution will, of course, depend on a boundary condition which we suppose to be given by a constant concentration $p_a = j/e\,\mu\,E_a$ at the anode. If we assume a depletion boundary layer at the anode ($p_a < p_0 + G\,\tau_0$) at low voltages, we will qualitatively find the curves given schematically in Fig. 4 and 5. Without entering into details, the following features are to be pointed out [14]:

1. If the boundary field E_a exceeds the quantity E_0, a region of increased quasi-constant field intensity will arise, beginning near the anode. The magnitude of this enhanced field strength tends toward a limit E_K; increasing the boundary field E_a by ΔE (and thereby increasing the current) causes a very large decline in potential within the region of high field strength. Thus, saturation is observed (point 1 of the introduction).

2. This critical field E_K cannot, for the present, be surpassed. The crystal is virtually divided into two parts with different field strengths (point 2 of the introduction).

3. If the boundary concentration p_a fulfills the condition $p_0 + G\tau_1 < p_a$, the boundary field intensity may reach the value [14]

$$E_K = \frac{E_0}{1 - \frac{p_0 + G\tau_0 - p_a}{G\Gamma}} \tag{16}$$

so that a constant field strength $E_a = E_K$ will be found all over the semiconductor (point 3 of the introduction).

4. As soon as the applied voltage exceeds $E_K L$, the field strength at the counterelectrode increases, whereas at the anode the field will practically retain its value E_K. The above-mentioned point 4 of the introduction can thus be understood even without carrier injection.

5. In the saturation region the boundary field intensity E_a will grow larger by an amount of $E_K - E_0$, while the applied voltage must increase by $E_K L$. If the interval (E_0, E_K) is comparatively small, the tendency of the enhanced field strength at the anode toward E_K may not be observable (cf. potential measurements of Kiess). In the opposite case the potential curves of Polke and Hasler will be found (point 5 of the introduction).

3.5 A theory for the onset of oscillations

As shown above, the high-field region near to the anode has a negative resistance. Hence, if we imagine the crystal to be divided into two domains of different but homogeneous field strengths, its resistance is to first approximation

$$R = r_1 (L - x_0) - |r_2|\, x_0 \,.$$

Consequently, there is a critical value x_k of the high-field region width x_0 which makes the overall resistance vanish. If we expect oscillations to occur as soon as this resistance turns negative, we have found a criterion for their setting-in.

It has been shown, however, that the field is by no means homogeneous. A rigorous study can be done by integrating equation (15) over x. The applied voltage

$$V = -\int_0^{x_0} E\, \mathrm{d}x - \int_{x_0}^{L} E\, \mathrm{d}x$$

is then evaluated as a function of x_0 and the boundary field intensities E_a, E_L at the anode and cathode. Retaining x_0 fixed, the resistance may be calculated by differentiation:

$$R = \frac{\partial V}{\partial j} \,.$$

This resistance vanishes if the following equation holds:

$$\left.\begin{aligned} \frac{x_0}{L} = \frac{x_k}{L} &= (1 - y)\left(1 - \frac{\mu\,\varepsilon\,\varepsilon_0\,E_L}{L}\frac{\partial E_L}{\partial j}\right) - y\,\frac{\mu\,\varepsilon\,\varepsilon_0\,E_a}{L}\frac{\partial E_a}{\partial j}\,; \\ y &= \frac{p_0 + G\tau_0}{G\Gamma} \approx \frac{\tau_0}{\Gamma} < 1\,. \end{aligned}\right\} \tag{17}$$

The width of the high-field region must be positive; hence we find as a necessary condition

$$1 > \frac{\mu \varepsilon \varepsilon_0}{L} E_L \frac{\partial E_L}{\partial j} \tag{18}$$

provided that the boundary field intensity E_a increases with increasing current density.

If the difference $L - x_k$ becomes considerably larger than one Debye length the electric field E_L can be replaced by

$$E_L = \frac{j}{e \mu (p_0 + G \tau_0)} .$$

Using the relation $G = \alpha B$ and $j = \beta B^c$, an inequality for the irradiation intensity B is found by substituting ($p_0 \ll G \tau_0$):

$$B^{2-c} L > \frac{\mu \varepsilon \varepsilon_0 \beta}{(e \mu \tau_0 \alpha)^2} . \tag{19}$$

Here the coefficient c has been determined experimentally for selenium [ ] ($1/3 < c < 1/2$). Hence, a low irradiation intensity is necessary in order to observe the oscillations (point 7 of the introduction).

If on the other hand the difference $L - x_k$ becomes smaller than one Debye length, the field strength E_L will increase far more than linearly with increasing current density because of the space charge region near the cathode. It may then become impossible to fulfil condition (18) (point 6 of the introduction).

4. The Non-Stationary System

When solving the non-stationary system (3 to 7), one encounters much more difficulties. In the following, we shall merely show that instabilities as observed by Böer and Wilhelm [1] can be understood within the framework of our model. For a more detailed discussion see [14].

From equations (3) to (7) we may eliminate the concentrations n and H^- as well as their derivatives. Then if we neglect the operator $\tau_1 \partial/\partial t$ against unity (low frequencies), we can introduce the effective lifetime (10). Thus we get a non-linear system of equations in which we shall neglect the hole diffusion current. Inserting the following variables:

$$j = j_s (1 + I) \qquad \text{und} \qquad E = E_s (1 + \varphi) ,$$

where the subscript "s" stands for "stationary", and carrying out the linearization, we obtain a system of equations that couples the stationary functions $E_s(x)$ and $p_s(x)$ with the unknown functions. But the former are almost constant in the high-field region and may therefore be replaced by an average value. Now, if we introduce the function (14) to approximate the effective lifetime, we are left with a system of linear equations with constant coefficients, valid in the high-field domain and soluble by the Ansatz

$$\varphi \sim \exp [j (k x - \omega t)]; \; k = \chi + j d , \quad \omega \text{ real} .$$

In general, there will be an attenuation in the direction of group velocity $v_g = \partial\omega/\partial\chi$. A local perturbation in homogeneous distribution is dissipated with a time constant $t_0 = 1/v_g d$. Instabilities cannot occur unless there are solutions with negative time constant t_0 [16], even though these may possibly exist only within a narrow frequency range. At low frequencies ($\omega \tau_0 \ll 1$), our evaluation yields some positive group velocity with minimum at $\omega = 0$ on the one hand and a negative absorption d on the other hand. Hence, there will happen instabilities at low frequencies.

As an interesting feature we emphasize that the minimum of group velocity does not depend on the hole mobility. Assuming the carrier concentration $G\,\tau_0 = 5\times10^{13}\ \text{cm}^{-3}$ and the ratio $\Gamma/\tau_0 = 10$, we find a minimum value of some cm/s.

The frequency cannot be figured out numerically unless the wave vector χ is known. If we suppose the wavelength to equal the distance of electrodes (i.e. 10^{-2} cm), we find an estimate of 500 Hz where a factor of 10 or even 30 is not excluded, however.

It should be pointed out that these numerical results are in good agreement with the experiments in point 8 of the introduction.

5. Summary

In the present paper it was attempted to reconcile experimental facts with the model of internal field emission of minority carriers. No effort was made to fit parameters to any particular measurement.

All results which have been derived from the stationary system of equations are in good agreement with experiment, though it should be mentioned that introduction and determination of the parameter Γ causes some difficulties from an experimental point of view.

For several reasons, the results obtained from the non-stationary system of equations cannot be exact. Moreover, a few questions have not yet been decided neither by theory nor by experiment. It is possible that the observed low-frequency current oscillations are connected with some unsolved motion of high-field regions.

Acknowledgement

Finally, the author wishes to express his gratitude to all those who contributed their aid during the course of this investigation. Thanks are due especially to Prof. Dr. O. Madelung for initiating and encouraging this work with continuous interest, and to Prof. Dr. F. Stöckmann, Prof. Dr. W. Ruppel, and Dipl.-Phys. K. Hasler for numerous enlightening and stimulating discussions.

References

[1] K. W. Böer and W. E. Wilhelm, phys. stat. sol. **4**, 237 (1964).
[2] A. Barraud, C. R. Acad. Sci. (France) **256**, 3632 (1963).
[3] C. Fischler-Hazoni and F. Williams, Phys. Rev. **139**, A583 (1965).
[4] B. K. Ridley and R. G. Pratt, Physics of Semiconductors, Dunod, Paris 1964 (p. 487).
[5] K. W. Böer and W. E. Wilhelm, phys. stat. sol. **3**, 1704 (1963).
[6] B. K. Ridley, Proc. Phys. Soc. **82**, 954 (1963).
[7] H. Kiess and F. Stöckmann, phys. stat. sol. **4**, 117 (1964).
[8] H. Kiess, phys. stat. sol. **4**, 107 (1964).
[9] M. Polke and K. Hasler, phys. stat. sol. **9**, 851 (1965).
[10] W. Franz, Ann. Phys. (Germany) **11**, 17 (1962).
[11] M. Polke, phys. stat. sol. **5**, 279 (1964).
[12] K. W. Böer, Phys. Rev. **139**, A1949 (1965).
[13] E. I. Adirovich, Z. Phys. **155**, 195 (1959).
[14] G. Döhler, Dissertation Marburg/Lahn, 1966.
[15] O. Madelung, Hdb. Phys. (Flügge), Vol. 20, Springer-Verlag, 1957 (p. 108).
[16] H. Rother, Ann. Phys. (Germany) (7), **5**, 203 (1960).

(Received November 11, 1966)

phys. stat. sol. 19, 565 (1967)

Subject classification: 10.2; 22.5.2

Department of Chemistry, University of Western Ontario, London, Ontario

Photochemical Properties of SH^- Ions in KCl and KBr Crystals[1)]

By

O. E. FACEY and P. W. M. JACOBS

The photochemistry of KCl and KBr containing SH^- impurity ions has been investigated. In addition to the formation of S^- and U_2-centres reported previously, U-centres are formed in high yield and F-centres in low yield. The thermal contribution to the photochemical process is slight (≈ 0.013 eV). The kinetics of the decay of U_2-centres has been shown to be bimolecular and to correspond mainly to the reaction $S^- + H_i \rightarrow SH^-$, where H_i denotes an interstitial hydrogen atom. A model involving the utilization of the thermal energy released during a radiationless transition from the excited state is successfully used to give a qualitative explanation of the data.

Es werden photochemische Prozesse in KCl- und KBr-Kristallen, die SH^--Fremdionen enthalten, untersucht. Zusätzlich zur Bildung von S^-- und U_2-Zentren, worüber schon früher berichtet wurde, werden mit hoher Ausbeute U-Zentren und mit niedriger Ausbeute F-Zentren gebildet. Der thermische Beitrag zum photochemischen Prozeß ist gering (0,013 eV). Es zeigt sich, daß die Kinetik des Zerfalls der U_2-Zentren bimolekular ist und hauptsächlich der Reaktion $S^- + H_i \leftarrow SH^-$ entspricht, wobei H_i ein Wasserstoffatom auf Zwischengitterplatz bezeichnet. Ein Modell, das die während des strahlungslosen Übergangs vom angeregten Zustand freiwerdende thermische Energie mit berücksichtigt, wird mit Erfolg zur qualitativen Deutung der Ergebnisse benutzt.

1. Introduction

Crystals grown from Analar grade KCl without pre-treatment with Cl_2 or HCl show a strong absorption band having a maximum at 186 nm at room temperature [1]. On irradiation with ultra violet light this band decreases in height and simultaneously a band develops at 235 nm [1, 2]. This is the well-known U_2-band [3] and is due to hydrogen atoms in interstitial positions in the crystal lattice [4, 5]. Rolfe [6] showed that the 186 nm band could be produced by adding sodium sulphide to the KCl before melting it. Because of the difficulty of producing anhydrous Na_2S, he considered that there would be sufficient water present to hydrolyze the sodium sulphide to sodium hydrosulphide and that his crystals, therefore, contained SH^- ions which were responsible for the 186 nm band. Concurrently at Göttingen, Fischer and Gründig [7] were growing KCl crystals under H_2S, a procedure which also results in the incorporation of SH^- ions by the reaction

$$KCl + H_2S \rightleftarrows KSH + KCl . \qquad (1)$$

These crystals, therefore, also show an SH^- band at 186 nm. On photolysis, they observed [7], in addition to the U_2-band, another of similar intensity at

[1)] Supported by the U.S. Department of the Navy, Office of Naval Research.

193 nm which was ascribed to S^- ions (on anion lattice sites) formed by the photochemical reaction

$$SH^- \rightarrow S^- + H_i \tag{2}$$

in which H_i denotes an interstitial hydrogen atom. A small band with an apparent maximum at 255 nm appears as a shoulder on the larger 235 nm absorption band [1, 2]. Fischer and Gründig suggest [7] that this band is also due to transitions within the S^- ion.

We have now investigated the photochemical properties of KCl + KSH and also of KBr + KSH in greater detail. It appears that both the primary photochemical process and the subsequent thermal processes which occur on annealing are more complicated than are implied by equation (2).

2. Experimental Procedure

Crystals doped with SH^- or SD^- ions were prepared by three different methods.

(i) Suitably sized cleaves of KCl crystals (either from Harshaw or grown from Analar KCl which had been melted under chlorine) were sealed off in silica ampoules under a pressure of H_2S gas in the range 60 to 160 Torr and annealed at 700 °C for times varying from 2 hours to 5 days. All such crystals showed the SH^- absorption band provided the specimen examined was cleaved from the outside of the crystal. This shows that the reaction (1) proceeds rapidly at the surface of the KCl but that diffusion of SH^- into the interior of the crystal is exceedingly slow. To obtain more uniform doping either of the following two methods was employed.

(ii) Crystals were grown from the melt containing 0.01 mol% anhydrous KSH. The latter salt was prepared by passing dry H_2S through a solution of potassium in absolute ethyl alcohol. The first-formed precipitate is a mixture of K_2S and KSH and this was discarded. On adding absolute ether to the filtrate, a precipitate of KSH forms. This was purified by re-precipitation from alcohol with ether and kept in a desiccator over SiO_2 gel + $CaCl_2$. All operations involved in the preparation and handling of KSH were performed in a dry box under nitrogen. The KCl was pre-melted under Cl_2 to remove OH^- ions and then cooled to room temperature before adding the KSH. This is probably the least ambiguous method of doping with SH^- ions.

(iii) Crystals were grown from the previously chlorinated melt under 0.33 atm pressure of hydrogen (or deuterium) after adding 3 mg of sulphur to the charge. For SD^- doping, the chlorinated melt was treated with deuterium at 1 atm pressure at 500 °C and allowed to cool in deuterium. This procedure minimized the possibility of contamination from hydrogen dissolved in the silica container.

For KBr method (iii) was used but pre-treatment with Cl_2 was omitted.

Spectra were run either on a Perkin-Elmer UV137 or on a Cary 14R spectrophotometer.

3. Results

3.1 KCl + KSH

The absorption spectra obtained with a 0.72 mm thick cleave from a KCl crystal doped with KSH are shown in Fig. 1. The large band at 186 nm (the peak of which is just off scale) is due to SH^-. This band always shows a long tail, on the long wavelength side, which includes a small peak in the region of

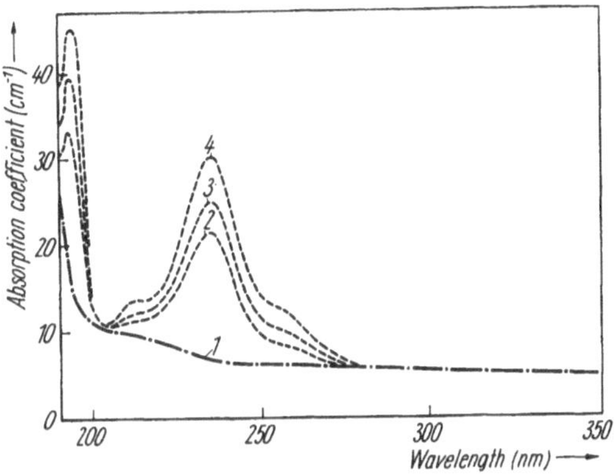

Fig. 1. Absorption spectra at 97 °K of KCl + KSH: (1) before irradiation, (2) after 10 min, (3) after 20 min, (4) after 41 min irradiation

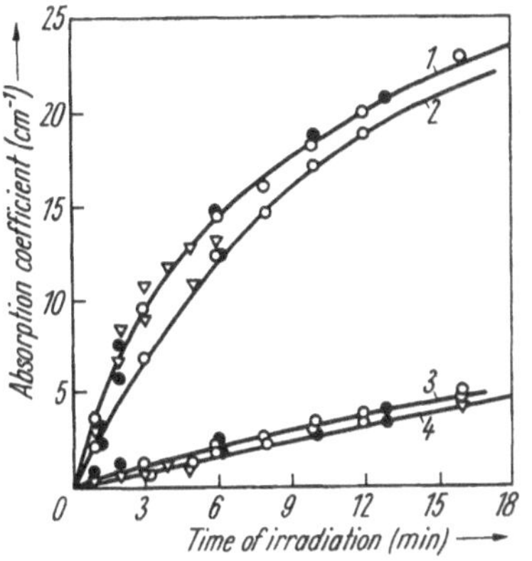

Fig. 2. Kinetics of formation of the absorption bands 193 nm (1), 235 nm (2), 212 nm (3), and 255 nm (4) in irradiated KCl + KSH. Results from three different runs at 90 °K are included

215 nm. There is no optical evidence for OH^- ion impurity which should have been completely removed by the chlorination procedure. On irradiation with uv light from a mercury low-pressure lamp, the SH^- band decreases, and bands develop at 193, 212, 235, and 255 nm. The kinetics of formation of these bands are shown in Fig. 2. For short irradiation times the intensity of the 193 and 235 nm bands are well correlated as are those of the weaker 212 and 255 nm bands. For long irradiations, however, the intensity of the 235 nm band passes through a maximum and then decreases, owing to the thermal instability of the interstitial H atoms. While the 255 and 193 nm bands saturate, the U-band at 212 nm grows steadily and eventually becomes as intense as the 193 nm band.

The sensitivity of the U_2-band at 235 nm to temperature is shown in Fig. 3. The thermal bleaching of this band may be followed in the temperature range 85 to 100 °K. At these temperatures the 212 and 255 nm bands do not bleach, but the S^- band decreases partially in intensity along with the U_2-band. An F-band is apparent in KCl + KSH only after very long irradiations lasting several hours. On warming the F-band intensity increases.

The kinetics of the bleaching of the U_2-band was investigated by measuring the absorbance as a function of time at constant temperature. Three possibilities seem likely, namely that the reaction should be unimolecular, bimolecular

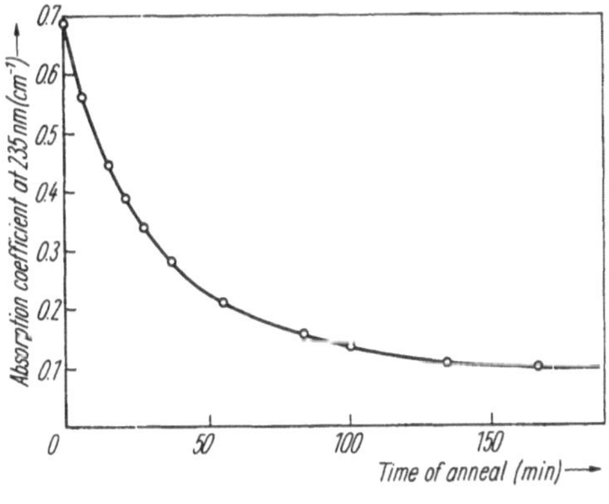

Fig. 3. Decrease in absorption coefficient at 235 nm with time on annealing a uv irradiated KSH-doped KCl crystal at 100 °K

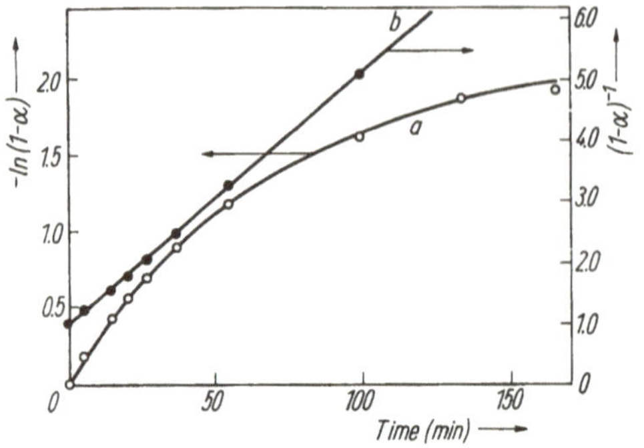

Fig. 4a. Kinetics of the thermal decay of U_2-centres (a) plotted according to equation (3), (b) according to equation (4)
b. Kinetics of the thermal decay of U_2-centres plotted according to equation (5)

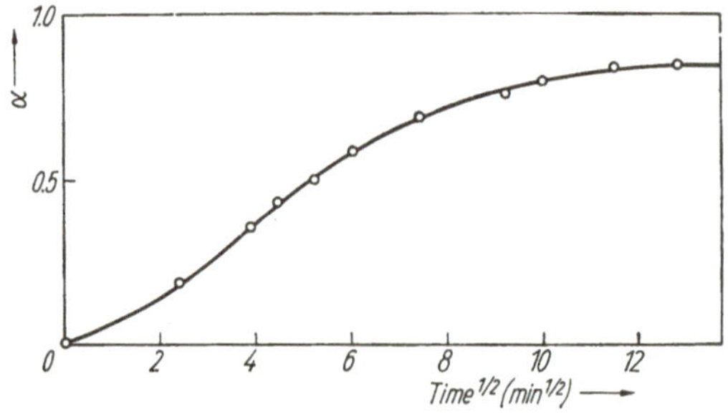

or diffusion controlled. Let α denote the fraction of H atoms which have reacted in time t; $1 - \alpha$ is given by the ratio $A(t)/A(0)$ where $A(t)$ is the absorbance (optical density) at time t. Then for a unimolecular process

$$- \ln (1 - \alpha) = k_1 t , \tag{3}$$

for a bimolecular process

$$(1 - \alpha)^{-1} = k_2 t , \tag{4}$$

and for a diffusion controlled process

$$\alpha = k_3 t^{1/2} . \tag{5}$$

Plots testing these kinetic equations are shown for a typical run in Fig. 4. Only the plot of $(1 - \alpha)^{-1}$ vs. t is linear so that the thermal decay of U_2-centres is a bimolecular reaction. The rate constants k_2 obey the Arrhenius equation, the activation energy being 0.2 eV.

3.2 KCl + KSD

Absorption spectra of KCl crystals containing KSD are very similar to those for KCl + KSH, and their behaviour on irradiation is analogous. In view of the danger of hydrogen contamination from the quartz, it was necessary to check that the crystal really did contain deuterium. The half-widths of the D^- and D_i bands were therefore measured by assuming their shape to be gaussian and the resulting values are compared in Table 1 with the corresponding values for KSH doped crystals. The deuterium bands are significantly sharper. The kinetics of formation of the four bands at 193, 212, 235, and 255 nm are shown in Fig. 5. These data show (i) the good reproducibility for two runs at the same temperature, (ii) that the 193 and 235 nm bands form at the same rate until the thermal bleaching of the latter becomes important, and (iii) the importance

Fig. 5. Kinetics of formation of the absorption bands at (1) 193 nm, (2) 235 nm, (3) 212 nm, and (4) 255 nm in irradiated KCl + KSD. Note the different time scale to Fig. 1. For both short and long irradiations KCl + KSH and KCl + KSD behave similarly

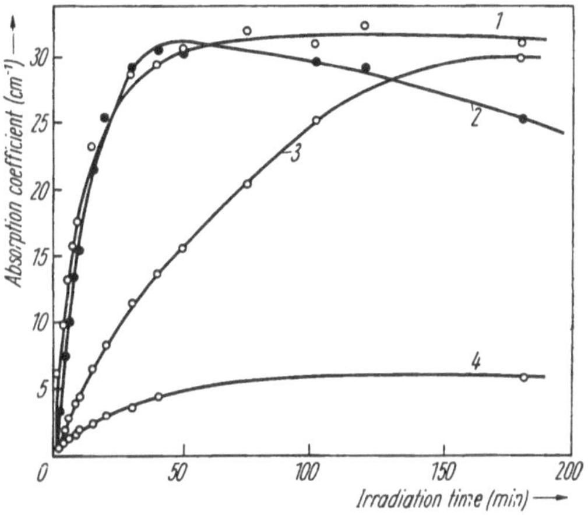

Table 1

Half-widths in eV for the U- and U_2-bands in KCl

	U-band	U_2-band
Hydrogen	0.35	0.39
Deuterium	0.27	0.30

of the U-band, which continues to grow after the 255 and 193 nm bands have saturated. This is also true for KSH doped crystals. At the doping levels used the absorption coefficient of the 235 nm depended on thickness unless thin crystals $\leqq$ 0.6 mm were used. This is because the intense absorption by SD^- ions causes so much attenuation of the light that the deuterium atoms are not produced at uniform concentration throughout the whole width of the crystal.

The kinetics of decay of the U_2-band again obey equation (4). The rate constants are slightly smaller than for the corresponding hydrogen band as would be expected for the heavier deuterium atoms, and the activation energy is slightly larger ($\approx$ 0.25 eV).

3.3 KBr + KSH

The KBr + KSH crystals display a large absorption band peaking at 194 nm at 80 °K. As in KCl there is a pronounced tail on the long wavelength side which can sometimes be resolved into a shoulder at 230 nm. Attempts to fit this SH^- band by an equation of gaussian form were not successful. The major part of the band is gaussian with a full width at half-height of 0.28 eV, but the long wavelength side cannot be fitted completely by either a single gaussian or by a superposition of gaussian bands. It was not possible to do a detailed study of the temperature variation of the half-width for above 123 °K the SH^- band disappears into the KBr band edge.

On irradiation the 194 nm SH^- band decreases and new bands form at 204 and 300 nm. The hydrogen bands also appear: the U-band at 225 nm and the U_2-band at 270 nm. The latter is somewhat less prominent than in KCl because the band bleaches thermally more easily than in KCl, and is hardly apparent

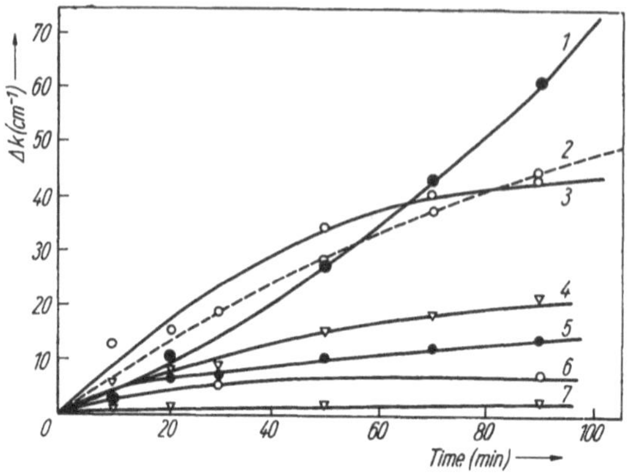

Fig. 6. Kinetics of decomposition of SH^- ions in KBr shown by plotting the change in absorption coefficient at various wavelengths. (1) U-band formation, (2) SH^- decomposition, (3) 204 nm band, (4) U_2-band, (5) F-band, (6) F-band in undoped KBr, (7) 300 nm band

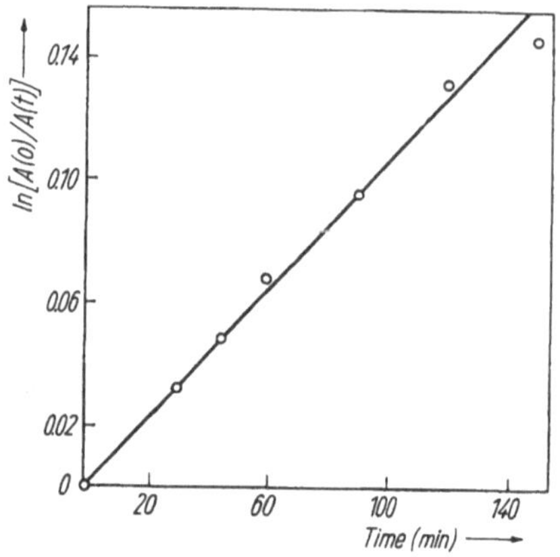

Fig. 7. Photochemical decomposition of SH^- band in KBr showing first order kinetics

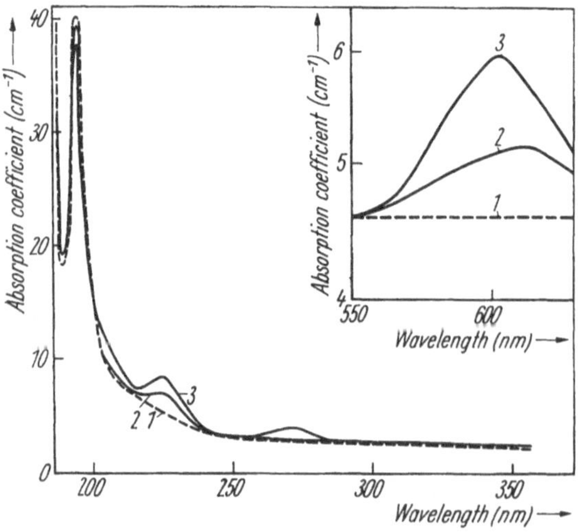

Fig. 8. Changes in the absorption spectrum of KBr + KSH on irradiation and annealing. (1) Unirradiated crystal at 88 °K, (3) after 90 min uv irradiation at 88 °K, (2) after overnight warm-up and recooling to 90 K

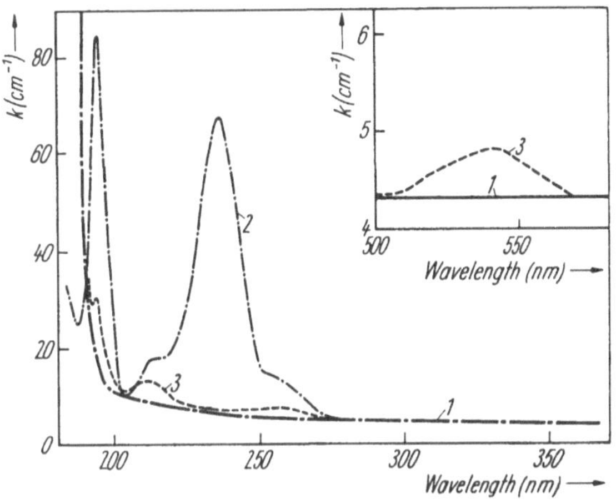

Fig. 9. Changes in the absorption spectrum of KCl + KSH on irradiation and annealing. (1) Unirradiated crystal at 85 °K, (2) after 2 h uv irradiation at 85 °K, (3) after 1 day at room temperature and recooling to 85 °K

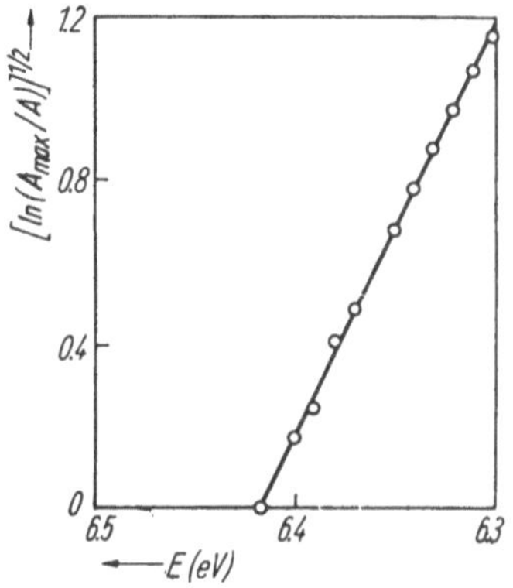

Fig. 10. Long wavelength side of the S^- band in KCl at liquid helium temperature

at temperatures > 90 °K. The F-band at 610 nm is also formed in variable amount. The formation of an F-band in undoped KBr was also observed (Fig. 6), but it is of smaller intensity than in KBr + KSH. Thus, F-centre formation in the latter is a genuine effect. No U- or U_2-centre formation was detected in the undoped KBr, but it remains an open question as to whether or not the F-centre formation is due to a small residual OH^- impurity. The decay of the SH^- band and the growth of the 204, 300, and 610 nm bands are shown in Fig. 6. The kinetics of decomposition of the SH^- centres follow a simple first order law (Fig. 7), i.e. the rate of photochemical decomposition is simply proportional to the amount undecomposed. The rate constant k increases slightly with temperature, this increase corresponding to a thermal activation energy of 300 cal/mol.

Thermal bleaching in irradiated KBr + KSH is shown in Fig. 8. On warming overnight the U_2-band bleaches and the 204 nm band almost disappears. The recovery of the SH^- band at 193 nm is, however, incomplete while the U- and F-bands show only an approximately 50% bleach under these conditions. This behaviour of the F-band contrasts with that in KCl + KSH for there the F-band at first increases on warming (Fig. 9). This figure also shows the slowness of the U-band bleach compared to that of the U_2-band and the extent of the S^- bleach which accompanies the decay of the U_2-band.

3.4 Band shapes

The 193 nm S^- band and the 186 nm SH^- band in KCl overlap to such an extent that a complete analysis of the bands is impossible. By making reasonable subtractions for SH^- background, however, we have been able to analyze the leading edge (long wavelength side) of the S^- band. Fig. 10 shows a plot of $[\ln (A_{max}/A)]^{1/2}$, where A is the absorbance at a photon energy E, against E. A reasonably good straight line is obtained from which one finds a half-width W of 0.173 eV. This actually represents twice the half-width on the low-energy side, i.e. the full width if the band is symmetric. The low-energy side of the

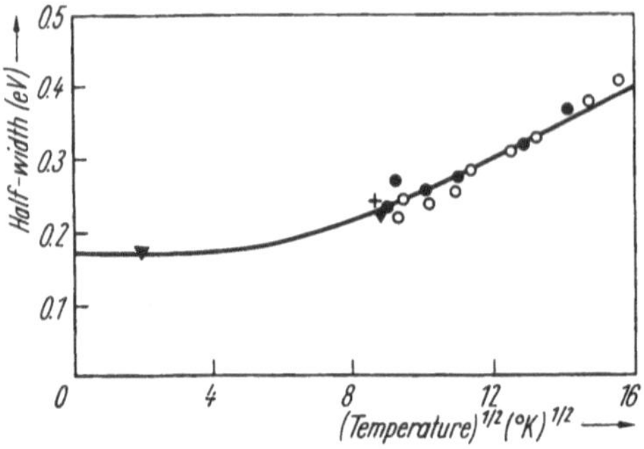

Fig. 11. Temperature dependence of the half-width of the 193 nm band in KCl. ○ and ● refer to two different runs; ▼ helium run, + Fischer and Gründig

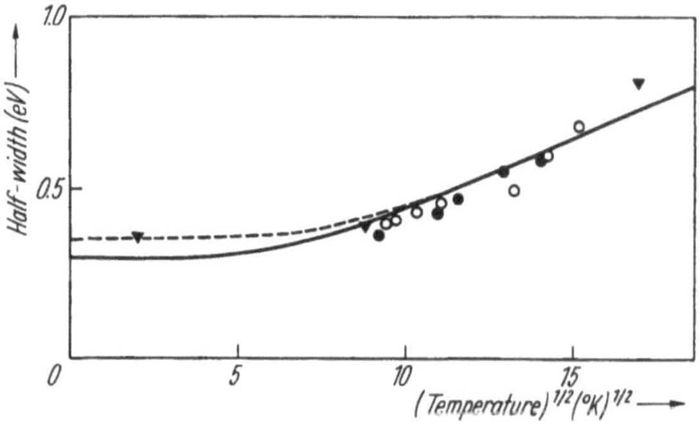

Fig. 12. Temperature dependence of half-width of 255 nm band. Symbols as for Fig. 11

255 nm band was analyzed similarly and the temperature dependences of W for the two bands are shown in Fig. 11 and 12, respectively. Because of the difficulties involved in the analysis of overlapping bands high accuracy is not expected. Nevertheless, both sets of data show the anticipated $T^{1/2}$ dependence at high temperatures i.e. > 80 °K. Both graphs show data obtained with four different crystals, namely two runs covering the temperature range 80 to 230 °K, a single point at liquid helium temperature and the liquid nitrogen point from the data of Fischer and Gründig [7]. The absorbance of the two bands was measured on different cleaves above 80 °K so as to allow increased accuracy by using longer irradiations when making measurements on the 255 nm band. The continuous lines in Fig. 11 and 12 were calculated from the equation

$$W(T) = W(0)\,[\coth(h\,\nu_g/kT)]^{1/2}\,, \tag{6}$$

where $W(T)$ is the half-width at temperature T and ν_g the vibrational frequency in the ground state. The values of ν_g used were 1.98×10^{12} s^{-1} for the S^- band and 2.09×10^{12} s^{-1} for the 255 nm band. The dotted line in Fig. 12 was calculated assuming no error in the helium point. The corresponding frequency is 2.99×10^{12} s^{-1}.

4. Discussion

Irradiation of KCl containing SH^- ions with uv light produces bands at 193, 212, 235, 255 nm together with a weak F-band which is only observed for long irradiation times. KBr behaves similarly, the positions of the four uv bands being 204, 226, 273, and 300 nm. The F-band is more prominent in KBr than in KCl. Four of these eight bands are the well-known hydrogen bands due to interstitial hydrogen atoms (235 and 273 nm) and hydride ions on lattice sites (212 and 226 nm). The kinetics of formation (Fig. 2, 5) and the bleaching experiments show that 193 and 235 nm bands are well correlated in KCl. Similarly the 204 and 273 nm bands are correlated in KBr. The decrease in the SH^- absorption when these bands form and the increase in the SH^- band when they decay makes the assignment of the 193 nm band in KCl and the 204 nm band in KBr to S^- ions on normal anion sites certain [7]. Thus one photochemical reaction is

$$SH^- + h\nu \rightarrow SH^{-*} \rightarrow S^- + H_i \,. \tag{7}$$

At temperatures above ≈ 85 °K the hydrogen atoms are mobile and the U_2-band bleaches. The kinetics of this process indicate a bimolecular reaction. Since the initial concentrations of S^- and H_i are equal, equation (4) does not distinguish between

$$H_i + H_i \rightarrow H_2 \tag{8}$$

and

$$H_i + S^- \rightarrow SH^- \,. \tag{9}$$

There is a third possible reaction

$$H_i + SH^- \rightarrow H_2S^- \,, \tag{10}$$

since H_2S^- has been identified by Hausmann [8] in KCl and KBr using ESR techniques. The H_2S^- molecular ion forms rapidly at 110 °K but is unstable above 150 °K. Reaction (10) would display unimolecular kinetics as the SH^- was present in large excess in our experiments. We conclude therefore that (9) is the major process with some contribution from (10) which is, however, insufficient in amount to upset the predominantly bimolecular kinetics. The H_2S^- ion thus acts as a trap for some of the H atoms and at higher temperatures these are given off again and reaction (9) concluded. This discovery of H_2S^- resolves an apparent anomaly in the data. When H atoms anneal out at 100 °K the S^- band only bleaches partially although the U_2-band will be 90% bleached in the time of the experiment. However, when an irradiated KBr + KSH crystal is allowed to warm up slowly over a period of several hours then the S^- bleaches completely along with the U_2-band (Fig. 8). This means that reaction (8) must occur to a very limited extent if at all in KBr, since the reaction $H_2 + S^- \rightarrow SH^- + H_i$ would require a good deal of thermal energy to split the strong H–H bond. The behaviour of KCl + KSH is a little different in that the S^- band bleach is incomplete even on warming to room temperature (Fig. 9) although it is more extensive than at 100 °K. Thus the formation of molecular hydrogen, reaction (8), occurs to some extent in KCl.

Fig. 8 shows that the recovery of the SH^- is incomplete when the S^- band has bleached. This fact, together with the absence of the U-band in irradiated undoped KBr, shows that the U-centres come from the SH^- ions and not from other impurities. Thus besides the reaction (7) there must be at least one more photochemical reaction:

$$SH^- \rightarrow SH^{-*} \rightarrow H^- + S_i \,. \tag{11}$$

There is no direct evidence or even indirect (optical) evidence for interstitial sulphur atoms but, since lattice hydride ions are formed, that is the only other possibility. If we assume that the SH^- ions are excited by the absorption of 185 nm photons to SH^{-*} and that the S–H bond is broken in the excited state, then the ejection of a hydrogen atom into an interstitial position seems much more likely than the ejection of a sulphur atom because of the greater mass of the latter. We thus reject this mechanism of bond fission in the excited state implied by (7) and (11) and consider two other possibilities. Both these are similar in principle to the two main theories of F-centre formation in alkali halides.

4.1 Ionization mechanism

If the excited SH^{-*} state lies either in or just below the conduction band, then the primary photochemical step, possibly with the assistance of a small amount of thermal energy, will be the formation of an SH radical and an electron in the conduction band. The neutral SH centre no longer has the benefit of the coulombic energy to keep it on an anion site and therefore moves easily into an interstitial position. The electron is then trapped by the resulting anion vacancy forming an F-centre with the SH radical close by. This is an unstable situation with respect to the formation of either S^- or H^- ions, so bond fission in the radical and the movement back into the anion site of either the H or S atoms occurs. Clearly the activation energy for this would be much smaller for the H atom and so S^- ions are formed initially at a greater rate than H^- ions. This process may be summarized:

$$SH^- + h\nu \rightarrow SH^{-*} \overset{?kT}{\rightarrow} SH + e \rightarrow SH_i + e + \boxed{-} \rightarrow SH_i + F \overset{kT}{\rightarrow} S^- + H_i$$

or

$$SH_i + F \overset{kT}{\rightarrow} H^- + S_i\,. \tag{12}$$

kT signifies the need for thermal energy. In KBr it was possible to measure the temperature dependence of the photochemical reaction though admittedly over a rather limited range. The thermal activation energy required turned out to be slight, only ≈ 0.013 eV or hardly more than kT at the temperatures under consideration. The annealing experiments show that the interstitial sulphur atoms move very slowly indeed even at room temperature (e.g. the slow bleaching of the U-band in Fig. 8). The annealing of the U_2-band is orders of magnitude more rapid, but even this requires an activation energy of 0.2 eV. Thus it seems that the ionization mechanism cannot be sustained because of the thermal energy needed to move the interstitial back into the anion site. However, this argument may not be quite correct since the activation energies may refer to the migration of the species from one interstitial site to the next whereas in (12) the moving atom benefits from the recombination energy with the electron. The main weakness in (12) seems to be that the energy available from the formation of the F-centre is simply dissipated instead of being usefully employed. This is avoided in the mechanism (ii).

4.2 Radiationless transition mechanism

Pooley [9] has shown the energy available from radiationless electron–hole recombination may be sufficient to result in the formation of a negative ion vacancy and a negative ion interstitial. The electron is first trapped in an excited state by a V_K centre and a radiationless transition from the excited state to the ground state results in a [110] replacement sequence. We consider that a similar mechanism may be operative in KCl and KBr containing SH^- ions. Photon absorption results in a bound excited state and a radiationless transition makes available nearly all the energy of the original quantum. The violent vibrations

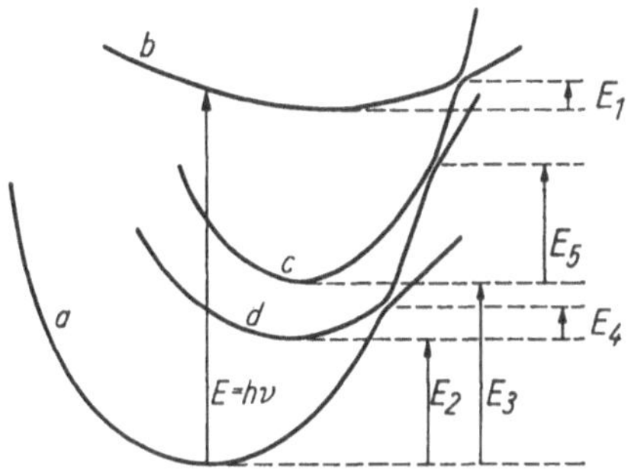

Fig. 13. Schematic representation of the energy of the $S + H + e$ system. In KCl $E = 6.68$ eV, $E_1 = 0.013$ eV, E_2 and E_3 are unknown, $E_4 = 0.2$ eV, and $E_5 \gg E_4$. In KBr $E = 6.08$ eV and other values are unknown.

(a) SH^-, (b) SH^{-*}, (c) $S_i + H^-$, (d) $S^- + H_i$

of the SH^- ion are now sufficient to split the S–H bond with consequent ejection of either the S or the H atom into an interstitial position. Schematic curves showing the energy of the $S + H + e$ system in various possible configurations are given in Fig. 13. Only E, E_1, and E_4 are known at present from experiment and the energy diagram is thus highly schematic. Nevertheless, we believe it represents the main features of the system in a qualitative manner. Precise computations would be rather difficult to achieve. It seems unlikely that F-centres are formed as part of the primary photochemical reaction if mechanism (ii) is valid. However, once U-centres are formed then absorption within the U-band would lead to the reactions

$$H^- \rightarrow \boxed{-} + H_i^- , \tag{13}$$

$$H_i^- \rightarrow \boxed{-}\, e + H_i . \tag{14}$$

The increase in the F-band on warming irradiated KCl + KSH is difficult to explain. The same phenomenon occurs in KCl + KOH where Kerkhoff [10] has suggested that it results from the reaction

$$2\, O^- \rightarrow O_2 + 2\, e + 2\boxed{-} . \tag{15}$$

Neither O^- nor S^- ions would have the requisite mobility for this process, however.[2])

The continued growth in the U-band after the S^- band saturates (Fig. 5, 6) requires comment. Because of the mobility of the interstitial H atoms at the irradiation temperature, a steady state is set up in which the back reaction (9) just balances the forward reaction (7). The H_i atom concentration actually decreases because of the operation of two further reactions, namely (10) and

$$H_i + F \rightarrow U . \tag{16}$$

However, the S_i atoms are far less mobile than H_i and so, the back reaction being much slower, the U-band continues to grow.

The assignment of the 255 nm band in KCl and the 300 nm band in KBr cannot be made with certainty. The bands form and decay along with the respective S^- bands and so could be due to transitions in the S^- centre. The similarity in the ground state local mode frequencies determined for the two bands tend to support this conclusion. The possibility that the 225 and 300 nm bands be due to charge transfer transitions from Cl^- to interstitial sulphur atoms has

[2]) A possible mechanism is the movement of H atoms from hydride ions to S^- ions located on near-neighbour sites: $S^- + H^- \rightarrow SH^- + F$.

not been overlooked. However, this is disproved by the kinetics which show the growth in the U-band after the 255 and 310 nm bands saturate (Fig. 5, 6). The growth curves would rather, as stated above, tend to support Fischer and Gründig's assignment [7]. Unfortunately it was not possible to determine the ground state vibrational frequency with sufficient accuracy to decide whether or not the two transitions have a common ground state. The two continuous lines in Fig. 10 und 11 both correspond to $\nu_g \approx 2 \times 10^{12}\ s^{-1}$, but a frequency of $3 \times 10^{12}\ s^{-1}$ would give almost as good a fit for the 255 nm band for which the experimental errors are greater.

Acknowledgement

We are indebted to Dr. A. Runciman for the use of a liquid helium cryostat.

References

[1] U. M. GRASSANO and P. W. M. JACOBS, J. appl. Phys. **35**, 2391 (1964).
[2] P. W. M. JACOBS and H. A. PAPAZIAN, Phys. Rev. **127**, 1567 (1962).
[3] H. THOMAS, Ann. Phys. (Germany) (5) **38**, 601 (1940).
[4] C. J. DELBECQ, B. SMALLER, and P. H. YUSTER, Phys. Rev. **104**, 599 (1956).
[5] F. KERKHOFF, W. MARTIENSSEN, and W. SANDER, Z. Phys. **173**, 184 (1963).
[6] J. ROLFE, Appl. Phys. Letters **6**, 66 (1965).
[7] F. FISCHER and H. GRUNDIG, Phys. Letters **13**, 113 (1964); Z. Phys. **184**, 299 (1965).
[8] A. HAUSMANN, Z. Phys. **192**, 313 (1966).
[9] D. POOLEY, Solid State Commun. **3**, 241 (1965).
[10] F. KERKHOFF, Z. Phys. **158**, 595 (1960).

(Received October 28, 1966)

phys. stat. sol. **19**, 577 (1967)

Subject classification: 13.1; 16; 22.2.1; 22.2.3

Department of Applied Mathematics and Mathematical Physics, University College, Cardiff

Recombination in Semiconductors by a Light Hole Auger Transition

By

A. R. BEATTIE and G. SMITH

The transition rates for an Auger collision process in semiconductors which involves the light hole band is calculated using a quantum mechanical perturbation method. Spherical energy surfaces are assumed although non parabolic energy bands are allowed for. The temperature dependence of the lifetime of excess carriers due to this process is investigated and the results applied to InSb and InAs in the temperature range 200 to 500 °K. The shape of the temperature dependence of this theoretical lifetime for InSb agrees well with experiment at room temperature and above, and when estimates of overlap parameters which occur in the theory are made the absolute magnitude of the lifetime also agrees with experiment. The probability per unit time that a light hole created by a photon of energy $h\nu$ will take part in an impact ionizing transition is also given as a function of $h\nu$. It is concluded that the transition rate for this process is at least comparable to that of the more usual Auger transitions involving the heavy hole and conduction bands only.

Die Übergangsraten für einen Augerkollisionsprozeß in Halbleitern werden unter Einbeziehung der Energiebänder leichter Löcher mit einer quantenmechanischen Störungsrechnung berechnet. Es werden sphärische Energieflächen angenommen, obwohl auch nichtparabolische Energiebänder erlaubt sind. Die Temperaturabhängigkeit der Lebensdauer der durch diesen Prozeß erzeugten überschüssigen Ladungsträger wird untersucht und die Ergebnisse werden auf InSb und InAs im Temperaturbereich 200 bis 500 °K angewendet. Die Temperaturabhängigkeit dieser theoretischen Lebensdauer stimmt für InSb bei Zimmertemperatur und darüber gut mit dem Experiment überein. Wenn die Überlappungsparameter der benutzten Theorie abgeschätzt werden, stimmt die absolute Größe der Lebensdauer ebenfalls mit dem Experiment überein. Die Wahrscheinlichkeit pro Zeiteinheit, daß ein leichtes Loch, durch ein Photon der Energie $h\nu$ erzeugt, an einem Stoßionisationsübergang teilnimmt, wird als Funktion von $h\nu$ gegeben. Es wird geschlossen, daß die Übergangsrate für diesen Prozeß mindestens vergleichbar ist mit dem gebräuchlicheren Augerübergang, an dem nur das Band der schweren Löcher und das Leitungsband beteiligt sind.

1. Introduction

If the conduction, heavy hole, and light hole bands in semiconductors are considered, there are ten Auger transitions which lead to recombination, or, if the reverse processes are considered, to impact ionization [1]. A theory for the two transitions involving the conduction and heavy hole bands only has been developed by Beattie and Landsberg [2, 3] who applied it to the problem of recombination of excess carriers. The theory for the transitions shown in Fig. 1 is now developed. In Fig. 1a two electrons in the heavy hole band interact via their Coulomb repulsion so that one electron drops into a hole in the light hole band while the other goes into the conduction band, conserving energy and

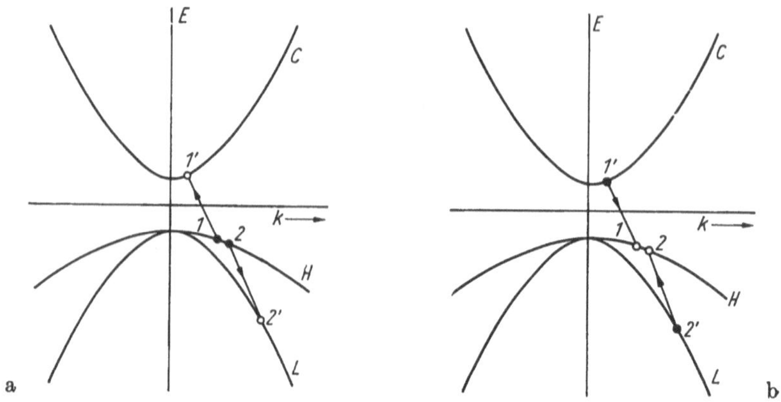

Fig. 1. Auger transitions involving the light hole band for a) recombination and b) impact ionization are shown. Curves C, H, L represent the conduction, heavy hole, and light hole bands, respectively. The initial and final states are 1, 2 and 1′, 2′, respectively

momentum in the process. Fig. 1b shows the reverse of Fig. 1a and leads to recombination.

The processes of Fig. 1 have been considered because their possible importance has been suggested by measurements of the quantum efficiency of incident photons in InSb [4, 5]. When Beattie [1] examined the threshold energies of all ten possible impact ionization transitions, he found that only two would be likely to cause a quantum efficiency greater than unity in the energy range of interest. These were the electron collision process of Beattie and Landsberg and the transition of Fig. 1a. In Tauc's [4] measurement of the quantum efficiency in InSb as a function of photon energy the two threshold energies which appeared agreed well with the theoretical values and, although the experimental error was such that there was some doubt about the existence of the second threshold (Antoncik [6]), its presence has since been confirmed by Nasledov [5]. Although the threshold photon energies are different for the two processes the threshold kinetic energies of the ionizing carriers are almost the same (1.069 × energy gap for the Beattie-Landsberg transition and 1.073 × energy gap for the present transition). Thus, if the transition of Fig. 1a is important in determining the quantum efficiency, then, the transitions of Fig. 1 should also be important in the recombination of excess carriers. The calculation of recombination lifetimes is preferred to a direct calculation of quantum efficiency which entails not only considering the impact ionization transitions but also all other energy loss mechanisms.

In calculating the transition probability in Section 2 a quantum mechanical perturbation treatment [2] is followed. The net transition rate, the difference between the recombination and impact ionization rates, is found for small departures from equilibrium by integrating over all allowed states the transition probability multiplied by the appropriate probabilities of occupancy and vacancy of the initial and final states. The overlap integrals which occur in the theory are discussed in Section 3 where their functional dependence on $\boldsymbol{k}$-vectors is given for small wave vector differences, and estimates of an overlap parameter are obtained by three methods which, in the case of InSb, agree within an order of magnitude. In the numerical work of Section 4, where the theory is used to evaluate intrinsic carrier lifetimes in InSb and InAs and to give the probability per unit time that a light hole of energy E will impact ionize, it is assumed that transitions between light and heavy holes are sufficiently rapid to permit the use of the same quasi-Fermi level for these two bands.

2. Transition Rates and Lifetimes

In calculating the net transition rate for the processes of Fig. 1 Bloch wave functions are used in the quantum mechanical perturbation treatment of [2]. Umklapp processes are neglected and the assumptions of spherical energy surfaces, quasi-Fermi levels and non-degenerate semiconductors are made. However, the possibility of non-parabolic energy bands is taken into account. The net transition rate per unit volume follows directly from equation (3.6) of [2] and is given by

$$R = \frac{e^4 t}{8 \hbar^2 \varepsilon^2 \pi^7} (\delta F_C + \delta F_L - 2\, \delta F_H) \int\int\int \theta(1, 1', 2, 2') \times$$
$$\times \left\{ \left| \frac{F(1,1')\, F(2,2')}{\lambda^2 + g^2} \right|^2 + \left| \frac{F(1, 2')\, F(2, 1')}{\lambda^2 + l^2} \right|^2 + \left| \frac{F(1, 1')\, F(2, 2')}{\lambda^2 + g^2} - \frac{F(1, 2')\, F(2, 1')}{\lambda^2 + l^2} \right|^2 \right\} \times$$
$$\times \frac{1 - \cos x}{x^2} \,\mathrm{d}\boldsymbol{k}_1\, \mathrm{d}\boldsymbol{k}_2\, \mathrm{d}\boldsymbol{k}_2' \,, \tag{1}$$

where e is the electronic charge, t is time, $\hbar$ is Dirac's constant, ε is the dielectric constant, δF_i is the departure of the quasi-Fermi level from its equilibrium value for the i-th band, subscripts C, H, L refer to the conduction, heavy hole, and light hole bands, θ (1, 1′, 2, 2′) is the appropriate probability of occupancy or vacancy of states 1, 1′, 2, 2′, $\boldsymbol{g} = \boldsymbol{k}_1 - \boldsymbol{k}_1'$, $\boldsymbol{l} = \boldsymbol{k}_2 - \boldsymbol{k}_1'$,
where

$$x = \frac{t}{\hbar} |E_1 + E_2 - E_1' - E_2'| \tag{2a}$$

and where

$$F(i, 1') = \int u_H^*(\boldsymbol{k}_i, \boldsymbol{r})\, u_C(\boldsymbol{k}_{1'}, \boldsymbol{r})\, \mathrm{d}\boldsymbol{r}$$

and

$$F(i, 2') = \int u_H^*(\boldsymbol{k}_i, \boldsymbol{r})\, u_L(\boldsymbol{k}_{2'}, \boldsymbol{r})\, \mathrm{d}\boldsymbol{r} \tag{2b}$$

are overlap integrals of the modulating parts of Bloch functions. The neglect of the screening parameter λ in (1) results in a slight overestimate of the transition rate. This effect, however, is small since the screening length in non-degenerate semiconductors is large and the correction to the Auger lifetime due to screening was found in [2] to be no more than 5%.

The overlap functions which are discussed in Section 3 are taken to be of the form

$$|F(i, j)|^2 = \alpha_{ij}\, | \boldsymbol{k}_i - \boldsymbol{k}_j|^2/(E_i - E_j) \,, \tag{3}$$

where E_i is the energy of the i-th state and α_{ij} is a parameter. Since the α_{ij} factors are slowly varying in comparison with the rest of the integrand they may be removed from the integral of (1) and evaluated for the most probable transition. For the most probable transition $\boldsymbol{k}_1 = \boldsymbol{k}_2$ (see [1]), and hence

$\alpha_{11'} = \alpha_{21'}$ and $\alpha_{22'} = \alpha_{12'}$ so that with (3) the net transition rate of (1) becomes

$$R = \frac{\alpha_{11'}\,\alpha_{22'}\,e^4\,t}{4\,\pi^7\,\varepsilon^2\,E_g^2\,\hbar^2} (\delta F_C + \delta F_L - 2\,\delta F_H) \exp(\eta_v - F_0) \times$$
$$\times \iiint \exp\{-(E_v - E_{2'})/kT\} \frac{1 - \cos x}{x^2} \,\mathrm{d}\boldsymbol{k}_1\,\mathrm{d}\boldsymbol{k}_2\,\mathrm{d}\boldsymbol{k}_{2'}\,. \quad (4)$$

It is convenient to write the band structure of the three bands in the form

$$E_{1'} - E_C = \frac{\hbar^2 k_{1'}^2}{2\,m_C}\,, \qquad E_v - E_2 = \frac{\hbar^2 k_2^2}{2\,m_H}\,, \qquad E_v - E_{2'} = \frac{\hbar^2 k_{2'}^2}{2\,m_L}\,, \quad (5)$$

where for non-parabolic bands the effective masses m_C, m_H, and m_L are, in general, energy dependent. However, since states, 1, 1′ and 2 of Fig. 1 are near their respective band edges their effective masses are taken to be constant with values m_{C0} and m_{H0}.

It is also convenient to define new variables of integration, $\boldsymbol{h}, \boldsymbol{j}$, and $\boldsymbol{k}_{2'}$, where

$$\boldsymbol{h} = \boldsymbol{k}_2 - \boldsymbol{k}_1\,, \qquad \boldsymbol{j} = (1 + \mu_H/2)^{1/2}\,(\boldsymbol{k}_1 + \boldsymbol{k}_2) - (1 + \mu_H/2)^{-1/2}\,\boldsymbol{k}_{2'} \quad (6)$$

so that (2a) becomes

$$x = \frac{t\,\hbar}{m_{C0}} \left(a\,k_{2'}^2 - j^2 - \frac{1}{2}\,\mu_H\,h^2 - k_g^2\right), \quad (7)$$

where

$$a = \mu_L - 1 + (1 + \mu_H/2)^{-1}\,; \qquad \mu_L = m_{C0}/m_L \quad \text{and} \quad \mu_H = m_{C0}/m_{H0}\,. \quad (8)$$

Thus with the use of (4), (5), (6), and (7) and when the changes in Fermi level are related to their excess carrier densities δn, δp_H, δp_L and corresponding thermal equilibrium values n_0, p_{H0}, p_{L0}, the recombination rate becomes

$$R = \frac{8\,e^4\,m_{C0}^{3/2}\,\alpha_{11'}\,\alpha_{22'}}{m_L^2\,\pi^{1/2}\,\varepsilon^2\,\hbar^7\,a^2(E_T)\,(2 + \mu_{H0})^{3/2}}\,\eta_T^{5/2}\,\frac{\delta n\,p_{H0}}{n_0} \left[1 + \frac{2\,n_0\,\delta p_H}{p_{H0}\,\delta n} - \frac{n_0\,\delta p_L}{p_{L0}\,\delta n}\right] I(\eta_T)\,, \quad (9)$$

where

$$I(\eta_T) = \int_1^\infty \mathrm{e}^{-\eta_T y}\,a^2\,m_L{}^{5/2} \left(m_L + y\,\frac{\mathrm{d}m_L}{\mathrm{d}y}\right) y^{1/2}\,(y - 1)^2\,\mathrm{d}y \quad (10)$$

and where $E_T (= \eta_T\,k\,T)$ is the threshold energy which is related to the energy gap E_g by solving the equation

$$E_T = \left[\frac{2\,m_{H0} + m_{C0}}{2\,m_{H0} + m_{C0} - m_L(E_T)}\right] E_g\,. \quad (11)$$

The integral of (10) can be computed numerically. However, for $\eta_T \gg 1$ the exponential term in (10) ensures that there is a sharp maximum in the integrand so that slowly varying functions may be evaluated at this maximum y_m and removed from the integral. Thus (10) may be approximated to

$$I(\eta_T) = \left\{a^2\,m_L^{5/2} \left(m_L + y\,\frac{\mathrm{d}m_L}{\mathrm{d}y}\right)\right\}_{y=y_m} [I_2(\eta_T) - 2\,I_1(\eta_T) + I_0(\eta_T)]\,, \quad (12)$$

where the $I_n(\eta_T)$ are related to incomplete gamma functions and are defined by

$$I_n(\eta_T) = \eta_T^{-(n+3/2)} \int_{\eta_T}^\infty \mathrm{e}^{-z}\,z^{n+1/2}\,\mathrm{d}z\,.$$

The lifetime of excess carriers which is given by $\tau = \delta n/R$ is then

$$\tau = D(E_{\mathrm{T}}) \left\{1 + \frac{2\, n_0\, \delta p_{\mathrm{H}}}{p_{\mathrm{H}\,0}\, \delta n} - \frac{\delta p_{\mathrm{L}}\, n_0}{\delta n\, p_{\mathrm{L}\,0}}\right\}^{-1} \frac{\eta_{\mathrm{T}}^{-3/2}}{I(\eta_{\mathrm{T}})}, \tag{13}$$

where

$$D(E_{\mathrm{T}}) = \frac{\pi^{1/2}\, \varepsilon^2\, \hbar^7\, m_{\mathrm{L}}^2(E_{\mathrm{T}})\, a^2(E_{\mathrm{T}})\, (2 + \mu_{\mathrm{H}\,0})^{3/2}\, n_0}{8\, m_{\mathrm{C}\,0}^{3/2}\, \alpha_{11'}\, \alpha_{22'}\, e^4\, p_{\mathrm{H}\,0}}. \tag{14}$$

In the case of an intrinsic semiconductor where the transition rate between the light and heavy hole bands is sufficiently great that they may be considered to have the same quasi-Fermi level, then (13) reduces to

$$\tau = \frac{D(E_{\mathrm{T}})\, \eta_{\mathrm{T}}^{-3/2}}{2\, I(\eta_{\mathrm{T}})} \tag{15}$$

and when asymptotic expressions for the integrals in (12) are used (15) reduces to

$$\tau = \frac{D(E_{\mathrm{T}})\, \mathrm{e}^{\eta_{\mathrm{T}}}\, \eta_{\mathrm{T}}^{3/2}}{4 \left\{a^2\, m_{\mathrm{L}}^{5/2} \left(m_{\mathrm{L}} + y\, \dfrac{\mathrm{d}m_{\mathrm{L}}}{\mathrm{d}y}\right)\right\}_{y=y_{\mathrm{m}}}} \left[1 + \frac{3}{2\,\eta_{\mathrm{T}}} - \frac{3}{2\,\eta_{\mathrm{T}}^2} + \cdots\right]^{-1}. \tag{16}$$

3. Overlap Functions

The overlap functions of the modulating parts of Bloch functions F_{ij} which occur in the matrix element of (1) and for which the states are in different bands, may be related to the matrix elements of the momentum operator [9, 10] by the equation

$$|F(i\, j)|^2 = \alpha_{ij}\, g^2/(E_i - E_j)\,, \tag{17}$$

where

$$\alpha_{ij} = \left(\frac{\hbar}{m}\right)^2 \frac{|\langle i|\, \boldsymbol{e} \cdot \boldsymbol{p}\, |\, j\rangle|^2}{(E_i'' - E_j'')^2}\, (E_i - E_j)\,, \tag{18}$$

where $E_i'' = E_i - \hbar^2\, k_i^2/2\, m$, $\boldsymbol{e}$ is the unit vector in the direction of $\boldsymbol{g}$ $(= \boldsymbol{k}_i - \boldsymbol{k}_j)$, m is the free electron mass, and $\boldsymbol{p}$ is the momentum operator. Thus to the first order in g^2 the overlap functions are obtained by evaluating α_{ij} for $\boldsymbol{k}_i = \boldsymbol{k}_j$, so that

$$\alpha_{ij} = \left(\frac{\hbar}{m}\right)^2 \frac{|\langle i|\, \boldsymbol{e} \cdot \boldsymbol{p} | j\rangle|^2}{E_i - E_j}\,; \qquad (\boldsymbol{k}_i = \boldsymbol{k}_j)\,. \tag{19}$$

Three ways of estimating values for α_{ij} are now considered.

3.1 Effective mass sum rule[1])

For spherical energy surfaces the effective mass sum rule [12] may be written in terms of α_{ij} as

$$\sum_{j \neq i} \alpha_{ij} = \frac{\hbar^2}{2\, m} \left(\frac{m}{m_i^*} - 1\right), \tag{20}$$

where

$$\frac{1}{m_i^*} = \hbar^{-2}\, \frac{\partial^2 E_i}{\partial k^2}\,.$$

---

[1]) See Antončik and Landsberg [9].

Although the summation is over all bands at the same wave-vector it is usually possible to include in the sum (20) only those bands which are closest in energy to the ones of interest. However, it should be noted that, if (20) is summed over all i, then, since $\alpha_{ij} = -\alpha_{ji}$,

$$\Sigma \left(\frac{m}{m_i^*} - 1\right) = 0 \tag{21}$$

and sufficient numbers of bands should be included so that (21) is at least approximately satisfied. Even when only the conduction, heavy hole, light hole, and split-off bands are taken into account, further approximation must be made. If, in general, the states which are closest in energy have the greatest value of overlap, then reasonable approximations[2])
are a) near $k = 0$:

$$\alpha_{\mathrm{CH}} = \alpha_{\mathrm{CL}} \gg \alpha_{\mathrm{CS}}$$

so that

$$\alpha_{\mathrm{CH}} = \frac{\hbar^2}{4\,m}\left(\frac{m}{m_{\mathrm{C}}^*} - 1\right) \tag{22}$$

and b) for large k:

$$\alpha_{\mathrm{LS}} \gg \alpha_{\mathrm{CS}}\,, \quad \alpha_{\mathrm{HS}} \text{ and } \alpha_{\mathrm{HL}} \gg \alpha_{\mathrm{CL}}$$

so that

$$\alpha_{\mathrm{HL}} = -\frac{\hbar^2}{2\,m}\left(\frac{m}{m_{\mathrm{L}}^*} - 1\right) - \frac{\hbar^2}{2\,m}\left(\frac{m}{m_{\mathrm{S}}^*} - 1\right). \tag{23}$$

Equation (22) is used for α_{CH} and (23) for α_{HL}.

3.2 Optical absorption

The optical absorption constant for transitions between bands i and j at a wave vector $\boldsymbol{k}$ is given by [13, 14] as

$$a_{ij} = \frac{4\,e^2\,\hbar^2\,M_{ij}^2\,k^2}{m^2\,c\,n\,(E_i - E_j)}\left|\frac{\mathrm{d}E_i}{\mathrm{d}k} - \frac{\mathrm{d}E_j}{\mathrm{d}k}\right|^{-1} \theta(i)\,\theta(j)\,, \tag{24}$$

where α_{ij} is the optical absorption coefficient, M_{ij}^2 is the square of the optical matrix element averaged over direction, $\theta(i)$ is the probability of vacancy of state i and $\theta(j)$ is the probability of occupation of state j, c is the speed of light and n is the refractive index. Thus from (19) and (24)

$$\alpha_{ij} = \frac{\hbar\,c\,n\left|\frac{\mathrm{d}E_i}{\mathrm{d}k} - \frac{\mathrm{d}E_j}{\mathrm{d}k}\right| a_{ij}}{4\,e^2\,k^2\,\theta(i)\,\theta(j)}\,. \tag{25}$$

3.3 Band structure calculations

If the wave functions for III–V compounds (equation (14) of [11]) are used the α_{ij} may be evaluated directly. When one of the states is in the heavy hole band the result is

$$|\alpha_{i\mathrm{H}}| = \left|\frac{2\,Q^2\,\Phi(E_i')}{9\,[E_i' - \hbar^2\,k^2/2\,m]}\right|\,, \tag{26}$$

where

$$\Phi(E_i') = \frac{E_i'\,(E_i' + \Delta)\,(E_i' + 2\,\Delta/3)}{[2\,E_i'^3 + (3\,\Delta - E_g)\,E_i'^2 + (4\,\Delta/3)\,(\Delta - E_g)\,E_i' - 2\,\Delta^2 E_g/3]}\,. \tag{27}$$

---

[2]) Different approximation to that of Antončik and Landsberg [9] are used.

E_i' is a solution of the equation

$$E_i' (E_i' - E_g) (E_i' + \Delta) - k^2 Q^2 (E_i' + 2 \Delta/3) = 0 \quad (E_C' > E_L' > E_S') . \quad (27\,a)$$

Δ is the spin–orbit splitting
and

$$Q^2 = \frac{3 E_g \hbar^2}{4 m} \left[1 + \frac{m}{m_{L0}}\right] . \quad (27\,b)$$

Strictly speaking m_{L0} in (27b) should be the effective mass corresponding to E_L' at $k = 0$. Thus, $\alpha_{11'}$ is evaluated from (26) at $\boldsymbol{k}_1 = 0$ and $\alpha_{22'}$ at $E_2' = E_T$.

There are difficulties in connection with all three methods. In Section 3.1 qualitative assumptions about relative magnitudes of the α_{ij} have to be made, in Section 3.2 the experimental optical absorption represents the total absorption and experiments have to be found where the transitions of interest are the dominant ones, and in Section 3.3 since a variational approach is used, the wave functions, although accurate at $k = 0$, may not be accurate for $k \neq 0$. However, all three methods should give an order of magnitude estimate for the overlap parameters.

4. Results

The general theory of Section 2 is applied to InSb and InAs. For III–V compounds the energies of the conduction, light hole, and split-off bands are given by the roots of the cubic equation (27a) so that

$$E_i - E_V = \frac{\hbar^2 k_i^2}{2 m} + E_i' + \delta(E_i') , \quad (28)$$

where $\delta(E_i')$ is a correction term due to interaction with bands other than the conduction, heavy hole, light hole, and split-off bands.

For the conduction, light hole, and split-off bands $\delta(E_i')$ is unimportant in InSb, but it is important in InAs. The results given below are appropriate for InSb where $\delta(E_i')$ is neglected but for InAs the correction term is taken into account. From (27a) the effective mass m_L defined in (5) is given by

$$\left(\frac{m}{m_L} - 1\right) = \left(\frac{m}{m_{L0}} - 1\right) \left(1 - \frac{E_L'}{E_g}\right)^{-1} (\Delta + E_L')^{-1} \; (\Delta + 3 E_L'/2) , \quad (29)$$

and when

$$m_L \ll m , \quad \Delta \gg |E_L'| ,$$

then

$$m_L = m_{L0} \{1 + (E_V - E_L)/E_g\} . \quad (30)$$

With the use of (29) or (30) the threshold energy is calculated from (11) and the lifetime is found from (15). The effective mass m_L^* is obtained from (27a) by finding the second derivative of the energy with respect to k. Since the conduction, heavy hole, light hole, and split-off bands only are included in the sum rules determination of α_{ij}, the correction term $\delta(E_i')$ is also omitted in finding m_L^* for InAs.

The material parameters which were used are summarized in Table 1, and the values of the overlap parameters corresponding to the threshold transition are given in Table 2. The total optical absorption near the band edge includes not

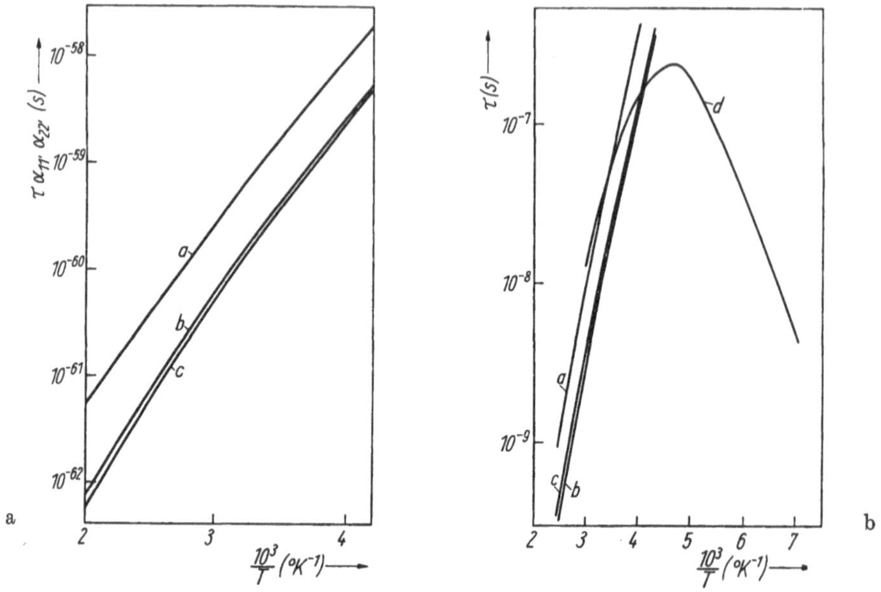

Fig. 2. Lifetimes for intrinsic InSb based on the theory of Section 2 are plotted as functions of inverse temperature. In a) where the lifetimes are multiplied by the overlap parameters, curve (a) is based on the assumption of parabolic bands, curve (b) represents the approximation to the non-parabolic case, and curve (c) gives the result using the computed integral. In b) the overlap parameters have been estimated by (a) sum rules, (b) optical absorption, and (c) band structure calculations. Curve (d) shows the experimental lifetimes from Zitter et al. [8]

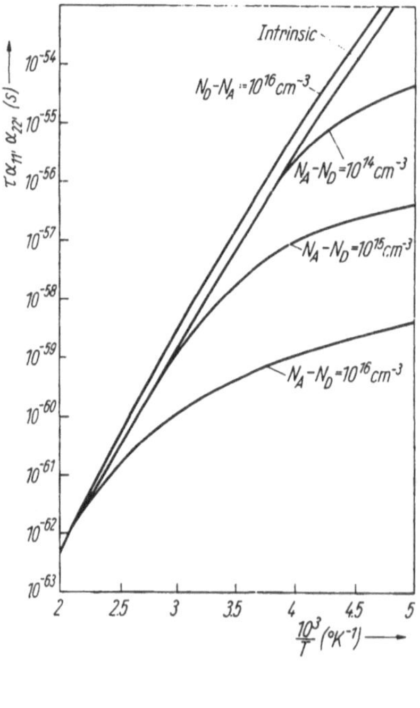

Fig. 3. Theoretical lifetimes multiplied by the overlap parameters for InAs are plotted as functions of inverse temperature. The donors and acceptors are assumed to be completely ionized

Fig. 4. The probability per unit time that a light hole of wave vector $\boldsymbol{k}$ will impact ionize is plotted as a function of the photon energy required to produce this light hole

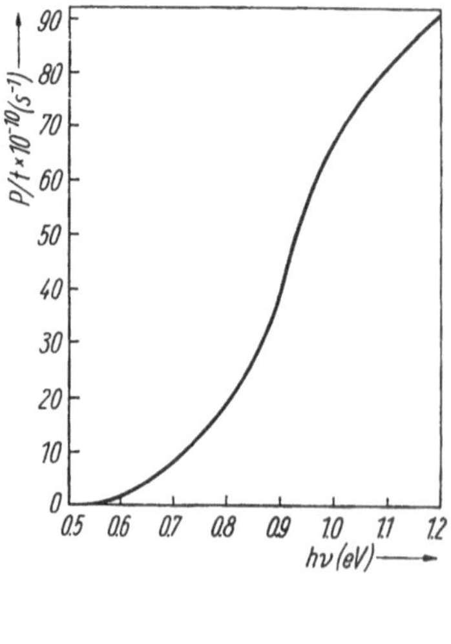

Table 1

Material parameters for InSb and InAs

Parameter	InSb	InAs
E_g	$(0.24-0.26\times10^{-3}\ T)$ eV	$(0.426-0.22\times10^{-3}\ T)$ eV
E_T	$1.073\ E_g$	$1.303\ E_g$
m_{C0}	$0.0114\ m$	$0.026\ m$
m_{H0}	$0.18\ m$	$0.4\ m$
m_{L0}	$0.0117\ m$	$0.025\ m$
ε	16.8	10.9
n	4.1	

Table 2

Overlap parameters for the threshold transition

	Sum rules	Optical absorption	Direct evaluation from wave function
InSb α_{CH}	2.57×10^{-26}	2.83×10^{-26}	2.42×10^{-26}
α_{HL}	0.24×10^{-26}	0.78×10^{-26}	0.71×10^{-26}
$\alpha_{CH}\,\alpha_{HL}$	0.62×10^{-52}	2.21×10^{-52}	1.71×10^{-52}
InAs α_{CH}	1.42×10^{-26}		1.12×10^{-26}
α_{HL}	1.48×10^{-28}		2.21×10^{-28}
$\alpha_{CH}\,\alpha_{HL}$	2.11×10^{-54}		2.48×10^{-54}

only the heavy hole–conduction band absorption, but also the light hole–conduction band absorption and this was allowed for.[3]) For the light hole–heavy hole transition, data on degenerate p-type material were used so that only this transition was being observed (Gobeli and Fan [7]).

The results of the lifetime calculations are shown in Fig. 2 and Fig. 3. It can be seen from graph a) of Fig. 2 that the assumption of parabolic bands leads to a lifetime which is about an order of magnitude greater than the computed lifetime, while with the approximation to the non-parabolic case there is a discrepancy of at most 14% which occurs at the higher temperatures ($\approx$500 °K).

On the basis of the theory of Section 2 the probability per unit time that a hole in the light hole band with a wave-vector $\boldsymbol{k}$ and energy E will take part in an impact ionization process (Fig. 1b) is

$$\frac{P}{t} = \frac{4\,\alpha_{11'}\,\alpha_{22'}\,e^4\,m_{H0}^{3/2}}{\varepsilon^2\,\hbar^7\,m_{C0}^{1/2}\,(2+\mu_{H0})^{3/2}}\,a^2(E)\,m_L^2\,\frac{(E_V - E - E_T)^2}{E_g^2} \tag{31}$$

and is shown in Fig. 4 as a function of the photon energy required to produce the light hole.

[3]) If the only difference between them were the density of states the ratio of the light hole optical absorption to the heavy hole optical absorption would be 1:2.

5. Conclusions

The recombination mechanism investigated here yields a temperature dependence which agrees with experimentally measured lifetimes in InSb at room temperature and above. When estimates of the overlap function are made the magnitude of the lifetime appears to give values which are somewhat smaller than the experimental lifetime (Fig. 2). The closest fit is obtained when the overlap parameters are estimated from the effective mass sum rules (a factor 1.4 too small), while the worst fit is obtained using optical absorption data (a factor 4 too small). However, all these methods of estimating the overlap parameters can only be hoped to give an order of magnitude estimate so that to this order it can be concluded that the magnitudes of the experimental and theoretical lifetimes are in agreement. In particular for the case of the optical absorption calculation an underestimate of the lifetime would be expected since included in the experimental optical absorption coefficient will be all allowed transitions and not just the direct transitions of interest.

The Auger theory of Beattie and Landsberg has an almost identical temperature dependence. Their calculation could be extended to non-parabolic energy bands, but direct comparison with the present process is difficult due to the overlap functions. For the dominant transitions in the present case the differences in $\boldsymbol{k}$-vectors are small, but for the other case these differences are large precluding the use of the results of Section 3. However, the process considered here seems comparable with the more usual Auger process for intrinsic semiconductors becoming less important for heavy n-type material and more important for heavy p-type material.

Acknowledgements

One of us (G. S.) is grateful to the Signals Research and Development Establishment, Christchurch, for support in this work. We should also like to thank Mr. G. Jones for his assistance in the computer programming.

References

[1] A. R. BEATTIE, J. Phys. Chem. Solids **24**, 1049 (1962).
[2] A. R. BEATTIE and P. T. LANDSBERG, Proc. Roy. Soc. **A 249**, 16 (1959).
[3] P. T. LANDSBERG and A. R. BEATTIE, J. Phys. Chem. Solids **8**, 73 (1959).
[4] J. TAUC, J. Phys. Chem. Solids **8**, 219 (1959).
[5] V. N. IVAKHNO and D. N. NASDELOV, Soviet Phys. — Solid State Phys. **6**, 1651 (1965).
[6] E. ANTONČIK, Proc. 7th Internat. Conf. Semiconductors, Paris 1964 (p. 473).
[7] G. W. GOBELI and H. Y. FAN, Phys. Rev. **199**, 613 (1960).
[8] R. N. ZITTER, A. J. STRAUSS, and A. E. ATTARD, Phys. Rev. **115**, 266 (1959).
[9] E. ANTONČIK and P. T. LANDSBERG, Proc. Phys. Soc. **82**, 337 (1963).
[10] W. BRAUER, phys. stat. sol. **5**, 136 (1964).
[11] E. O. KANE, J. Phys. Chem. Solids **1**, 249 (1957).
[12] D. J. MORGAN and P. T. LANDSBERG, Proc. Phys. Soc. **86**, 261 (1965).
[13] J. BARDEEN, F. J. BLATT, and L. H. HALL, Proc. Photoconductivity Conf. Atlantic City 1954, Wiley 1956 (p. 146).
[14] D. L. DEXTER, ibid. (p. 155).

(Received November 11, 1966)

phys. stat. sol. 19, 587 (1967)

Subject classification: 13.3; 13.4

Theoretical Chemistry Institute, University of Wisconsin, Madison, Wisconsin (a) and Department of Mathematics, University of Nottingham (b)

On the Theory of Surface and Impurity States

By

R. D. LEVINE (a) and A. T. AMOS (b)

A formalism is developed for perturbation theory in which the Hamiltonian is resolved as $H = H_0 + V_1 + V_2$ and the two perturbing potentials V_1 and V_2 are not treated in a symmetric fashion. All the operators are expressed in the H_0 representation so that the formalism has particular applications in the theory of impurity states and surface and chemisorption states since H_0 can be taken as that part of the Hamiltonian which acts on individual atomic sites, V_1 the interatomic interaction and V_2, which will have only a small number of non-zero matrix elements, the effect of the surface or impurity atoms. These applications are illustrated using the linear chain as a model for a one-dimensional crystal, and the present theory allows the results to be obtained in a very simple way.

Ein Formalismus für Störungstheorie wird entwickelt, in der der Hamiltonoperator gemäß $H = H_0 + V_1 + V_2$ aufgelöst wird und die beiden Störungspotentiale V_1 und V_2 nicht in symmetrischer Weise behandelt werden. Alle Operatoren werden in der H_0-Darstellung ausgedrückt, so daß der Formalismus insbesondere Anwendungen in der Theorie der Fremdzustände, der Oberflächen- und der Chemisorptionszustände findet, da H_0 als der Teil des Hamiltonoperators, der auf individuelle Atomplätze wirkt, V_1 als interatomare Wechselwirkung und V_2, mit einer nur kleinen Anzahl von Null verschiedener Matrixelemente, als Oberflächen- oder Fremdatomeffekt angesehen werden kann. Diese Anwendungen werden mit der linearen Kette als Modell für einen eindimensionalen Kristall illustriert. Die hier dargestellte Theorie liefert die Ergebnisse in sehr einfacher Weise.

1. Introduction

The aim of this paper is twofold. On the formal side it introduces the use of the partitioning technique and Green's operators and associated quantities in problems involving two potentials, with particular emphasis on the case where the two potentials are not treated in a symmetric fashion. The actual Hamiltonian H is resolved as

$$H = H_0 + V_1 + V_2 = H_1 + V_2 = H_0 + V \tag{1.1}$$

and all operators are expressed in the H_0 representation. The non-symmetric role of V_1 and V_2 is to be contrasted with the formalism of multiple scattering [1], where the interaction of a particle with a homogeneous many-body system is discussed and V is resolved as a sum of equivalent terms.

Physically we are interested in those situations where both V_1 and V_2 have off-diagonal elements in the H_0 representation, and where the Hamiltonian H_1 has at least an intuitive physical significance. The simplest physical model is that of an 'impure' linear chain, where H_1 corresponds to the pure chain, and our second aim is to use the formal theory to deal with this physical model in some detail. This problem is of some considerable theoretical and practical significance since, for example, it gives insight into the properties of impurity states and surface and chemisorption states in real crystals [2 to 8, 30]. The present approach differs from previous work, however, since apart from the computational simplicity of the method it can be extended to analyse explicitly the time evolution of the system by appropriate summations of perturbation theory [9]. In addition, we wish to concentrate particularly on those cases where the energies of interest fall within the continuous spectra of the pure system. In these cases the formalism of Section 2 is not merely a compact representation of perturbation theory for two potentials but an essential tool.

In addition to the problems to be discussed here the formalism can be applied to the theory of the linear response of a system to an external perturbation and the theory of reaction rates. It can also be used to investigate the possibility of defining an 'adsorption index' which is a measure of the adsorption properties of a given type of surface towards a series of similar molecules. These topics will be taken up by one of us (R. D. L.) in a separate publication.

2. Theory

Let H_0 be the unperturbed Hamiltonian for a crystal with $|k\rangle$ as an eigenstate so that

$$\big(E(k) - H_0\big)\,|k\rangle = 0\,. \tag{2.1}$$

For a perturbed Hamiltonian $H_0 + V$ the Schrödinger equation is

$$(E - H_0)\,|\psi\rangle = V\,|\psi\rangle\,. \tag{2.2}$$

An explicit formal expression for $|\psi\rangle$ can be written in two ways. If $|\psi\rangle$ and $|k\rangle$ have the same spatial behaviour, i.e. if they are both delocalized, then the Lippmann-Schwinger equation [12, 13] holds

$$|\psi\rangle = |k\rangle + G_0\,V|\psi\rangle \tag{2.3}$$

with G_0 the Green's operator $(E - H_0)^{-1}$. However, the well-known Fredholm alternative implies that there may also be localized solutions $|\psi_l\rangle$ for which $\langle k|\psi_l\rangle$ is zero for all k and which satisfy the homogeneous equation

$$|\psi_l\rangle = G_0\,V|\psi_l\rangle\,. \tag{2.4}$$

The solutions of (2.4) are, in the case of the impurity states, the localized states discussed by many authors [2, 14, 15].

The symbolic operator equations (2.2) to (2.4) in practice have to be written in a convenient representation. Choosing one which diagonalizes H_0 we can put

$$G_0(E) = \sum_j \frac{P_j}{E - E_j} + \int\limits_{\mathrm{Sp}} \frac{P(\varepsilon)}{E - \varepsilon}\,\mathrm{d}\varepsilon\,, \tag{2.5}$$

where P_j and $P(\varepsilon)$ are projection operators on the discrete states and states of energy ε in the continuous spectra, respectively. The integration has to be carried out around the continuous spectra. If E as defined by (2.2) can take up values in the continuous spectra of H_0, the integration has to be carried out with care and is done using the identity

$$\lim_{\gamma \to +0} \int \frac{P(\varepsilon)}{E + i\gamma - \varepsilon} \mathrm{d}\varepsilon = \mathrm{Pv} \int \frac{P(\varepsilon)}{E - \varepsilon} \mathrm{d}\varepsilon - i\pi\delta(E - \varepsilon), \tag{2.6}$$

where Pv is Cauchy's principal value of the integral. Equation (2.6) corresponds to 'outgoing' solutions at an energy we denote as $E^+ = E + i\gamma$. See [9] for the equivalent formulation for 'ingoing' and 'standing' wave solutions.

The energy levels corresponding to (2.3) and (2.4) are given by the solutions of

$$\det D_l(E_l) = 0 \tag{2.7}$$

for the localized state, where $D_l(E) = 1 - G_0(E)\,V$ is evaluated in some convenient representation while for the delocalized states the real parts of the in-band energies are given by

$$\det |D(E)| = 0\,, \tag{2.8}$$

where

$$D(E) = 1 - \mathrm{Pv}\; G_0^+(E)\; V\,.$$

The operator G_0 is not the only useful one. Suppose $|0\rangle$ is an eigenstate of H_0 and let $P = 1 - |0\rangle\langle 0|$ be a projection operator on the spectrum of H_0. The Brillouin-Wigner wave operator W can be defined by [9, 16, 17]

$$|\psi\rangle = W|0\rangle \tag{2.9}$$

or

$$W = 1 + G_0\, P\, V\, W\,. \tag{2.10}$$

The state $|\psi\rangle = W|0\rangle$ is an eigenstate of H, for times $t > 0$ which evolved from the state $|0\rangle$ provided the latter was occupied at $t = 0$. The difference in energy between $|0\rangle$ and $|\psi\rangle$ is given by Re $\langle 0|\Gamma|0\rangle$ where in terms of W the rate operator Γ is

$$\Gamma = V\, W\,. \tag{2.11}$$

The Møller wave operator Ω defined by [12, 13]

$$(E^+ - H)\,\Omega = (E^+ - H_0) \tag{2.12}$$

and the transition operator $T = V\,\Omega$ are other useful operators. Using equation (22b) of [9], i.e.

$$T = \Gamma + \Gamma\, Q\, G\, Q\, \Gamma, \tag{2.13}$$

where $Q = 1 - P$ and G is the Green's operator, for H gives

$$\Omega = W + W\, Q\, G\, Q \tag{2.14}$$

It is clearly most convenient to use these operators and operator relations with $|0\rangle$ an eigenstate of H_0. At the same time, for a discussion of impurity states and other problems it is convenient for $|0\rangle$ to refer to one atomic site so that, for example, the order of the determinants in (2.6) and (2.7) is small [18]. We must, therefore, take H_0 to be that part of the total Hamiltonian which acts on individual atomic sites. The terms which give rise to interatomic interaction must be included as part of V, say V_1. The remaining part V_2 refers to the effect of the impurity atom. Thus, we are led naturally to a division of the Hamiltonian into the terms indicated by equation (1.1), and in the remainder of this section we revise the previous equations to take account of this circumstance [19].

Let W_1 be the Brillouin-Wigner wave operator for H_1, i.e.

$$W_1 = 1 + G_0 P V_1 W_1 . \tag{2.15}$$

From equation (2.10)

$$W = W_1 + W_1 G_0 P V_2 W \tag{2.16}$$

since

$$W = 1 + G_0 P V_1 W_1 + G_0 P V_2 W_1 + G_0 P V (W - W_1)$$

so that

$$W - W_1 = G_0 P V_2 W + G_0 P V_1 (W - W_1) .$$

From (2.16) since $\Gamma_1 = V_1 W_1$

$$V_1 W = \Gamma_1 + \Gamma_1 G_0 P V_2 W$$

and therefore

$$\Gamma = \Gamma_1 + [1 + \Gamma_1 G_0 P] V_2 W = \Gamma_1 + W_1^{\mp} V_2 W , \tag{2.17}$$

where $W_1^{\mp} = 1 + \Gamma_1 G_0 P$ and the operation $\mp$ is defined as

$$A^{\mp}(\lambda) = A^{\dagger}(\bar{\lambda}) ,$$

where the bar denotes the complex conjugate.

In exactly the same fashion

$$T = T_1 + \Omega_1^{\mp} V_2 \Omega , \tag{2.18}$$

where $T_1 = V_1 \Omega_1 = V_1 + V_1 G_0 T_1$ is the transition operator due to V_1 only. Equation (2.18) is the known result for scattering by two potentials [13]. Also corresponding to (2.16)

$$\Omega = \Omega_1 + (E^+ - H_1)^{-1} V_2 \Omega . \tag{2.19}$$

Using equation (2.14) or (2.13), equation (2.19) can be solved in the H_0 representation. If we specialize the discussion to impurity states in the linear chain we recover essentially the formalism of Slater and Koster [11], except that here we impose outgoing boundary conditions, a distinction which was recognized of course, by these authors.

As a simple example let $|k\rangle$ be a solution of the Lippmann-Schwinger equation for H_1 and let $|n\rangle$ be a site of the linear chain. Equation (2.19) reads

$$\langle n|\psi\rangle = \langle n|k\rangle + \sum_{m,l} \langle n|G_1|m\rangle \langle m|V_2|l\rangle \langle l|\psi\rangle \tag{2.20}$$

or

$$\langle n|\psi\rangle = \langle n|k\rangle + \langle n |G| n\rangle \langle n |\Gamma_2| k\rangle , \tag{2.21}$$

where we have used equation (2.14) with $Q = |n\rangle \langle n|$ and $\Gamma_2 = V_2 W = W^{+} V_2$. Equation (2.21) is an explicit solution of the Lippmann-Schwinger equation.

Using the wave and rate operators one can thus generate a convergent perturbation expansion for the solution of the Lippmann-Schwinger equation. This is of some importance since an iteration of equation (2.3) in the form

$$|\psi\rangle = |k\rangle + G_0 V|k\rangle + G_0 V G_0 V|k\rangle + \cdots$$

is likely to diverge in the presence of impurity states. On the other hand, using an iterated value of the rate operator leads to a convergent expansion since the denominator of the Green's function $\langle n|G|n\rangle$ has the correct analytic behaviour.

3. Internal Structure in the Q-Subspace

Up to now we have discussed the subspace Q as if it were one-dimensional so that no internal structure in Q can be observed. In the study of impurity states it is convenient to explicitly indicate such a structure in order to deal with the interaction between impurities and between different excited states when each site has several excited states. The general formalism is also relevant to semi-empirical theories of unimolecular breakdown [20].

Let $L = H_0 + \Gamma$ so that (equation (12b) of [9])

$$(E - Q\,L\,Q)\,Q\,G\,Q = Q\,. \tag{3.1}$$

Suppose Q is divided into non-overlapping subspaces so that

$$Q = \sum_i Q_i \quad \text{and} \quad Q_i\,P = 0 \quad \text{for all } i\,. \tag{3.2}$$

If F is the wave operator in the Q-subspace satisfying $F\,Q_i\,G\,Q_i = Q\,G\,Q_i$, so that

$$(E - Q_i\,L\,F\,Q_i)\,Q_i\,G\,Q_i = Q\,, \tag{3.3}$$

then defining $Q^i = Q - Q_i$, we have that

$$F = Q_i + (E - Q^i\,L\,Q^i)^{-1}\,Q^i\,L\,F\,Q_i\,. \tag{3.4}$$

Let Q be the projection on the sites of the impurities so that $Q_i = |i\rangle\,\langle i|$ projects on the impurity at site i. Equation (3.7) is then of considerable interest in that it shows that even though $Q_i\,V\,Q_j$ may vanish (and is assumed to vanish for very diluted impure systems) the various orthogonal components of the Q-subspace can interact indirectly due to terms of the type $Q_i\,V\,P$.

In the context of the linear chain, equations (3.1) to (3.7) enable us to remove the most serious limitation of the model, namely rigidity. In this form Q is the projection on the electronic component of the lattice electronic Hamiltonian, $P\,H_0\,P$ is the Hamiltonian for the site vibrations of the unexcited lattice and V is the coupling between electronic and nuclear motion, $Q\,V\,Q$ leads to a uniform shift of the electronic band, and similarly for $P\,V\,P$. The interaction of different sites can be mediated by the lattice vibrations.

4. Impurity States

Consider a single impurity located at the origin of an infinite linear chain. The Hamiltonian of the system is taken as the sum of H_1, the ordinary tight binding Hamiltonian, and V_2, which represents the modifications due to the presence of the impurity. This specification is thus also relevant to the interaction of Frenkel excitons with impurity centres. In terms of the site representation with the basis set $|n\rangle$ referring to the n-th site of the chain we have that

$$\left.\begin{aligned} \langle r|H_0|s\rangle &= \alpha\,\delta_{rs}\,,\\ \langle r|V_1|s\rangle &= \beta\,\delta_{r,\,s\pm 1}\,,\\ \langle 0|V_2|0\rangle &= y\,,\\ \langle 0|V_2|\pm 1\rangle &= \beta\,x \end{aligned}\right\} \tag{4.1}$$

on the assumption that the effect of the impurity changes only the matrix elements $\langle 0|H|0\rangle$ and $\langle 0|H|\pm 1\rangle$. It follows, therefore, that V_2 is of the type $P\,V_2\,P = 0$, where $P = 1 - Q$ and Q is the projection operator $|0\rangle\,\langle 0|$.

For such a localized perturbation the Fredholm determinant is of the third order. Putting $g = \langle 0|G_1|0\rangle$ and $E = \alpha + \beta\,(a + a^{-1})$ and using equations (19) and (23) of [16] it becomes

$$\begin{vmatrix} 1 - a\,g\,\beta\,x & -(1 + a^2)\,g\,\beta\,x - a\,g\,y & -a\,g\,\beta\,x \\ -g\,\beta\,x & 1 - 2\,a\,g\,\beta\,x - g\,y & -g\,\beta\,x \\ -a\,g\,\beta\,x & -(1 + a^2)\,g\,\beta\,x - a\,g\,y & 1 - a\,g\,\beta\,x \end{vmatrix} . \tag{4.2}$$

The reflection symmetry of the system implies that

$$\det D = \det D_{\mathrm{sym}} \det \mathrm{D}_{\mathrm{anti}} , \tag{4.3}$$

where D_{sym} is the Fredholm matrix in the subspace of even functions with respect to the origin, and conversely for D_{anti}. The latter has a node at the origin and so is the identity in the present case. Hence (4.2) reduces to

$$\begin{vmatrix} 1 - 2\,a\,g\,\beta\,x & -g\,(1 + a^2)\,\beta\,x - a\,g\,y \\ -2\,a\,g\,\beta\,x & 1 - 2\,a\,g\,\beta\,x - g\,y \end{vmatrix} . \tag{4.4}$$

A special case of (4.4) is when $y = 0$ so that

$$\left(\frac{\mathrm{d}}{\mathrm{d}x} \det D\right)_{x=0} = -\,4\,a\,g\,\beta , \tag{4.5}$$

which is the extended Hellmann-Feynman theorem [16] for this case.

Considering outgoing boundary conditions and defining

$$E^+ - \alpha = \beta\,(a + a^{-1}) + i\,\varepsilon , \tag{4.6}$$

we can use the condition $\langle 0|W|0\rangle = 1$ and the result

$$\langle n \pm l|\,W_1(E^+)|n\rangle = a^{|l|} \tag{4.7}$$

obtained by Levine [9] to deduce from (2.16) that

$$\langle n|W|0\rangle = (1 + x)\,a^{|n|} . \tag{4.8}$$

Using (4.8) and (2.17) or (2.11) gives

$$\langle 0|\Gamma^+|0\rangle = y + 2\,\beta\,(1 + x)^2\,a . \tag{4.9}$$

This result has previously been derived by the direct integration of the differential form of equation (2.17) [16].

The zeros of $\langle 0|G(E^+)|0\rangle^{-1}$, i.e. the zeros of $\{E^+ - \alpha - \langle 0|\Gamma^+|0\rangle\}$ determine the energy levels of the system or from equations (4.7) and (4.9)

$$\beta\,(a + a^{-1}) - y - 2\,\beta\,(1 + x)^2\,a = 0 . \tag{4.10}$$

This could equally well have been obtained by setting (4.4) equal to zeo. For standing wave solutions we use the standing solution of equation (4.9) which gives

$$\beta\,(a + a^{-1}) - y - \beta\,(1 + x)^2\,(a + \bar{a}) = 0 . \tag{4.11}$$

Equation (4.11) shows that since if all the solutions are inside the band of the pure system, $|a + a^{-1}| \leqq 1$, while, if two solutions are outside the band, one at the top and one at the bottom, $|a + a^{-1}| > 1$, the occurrence of localized states depends on the magnitude of

$$\frac{y}{2\,\beta\,[1-(1+x)^2]}\,.$$

These results are equivalent to the conclusions derived from the study of the limiting behaviour of the finite chain [6, 21].

For the solutions with outgoing boundary conditions

$$\begin{aligned}\langle 0|G(E^+)|0\rangle &= [\beta\,(a+\bar{a})\,(1-(1+x)^2) - y - \beta\,(a-\bar{a})\,(1+x)^2]^{-1}\\ &= [\,2\,\beta\,(1-c^2)\cos\varphi - y - 2\,i\,\beta\,c^2\sin\varphi\,]^{-1}\,,\end{aligned} \tag{4.12}$$

where $a = \exp i\,\varphi$, $c = (1+x)$, so that the phase of the Green's function is given by

$$\tan\delta = \frac{-\,2\,\beta\,c^2\sin\varphi}{2\,\beta\,(1-c^2)\cos\varphi - y}\,. \tag{4.13}$$

The unperturbed ($c = 0$) impurity clearly absorbs at $E = y + \alpha$, while from equation (4.13) the absorption in the chain occurs at a shifted frequency

$$E' = E + \frac{c^2}{1-c^2}\,y \tag{4.14}$$

and the wave functions are given by

$$W|0\rangle = \sum_{n\neq 0}(1+x)\,a^{|n|}\,|n\rangle + |0\rangle\,. \tag{4.15}$$

If $|y| > (1-c^2)\,2\,\beta$, the impurity absorption is outside the band ($c = 0$ is the case studied by Rashba). In this case $a = \exp(-\varkappa)$ and the wave function decays rapidly in space unless $\alpha + y$ is just outside the band.

From the relation $P\,G\,P = (E - P\,H\,P)^{-1} + P\,W\,Q\,G\,Q\,W^{+}\,P$ and using the fact that the change in the density of states is $\mathrm{Im}(\mathrm{Tr}\,P\,G\,P)$ it is obvious that the change in the density of states due to the impurity is compensated by the same amount in the density of the remainder. The impurity density of states vanishes at the band edges, as expected (compare [22 to 24]).

The lifetime for the transfer of excitation among sites is unchanged as the phase of the wave operator remains the same. (For the connection between the phase and the lifetime see [9, 25].) The change in the imaginary part of $\langle 0|\Gamma|0\rangle$ is thus entirely due to the further delay experienced by the excitation at the impurity which cannot permanently trap the excitation. This is in contrast with the pure chain in which there is no delay due to the occupation of the site. The phenomenon is rather similar to the delay of a plane wave at a point of discontinuity of the potential, and may explain the observation that minute quantities of impurities are enough to change the emission spectra to that of the impurity [26]. Since the phenomenon is exhibited even by molecules rather similar to that of the host crystal, it is perhaps best explained as due to the enhanced probability of the excitation to be found on the site of the impurity.

5. Surface and Chemisorption States

Theoretical studies of surface and chemisorption states have generally concentrated on localized states in which the electrons of the foreign atom remain near the surface. It seems to us that it is by no means clear that these are the most important. Indeed, for surface states it seems that the existence condition for localized states is, in practice, too restrictive to be of general significance, while in the case of chemisorption states, if the bond order between the adsorbed atom and the surface is considered as a measure of the effective surface interaction then simple arguments show that delocalized states can be just as important as the localized ones. Indeed, there are cases when only the delocalized solutions are bonding. In this section, therefore, we shall consider surface and chemisorption states corresponding to propagating solutions inside the band of the unperturbed chain.

Following Koutecky we use a tight-binding Hamiltonian for the linear chain with alternating Coulomb and exchange integrals and β as the unit of energy, i.e.

$$\langle n|H_0|m\rangle = \delta_{n,m}\{\alpha + y(-1)^n\} \tag{5.1}$$

and

$$\langle n|V_1|n-1\rangle = \begin{cases} 1 & \text{for even } n\,, \\ x & \text{for odd } n\,. \end{cases}$$

By suitable choice of x and y we can use (5.1) to treat several different models. For example, with $x = 1$ we can apply the results to mixed crystals [27] or deal with correlation effects [28], while with $y = 0$ we can deal with alternating bond lengths [29].

The recurrence relation for the matrix elements $W_r = \langle r|W_1|0\rangle$ of the wave operator is

$$W_{2n} = (E^+ - \alpha - y)^{-1}(W_{2n+1} + x\,W_{2n-1})\,. \tag{5.2}$$

Iterating once:

$$W_{2n} = (E^+ - \alpha - y)^{-1}(E^+ - \alpha + y)^{-1}[x(W_{2n+2} + W_{2n-2}) + (1 + x^2)\,W_{2n}]\,. \tag{5.3}$$

The translational symmetry of the wave operator is obvious from (5.3), so that we can put

$$W_{2n-1} = k\,W_{2n} \quad \text{and} \quad W_{2n+1} = k\,z\,W_{2n}\,. \tag{5.4}$$

From (5.3) and (5.4)

$$z + z^{-1} = x^{-1}(\alpha^2 - 1 - x^2 - y^2) \qquad (\text{say}, = U) \tag{5.5}$$

where

$$z_{\pm} = \frac{U}{2}[1 \mp (1 - 4\,U^{-2})^{1/2}]\,.$$

Koutecky investigated the problem subject to $U^2 > 4$. Here we are interested in the complementary case $U^2 \leqq 4$ which corresponds to propagating solutions inside the band.

We also need to know $\langle 0|\Gamma_1|0\rangle$, and this is given by

$$\langle 0|\Gamma_1|0\rangle = \langle 0|V_1|+1\rangle\langle +1|W_1|0\rangle + \langle 0|V_1|-1\rangle\langle -1|W_1|0\rangle = k + x\,k\,z\,. \tag{5.6}$$

The formation of the surface is described by the breaking of the chain between the atoms at sites -1 and 0. It is accompanied by an increase of the Coulomb integral of the surface atom, 0, by b. Thus

$$\langle 0|V_2|0\rangle = b \quad \text{and} \quad \langle -1|V_2|0\rangle = -1\,. \tag{5.7}$$

Hence

$$\langle 0|\Gamma|0\rangle = \langle 0|\Gamma_1|0\rangle + b - k = b + x\,k\,z\,. \tag{5.8}$$

Let $e = E - \alpha$ and $e - y = \mathrm{Re}\,\langle 0|\Gamma_2|0\rangle = \mathrm{Re}\,\langle 0|\Gamma_1|0\rangle$ so that

$$\mathrm{Re}\,k = b \tag{5.9}$$

which is the statement that the change in the Coulomb integral of the surface atom compensates for the change in energy due to surface formation.

Let $b = 0$, then $\mathrm{Re}\,k = 0$ and from (5.2) the state is in the band so that $z = \exp(\mathrm{i}\,\varphi)$ and

$$x \geqq 1 \tag{5.10}$$

which is the opposite condition derived by Koutecky for the state outside the band. Equation (5.10) corresponds to states inside the band which are bonding contrary to the non-bonding character of the state outside the band. From equation (5.8), using the result [16] that

$$\frac{\mathrm{d}}{\mathrm{d}\beta_{rs}}\langle 0|\Gamma|0\rangle = \langle 0|W^{+}|r\rangle\,\langle s|W|0\rangle\,, \tag{5.11}$$

it follows that the bond order 0–1 is $|k|\,|z|$ and one should note that inside the band $|k| \neq \mathrm{Re}\,k$, so that while those surface states outside the band for which $\mathrm{Re}\,k = 0$ are non-bonded, their counterparts inside the band are definitely bonded.

Clearly the 'surface states' refer to the energy levels of the 'clean' surface. The chemisorption states, corresponding to atom -1, being a foreign atom, can be treated by the present approach if we let the adsorption be described by V, where

$$\langle -1|V_3|-1\rangle = c + \alpha \quad \text{and} \quad \langle -1|V_3|0\rangle = d\,. \tag{5.12}$$

Then $\langle -1|W|0\rangle = (e - c)^{-1}\,d$ and an extra term is present in Γ:

$$\langle 0|V_3|-1\rangle\,\langle -1|W|0\rangle = (e - c)^{-1}\,d^2\,. \tag{5.13}$$

Proceeding as in equations (5.8), (5.9), and (5.11) this makes a further contribution to $\mathrm{Re}\,k$. We are not able, therefore, to explain the selective properties of surfaces as consistency conditions on the parameters of the tight-binding Hamiltonian, since there is no reason to exclude the in-band solutions of the wave equation in the absence of experimental support.

Using the extended Hellmann-Feynman theorem [16] it is possible to evaluate the dependence of all relevant expectation values on the matrix elements of V_2 or V_3, in a manner analogous to equation (5.11). The success of similar methods in rationalizing reactivity data in a series of similar organic molecules suggests that it may be possible to correlate data on the adsorption states of a series of similar molecules on the same surface [19].

Acknowledgements

One of us (R.D.L.) would like to thank the Friends of the Hebrew University and the Humanitarian Trust for financial support, and Professor C. A. Coulson, F.R.S., for his hospitality in the Mathematical Institute, Oxford, where most of the work was carried out.

References

[1] K. M. WATSON, Phys. Rev. **105**, 1388 (1957).
K. M. WATSON and M. H. MITTLEMAN, Phys. Rev. **113**, 198 (1959).
[2] J. KOUTECKY, Adv. chem. Phys. **9**, 85 (1965).
[3] J. CALLAWAY, J. math. Phys. **5**, 784 (1964).
[4] S. TAKENO, J. chem. Phys. **44**, 853 (1966).
[5] YU. A. IZYNMOV, Adv. Phys. **14**, 569 (1965).
[6] K. KATASURA and M. INOKUTI, J. Phys. Soc. Japan **18**, 1486 (1963).
[7] H. MATSUDA and K. OKADA, Progr. theor. Phys. **34**, 539 (1965).
[8] E. I. RASHBA, Soviet Phys. — Solid State **4**, 2417 (1963).
[9] R. D. LEVINE, J. chem. Phys. **44**, 2035 (1966).
[10] A. MESSIAH, Quantum Mechanics, Part II, North-Holland Publ. Comp. 1962.
[11] G. F. KOSTER and J. C. SLATER, Phys. Rev. **95**, 1167 (1954).
G. F. KOSTER, Phys. Rev. **95**, 1436 (1954).
[12] B. A. LIPPMANN and J. SCHWINGER, Phys. Rev. **79**, 469 (1950).
[13] M. GELL-MANN and M. L. GOLDBERGER, Phys. Rev. **91**, 398 (1953).
[14] T. B. GRIMLEY, Adv. Catalysis **12**, 1 (1960).
[15] A. T. AMOS, Modern Quantum Chemistry III, Academic Press, 1965 (p. 265).
[16] R. D. LEVINE, Proc. Roy. Soc. **A294**, 467 (1966).
[17] P. O. LOWDIN, J. math. Phys. **3**, 969 (1962).
[18] C. A. COULSON and H. C. LONGUET-HIGGINS, Proc. Roy. Soc. **A 191**, 39 (1947).
[19] R. D. LEVINE, D. P. Thesis, University of Oxford, 1966, and to be published.
[20] R. D. LEVINE, Kinetics of Unimolecular Breakdown, VI, to be published.
[21] R. E. MERRIFIELD, J. chem. Phys. **38**, 920 (1963).
[22] P. W. ANDERSON, Phys. Rev. **124**, 41 (1961).
[23] P. A. WOLFF, Phys. Rev. **124**, 1030 (1961).
[24] A. M. CLOGSTON, Phys. Rev. **125**, 439 (1962).
[25] R. D. LEVINE, J. chem. Phys. **44**, 2029 (1966).
[26] A. S. DAVYDOV, Soviet Phys. — Uspekhi **7**, 145 (1964).
[27] A. T. AMOS and S. G. DAVISON, Physica **30**, 905 (1964).
[28] S. G. DAVISON and A. T. AMOS, J. chem. Phys. **43**, 2223 (1965).
M. TOMASEK, Surface Sci. **4**, 471 (1966).
[29] J. A. POPLE and S. H. WALMSLEY, Mol. Phys. **5**, 15 (1962).
[30] V. HEINE, Phys. Rev. **138**, A 1689 (1965).
J. D. LEVINE, Phys. Rev. **140**, A 586 (1965).

(Received November 11, 1966)

phys. stat. sol. 19, 597 (1967)

Subject classification: 13; 14.3; 22.1.1

Institut für Theoretische Physik der Universität Münster

Verteilungsfunktion und Driftgeschwindigkeit warmer Elektronen in Germanium[1]

Von

W. FRANZ und P. MIOSGA

Das in [1] und [2] beschriebene Verfahren, nach welchem die Verteilungsfunktion heißer Elektronen durch iterative Quadraturen bestimmt wurde, wird für den Fall kleiner elektrischer Felder modifiziert. Die Ergebnisse einer Rechnung für Germanium zeigen bei höheren Temperaturen befriedigende Übereinstimmung mit Messungen.

The iterative method described in [1] and [2] is modified for the special case of low electric fields. Numerical results for germanium agree fairly well with measurements made at moderate temperatures.

1. Berechnung der Verteilungsfunktion

Das in [1] und [2][2]) aufgestellte Gleichungssystem soll im folgenden verwendet werden, um die ersten Abweichungen vom Ohmschen Verhalten bei kleinen Feldstärken F zu berechnen. Im Gleichungssystem (I, 6) bis (I, 14) ist F ausschließlich in den Parametern B_i von Gleichung (I, 8) enthalten, welche zu F^2 proportional sind. Für $F = 0$ werden die Gleichungen durch die Boltzmannverteilung gelöst, wobei die Größen $S_i(x)$ und $\sigma_i(x)$ identisch Null werden. Bei kleinen Werten von F ergeben sich Abweichungen vom thermischen Gleichgewicht, welche zu F^2 proportional sind, was den Ansatz nahelegt

$$\chi_i(x) = \chi_0(x)\,[1 + \xi_i(x)]\,, \tag{1}$$

worin

$$\chi_0(x) = \frac{2}{\sqrt{\pi}\, t_0^{3/2}} \exp\left(-\frac{x}{t_0}\right) \tag{2}$$

die Boltzmannverteilung ist, so normiert, daß

$$\int_0^\infty \mathrm{d}x\,\sqrt{x}\,\chi_0(x) = 1\,.$$

Da die Größen S_i und σ_i proportional F^2 sind, kann man in Gleichung (I, 11) χ_i durch χ_0 ersetzen, so daß man hat

$$S_i(x) = \sigma_i(x)\,\chi_0(x)\,. \tag{3}$$

Setzt man Gleichung (1) und (3) in Gleichung (I, 6) ein und vernachlässigt

---

[1]) Auszug aus der Diplomarbeit von P. Miosga, Münster 1965.

[2]) Im folgenden bei Gleichungen zitiert als I, z. B. (I, 6).

Größen, die in F vom vierten Grade sind, so ergibt sich

$$\xi_i(x) = \int\limits_0^x \mathrm{d}y \left(\frac{w_i \tau \sqrt{y}}{t_0^2 y} B_i - \frac{\sigma_i}{t_0 y^2} \right) + C_i . \tag{4}$$

Aus Gleichung (I, 10) andererseits folgt (beachte, daß $\sigma_i(0) = 0$)

$$\chi_0(x)\, \sigma_i(x) = \Omega \sum_{\nu=2}^{3} \frac{w_\nu n_\nu}{w_2} \int\limits_{x-\omega_\nu}^{x} \mathrm{d}y \sqrt{y\,(y+\omega_\nu)}\, \chi_0(y)\, [\xi_i\,(y+\omega_\nu) - \xi_i(y)] -$$

$$- \Omega \frac{n_3 w_3}{(l-1)\, w_2} \left\{ \int\limits_0^x \mathrm{d}y \sqrt{y\,(y+w_3)}\, \chi_0(y) \left[l\, \xi_i\,(y+\omega_3) - \sum_{j=1}^{l} \xi_j\,(y+\omega_3) \right] + \right.$$

$$\left. + \int\limits_0^{x-\omega_3} \mathrm{d}y \sqrt{y\,(y+\omega_3)}\, \chi_0(y) \left[l\, \xi_i(y) - \sum_{j=1}^{l} \xi_j(y) \right] \right\} . \tag{5}$$

Mittelt man (4) und (5) über sämtliche Täler, so erkennt man, daß sich die Mittelwerte der Größen $\xi_i(x)$ und $\sigma_i(x)$ allein durch den Mittelwert der Größen B_i bestimmen, und zwar zu ihm proportional sind. Genauso ergibt sich, daß auch die Abweichung des Wertes ξ_i, σ_i vom Mittelwert proportional zu der Abweichung der Größe B_i vom Mittelwert ist. Man kann deshalb zur Lösung des Gleichungssystems den folgenden Ansatz machen:

$$\left. \begin{aligned} \xi_i(x) &= F^2 \left(\xi^{(1)}\,(x) + \varphi_i\, \xi^{(2)}(x)\right) , \\ C_i &= F^2 \left(C^{(1)} + \varphi_i\, C^{(2)}\right) , \\ \sigma_i(x) &= F^2 \left(\sigma^{(1)}(x) + \varphi_i\, \sigma^{(2)}(x)\right) . \end{aligned} \right\} \tag{6}$$

Darin ist

$$\varphi_i \equiv \frac{B_i}{\overline{B}} - 1 ; \qquad \overline{B} \equiv \frac{1}{l} \sum_{j=1}^{l} B_i \equiv B_0\, F^2 . \tag{7}$$

Die in Gleichung (7) eingeführten Abkürzungen bedeuten

$$B_0 \equiv \frac{e^2}{3\, \overline{m}\, u^2\, w_1^2} \frac{1}{3} \left(\frac{2}{m_{\mathrm{tr}}} + \frac{1}{m_{\mathrm{l}}} \right) = 3{,}5 \left(\frac{\mathrm{cm}}{\mathrm{kV}} \right)^2 , \tag{8}$$

$$\varphi_i = \frac{m_{\mathrm{l}} - m_{\mathrm{tr}}}{2\, m_{\mathrm{l}} + m_{\mathrm{tr}}} \frac{\boldsymbol{F}}{F} (I - 3\, \boldsymbol{e}_i\, \boldsymbol{e}_i) \frac{\boldsymbol{F}}{F} ; \quad \frac{m_{\mathrm{l}} - m_{\mathrm{tr}}}{2\, m_{\mathrm{l}} + m_{\mathrm{tr}}} = 0{,}46 . \tag{9}$$

Die beiden Gleichungssysteme für die Variabelnpaare $\xi^{(\nu)}$, $\sigma^{(\nu)}$ sind aus (4) und (5) unmittelbar abzulesen. Sie enthalten Feldstärke und Feldrichtung nicht mehr. — Die Konstante $C^{(1)}$ bestimmt sich aus der Forderung, daß die Gesamtzahl der Elektronen durch das Feld nicht verändert wird:

$$\int \mathrm{d}x \sqrt{x}\, \chi_0(x)\, \xi^{(1)}(x) = 0 . \tag{10}$$

Zur Bestimmung von $C^{(2)}$ hat man Gleichung (I, 13) zu benutzen:

$$\int\limits_0^\infty \mathrm{d}x \sqrt{x\,(x+\omega_3)}\, \chi_0(x) [\xi^{(2)}\,(x+\omega_3) + \xi^{(2)}(x)] = 0 . \tag{11}$$

Die numerischen Ergebnisse für die $\xi^{(\nu)}$ zeigt für einige Temperaturen Fig. 1.

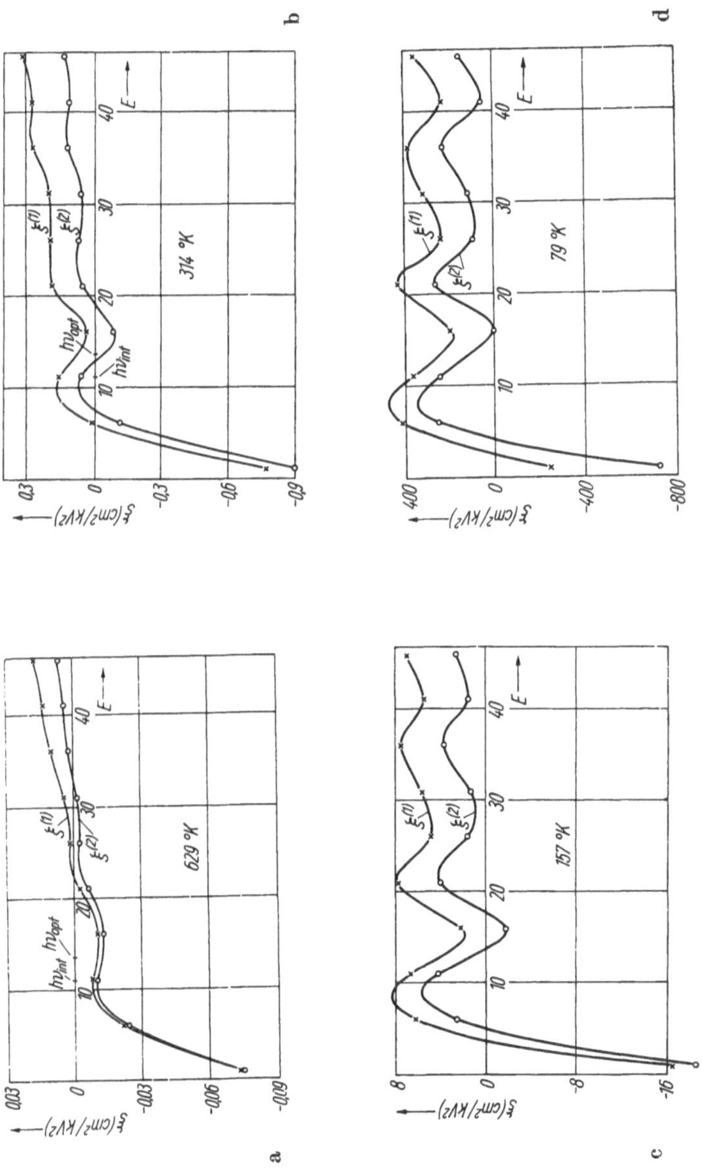

Fig. 1a, b, c, d. Korrekturen der Verteilungsfunktion

Aus den Gleichungen (6) und (9) ersieht man, daß sich für ein kaltes Tal ($\boldsymbol{F} \| \boldsymbol{e}_i$) die Beiträge von $\xi^{(1)}$ und $\xi^{(2)}$ zum großen Teil kompensieren, während sie sich für ein heißes Tal ($\boldsymbol{F} \perp \boldsymbol{e}_i$) verstärken.

2. Korrekturen an der Driftgeschwindigkeit

Setzt man Gleichung (6) in Gleichung (I, 45) ein und multipliziert mit der Feldstärke, so ergibt sich für die Driftgeschwindigkeit

$$\boldsymbol{v} = \frac{1}{l} \sum_{i=1}^{l} \boldsymbol{v}_i \tag{12}$$

mit

$$\boldsymbol{v}_i = P_0 \boldsymbol{F} \cdot \frac{m}{m_i^*} [1 + F^2 P^{(1)} + F^2 \varphi_i P^{(2)}] \,. \tag{13}$$

Dabei bedeutet

$$P_0 \equiv - \frac{2e}{m\,kT} \, \frac{\int\limits_0^\infty \mathrm{d}E \, E^{3/2} \, \tau \, \chi_0}{\int\limits_0^\infty \mathrm{d}E \sqrt{E} \, \tau \, \chi_0} \,, \tag{14}$$

$$P^{(\nu)} \equiv \frac{\int\limits_0^\infty \mathrm{d}E \, E^{3/2} \, \tau \, \chi_0 \left[\xi^{(\nu)} - kT \, \frac{\mathrm{d}\xi^{(\nu)}}{\mathrm{d}E} \right]}{\int\limits_0^\infty \mathrm{d}E \, E^{3/2} \, \tau \, \chi_0} \,. \tag{15}$$

Setzt man (13) in (12) ein und führt die Summation über die Täler explizit aus, so ergibt sich

$$\boldsymbol{v} = \beta_0 \left\{ (1 + F^2 P^{(1)}) \boldsymbol{F} + 2 \left(\frac{m_l - m_{tr}}{2 m_l + m_{tr}} \right)^2 P^{(2)} [F^2 \boldsymbol{F} - (F_x^3, F_y^3, F_z^3)] \right\} . \tag{16}$$

Für Germanium ist

$$2 \left(\frac{m_l - m_{tr}}{2 m_l + m_{tr}} \right)^2 = 0{,}43 \,. \tag{17}$$

β_0 ist die Ohmsche Beweglichkeit. Die eigentlich interessanten Größen sind die Komponente von $\boldsymbol{v}$ parallel zum Feld, $v_{||}$, und der Winkel φ zwischen Feld und Driftgeschwindigkeit:

$$v_{||} = \beta_0 F \left[1 + F^2 P^{(1)} + 4 \left(\frac{m_l - m_{tr}}{2 m_l + m_{tr}} \right)^2 F^2 P^{(2)} (\alpha^2 \beta^2 + \beta^2 \gamma^2 + \gamma^2 \alpha^2) \right] , \tag{18}$$

$$\varphi = 2 \left(\frac{m_l - m_{tr}}{2 m_l + m_{tr}} \right)^2 P^{(2)} F^2 \sqrt{\alpha^2 \beta^2 (\alpha^2 - \beta^2)^2 + \beta^2 \gamma^2 (\beta^2 - \gamma^2)^2 + \gamma^2 \alpha^2 (\gamma^2 - \alpha^2)^2} \,. \tag{19}$$

α, β und γ sollen hierin die Richtungskosinusse des Feldes gegen die kubischen Hauptachsen sein. φ verschwindet ersichtlich für die Richtungen (100), (110), (111), und nur für diese. Die Wurzel nimmt ihren Maximalwert von 0,28

Tabelle 1

Beweglichkeitskorrekturen

T (°K)	79	157	236	314	393	472	550	629
$P^{(1)} \left(\frac{\text{cm}}{\text{kV}}\right)^2$	97	4,67	0,684	0,203	0,088	0,0483	0,0309	0,0220
$P^{(2)} \left(\frac{\text{cm}}{\text{kV}}\right)^2$	328	8,44	1,134	0,324	0,135	0,0698	0,0440	0,0271

$\left(= \frac{1}{96} \sqrt{\frac{827 + 73\sqrt{73}}{2}}\right)$ etwa in der Mitte zwischen den Richtungen (100) und (111) an, ist aber auf einem Kegel vom Öffnungswinkel 22,5° um die (100)-Richtung stets $> 0{,}25$. Zwischen den Richtungen (110) und (111) bleibt die Wurzel $< 0{,}105 \left(= \frac{1}{96} \sqrt{\frac{827 - 73\sqrt{73}}{2}}\right)$. — Die errechneten Werte der beiden Korrekturgrößen $P^{(1)}$, $P^{(2)}$ zeigt Tabelle 1. Von T hängen diese Größen etwa wie T^{-4} ab. Die Rechnungen wurden so weit getrieben, daß die Prüfgröße (quadratisch gemittelte Abweichung der iterierten Lösung von der Lösung) $< 10^{-6}$ wurde. Lediglich bei der tiefsten Temperatur von 79 °K konvergierte das Verfahren so langsam, daß bei einer Prüfgröße von $0{,}6 \times 10^{-2}$ die Rechnung abgebrochen wurde, was auch deshalb sinnvoll erschien, weil bei diesen tiefen Temperaturen auch bei sehr reinem Material die Verunreinigungsstreuung nicht mehr vernachlässigt werden darf.

3. Vergleich mit Messungen

Die von uns berechneten und in [5] gemessenen Abweichungen vom Ohmschen Verhalten sind in Fig. 2 für 298 °K aufgetragen. Die Abweichung bei höheren Feldstärken erklärt sich daraus, daß dort F^4 nicht mehr vernachlässigt werden darf. Bei tieferen Temperaturen zeigen sich Abweichungen zwischen Theorie und Experiment, welche vermutlich darauf zurückzuführen sind, daß bei unserer Rechnung die Verunreinigungsstreuung nicht berücksichtigt wurde. Der Verlauf

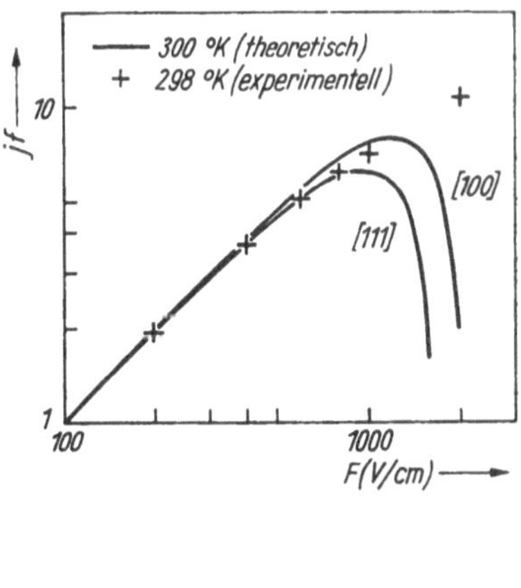

Fig. 2. Driftgeschwindigkeitskomponente in Feldrichtung für n-Ge. Messung nach [5; Fig. 26]

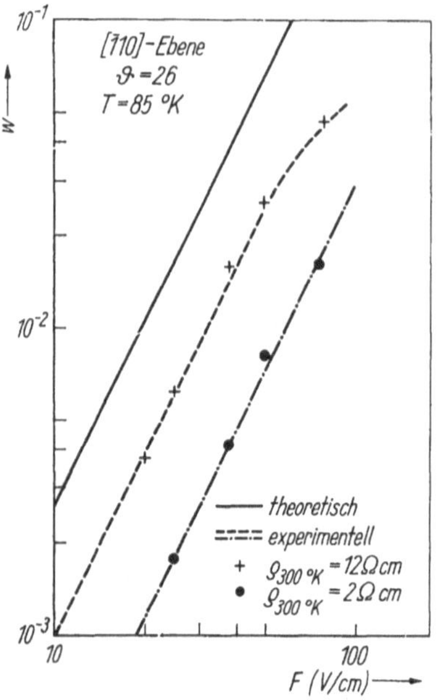

Fig. 3. Anisotropiewinkel für n-Ge. $T = 85$ °K, Meßwerte nach [5; Fig. 30a]

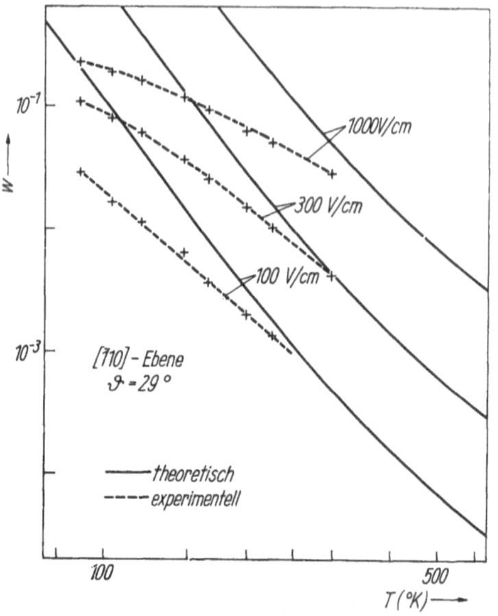

Fig. 4. Anisotropiewinkel für n-Ge. Meßwerte nach [5; Fig. 30c]

des Anisotropiewinkels als Funktion der Feldstärke (Fig. 3) zeigt, daß die Abweichung um so größer ist, je unreiner das Material ist. Aus Fig. 4 entnimmt man, daß — wie zu erwarten — die Abweichungen bei höherer Temperatur kleiner werden.

Die numerischen Rechnungen wurden auf den Anlagen Zuse Z23 der Universität Münster und Siemens S2002 des Instituts für Angewandte Mathematik der Universität Mainz durchgeführt. Für ihre Hilfe bei der Abwicklung der Rechnungen sei hier den Herren Dr. Roch, Durchholz (Münster) und besonders Herrn Braun (Mainz)gedankt.

Literatur

[1] W. FRANZ, Z. Naturf. **15a**, 366 (1960).
[2] W. FRANZ, phys. stat. sol. **1**, 600 (1961) (zitiert als I).
[3] W. FRANZ, phys. stat. sol. **3**, 1260 (1963).
[4] H. G. REIK, in Festkörperprobleme I, Vieweg, Braunschweig 1962.
[5] K. J. SCHMIDT-TIEDEMANN, in Festkörperprobleme I, Vieweg, Braunschweig 1962.

(Received November 11, 1966)

phys. stat. sol. **19**, 603 (1967)

Subject classification: 13.1

Institut für Theoretische Physik (II) der Universität Marburg/Lahn

Green's Function Method for Energy Band Calculations Including Spin–Orbit Coupling

By

J. TREUSCH

The Green's function method proposed by Korringa, and Kohn and Rostoker, is modified to include spin-orbit coupling. The formalism follows Korringa's work, and involves solving the two-dimensional Pauli equation with the condition that the wave function be continuous. The computer time needed is comparable to the case in which spin is not included, as the structure constants are the same in both cases.

Die von Korringa sowie Kohn und Rostoker entwickelte Methode der Green'schen Funktion zur Berechnung von Bandstrukturen wird so erweitert, daß die Spin-Bahn-Kopplung berücksichtigt wird. Ausgangspunkt der Rechnung ist die Pauli-Gleichung, die unter der Bedingung, daß die Wellenfunktion stetig sein soll, gelöst wird. Der Rechenaufwand ist nicht wesentlich höher als im spinlosen Fall, da dieselben Strukturkonstanten verwendet werden können.

1. Introduction

Lately the orthogonalized-plane-wave method [1], the augmented-plane-wave method [2], and the pseudopotential method [3] have been modified in order to include spin–orbit coupling or — more generally — relativistic effects into band-structure calculations. Since in future a comparison of methods will strongly depend on how rigorously they treat such effects, it seemed worthwhile to modify the Green's function method by Korringa [4] and Kohn and Rostoker [5], too.

Most recently and nearly at the same time Onodera and Okazaki [6] and Takada [7] published papers concerning a solution of the crystal Dirac equation by means of a Green's function. It is felt that the work presented here is a fruitful supplement to their rather laborious derivations. In contrast to them we start with the two-component wave equation derived from the Dirac equation by twice applying the Foldy-Wouthuysen transformation [8]. In this way we are enabled to treat the spin–orbit coupling, the Darwin term, and the mass-velocity term separately, thus getting better insight into the nature of bands. Moreover, our derivation runs along the lines of Korringa's scattering method which is advantageous in two respects: Firstly, the theory is by far more perspicuous, and secondly, we get the additional result that the wave function is continuous with continuous derivatives.

The mathematical formalism of the KKR method shall not be repeated here. The following work is to be seen as a continuation of the paper by Treusch and Sandrock [9] (hereafter referred to as TS); therefore and to facilitate reading Korringa's and Kohn and Rostoker's formulae used here are taken from TS.

2. Theory

As usual in the KKR method a muffin-tin potential shall be assumed. The two-component wave equation to be solved is then [8, 10] (in atomic units; correct to order α^2, where α is the fine-structure constant)

$$\left[-\Delta + V(r) - \frac{\alpha^2}{4}\nabla^4 + \frac{\alpha^2}{8}\Delta V(r) + \frac{\alpha^2}{4}\frac{1}{r}\frac{\mathrm{d}V}{\mathrm{d}r}\boldsymbol{L}\cdot\boldsymbol{\sigma}\right]\Psi(\boldsymbol{r}) = E\,\Psi(\boldsymbol{r}) \tag{1}$$

with

$$\boldsymbol{L} = -i\,(\boldsymbol{r}\times\nabla)\,. \tag{2}$$

We will denote

$$\frac{\alpha^2}{4}\frac{1}{r}\frac{\mathrm{d}V}{\mathrm{d}r} = \Phi(r)\,, \tag{3}$$

where

$$\Phi(r) = 0 \quad \text{for} \quad r > \varrho \tag{4}$$

because of the muffin-tin form of the potential.

For sake of clarity and since they do not split terms, the mass-velocity term and the Darwin term are suppressed for the present. Later on they can easily be included by substituting ∇^4 through $(E_0 - V(r))^2$, where E_0 is the nonrelativistic energy, the substitution being correct to order α^2 [10]. Both terms only affect the radial part of the wave function and, therefore, yield no complication.

A solution of equation (1) is given by

$$\Psi_{\boldsymbol{k}}(\boldsymbol{r}) = \int_{r'\leqq\varrho} G_{\boldsymbol{k}}(\boldsymbol{r},\boldsymbol{r}')\,[V(r') + \Phi(r')\,\boldsymbol{L}\cdot\boldsymbol{\sigma}]\,\Psi_{\boldsymbol{k}}(\boldsymbol{r}')\,\mathrm{d}\boldsymbol{r}'\,, \tag{5}$$

with

$$(\Delta + E)\,G_{\boldsymbol{k}}(\boldsymbol{r},\boldsymbol{r}') = \sum_n \exp(i\,\boldsymbol{k}\,\boldsymbol{R}_n)\,\delta(\boldsymbol{r}-\boldsymbol{r}'-\boldsymbol{R}_n)\,. \tag{6}$$

Following the lines of Korringa [4] this yields

$$\Psi_{\text{out}}(\boldsymbol{r}) = \int_{r'=\varrho}\left[G(\boldsymbol{r},\boldsymbol{r}')\,\frac{\mathrm{d}\Psi_{\text{in}}}{\mathrm{d}r'}(\boldsymbol{r}') - \frac{\mathrm{d}G(\boldsymbol{r},\boldsymbol{r}')}{\mathrm{d}r'}\,\Psi_{\text{in}}(\boldsymbol{r}')\right]\mathrm{d}\boldsymbol{r}'\,. \tag{7}$$

A suitable trial function solving equation (1) exactly is [8, 11]

$$\Psi_{\text{in}}(\boldsymbol{r}) = \sum_{\lambda,\mu} C_{\lambda\mu}\,R_\lambda(r)\,\chi^\mu_\lambda(\boldsymbol{r})\,, \tag{8}$$

where the central field orbitals are given by

$$\chi^\mu_\lambda(\boldsymbol{r}) = \sum_{m_s=\pm 1/2} C\left(l\,\frac{1}{2}\,j;\,\mu - m_s,\,m_s\right) Y_l^{\mu-m_s}(\boldsymbol{r})\,\chi^{m_s} \tag{9}$$

with

$$j = l - \frac{1}{2}\operatorname{sign}\lambda \tag{10a}$$

and

$$\begin{aligned} l &= \lambda \qquad &&\text{for}\quad \lambda > 0\,,\\ l &= -\lambda - 1 \qquad &&\text{for}\quad \lambda < 0\,. \end{aligned} \tag{10b}$$

μ is the eigenvalue of j_z, the $\chi^{\pm 1/2}$ are the Pauli spinors, and $R_\lambda(r)$ solves the radial equation

$$\left[\Delta_r + E - V(r) - \frac{l(l+1)}{r^2} + \Phi(r)(1+\lambda)\right] R_\lambda(r) = 0\,, \tag{11}$$

since

$$\boldsymbol{L}\cdot\boldsymbol{\sigma}\,\chi_\lambda^\mu = -(1+\lambda)\,\chi_\lambda^\mu\,. \tag{12}$$

The $C\left(l\,\frac{1}{2}\,j;\,\mu - m_s, m_s\right)$ are the well-known Clebsch-Gordan coefficients. For the same reason as in TS (equation (18), etc.) the trial function is written as

$$\Psi_{\text{in}}(\boldsymbol{r}) = \sum_{\substack{-N-1\leq\lambda\leq N\\ \lambda\neq 0\\ \mu}} C_{\lambda\mu}\,R_\lambda(r)\,\chi_\lambda^\mu(\boldsymbol{r}) + \sum_{\substack{\lambda<-N-1\\ \lambda>N\\ \mu}} C_{\lambda\mu}\,j_{l(\lambda)}(\varkappa r)\,\chi_\lambda^\mu(\boldsymbol{r})\,. \tag{13}$$

This expansion is justified by the fact that there exists an N such that for $l > N$

$$[V(r) - \Phi(r)(1+\lambda)] \ll \frac{l(l+1)}{r^2}\,, \tag{14}$$

the radial solutions then being spherical Bessel functions $j_l(\varkappa r)$ (with $\varkappa = \sqrt{E}$) independent of whether λ is positive or negative.[1])

The Green's function can be expanded as

$$G(\boldsymbol{r},\boldsymbol{r}') = \sum_{l\,m}\sum_{l'\,m'} [A_{lml'm'}\,j_l(\varkappa r)\,j_{l'}(\varkappa r') + \varkappa\,\delta_{ll'}\,\delta_{mm'}\,n_l(\varkappa r)\,j_l(\varkappa r')]\;Y_{lm}(\boldsymbol{r})\;Y^*_{l'm'}(\boldsymbol{r}') \tag{15}$$

for $r > r'$.

In order to make compatible the expansions (13) and (15), the summation over λ and μ in (13) is replaced by a summation over l and m, making use of (10b) and setting $\mu - m_s = m$. $C_{\lambda\mu}$ and $R_\lambda(r)$ are called C^j_{lm} and $R^j_l(r)$ with $j = 1, 2$ for $\lambda \lessgtr 0$, respectively, the m-summation runs over

$$\begin{aligned} -l-1 &\leqq m \leqq l \qquad &&\text{for}\quad j = 1\,,\\ -l\quad &\leqq m \leqq l-1 &&\text{for}\quad j = 2\,. \end{aligned} \tag{16}$$

The spherical harmonics belonging to one spinor are Y_l^m for the upper, Y_l^{m+1} for the lower component.

Substituting (15) and (13) in the rewritten form into (7), and then using the condition (TS, equation (16)) of continuity of the wave function, one gets a system of linear equations for the $C^j_{l'm}$, which is soluble, if the following determinant equals zero:

$$\det\left|\left\{A_{lm;\,l'm'-m_s+1/2} + \varkappa\,\delta_{ll'}\,\delta_{m\,m'-m_s+1/2}\,\frac{n_l'(\varkappa r) - n_l\,L_l^j(r)}{j_l'(\varkappa r) - j_l\,L_l^j(r)}\bigg|_{r=\varrho}\right\}\times\right.$$
$$\left.\times\; C\left(l'\,\frac{1}{2}\,j;\,m', m_s\right)\right| = 0\,. \tag{17}$$

---

[1]) Since $\Phi(r)$ is of the character $1/r^3$ that seems impossible, but one most keep in mind that there is no sense in taking r smaller than e.g. the Compton wavelength, due to physical and computational reasons.

$A_{lm;\,l'm'}$ are the structure constants of the spinless case, $L_l^j(r) = (\mathrm{d}R_l^j(r)/\mathrm{d}r)/R_l^j(r)$. l' runs from 0 to N, the m'-summation obeying (16), l and m run in the usual sense with $l \leqq N$, j takes the values 1 and 2, indicating whether $\lambda < 0$ or $\lambda > 0$, and m_s takes the values $+1/2$ and $-1/2$, denoting the upper and lower component of the wave function. Thus, by the double-valued indices j and m_s, the dimension of the secular determinant is doubled as compared to the spinless case. The C_{lm}^j for $l > N$ are given by two linear equations corresponding to (TS equation (21)). They couple C_{lm}^1 and C_{lm}^2 and can easily be set up following equations (20) and (21) of TS.

3. Conclusion

The main results of this paper are 1. that the KKR method — esteemed because of its mathematical rigor and good convergence — is able to include spin-orbit coupling from first principles, 2. that the computational effort is not much larger than that for the spinless calculation, since the same structure constants can be used.

To show the connexion to the work of Onodera [6] and Takada [7], we transform our resultant secular equation (17) as follows: Multiplying the (l, m, m_s) row by $C\left(l\,\frac{1}{2}\,j;\, m,\, m_s\right)$ and adding to it the $(l, m + 2\,m_s, -m_s)$ row times $C\left(l\,\frac{1}{2}\,j;\, m,\, -m_s\right)$ with $j = 1, 2$ for $m_s = \pm\frac{1}{2}$, respectively, one gets an equivalent secular equation of the form

$$\det\left|B_{\lambda\mu\lambda'\mu'} + \varkappa\;\delta_{\lambda\lambda'}\,\delta_{\mu\mu'}\,\frac{n_l' - n_l\,L_\lambda}{j_l' - j_l\,L_\lambda}\right| = 0\,, \tag{18}$$

with

$$\mu - m_s = m \qquad \text{and} \qquad l = l(\lambda) \quad \text{as equation} \quad (10\,\text{b}).$$

The "compound" structure constants $B_{\lambda\mu\lambda'\mu'}$ are the same as in [6; equation (A14)] and in [7; equation (35)]. In view of the different starting points which led us to a scalar Green's function in contrast to Onodera's and Takada's four-component spin-dependent Green's function, this equality might seem surprising. It is a consequence of the fact that terms of order c^{-2} were neglected in the calculation of $B_{\lambda\mu\lambda'\mu'}$ by Onodera and Takada. The difference between the secular equations is to be found in the radial parts of the wave function. In [6] and [7] these are the solutions of two coupled differential equations originating from the Dirac equation. Therefore, they do not suffer from the neglects made in connexion with the Foldy-Wouthuysen transformation. But we doubt if the accuracy obtained there is not destroyed by the neglect mentioned above concerning the compound structure constants.

Moreover, there is a great practical advantage in making use of equation (17). As described earlier [5, 9], the computing technique runs as follows: 1. calculation and storage of structure constants which are independent of the potential used, 2. calculation and storage of radial solutions, 3. setting up and evaluating the secular determinants. The second and third step take considerably less computer time than the first. Therefore, it is easy to vary the radial solution and to produce the corresponding band structures. This can be done by changing the potential and, as already mentioned, by taking into account spin-orbit coupling, the Darwin term and the mass-velocity term separately, thus including straightforwardly the comparison to the spinless energy bands. This way one will get a better understanding of the influence that the potential and the relativistic terms have on certain properties of the band structure. Starting with the Dirac equation one looses these possibilities.

At any rate we are sure that the KKR method in either modification to include spin will bear comparison with the other spin-including methods developed recently.

Acknowledgement

The author wishes to thank Professor O. Madelung for his continuous support and encouragement.

References

[1] P. Soven, Phys. Rev. **137**, A 1706 (1965).
[2] T. L. Loucks, Phys. Rev. **139**, A 1333 (1965).
[3] A. O. E. Animalu, Phil. Mag. **13**, 53 (1966).
[4] J. Korringa, Physica **13**, 392 (1947).
[5] W. Kohn and N. Rostoker, Phys. Rev. **94**, 1111 (1954).
[6] Y. Onodera and M. Okazaki, J. Phys. Soc. Japan **21**, 1273 (1966).
[7] S. Takada, Progr. theor. Phys. **36**, 224 (1966).
[8] M. E. Rose, Relativistic Electron Theory, Chap. 7, 8, 22, Wiley, New York 1961.
[9] J. Treusch and R. Sandrock, phys. stat. sol. **16**, 487 (1966).
[10] F. Herman, C. D. Kuglin, K. F. Cuff, and R. L. Kortum, Phys. Rev. Letters **11**, 541 (1963).
[11] M. E. Rose, Elementary Theory of Angular Momentum, Chap. 10, 30, Wiley, New York 1957.

(Received November 3, 1966)

phys. stat. sol. 19, 609 (1967)

Subject classification: 1.2; 8; 14.3; 22.8;

Institute of Applied Physics of the Academy of Sciences of the MSSR,
Kishinev Polytechnical Institute

Influence of Ordering on the Properties of Solid Solutions of the System $(HgSe)_{3x}$–$(In_2Se_3)_{1-x}$

By

E. I. Gavrilitsa and S. I. Radautsan

Some results are given of an investigation of the transition effect in solid solutions of the system $(HgSe)_{3x}$-$(In_2Se_3)_{1-x}$. The concentration and thermal ordering of the atomic components and defects in the cation sublattice are found to have an important effect on the electron mobility and lattice thermal conductivity.

Приводятся некоторые результаты исследования явлений переноса в твёрдых растворах системы (($(HgSe)_{3x}$-$(In_2Se_3)_{1-x}$. Обнаружено существенное влияние концентрационного и температурного упорядочения атомов компонентов и дефектов в катионной подрешётке на подвижность электронов и решёточную теплопроводность.

For the system $(HgSe)_{3x}$–$(In_2Se_3)_{1-x}$ the authors have shown, by X-ray, microstructural, and thermal studies, that a homogeneous region exists in the composition range $x = 1$ to 0.5, its alloys having diamond-like structure with varying defect concentration [1, 2]. Two systems of lines in X-ray powder photographs corresponding to $Hg_3In_2Se_6$ ($x = 0.5$) and $HgIn_2Se_4$ ($x = 0.25$) show that samples with $x = 0.49$ to 0.3 contain two phases. For alloys with more than 75 mol% In_2Se_3 the X-ray photographs exhibit a large number of lines which do not correspond to any diamond-like structure. Thermal and microstructural analyses confirm that two phases occur in the above mentioned regions.

Later on physical properties have been studied only for samples with $x = 1$ to 0.5 and 0.25.

Ordering solid solutions have been found in the homogeneous region $x = 1.0$ to 0.5 (Fig. 1). Slowly cooled samples of composition 25 to 45 mol% In_2Se_3 have an ordered structure with a cubic unit cell whose edges are twice as long as those of the unit cell of zinc blende. Maximum intensity of the superstructural lines of ordering is observed in the X-ray powder photographs for the composition $Hg_5In_2Se_8$ ($x = 0.625$). The symmetry group of this alloy is found to be $F\bar{4}3m$, with the reliability factor of 15.3%.

The samples of the phase $HgIn_2Se_4$, for any method of preparation used by the authors, possess a thiogalate type ordered structure. In both types of ordering Se atoms are slightly shifted from the positions of closest packing towards vacancies [2, 3].

In the present paper the authors give some results of the investigation of transport phenomena in these alloys which were carried out to reveal the influence of the concentration and temperature ordering of the atoms and of the

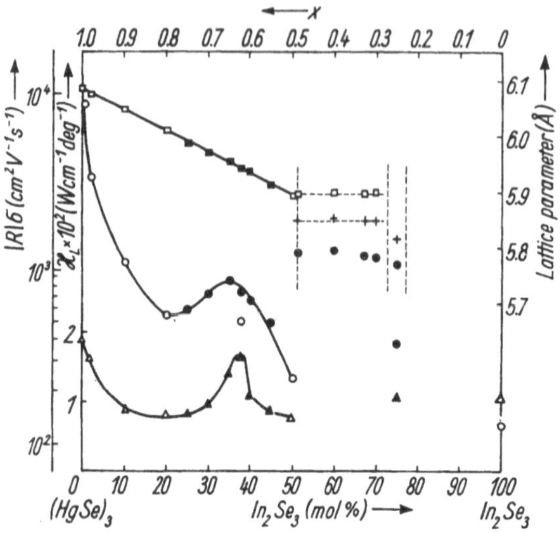

Fig. 1. Dependence of lattice parameters, electron mobility, and lattice thermal conductivity on the composition of $(HgSe)_{3x}$-$(In_2Se_3)_{1-x}$ solid solutions. Electron mobility ($R\,\sigma$) at $T = 100$ °K: ● ordered, ○ disordered solid solutions; lattice thermal conductivity (χ_L) at $T = 300$ °K: ▲ ordered, △ disordered solid solutions; lattice parameters: □ ZnS type phase, ■ ordered cubic phase, + a thiogalate phase, ⊕ $c/2$ thiogalate phase

defects in the cation sublattice on the physical properties of these substance [4, 5].

All the measurements were made on monocrystalline samples prepared by a modification of the Bridgman method.

All samples showed n-type conductivity. The electron concentration slowly decreases with x in the range 1.0 to 0.8 (2.8 to 1.7×10^{19} cm^{-3}). The samples with ordered structure ($x = 0.75$ to 0.55) have approximately the same electron concentration, which is of the order 5×10^{18} cm^{-3}. The compositions $x = 0.5$ and $x = 0.25$ were studied for samples with electron concentrations 8×10^{17} cm^{-3} and 1.3×10^{19} cm^{-3}, respectively.

The dependence of the Hall mobility ($R\,\sigma$) at $T = 100$ °K and the lattice component of the thermal conductivity (χ_L) at $T = 300$ °K, which characterize the influence of concentration ordering on the properties of the studied alloys [6], are shown in Fig. 1.

The rapid decrease of the electron mobility in the region $x = 1.0$ to 0.8 is obviously determined mainly by the electron scattering on randomly distributed atoms and intrinsic defects. In addition, this decrease is partly due to the fact that the electron mobility in In_2Se_3 is considerably less than in HgSe.

The dependence of the electron mobility on composition in the ordered region shows a maximum corresponding to the alloys with the highest (thermodynamically possible) degree of order. Thus ordering in the lattice leads to an increase in electron mobility.

The observed dependence of the lattice thermal conductivity on the compositions of the solid solutions (Fig. 1) may be explained, in general, by the existence of two mechanisms of phonon scattering acting simultaneously. Phonon scattering by defects is dominant. The second mechanism is due to perturbations of short-range order which take place in disordered solid solutions as a result of the random distribution of the atomic components. These perturbations may greatly contribute to the scattering of phonons because their wavelengths at room temperature are equal to several interatomic distances. The large increase

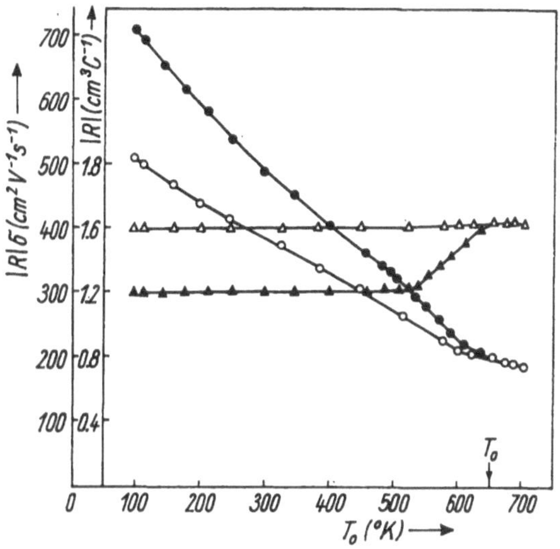

Fig. 2. Temperature dependence of the electron mobility ($R\sigma$) and Hall constant (R) of $Hg_5In_2Se_8$ crystal

$R\sigma$ {● ordered state, ○ disordered state}

R {▲ ordered state, △ disordered state}

of the lattice thermal conductivity in the ordered region, in which a high degree of short-range and long-range order is established, may serve as an evidence for this explanation.

To investigate the influence of thermal ordering we studied samples having the same composition but different degrees of order which were obtained by different heat treatments. Thus, for example, samples with $x = 0.625$ were ordered by slow cooling (at a rate of 10°/h) and were disordered by quenching from 420 °C.

Fig. 2 shows the temperature dependences of $R\sigma$ and R for one sample of given composition before and after quenching . The measurements of the electrical properties of the specimen before quenching were performed in such a way that equilibrium of short- and long-range order was established at each temperature. In contrast to this, the measurements after quenching were performed rapidly (using a semi-automatic arrangement). One may determine the temperature of order–disorder transition (T_0) from the intersection points of the curves measured after different heat-treatments. From Fig. 2 the transition temperature is found to be 650 °K. At this temperature, in the curves for $x = 0.625$ a very weak thermal effect is observed.

References

[1] S. I. Radautsan and E. I. Gavrilitsa, Izv. Akad. Nauk Mold SSR **10**, 95 (1961).

[2] S. I. Radautsan, E. I. Gavrilitsa, O. P. Derid, M. M. Markus, and R. A. Maslyanko, Proc. XX. Congress IUPAC, D 47, 93, Moscow 1965.

[3] H. Hahn, G. Frank, W. Klingler, A. D. Störger, and G. Störger, Z. anorg. allg. Chem. **279**, 241 (1955).

[4] T. Muto and Y. Takagi, The Theory of Order-Disorder Transitions in Alloys, Izd. IL, Moscow 1959 (in Russian).

[5] M. A. Krivoglaz and A. A. Smirnov, The Theory of Ordering Alloys, Izd. fiziko-matem. lit., Moscow 1958 (in Russian).

[6] Z. G. Pinsker, Kristallografiya **5**, 627 (1960).

(Received July 23, 1966)

phys. stat. sol. 19, 613 (1967)

Subject classification: 18.2

Polytechnical Institute, Timisoara, and Agronomic Institute, Timisoara

Contributions to the Mechanism of the Procopiu and Inverse Wiedemann Effects

By

B. F. Rothenstein, A. Policec, C. Anghel, and M. Lupulescu

A study is made of the form of the electromotive force induced by the Procopiu effect or by the inverse Wiedemann effect in a search coil surrounding the ferromagnetic specimen. This provides information regarding the relative positions of the specimen axis and easy axis, and the torsion of the specimen.

Durch den Procopiu-Effekt oder den inversen Wiedemann-Effekt wird in einer die Probe umgebenden Sondenspule eine EMK induziert. Es wird gezeigt, daß die Form dieser Spannung Aufschluß gibt über die relative Lage von Probenachse und Richtung leichter Magnetisierung, sowie über die Torsionsverformung der Probe.

1. Introduction

The Procopiu effect consists in the induction of an electromotive force E_i in a coil which surrounds a cylindrical ferromagnetic specimen carrying an alternating current, and which is subjected to a stationary longitudinal magnetic field and zero external strain [1 to 5]. The dependence of the induced e.m.f. E_i upon the intensity of the stationary longitudinal magnetic field $\boldsymbol{H}_y$ and upon the amplitude $\boldsymbol{H}_{0,x}$ of the oscillatory circular magnetic field caused by the alternating current in the wire [6 to 10], the sensibility of the effect to applied strain [2, 3, 11], to temperature [12], to the parameters of the circuit which picks up the induced e.m.f. [13], and to the thickness in the case of thin ferromagnetic films [14] has been studied.

We know two recent attempts to give a theory of the Procopiu effect [10, 15] by means of the Akulov formula [16]

$$\boldsymbol{J} = \boldsymbol{J}_s\left(1 - \frac{a_1}{H} - \frac{a_2}{H^2}\right) \tag{1}$$

which expresses the dependence of the magnetization vector $\boldsymbol{J}$ upon the intensity of the magnetic field $\boldsymbol{H}$ by means of the constants a_1 and a_2. $\boldsymbol{J}_s$ represents the saturation value of $\boldsymbol{J}$.

The effect can be studied using a very sensitive galvanometer or an oscilloscope which records the time variation of the induced e.m.f. or the time variation of the magnetic flux through the surface limited by the coil.

The inverse Wiedemann effect consists in the induction of an electromotive force in a coil which surrounds a ferromagnetic wire performing torsional oscillations in a static axial magnetic field [17 to 19].

2. Mechanism of the Procopiu and Inverse Wiedemann Effects

2.1 The easy axis coincides with the geometrical axis

We admit that the studied sample is a cylindrical film and presents an easy axis which coincides with its geometrical axis. The anisotropy of the sample can be characterized by the anisotropy constant K. The energy stored at a given moment in the volume unit of the specimen is given by [20]

$$W = K \sin^2 \alpha - \boldsymbol{J}_s \boldsymbol{H}_y \cos \alpha - \boldsymbol{J}_s \boldsymbol{H}_x \sin \alpha \tag{2}$$

in the case of the Procopiu effect and by

$$W = K \sin^2 \alpha + \frac{3}{2} \lambda_s \sigma \sin^2 (45^\circ - \alpha) - \boldsymbol{J}_s \boldsymbol{H}_y \cos \alpha \tag{3}$$

in the case of the inverse Wiedemann effect, where $\boldsymbol{H}_y$ is the intensity of the stationary magnetic field, $\boldsymbol{H}_{x,0}$ the amplitude of the oscillatory magnetic field with frequency f, $\boldsymbol{H}_x = \boldsymbol{H}_{x,0} \sin 2\pi f t$ its momentary value, σ the applied shear strain, and λ_s the saturation value of the magnetostriction.

The equilibrium value of the angle α is determined by $\partial W/\partial\alpha = 0$ and $\partial^2 W/\partial\alpha^2 > 0$. $\partial W/\partial\alpha = 0$ and $\partial^2 W/\partial\alpha^2 = 0$ represent straight lines. Their crossing point separates the region of stable equilibrium from the region of unstable equilibrium. The geometrical locus (critical curve) of the crossing points is an astroide in the case of the Procopiu effect [21]:

$$h_x^{2/3} + h_y^{2/3} = 1\,, \tag{4}$$

where $h_x = \boldsymbol{H}_x \boldsymbol{J}_s/2K$ and $h_y = \boldsymbol{H}_y \boldsymbol{J}_s/2K$. In the case of the inverse Wiedemann effect the critical curve, represented in Fig. 1, is given by

$$h_s = \frac{\sin^3 \alpha}{\cos \alpha\,(1 + 2\sin^2 \alpha)}\,, \qquad h_y = -\frac{1}{\cos \alpha\,(1 + 2\sin^2 \alpha)}\,, \tag{5}$$

where

$$h_s = \frac{3}{2} \frac{\lambda_s \sigma}{2K}\,.$$

In the Procopiu effect the intensity of the stationary longitudinal magnetic field h_y is maintained at a constant value. $h_{x,c}$ is the intensity of the oscillatory circular magnetic field which determines with h_y a point of the astroide.

As long as $h_x < h_{x,c}$, $\boldsymbol{J}_s$ can be oriented along the directions $90^\circ > \alpha > 0^\circ$ and $180^\circ > \alpha > 90^\circ$ with the specimen axis. As along as $h_x > h_{x,c}$, $\boldsymbol{J}_s$ can be oriented after a unique direction with the specimen axis $90^\circ > \alpha > 0^\circ$. If the value $h_{x,c}$ is exceeded, the system fails to return in the state with two possible orientations of $\boldsymbol{J}_s$ towards the specimen axis, and the time variation of h_x determines the time variation of the magnetic flux Φ_y through the surface limited by the coil.

The time variations of h_x and Φ_y, plotted on the basis of the proposed

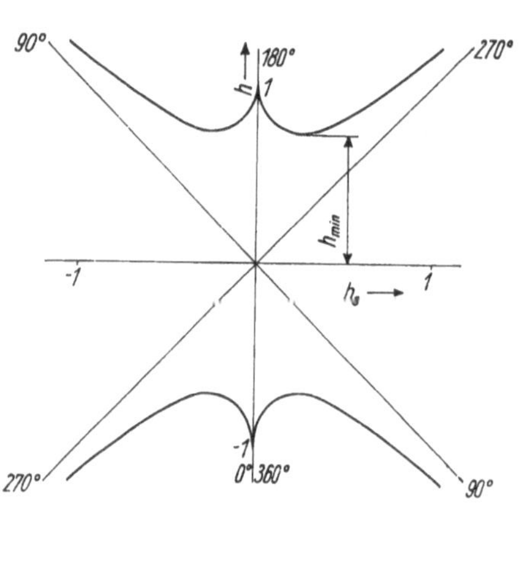

Fig. 1. The critical curve in the case that the easy axis coincides with the geometrical axis of the sample (inverse Wiedemann effect)

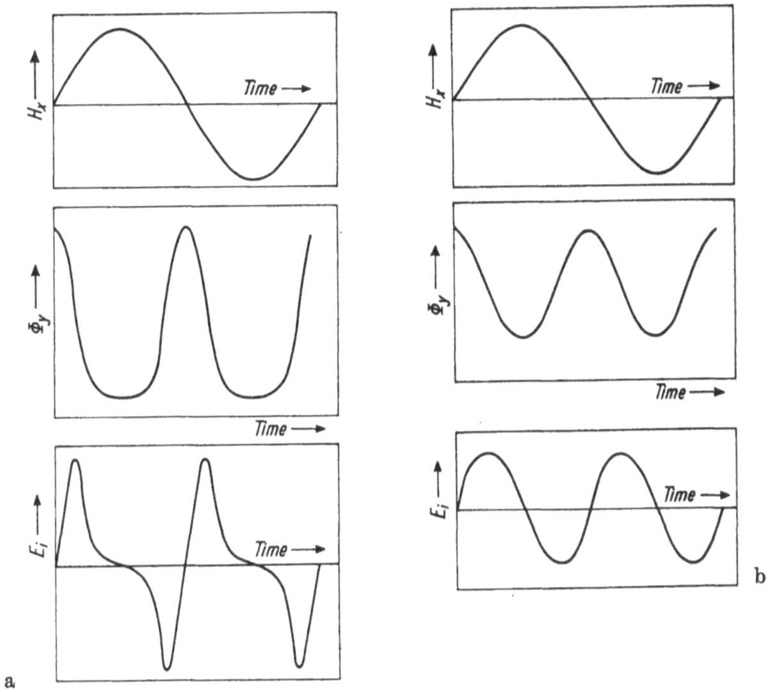

Fig. 2. Time variation of the intensity of the oscillatory circular magnetic field h_x, of the magnetic flux Φ_y through the surface limited by the coil, plotted on the basis of the proposed model, and of the e.m.f. E_i induced in the coil obtained by graphic derivation of Φ_y (Procopiu effect).

a) $h_{x,0} = 2$, $h_y = 0.1$; b) $h_{x,0} = 1$, $h_y = 0.3$

model as well as of E_i obtained by graphic derivation of Φ_y together with the phase relations are given in Fig. 2.

In the inverse Wiedemann effect we distinguish the following characteristic situations:

The ferromagnetic sample is in a demagnetized state. A periodic external shear strain of amplitude $h_{s,0}$ and frequency f_s, $h_s = h_{s,0} \sin 2\pi f_s t$, is applied. The phenomenon does not induce an e.m.f. in the search coil, because the macroscopic variation of the magnetic flux is equal to zero.

The ferromagnetic sample is in a state of remanent magnetization. The magnetization vector is displaced along the axis of the sample. The periodic external shear strain causes a change in the magnetic flux which induces an e.m.f. in the coil.

Starting from the demagnetized state, an axial static magnetic field $h_y < h_{min}$ is applied. If a periodic external shear strain is applied, the point which characterizes the magnetization states of the sample is always located on a straight line which is parallel to the h_s-axis and does not intersect the critical curve. In each point of the straight line the magnetization vectors can be oriented along two different directions inclined at ($0^\circ < \alpha < 90^\circ$) and ($180^\circ < \alpha < 270^\circ$) respectively to the specimen axis. The time variation of h_s determines the time variation of the magnetic flux through the surface limited by the coil.

Starting from the demagnetized state, an axial magnetic field $h_y > h_{min}$ is applied. Let $h_{s,c}$ be the value of the shear strain which determines with h_y

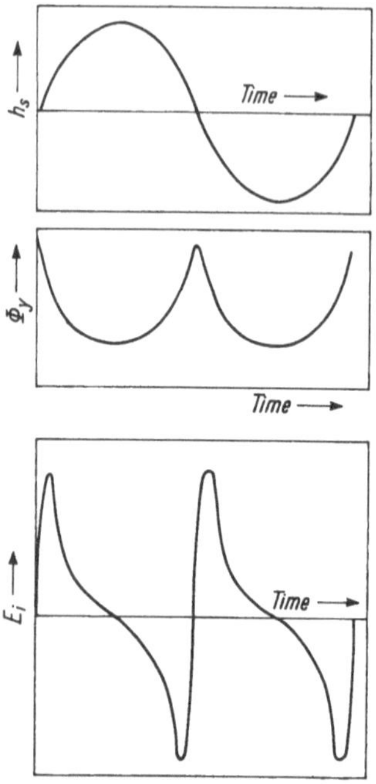

Fig. 3. Time variation of the shear strain h_s, of the magnetic flux Φ_y through the surface limited by the coil, plotted on the basis of the proposed model, and of the e.m.f. E_i induced in the test coil obtained by graphic derivation of Φ_y, in the case that $h_{s,0} > h_{s,c}$ and $h_y > h_{min}$ (inverse Wiedemann effect)

a point of the critical curve. If the value $h_{s,c}$ is exceeded, the system fails to return in the state with two possible orientations of $\boldsymbol{J}_s$ towards the specimen axis and the time variation of h_s determines the time variation of the magnetic flux through the surface limited by the coil. The time variation of h_s, of the magnetic flux Φ_y through the surface limited by the coil plotted on the basis of the proposed model, and of the e.m.f. E_i induced in the test coil obtained by graphic derivation of Φ_y are shown in Fig. 3 in the case that $h_{s,0} > h_{s,c}$ and $h_y > h_{min}$.

2.2 The easy axis is inclined at an angle ψ towards the specimen axis

We admit now that the easy axis is inclined at an angle ψ and presents cylindrical symmetry towards the specimen axis. The energy stored at a given moment in the volume unit of the specimen is given by

$$W = K \sin^2 (\alpha - \psi) - \boldsymbol{H}_y \boldsymbol{J}_s \cos \alpha - \boldsymbol{H}_x \boldsymbol{J}_s \sin \alpha \tag{6}$$

in the case of the Procopiu effect and by

$$W = K \sin^2 (\alpha - \psi) - \boldsymbol{H}_y \boldsymbol{J}_s \cos \alpha + \frac{3}{2} \lambda_s \sigma \sin^2 (45^\circ - \alpha) \tag{7}$$

in the case of the inverse Wiedemann effect.

In the case of the Procopiu effect the critical curve is an astroide as in Fig. 4. In the case of the inverse Wiedemann effect, the critical curve represented in Fig. 5 is given by

$$h_s = \frac{\cos 2\psi \sin^3 \alpha}{\cos \alpha \, (1 + 2 \sin^2 \alpha)} - \frac{\sin 2\psi}{2}, \qquad h_y = - \frac{\cos 2\psi}{\cos \alpha \, (1 + 2 \sin^2 \alpha)}. \tag{8}$$

In order to deduce the possible features of the induced e.m.f. in the Procopiu effect we consider the following cases:

a) $0 \leqq h_y < \sin \psi$

Due to the rotation of the critical curve we have the following sequence of the phenomena:

If h_x decreases from positive values $h_x > h_{x,c1}$ the angle between $\boldsymbol{J}_s$ and the specimen axis ($0° < \alpha \leqq 90°$) decreases and passes through zero only after a change in sign of h_x and reaches a value corresponding to the point "a" in Fig. 4. If h_x continues to rise in absolute value, the angle between $\boldsymbol{J}_s$ and the

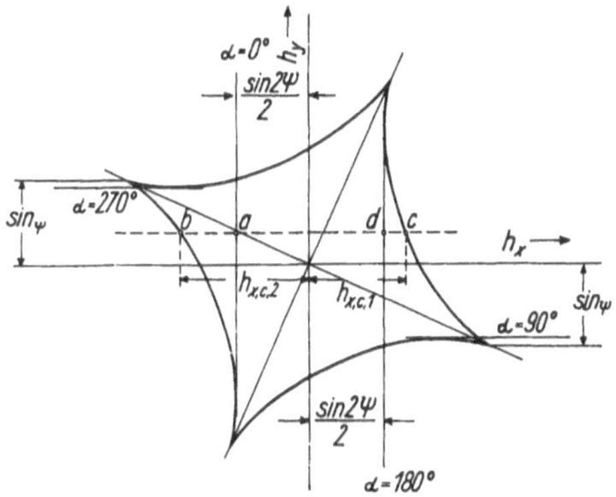

Fig. 4. Critical curve in the case that the easy axis is inclined by an angle ψ towards the specimen axis (Procopiu effect)

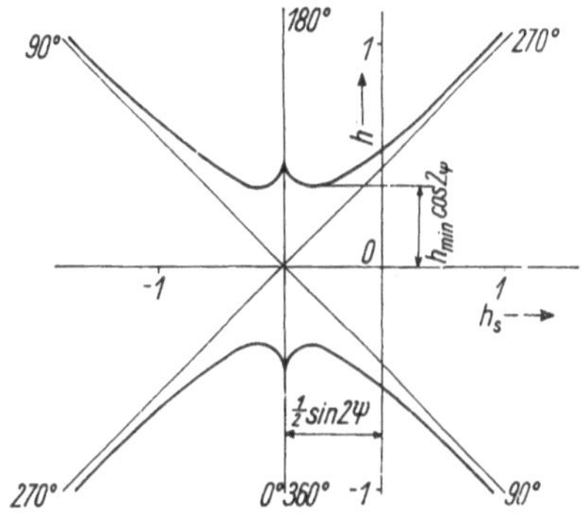

Fig. 5. Critical curve in the case that the easy axis is inclined by an angle ψ towards the specimen axis (inverse Wiedemann effect)

specimen axis continues to decrease ($360° \geqq \alpha > 270°$) and by crossing the critical curve in point "b" ($h_x = -h_{x,\,c\,2}$) the magnetization vector partially switches to a direction which forms an angle ($270° > \alpha > 180°$) with the specimen axis.

If the intensity of the circular magnetic field decreases in its absolute value, the angle between $\boldsymbol{J}_s$ and the specimen axis decreases continuously. By crossing the critical curve in point "d" ($h_x = h_{x,\,c\,1}$) the magnetization vector partially switches to a direction which forms an angle ($90° > \alpha \geqq 0°$) with the specimen axis. The switching conditions in the two intersection points of the straight line h_y = const. with the astroide may be very different, and the e.m.f. induced in the search coil presents a great variety of shapes depending on h_y, but two successive tension pulses, though inequal in size, are opposite in sign.

b) $h_y > \sin\psi$

Now, only continuous rotations of $\boldsymbol{J}_s$ take place and the discussed case does not differ from the case presented in Fig. 2.

Taking into account the wire form of the specimens used in the Procopiu effect, the studied cases exhaust the number of possible cases.

We remark that if $\psi \neq 0$ the size of two successive tension pulses and the time interval between two successive tension pulses are not equal in the inverse Wiedemann effect.

3. The Influence of Torsion on the Procopiu Effect

If $\sigma \neq 0$ the energy stored at a given moment in the volume unit of the specimen is given by

$$W = K \sin^2 \alpha + \frac{3}{2} \lambda_s \sigma \sin^2 (45° - \alpha) - \boldsymbol{H}_x \boldsymbol{J}_s \sin \alpha - \boldsymbol{H}_y \boldsymbol{J}_s \cos \alpha \,. \tag{9}$$

For a constant value of $h_s = \frac{3}{2} \frac{\lambda_s \sigma}{2K}$ the critical curve is an astroide given by

$$\left.\begin{aligned} h_x &= \sin^3 \alpha - \frac{h_s}{2} \cos \alpha \,(1 + 2 \sin^2 \alpha)\,, \\ h_y &= -\cos^3 \alpha - \frac{h_s}{2} \sin \alpha \,(1 + 2 \cos^2 \alpha) \end{aligned}\right\} \tag{10}$$

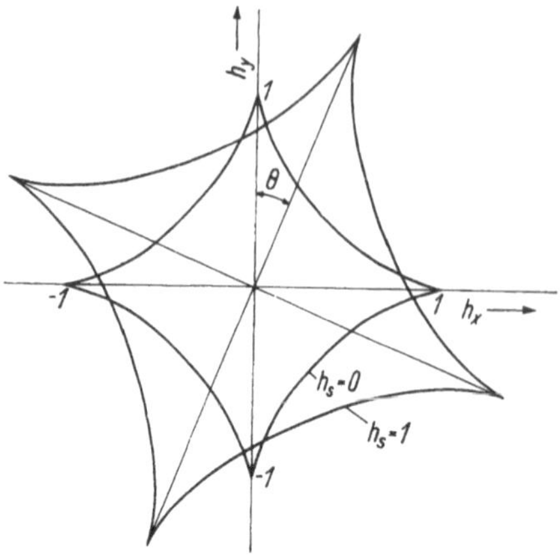

Fig. 6. Critical curve for $\psi = 0$, $h_s = 0$, and $h_s = 1$ (Procopiu effect)

and represented in Fig. 6 for different values of h_s. The astroide corresponding to $h_s \neq 0$ is rotated by an angle θ given by $\operatorname{tg} 2\,\theta = h_s$ towards the astroide corresponding to $h_s = 0$. The semiaxis of the astroide is given by $h_k = \sqrt{1 + h_s^2}$.

4. Conclusions

4.1 The Procopiu effect

The proposed model explains and emphasizes the following features of the Procopiu effect:

If $h_s = 0$ and $\psi = 0$ the double frequency of E_i as compared to the frequency of h_x is determined by the symmetry to the specimen axis of the phenomena released by the positive and the negative alternation of the circular magnetic field.

The presence of a shear strain is followed by the equalization of the frequencies of h_x and E_i.

In the Procopiu effect continuous rotations of $\boldsymbol{J}_s$ occur except for the cases $\psi \neq 0$ and $0 < h_y < \sin\psi$ or $0 \leq h_y < \sqrt{1 + h_s^2}\,\sin\theta$ and $h_s \neq 0$ in which in addition partial switchings of $\boldsymbol{J}_s$ also take place. The presence of continuous rotations determines the dependence of the tension induced in the coil upon the frequency of the circular magnetic field.

If the shear strain changes its sign and if $\psi = 0$ we find a change in the sign of the tension pulses which reflect partial switchings of $\boldsymbol{J}_s$. In case that $\psi \neq 0$ we obtain, as a function of sign of the shear strain, a modification of the tension pulses reflecting the switching of $\boldsymbol{J}_s$ without a change in its sign or a gradual decrease until zero without a change in sign, followed by a change in the sign of two successive tension pulses.

4.2 The inverse Wiedemann effect

The proposed model explains the following features of the inverse Wiedemann effect:

In the inverse Wiedemann effect only continuous rotations of $\boldsymbol{J}_s$ occur. The tension induced in the search coil depends upon the frequency of the torsional oscillations.

The double frequency of E_i as compared to the frequency of h_s is determined by the symmetry to the specimen axis of the phenomena released by the positive and the negative alternation of the shear strain.

The tension induced in the coil depends upon the intensity of the axial static magnetic field and upon the amplitude of the shear strain.

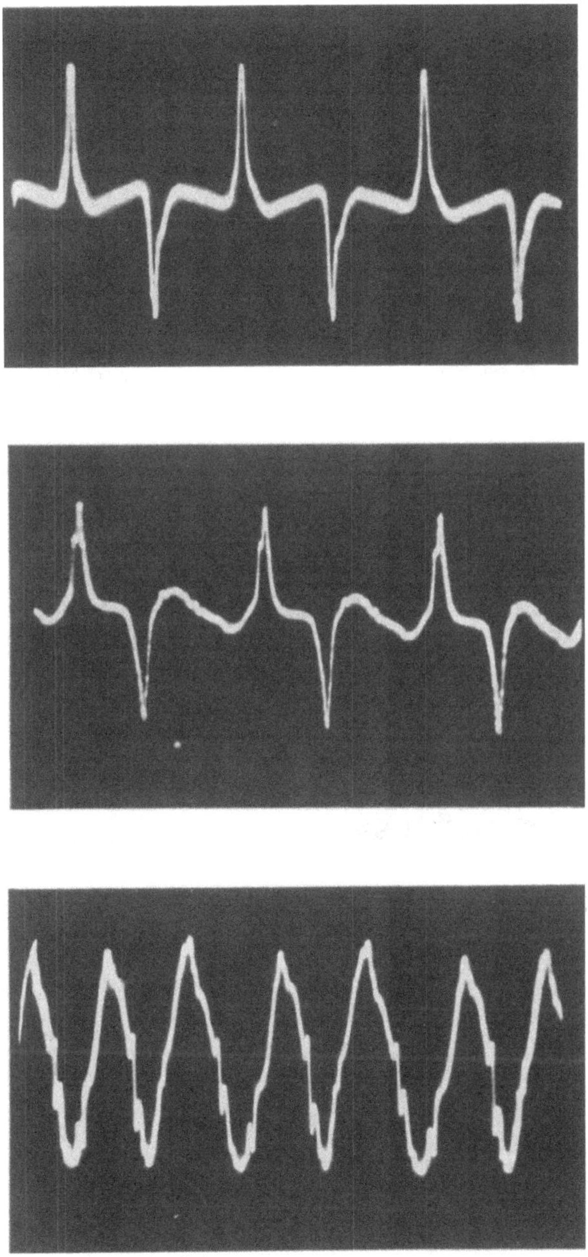

a

b

c

Fig. 7. Time variation of the e.m.f. induced in the coil in the Procopiu effect with a copper wire electroplated with permalloy (a, b, c) and the e.m.f. induced in the coil as a function of the oscillatory circular magnetic field (a', b', c'). Amplitude of the intensity of the oscillatory circular magnetic field $H_{x,0} = 10$ Oe, intensity of the stationary longitudinal magnetic field $H_y = 0$ Oe (a, a'), 0.08 Oe (b, b'), and 5 Oe (c, c'), respectively

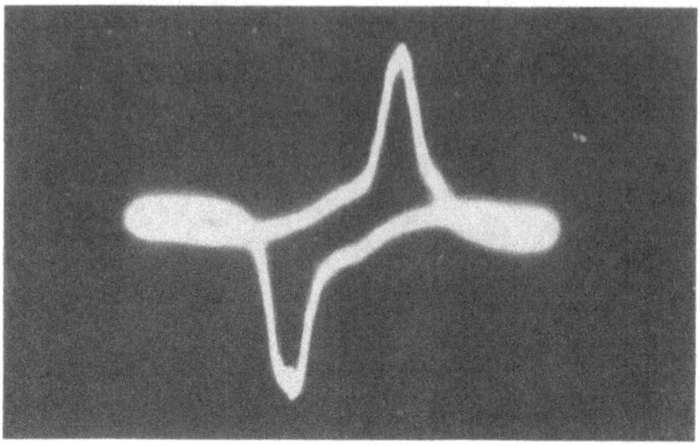

a′

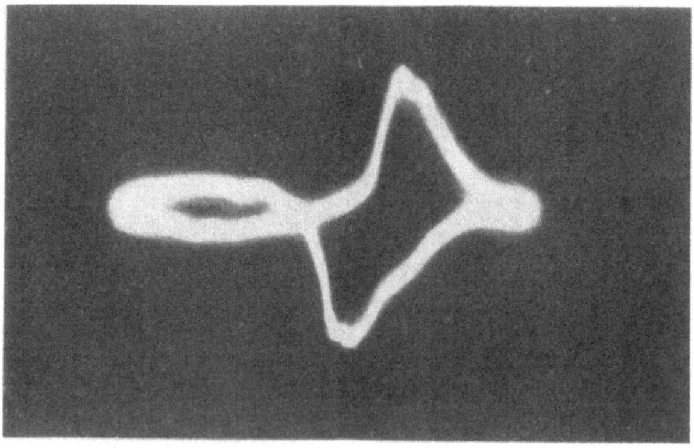

b′

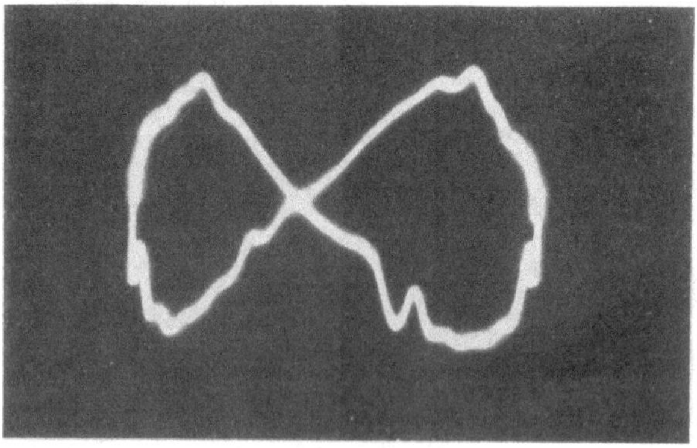

c′

Fig. 7 a′, b′, c′

5. Experimental Results

5.1 Procopiu effect

For the verification of the concordance between the shapes of E_i deduced on the basis of the proposed model and the experience, results published in current literature are available.

The induced e.m.f. as shown in Fig. 2 has been obtained by Procopiu and Viscrian [8] with nickel and iron, by Koch [12] and Sorohan [13] with iron, and by Skorski and Duracz [10] with permalloy.

Very good agreement between the form of the induced e.m.f. plotted on the basis of the proposed model and experiment has been obtained with copper wires (1.4 mm in diameter) which were electroplated with permalloy and had a preferred orientation.

The oscillograms (Fig. 7) reproduce the time variation of the e.m.f. induced in the coil as a function of the oscillatory circular magnetic field. The shape of the e.m.f. induced in the coil, in the case that the intensity of the stationary longitudinal magnetic field is equal to zero (Fig. 7a and 7a′), shows that the studied specimen presents an easy axis which does not coincide with the specimen axis. A partial switching of the magnetization vector takes place and the frequency of the e.m.f. induced in the coil is equal to the frequency of the oscillatory circular magnetic field. Increasing the intensity of the stationary longitudinal magnetic field, we come out from the field interval in which switchings occur, and the frequency of the tension induced in the coil is twice higher than the frequency of the circular oscillatory magnetic field (Fig. 7b and 7b′, Fig. 7c and 7c′).

5.2 Inverse Wiedemann effect

The oscillogram (Fig. 8) reproduces the time variation of the e.m.f. induced in the coil, obtained with copper wire (1.4 mm in diameter) which were electroplated with nickel and had a preferred orientation. Its shape shows that the studied specimen presents an easy axis which does not coincide with the specimen axis.

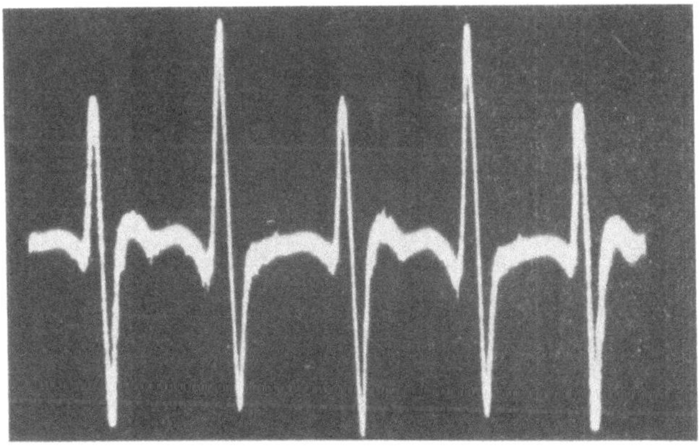

Fig. 8. Time variation of the e.m.f. induced in the coil in the inverse Wiedemann effect with copper wire electroplated with nickel. Intensity of the stationary longitudinal magnetic field $H_y = 10$ Oe, amplitude of the surface shear strain 2×10^{-5}

References

[1] S. Procopiu, J. Phys. Radium **1**, 306 (1930).
[2] S. Procopiu and G. Vasiliu, C. R. Acad. Sci. (France) **204**, 673 (1937).
[3] S. Procopiu and G. Vasiliu, C. R. Acad. Sci. (France) **204**, 971 (1937).
[4] S. Procopiu, C. R. Acad. Sci. (France) **205**, 137 (1937).
[5] S. Procopiu, C. R. Acad. Sci. (France) **208**, 1212 (1939).
[6] R. K. Bozorth and J. F. Dillinger, Phys. Rev. **41**, 345 (1932).
[7] T. Hofbauer and K. M. Koch, Z. Phys. **130**, 409 (1951).
[8] S. Procopiu and I. Viscrian, Studii Cercetări ştiinţ., Fiz. şi ştiinţe Tehn. **14**, 13 (1963).
[9] S. Procopiu and I. Viscrian, Studii Cercetări ştiinţ., Fiz. şi ştiinţe Tehn. **14**, 307 (1963).
[10] R. Skorski and A. Duracz, J. appl. Phys. **36**, 511 (1965).
[11] S. Procopiu and I. Viscrian, Studii Cercetări ştiinţ., Fiz. şi ştiinţe Tehn. **15**, 205 (1964).
[12] F. A. Koch, Z. Phys. **155**, 475 (1959).
[13] M. Sorohan, Studii Cercetări ştiinţ., Fiz. şi ştiinţe Tehn. **10**, 211 (1959).
[14] V. Tutovan and I. B. Bursuc, C. R. Acad. Sci. (France) **253**, 405 (1961).
[15] M. Sorohan, phys. stat. sol. **6**, K81 (1964).
[16] N. S. Akulov and T. Z. Miriasov, Dokl. Akad. Nauk SSSR **66**, 29 (1949).
[17] B. F. Rothenstein and A. Policec, Z. Phys. **183**, 234 (1965).
[18] B. F. Rothenstein and A. Policec, J. appl. Phys. **36**, 1808 (1965).
[19] R. C. Frank and J. W. Ferman, J. appl. Phys. **36**, 2235 (1965).
[20] E. C. Stoner and E. P. Wolfarth, Trans. Roy. Soc. **240**, 599 (1948).
[21] E. Kneller, Ferromagnetismus, Springer-Verlag, Berlin/Göttingen/Heidelberg 1962.

(Received October 29, 1966)

phys. stat. sol. 19, 623 (1967)

Subject classification: 20; 18; 14.3

Institute of Physics, Polish Academy of Sciences, Warsaw

Free Carrier Electro-Magneto-Optical Phenomena in Semiconductors

By

J. KOŁODZIEJCZAK and E. KIERZEK-PECOLD

Calculations are made of free carrier magneto-optical phenomena in the presence of an external d.c. electric field assuming a quadratic deviation from Ohm's law. The calculations are based on the assumption of non-parabolic energy bands with spherical symmetry. Expressions are obtained for the Faraday effect, and the Voigt effect in a d.c. electric field as well as for the electro-optical Kerr effect in a magnetic field.

Es werden magneto-optische Phänomene freier Ladungsträger bei äußeren elektrischen Gleichfeldern berechnet, wobei quadratische Abweichung vom Ohmschen Gesetz angenommen wird. Die Berechnungen basieren auf der Annahme nichtparabolischer Energiebänder mit sphärischer Symmetrie. Es werden Ausdrücke für den Faradayeffekt und Voigteffekt in einem elektrischen Gleichfeld und für den elektro-optischen Kerreffekt im Magnetfeld erhalten.

1. The Conductivity Tensor

In a previous paper [1] the Boltzmann equation was analysed from the point of view of its application for the calculation of free carrier electro-magneto-optical effects, i.e. optical effects in the presence of external d.c. electric and magnetic fields. By solving the nonlinear Boltzmann equation the general expression for the electric current density has been obtained in the approximation of a quadratic deviation from Ohm's law.

The calculations were carried out assuming ellipsoidal energy surfaces and a nonparabolic dependence of energy on the absolute value of the wave vector. Using the electric current density obtained in this way and solving the Maxwell equations, the following equation for the propagation vector was obtained:

$$\begin{vmatrix} \left(N^2 + \frac{4\pi i}{\omega}\sigma_{11}^{\text{eff}} - 1\right) & \frac{4\pi i}{\omega}\sigma_{12}^{\text{eff}} & \frac{4\pi i}{\omega}\sigma_{13}^{\text{eff}} \\ \frac{4\pi i}{\omega}\sigma_{21}^{\text{eff}} & \left(N^2 + \frac{4\pi i}{\omega}\sigma_{22}^{\text{eff}} - 1\right) & \frac{4\pi i}{\omega}\sigma_{23}^{\text{eff}} \\ \frac{4\pi i}{\omega}\sigma_{31}^{\text{eff}} & \frac{4\pi i}{\omega}\sigma_{32}^{\text{eff}} & \left(\frac{4\pi i}{\omega}\sigma_{33}^{\text{eff}} - 1\right) \end{vmatrix} = 0\,, \tag{1}$$

where the tensor $\sigma_{\beta\mu}^{\text{eff}}$ is defined as follows:

$$\sigma_{\beta\mu}^{\text{eff}} = \sigma_{\beta\mu}^{*} + (\sigma_{\alpha\varrho\beta\mu}^{(0)*} + \sigma_{\alpha\mu\beta\varrho}^{(1)*})\, E_{\alpha}^{(0)}\, E_{\varrho}^{(0)}\,. \tag{2}$$

$\boldsymbol{E}^{(0)}$ is the d.c. electric field that gives a nonlinear effect. The tensors $\sigma^{*}_{\beta\mu}$, $\sigma^{(0)*}_{\alpha\varrho\beta\mu}$,and $\sigma^{(1)*}_{\alpha\mu\beta\varrho}$ are given by the following expressions:

$$\left.\begin{aligned}\sigma^{*}_{\beta\mu} &= q\,N\,\frac{\langle f_0'\,G^{*}_{\beta\mu}\rangle}{\langle f_0'\rangle}\,,\\ \sigma^{(0)*}_{\alpha\,\varrho\beta\mu} &= q\,N\,\frac{\langle (S^{(0)}_{\alpha\varrho})'\,G^{*}_{\beta\mu}\rangle}{\langle f_0'\rangle}\,,\\ \sigma^{(1)*}_{\alpha\,\mu\beta\varrho} &= q\,N\,\frac{\langle (S^{(1)}_{\alpha\,\mu})'\,G^{*}_{\beta\varrho}\rangle}{\langle f_0'\rangle}\,,\end{aligned}\right\}\tag{3}$$

where q is the charge of the carrier, N is the carrier concentration, and the bracket $\langle\ \rangle$ is defined as

$$\langle A\rangle = \int_0^\infty A(\varepsilon)\,\gamma^{3/2}\,\mathrm{d}\varepsilon\,. \tag{4}$$

The primes in equations (3) denote the derivatives with respect to energy, f_0 is the Maxwell-Boltzmann distribution function. The function γ in (4) defines the structure of the energy band [1]. According to the previous paper [1], for energy bands of spherical symmetry and for weak magnetic fields the tensor $G^{*}_{\beta\mu}$ can be written in the form

$$G^{*}_{\beta\mu} = u^{*}\,\delta_{\beta\mu} - \frac{u^{*2}}{c}\,\varepsilon_{\alpha\mu\beta}\,H_\alpha + \frac{u^{*3}}{c^2}\,[H_\beta\,H_\mu - H^2\,\delta_{\beta\mu}] \tag{5}$$

and

$$u^{*} = u\,\frac{1}{1+i\,\omega\,\tau}\,, \tag{6}$$

where u is the microscopic mobility equal to $(q\,\tau)/m^{*}$, τ is the momentum relaxation time, and m^{*} is the effective mass which for a nonparabolic band of spherical symmetry is a function of energy and is defined by [1]

$$m^{*} = m_0^{*}\left(\frac{\mathrm{d}\gamma}{\mathrm{d}\varepsilon}\right), \tag{7}$$

where m_0^{*} is the effective mass at the bottom of the band.

Basing on the results of the previous paper [1] (equations (32) and (33)) and equation (5) of the present paper, one can find the tensors $S^{(0)}_{\alpha\mu}$ and $S^{(1)}_{\alpha\mu}$ in the following form:

$$S^{(0)}_{\alpha\mu} = \Phi_1\,\delta_{\alpha\mu} - \frac{1}{c}\,\Phi_2\,\varepsilon_{\varrho\mu\alpha}\,H_\varrho + \frac{1}{c^2}\,\Phi_3\,[H_\alpha\,H_\mu - H^2\,\delta_{\alpha\mu}]\,, \tag{8}$$

$$S^{(1)}_{\alpha\mu} = \Psi_1\,\delta_{\alpha\mu} - \frac{1}{c}\,\Psi_2\,\varepsilon_{\varrho\mu\alpha}\,H_\varrho + \frac{1}{c^2}\,\Psi_3\,[H_\alpha\,H_\mu - H^2\,\delta_{\alpha\mu}]\,, \tag{9}$$

where the functions Φ_k and Ψ_k depend on energy and can be found from the equations

$$\hat{C}\,\Phi_k = -\frac{2}{3}\,q\,\gamma^{-1/2}\left(\frac{\mathrm{d}\gamma}{\mathrm{d}\varepsilon}\right)^{-1}\frac{\mathrm{d}}{\mathrm{d}\varepsilon}\left[\gamma^{3/2}\,f_0'\,u^k\right] \tag{10}$$

and

$$(\hat{C} - i\,\omega)\,\Psi_k = -\frac{2}{3}\,\gamma^{-1/2}\left(\frac{\mathrm{d}\gamma}{\mathrm{d}\varepsilon}\right)^{-1}\frac{\mathrm{d}}{\mathrm{d}\varepsilon}\left\{\gamma^{3/2}\,f_0'\left[u^{*k} - (-1)^k\,u^k\right]\right\}. \tag{11}$$

Equation (8) can be simplified since the tensor $S^{(0)}_{\alpha\mu}$ is always associated with the product $E^{(0)}_\alpha\,E^{(0)}_\mu$ [1]. It is obvious that $\varepsilon_{\varrho\mu\alpha}\,H_\varrho\,E^{(0)}_\alpha\,E^{(0)}_\mu = 0$, thus we can write

$$S^{(0)}_{\alpha\mu} = \Phi_1\,\delta_{\alpha\mu} + \frac{1}{c^2}\,\Phi_3\left[H_\alpha\,H_\mu - H^2\,\delta_{\alpha\mu}\right]. \tag{12}$$

It is seen from equation (11) that the functions Ψ_k are complex and thus can be written in the form

$$\Psi_k = X_k + i\,Y_k\,. \tag{13}$$

Making use of equations (10), (12), and (13) and assuming high frequencies ω such that $\tau^2\,\omega^2 \gg 1$ and $\omega^2\,X_k \gg \hat{C}\,(\hat{C}\,X_k)$, one can find the following results for X_k and Y_k:

$$X_1 = \frac{1}{\omega^2}\left\{\hat{C}\,(\hat{C}\,\Phi_1) - \frac{2}{3}\,q^2\,\gamma^{-1/2}\left(\frac{\mathrm{d}\gamma}{\mathrm{d}\varepsilon}\right)^{-1}\frac{\mathrm{d}}{\mathrm{d}\varepsilon}\left[\gamma^{1/2}\,f_0'\,\frac{1}{m^*}\right]\right\}, \tag{14}$$

$$X_2 = -\frac{1}{\omega^2}\,\hat{C}(\hat{C}\,\Phi_2)\,, \qquad X_3 = \frac{1}{\omega^2}\,\hat{C}(\hat{C}\,\Phi_3)\,, \tag{15}$$

$$Y_1 = \frac{1}{\omega}\,\hat{C}\,\Phi_1\,, \qquad Y_2 = -\frac{1}{\omega}\,\hat{C}\,\Phi_2\,, \qquad Y_3 = \frac{1}{\omega}\,\hat{C}\,\Phi_3\,, \tag{16}$$

with $\hat{C}\,\Phi_k$ given by the right-hand side of equation (10). It was proved that for interactions with lattice vibrations and for temperatures above the Debye temperature, the inequality $\hat{C}\,\tau \ll 1$ is satisfied. From the physical point of view it means that the momentum scattering is much more effective than the energy scattering. The momentum relaxation time τ and the operator $\hat{C}$ describing the energy scattering were calculated in an earlier paper [2] for different modes of lattice vibration. If the inequality $\hat{C}\,\tau \ll 1$ is satisfied the first term in (14) can be neglected and X_1 can be simplified to

$$X_1 = -\frac{2\,q^2}{3\,\omega^2}\,\gamma^{-1/2}\left(\frac{\mathrm{d}\gamma}{\mathrm{d}\varepsilon}\right)^{-1}\frac{\mathrm{d}}{\mathrm{d}\varepsilon}\left[\gamma^{3/2}\,f_0'\,\frac{1}{m^*}\right]. \tag{17}$$

Making use of equations (3), (5), (6), (9), and (12), the components of the tensor $\sigma^{\mathrm{eff}}_{\beta\mu}$ can be calculated for different configurations of the magnetic field, electric field $\boldsymbol{E}^{(0)}$, and propagation vector $\boldsymbol{N}$.

It is seen that in this case $\sigma_{11}^{\text{eff}} = \sigma_{22}^{\text{eff}}$ and according to (19) and (20) only the Faraday effect will exist. The Faraday rotation in this case is equal to

$$\theta_{\text{F}}(E) = -\frac{2\pi d q N H}{n c^2} \text{Re}\left\{\frac{\langle f_0' u^{*2}\rangle}{\langle f_0'\rangle} + \frac{\langle \Phi_1' u^{*2}\rangle}{\langle f_0'\rangle} E^2\right\} \tag{31}$$

and the ellipticity is

$$\Delta_{\text{F}}(E) = i\frac{2\pi d q N H}{n c^2} \text{Im}\left\{\frac{\langle f_0' u^{*2}\rangle}{\langle f_0'\rangle} + \frac{\langle \Phi_1' u^{*2}\rangle}{\langle f_0'\rangle} E^2\right\}. \tag{32}$$

For vanishing electric field E the above equations become identical with classical formulas for the Faraday rotation and ellipticity.

3. Voigt Configuration $\vec{N}(0, 0, N)$, $\vec{H} \perp \vec{N}$ for $\vec{H}(0, H, 0)$

3.1 Case 1: $\vec{E}^{(0)} \parallel \vec{H}$ i.e. $\vec{E}^{(0)}(0, E, 0)$

Neglecting the terms proportional to $E^2 H^2$, the components of the tensor $\sigma_{\beta\mu}^{\text{eff}}$ can be written in the following form:

$$\left.\begin{aligned}
\sigma_{11}^{\text{eff}} &= \sigma_{33}^{\text{eff}} = \frac{q N}{\langle f_0'\rangle}\left\{\langle f_0' u^*\rangle - \langle f_0' u^{*3}\rangle \frac{H^2}{c^2} + \langle \Phi_1' u^*\rangle E^2\right\}, \\
\sigma_{22}^{\text{eff}} &- \frac{q N}{\langle f_0'\rangle}\{\langle f_0' u^*\rangle + \langle(\Phi_1' + \Psi_1') u^*\rangle E^2\}, \\
\sigma_{31}^{\text{eff}} &= -\sigma_{13}^{\text{eff}} = \frac{q N}{\langle f_0'\rangle}\frac{H}{c}\{\langle f_0' u^*\rangle + \langle \Phi_1' u^{*2}\rangle E^2\}, \\
\sigma_{12}^{\text{eff}} &= \sigma_{21}^{\text{eff}} = \sigma_{23}^{\text{eff}} = \sigma_{32}^{\text{eff}} = 0\,.
\end{aligned}\right\} \tag{33}$$

Substituting these components in equation (1) one can find the solutions

$$N_{\parallel}^2 - 1 = -\frac{4\pi i}{\omega}\sigma_{22}^{\text{eff}}, \tag{34}$$

$$N_{\perp}^2 - 1 = -\frac{4\pi i}{\omega}\sigma_{11}^{\text{eff}} - \left(\frac{4\pi}{\omega}\right)^2 \frac{\sigma_{13}^{\text{eff}}\sigma_{31}^{\text{eff}}}{\frac{4\pi i}{\omega}\sigma_{33}^{\text{eff}} - 1}, \tag{35}$$

where $N_{\parallel}$ denotes the propagation vector corresponding to the wave polarized parallel to the magnetic field, while $N_{\perp}$ denotes the propagation vector of the wave polarized perpendicular to the magnetic field. For very high frequencies, the second term in (35) can be neglected.

Making use of the usual procedure we find the phase shift associated with birefringence due to magnetic and electric fields:

$$\varphi = i\frac{2\pi d q N}{n c} \text{Im}\left\{\frac{\langle f_0' u^{*3}\rangle}{\langle f_0'\rangle}\frac{H^2}{c^2} + \frac{\langle \Psi_1' u^*\rangle}{\langle f_0'\rangle} E^2\right\}. \tag{36}$$

The first term of (36) determines the classical Voigt effect, while the second term determines the classical Kerr effect. The whole phase shift is a sum of these two effects. The dichroism is given by the following formula:

$$\delta = -\frac{2\pi d q N}{n c} \operatorname{Re}\left\{\frac{\langle f_0' u^{*3}\rangle}{\langle f_0'\rangle} \frac{H^2}{c^2} + \frac{\langle \Psi_1' u^*\rangle}{\langle f_0'\rangle} E^2\right\} \tag{37}$$

and is also equal to the sum of the dichroisms associated with Voigt and Kerr effects.

3.2 *Case 2:* $\vec{E}^{(0)} \perp \vec{H}$ *i.e.* $\vec{E}^{(0)}(E, 0, 0)$

Neglecting the terms proportional to $E^2 H^2$ we find the components of the tensor $\sigma_{\beta\mu}^{\text{eff}}$:

$$\left.\begin{aligned}
\sigma_{11}^{\text{eff}} &= \frac{q N}{\langle f_0'\rangle}\left\{\langle f_0' u^*\rangle - \langle f_0' u^{*3}\rangle \frac{H^2}{c^2} + E^2 \langle(\Phi_1' + \Psi_1') u^*\rangle\right\}, \\
\sigma_{22}^{\text{eff}} &= \frac{q N}{\langle f_0'\rangle}\{\langle f_0' u^*\rangle + E^2 \langle \Phi_1' u^*\rangle\}, \\
\sigma_{33}^{\text{eff}} &= \frac{q N}{\langle f_0'\rangle}\left\{\langle f_0' u^*\rangle - \langle f_0' u^{*3}\rangle \frac{H^2}{c^2} + E^2 \langle \Phi_1' u^*\rangle\right\}, \\
\sigma_{31}^{\text{eff}} &= \frac{q N}{\langle f_0'\rangle}\{\langle f_0' u^{*2}\rangle + E^2 \langle(\Phi_1' + \Psi_1') u^{*2}\rangle\} \frac{H}{c}, \\
\sigma_{13}^{\text{eff}} &= -\frac{q N}{\langle f_0'\rangle}\{\langle f_0' u^{*2}\rangle + E^2 \langle \Phi_1' u^{*2} + \Psi_2' u^*\rangle\} \frac{H}{c}, \\
\sigma_{12}^{\text{eff}} &= \sigma_{21}^{\text{eff}} = \sigma_{13}^{\text{eff}} = \sigma_{31}^{\text{eff}} = 0 .
\end{aligned}\right\} \tag{38}$$

The phase shift φ is now determined as a difference of the phase shift due to the Voigt and Kerr effects and is equal to

$$\varphi = i \frac{2\pi q d N}{n c} \operatorname{Im}\left\{\frac{\langle f_0' u^{*3}\rangle}{\langle f_0'\rangle} \frac{H^2}{c^2} - \frac{\langle \Psi_1' u^*\rangle}{\langle f_0'\rangle} E^2\right\}. \tag{39}$$

Similarly the dichroism is now given by

$$\delta = -\frac{2\pi q d N}{n c} \operatorname{Re}\left\{\frac{\langle f_0' u^{*3}\rangle}{\langle f_0'\rangle} \frac{H^2}{c^2} - \frac{\langle \Psi_1' u^*\rangle}{\langle f_0'\rangle} E^2\right\}. \tag{40}$$

It is worth mentioning here that in both Voigt configurations (Sections 3.1 and 3.2) a longitudinal component of the electromagnetic wave is induced. Making use of the fact that $\sigma_{12}^{\text{eff}} = \sigma_{21}^{\text{eff}} = \sigma_{23}^{\text{eff}} = \sigma_{32}^{\text{eff}} = 0$ we can write a system of uniform equations for the components of the electric vector $\boldsymbol{E}^{(1)}$ and find $E_3^{(1)}$ in the form

$$E_3^{(1)} = \frac{\dfrac{4\pi i}{\omega} \sigma_{31}^{\text{eff}}}{1 - \dfrac{4\pi i}{\omega} \sigma_{33}^{\text{eff}}} E_1^{(1)} . \tag{41}$$

$E_3^{(1)}$ is the longitudinal component of the electromagnetic wave since it is parallel to the propagation vector $\boldsymbol{N}$. Since σ_{31}^{eff} vanishes only for vanishing magnetic field the longitudinal component of the electromagnetic wave is associated with the magnetic field only.

References

[1] J. Kołodziejczak and H. Stramska, phys. stat. sol. **17**, 701 (1966).
[2] J. Kołodziejczak, phys. stat. sol. **19**, 231 (1967).

(Received November 21, 1966)

phys. stat. sol. **19**, 631 (1967)

Subject classification: 10; 4; 22.6

Department of Physics, Monash University, Victoria (a), and Institute of Geophysics and Planetary Physics, University of California, Los Angeles (b)

Transmission Electron Microscope Study of Brazil Twins and Dislacations Experimentally Produced in Natural Quartz [1)]

By

A. C. McLaren (a), J. A. Retchford (a), D. T. Griggs (b), and J. M. Christie (b)

Transmission electron microscopy has been used to examine single crystals of natural quartz which have been plastically deformed in compression at high temperature (500 to 700 °C) and confining pressure (15 to 20 kbar) with high shear stress on (0001). In specimens deformed at 500 °C, the only features found are closely spaced pairs of planar defects parallel to {0001}. Study of the diffraction contrast fringes reveals that each pair of defects bounds a Brazil twin band which has been produced by the high shear stress. At the higher temperature there are dislocations in the twin bands and other regions of the crystals with dislocations only. The dislocations are almost invariably either pure screw ($\boldsymbol{b} = a \langle 11\bar{2}0 \rangle$) or of a particular mixed character (predominantly edge, but with a 30° screw component). Arrays of the mixed dislocations are observed which would produce stress fields consistent with the stress-optical effects observed by optical microscopy.

Elektronenmikroskopische Durchstrahlungsmethoden werden zur Untersuchung von natürlichen Quarzeinkristallen benutzt, die in Kompression bei hohen Temperaturen (500 bis 700 °C) im Druckbereich 15 bis 20 kbar mit hoher Schubspannung auf (0001) plastisch deformiert wurden. In Proben, die bei 500 °C verformt wurden, sind die einzigen Merkmale eng beieinander liegende Paare von ebenen Defekten parallel zu {0001}. Eine Untersuchung der Beugungskontrastbilder zeigt, daß jedes Defektpaar ein Zwillingsband vom Brazil-Typ begrenzt, das durch die hohe Schubspannung erzeugt wurde. Bei den höheren Temperaturen findet man Versetzungen in den Zwillingsbändern und andere Kristallbereiche, die nur Versetzungen enthalten. Die Versetzungen sind meist gleichbleibend von reinem Schrauben- ($\boldsymbol{b} = a \langle 11\bar{2}0 \rangle$) oder von einem speziell gemischten Charakter (hauptsächlich Stufencharakter aber mit einem 30° Schraubenanteil). Anordnungen der gemischten Versetzungen werden beobachtet, die Spannungsfelder hervorrufen könnten in Übereinstimmung mit dem spannungsoptischen Effekten, die lichtmikroskopisch beobachtet werden.

1. Introduction

Early experimental investigations of the mechanical deformation of quartz showed that it fractures in a brittle manner without evidence of plastic flow [1]. However, it is well known that high (hydrostatic) confining pressure tends to inhibit the fracture of materials thus enhancing the possibility of producing plastic deformation. In view of this Christie, Heard, and La Mori [2] attempted to deform quartz plastically at confining pressures in the region of 25 to 30 kbar at room temperature. In all experiments the quartz fractured as it does at lower confining pressures.

---

1) Publication No. 532, Institute of Geophysics and Planetary Physics, University of California, Los Angeles.

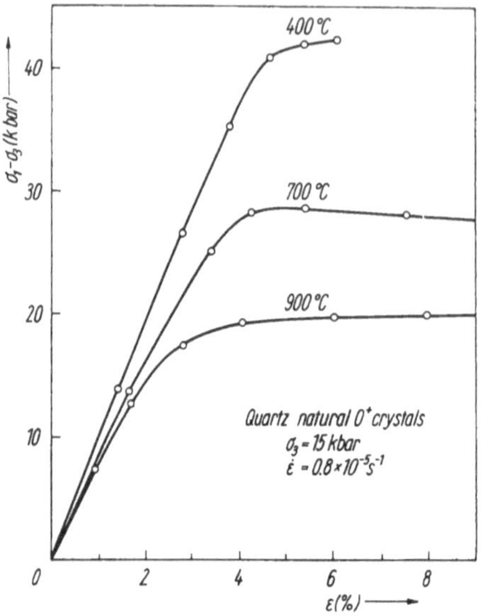

Fig. 1. Stress-strain curves for single crystals of natural quartz. σ_1 is the maximum (compressive) principal stress, σ_3 the minimum (= confining pressure). Unpublished experiments by J. D. Blacic and D. T. Griggs

However, more recently Christie, Griggs, and Carter [3] and Carter, Christie, and Griggs [4] obtained clear evidence for plastic flow in single crystals of natural quartz deformed in compression at high temperature (500 to 900 °C) and confining pressure (15 to 20 kbar).

All the deformed crystals discussed in this paper were oriented to give high shear stress on the basal plane. Typical stress–strain curves at three different temperatures are shown in Fig. 1. The initial slope of the stress–strain curve at 400 °C corresponds closely to that expected from the elastic constants of quartz. It will be seen that for this curve there is nearly 5% elastic deformation before the yield point at a compressive stress of about 40 kbar, corresponding to a shear stress on (0001) of approximately 20 kbar. The yield stress decreases with increasing temperature. The shear modulus μ of quartz for shear in the basal plane is 570 kbar at room temperature and pressure. MacKenzie [5] has shown that the theoretical shear strength of a crystal is about $\mu/15$ which is roughly twice the observed shear strength.

Christie, Griggs, and Carter [3] found that crystals deformed under these conditions developed slip bands on their cylindrical surface which indicated that the operating slip system was predominantly $\langle 11\bar{2}0\rangle$ {0001}. Thin sections of the deformed crystals were examined by optical microscopy and showed thin bands, referred to as *deformation lamellae*, parallel to the slip plane. It was also shown that the optical properties of the deformation lamellae are consistent with the photoelastic effects which would be expected from an array of edge dislocations of Burgers vector $\boldsymbol{b} = \boldsymbol{a}$ in the basal plane. Calculations of the stress field due to such an array show that a dislocation density of 8×10^4 cm^{-1} in average lamellae is required to produce the observed optical effects. The validity of this model was supported by electron micrographs of replicas of etched polished surfaces of crystals containing experimentally produced deformation lamellae. Pyramidal etch pits, the local density of which varied from 5×10^4 to 13×10^4 cm^{-1}, were observed in rows parallel to the trace of {0001}. Etched grooves parallel to the same direction were also observed, but their origin was not discussed.

Recently McLaren and Phakey [6] developed a simple method of preparing thin specimens of quartz suitable for transmission electron microscopy and have used this technique to examine the defect structures of a number of quartz minerals [7, 8, 9].

In order to examine in more detail the dislocation mechanism of the deformation of natural quartz suggested by Christie, Griggs, and Carter [3], we have now used transmission electron microscopy and selected area electron diffraction to study samples from the same suite of experimentally deformed quartz crystals.

2. Experimental

Thin specimens suitable for transmission electron microscopy were prepared by the method previously described [6] from the samples listed in Table 1. It will be seen from this table that the conditions of temperature and confining pressure placed all the samples well within the α-quartz stability field.

The thin specimens were examined in a JEM-6A electron microscope fitted with a goniometer stage.

Unfortunately, these specimens became damaged to some extent by the electron beam during observation [6] and this seriously limited the time available for tilting the specimen into various diffracting conditions.

Table 1

Details of the natural quartz samples deformed in compression with high shear stress on the basal plane

Sample	Orientation *)	Direction of highest resolved shear stress in (0001)	Temperature (°C)	Confining pressure (kbar)	Duration of experiment (min)	Strain (% shortening)
C-247	0^+	$[2\bar{1}\bar{1}0]$	500	20	33	10
C-476	0^+	$[2\bar{1}\bar{1}0]$	700	15	42	24
C-253	⊥ r	$[0\bar{1}10]$	500	20	35	6
C-478	⊥ r	$[0\bar{1}10]$	700	15	62	13

*) For definition of these symbols see [4].

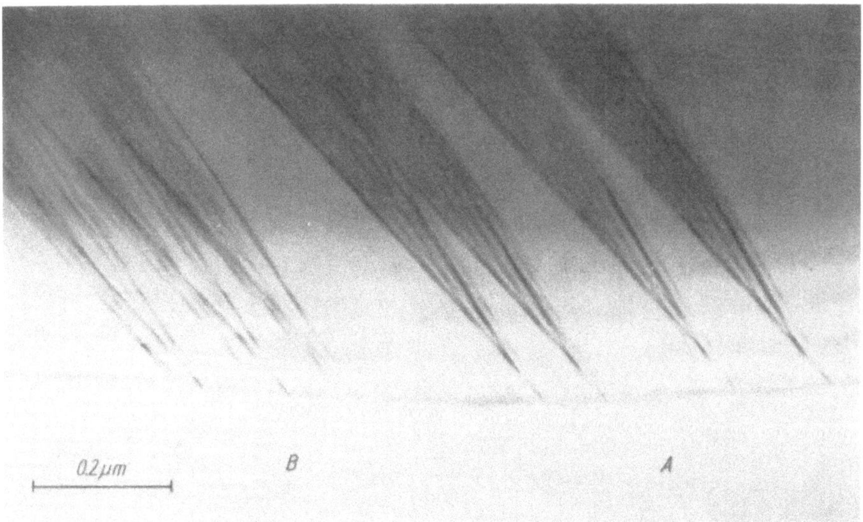

Fig. 2. Diffraction contrast fringe patterns observed in bright field. $g = \{10\bar{1}1\}$. Specimen C-247

3. Electron Microscope Observations

In samples C-247 and C-476 the highest resolved shear stress in (0001) was along one of the a-axes, say $[2\bar{1}\bar{1}0]$, and the deformations were 10 and 24% respectively. The only features consistently observed in C-247 were diffraction contrast fringe patterns due to some kind of planar defects which were always parallel to (0001). Examples are shown in Fig. 2. Similar fringe patterns from basal planar defects were consistently observed in C-476 as can be seen in Fig. 3 and 4. However, it will be seen from these micrographs that this sample also contained a relatively high density of dislocations, of the order of 10^9 cm^{-2}. Other examples of the dislocations in a region of the crystal free of the planar defects are shown in Fig. 5. It will be seen that the dislocations are almost invariably straight and parallel to the crystallographic a-axes. Usually one or two directions predominate, and occasionally a dislocation consisting of two or more straight segments, each parallel to an a-axis, will be seen.

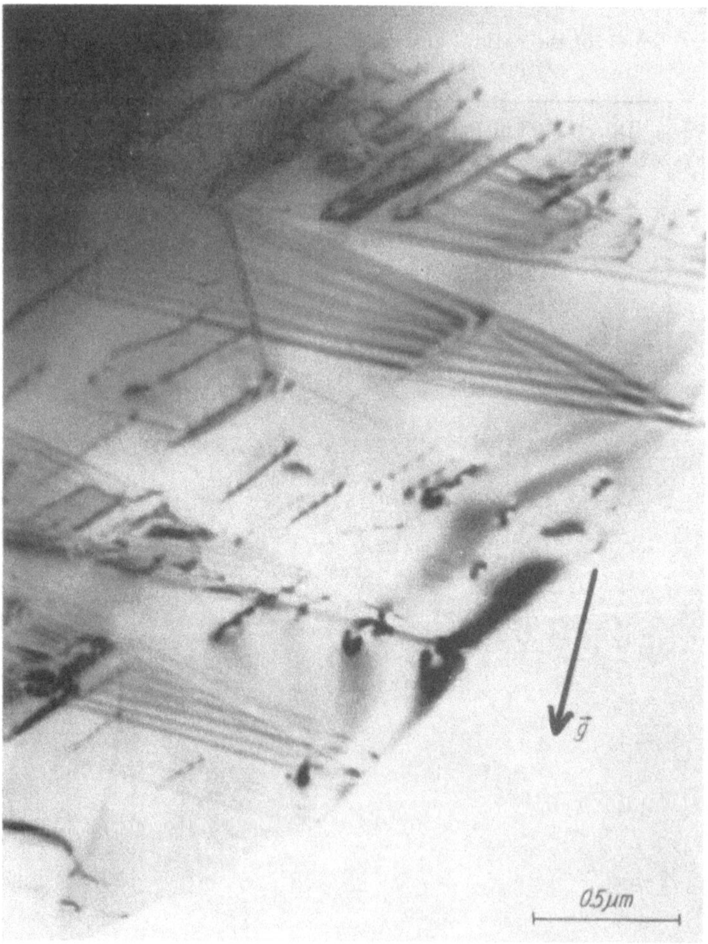

Fig. 3. Dislocations parallel to the crystallographic a-axes and diffraction contrast fringe patterns. Bright field, $\boldsymbol{g} = \{10\bar{1}1\}$. Specimen C-476

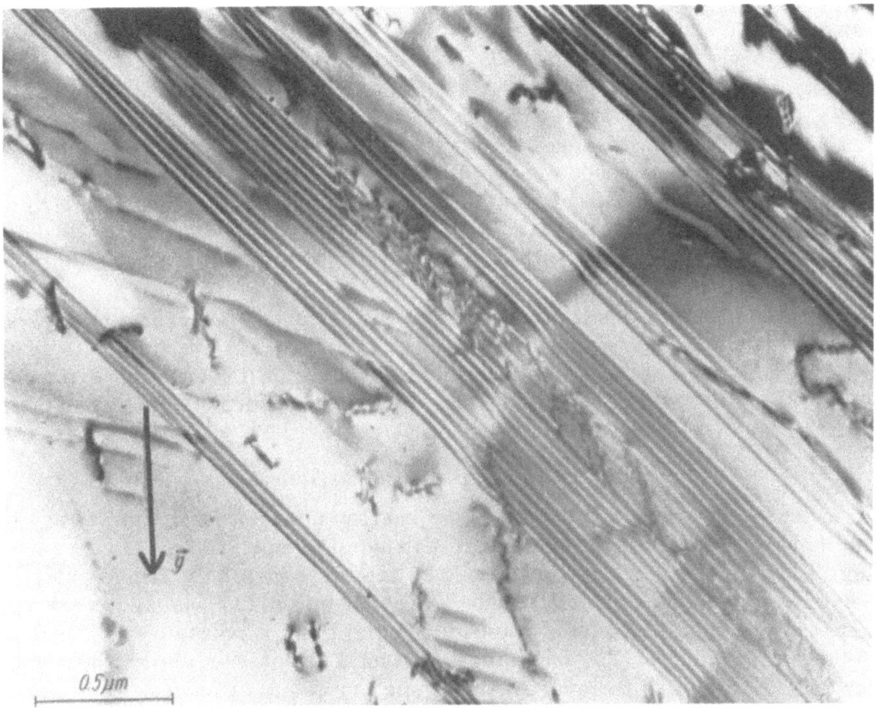

Fig. 4. Dislocations parallel to the crystallographic a-axes and diffraction contrast fringe patterns. Note the asymmetrical nature of the fringe patterns. Bright field, $\boldsymbol{g} = \{10\bar{1}1\}$. Specimen C-476

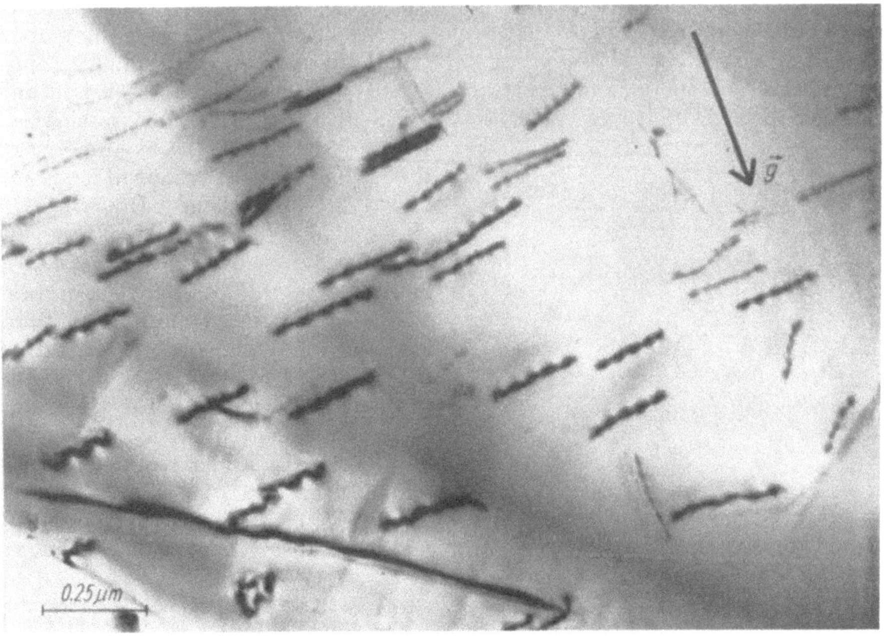

Fig. 5. Dislocations parallel to the crystallographic a-axes in a region of the crystal free of fringe patterns. Bright field, $\boldsymbol{g} = \{10\bar{1}1\}$. Specimen C-476

Samples C-253 and C-478 were deformed 6 and 13% respectively and the highest resolved shear stress in (0001) was along $\langle 10\bar{1}0 \rangle$. Fringe patterns from basal planar defects were consistently observed in C-253, but no dislocations were seen. On the other hand, sample C-478 contained a relatively high density of dislocations similar to those observed in sample C-476, but basal planar defects were only very rarely observed.

Occasionally Dauphiné twin boundaries were observed. The diffraction contrast from these boundaries has recently been studied by McLaren and Phakey, and the results will be published later. Dauphiné twins will not contribute to deformation.

From these observations it appears that the deformation mechanism involves both basal planar defects and dislocations. (Features corresponding to both types of defect were also observed in replicas from etched surfaces of deformed crystals [3].) In the following two sections these crystal defects are discussed in detail.

4. Nature of the Basal Planar Defects

In order to determine the nature of the basal planar defects it is necessary to examine in detail the diffraction contrast fringe patterns.

Fig. 2, 3, and 4 are bright-field electron micrographs of the fringe patterns observed with $\boldsymbol{g} = \{10\bar{1}1\}$. It will be seen that across the planar defects the thickness extinction contours are continuous and there is no change of contrast. It follows that there can be no change of crystal orientation across the planar defects, and this is confirmed by the single crystal nature of the associated selected area diffraction patterns. There is no evidence for a second phase of silica or for inclined axis twins. It might be expected, therefore, that the planar defects would give rise to α-fringe patterns which have symmetrical profiles, i.e. the outermost fringes of the pattern are either bright or dark depending upon the sign of α [10, 11]. However, it was found that for $\boldsymbol{g} = \{10\bar{1}1\}$ the fringe profiles were, in general, asymmetrical, i.e. one outermost fringe was bright and the other dark. The fringe patterns shown in Fig. 4 are clearly asymmetrical. On the other hand, the asymmetry of the fringe patterns shown in Fig. 2 and 3 is less obvious. However, on closer examination of these fringe patterns (in particular those marked A in Fig. 2) it will be seen that on passing a thickness extinction contour, the additional fringe is created within the fringe pattern, but it is not the central fringe which forks as in the case of α-fringe patterns [12] when $\alpha \neq \pm \pi$. This behaviour is shown schematically in Fig. 6a for greater clarity. The group of fringe patterns marked B in Fig. 2 provide the clue to

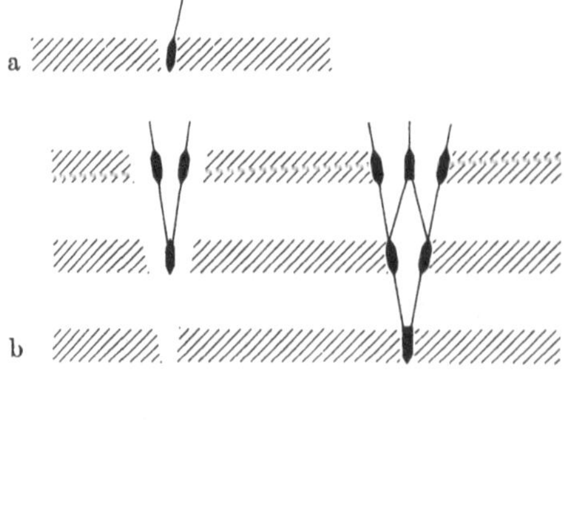

Fig. 6. a) Schematic diagram of the asymmetrical fringe patterns of the type shown at A in Fig. 2. b) Schematic diagram of α-fringes of opposite contrast. Compare with the fringes shown at B in Fig. 2

the real nature of the asymmetrical fringes. Near the edge of the crystal it is clear that in this group there are three pairs of fringe patterns. Each pair consists of two *symmetrical* fringe patterns of opposite contrast and separated by about 200 to 300 Å. This reversal of contrast, which occurs at each successive symmetrical fringe pattern, can be seen easily if the fringe patterns are compared with the schematic diagram shown in Fig. 6b. The features of these fringe patterns are typical of α-fringes with $\alpha = \pm\pi/3$ or $\pm 2\,\pi/3$ for the exact Bragg condition [12, 13]. Now it is clear that on looking down upon two closely spaced partially overlapping α-fringe patterns of opposite contrast on parallel planes, the combined fringe pattern will appear asymmetrical. Asymmetrical fringe patterns due to intrinsic-extrinsic stacking fault pairs have been observed in silicon by Booker and Howie [14]. The fringe profiles were always asymmetrical in bright field and the nature of the profiles changed progressively with the distance between the two faults.

We interpret, therefore, the asymmetrical fringe patterns observed in the deformed quartz as closely spaced partially overlapping α-fringe patterns of opposite contrast. The separation between the fringe patterns generally appears to be less than 300 Å. This interpretation is supported by micrographs such as Fig. 7 in which the basal planar defects are viewed edge on. It will be seen that the planar defects occur in pairs, the smallest separation being about 50 Å.

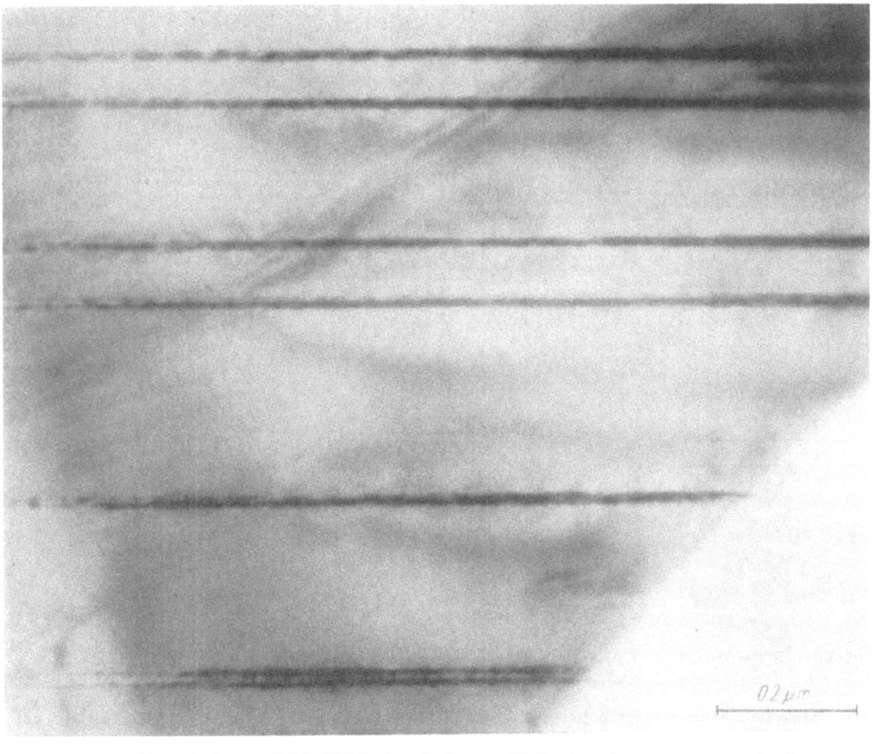

Fig. 7. Six pairs of boundaries parallel to (0001) viewed edge on. Notice that the average separation between the individual boundaries of a pair is of the order of 100 Å. Specimen C-476

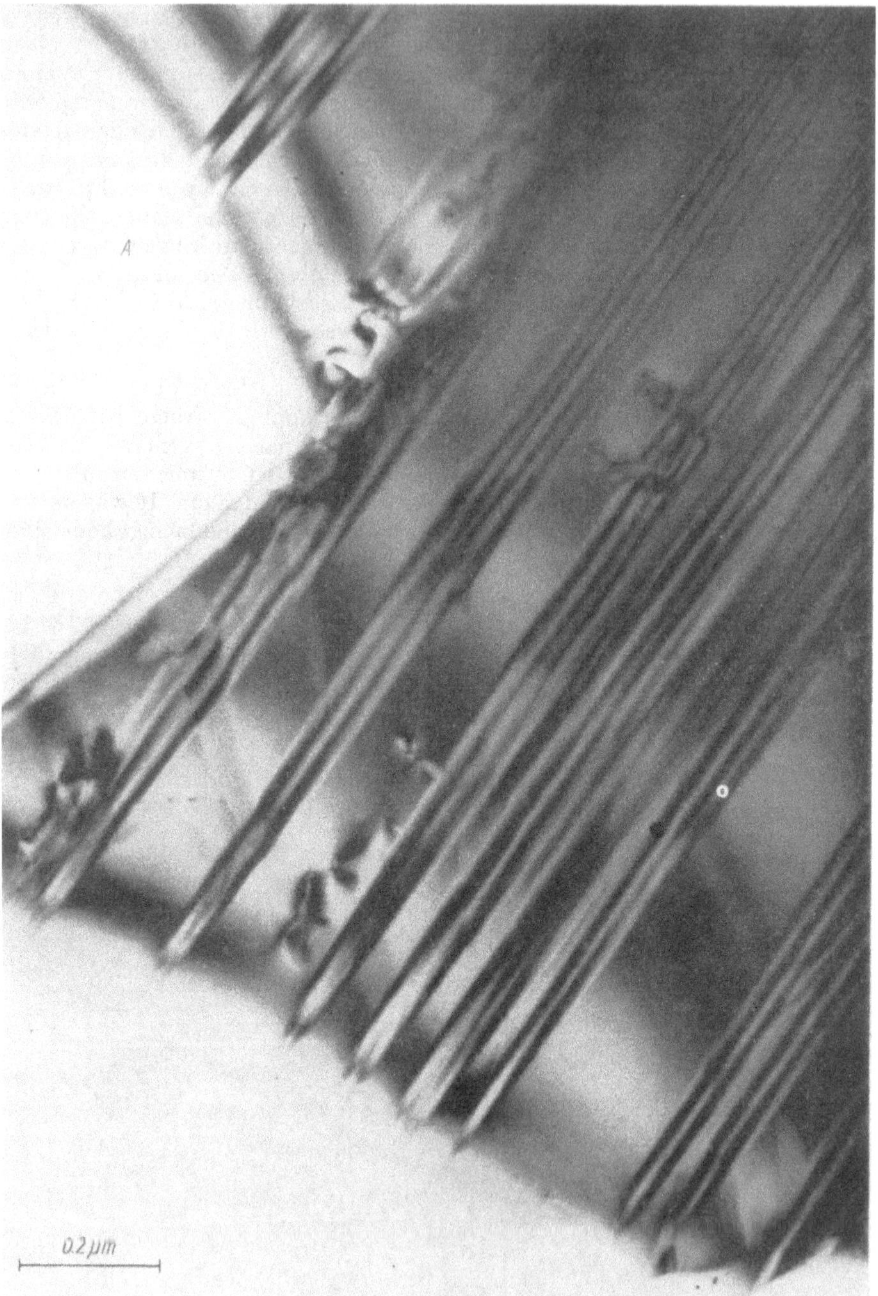

Fig. 8. Diffraction contrast fringe patterns observed in bright field. $\boldsymbol{g} = \{10\bar{1}0\}$. Note the symmetrical nature of the fringe patterns. Specimen C-476

All the fringe patterns discussed so far were imaged with $\boldsymbol{g} = \{10\bar{1}\bar{1}\}$. However, when $\boldsymbol{g} = \{10\bar{1}0\}$ the fringe patterns are always symmetrical, but there is no reversal of contrast (and, therefore, no change in the sign of α) at successive fringe patterns. Also, on passing a thickness extinction contour the additional fringes are created on the outside of the fringe pattern. Examples of these fringes are shown in Fig. 8. If these are α-fringes from single planar defects then the absence of reversal of contrast at each successive fringe pattern indicates that $\alpha = \pm\pi$. The manner in which the additional fringes are created on passing a thickness extinction contour is also characteristic of α-fringes with $\alpha = \pm\pi$ [12]. However, it has been shown above that the fringe patterns generally arise from closely spaced pairs of planar defects. It is clear that two closely spaced partially overlapping α-fringe patterns with $\alpha = \pm\pi$ will give rise to a symmetrical fringe pattern. However, van Landuyt, Gevers, and Amelinckx [12] have shown that for the exact Bragg condition ($s = 0$) the contrast is uniform over the actual area of overlap, but if $s \neq 0$ weak fringes will be seen. A pair of α-fringe patterns which are sufficiently far apart (about 500 Å) not to overlap near the edge of the crystal can be seen at A in Fig. 8. It will be seen that these fringe patterns have all the properties of α-fringes with $\alpha = \pi$ for $s = 0$ [12]. On the other hand, most of the other fringe patterns in this micrograph are sufficiently close together to overlap and the effects of this are clearly seen. We conclude, therefore, that the fringes imaged with $\boldsymbol{g} = \{10\bar{1}0\}$ have $\alpha = \pm\pi$.

The fringe patterns produced by the individual planar defects have precisely the same diffraction contrast characteristics as the fringe patterns due to Brazil twin boundaries observed in amethyst quartz by McLaren and Phakey [8]. It is suggested, therefore, that across each single basal planar defect in the deformed quartz there is a change of hand of the structure, that is, the thin region between two closely-spaced planar defects is of opposite hand to the quartz on either side of it. The composition plane of natural Brazil twins is usually $\{10\bar{1}1\}$, but a change of hand across (0001) is occasionally observed [15].

We now suggest a mechanism that approximates to simple shear by which thin Brazil twin bands may be produced by high shear stress on the basal plane parallel to an a-axis.

Assume the structure is initially left-handed. A projection onto (0001) of this structure[2]) is shown in Fig. 9a. A similar projection of the right-handed structure mirrored about $(2\bar{1}\bar{1}0)$ is shown in Fig. 9b. Fig. 9c shows the transformation of the left-handed structure to the right-handed structure by basal shear parallel to $\boldsymbol{a}_1$, that is parallel to $[2\bar{1}\bar{1}0]$.

The Si atoms at $z = 1.0$ and $z = 0.67$ are translated $a\,[2\bar{1}\bar{1}0]$ into register with similar atoms at lower levels in the left-handed structure. There are no intermediate positions into which these atoms could be displaced without severe misfit, that is without leaving broken Si–O bonds. The Si atoms at $z = 0.33$ are translated $0.395\,a\,[2\bar{1}\bar{1}0]$. For a shear stress in the opposite direction these Si atoms are translated $0.605\,a$. The translation is therefore approximately

[2]) In [8] the left-handed and right-handed structures were, following Frondel [1], incorrectly given the space groups $P3_121$ and $P3_221$ respectively. The space group of the right-handed structure should have been written $P3_121$. The mnemonic given by Lang [16] was, therefore, incorrectly applied. To correct this, the a-axes in Fig. 1 of [8] should be rotated through 60° as in Fig. 9 of the present paper.

(1/2) a and places the Si atoms at $z = 0.33$ in the c-axis channels of the left-handed structure. The tetrahedra centred on the Si atoms at $z = 0.67$ and $z = 1.0$ rotate in opposite directions about an axis which is approximately normal to $(0\bar{1}11)$. The tetrahedra centred at $z = 0.33$ invert parallel to $[2\bar{1}\bar{1}0]$ with very slight motion of the oxygen atoms at $z = 0.21$ and $z = 0.45$. In this model only two Si–O bonds must be broken and reformed. These are either Si(0)–O(0.12) or Si(0.33)–O(0.12) and Si(0.33)–O (0.45) or Si(0.67)–O(0.45). In Fig. 9c

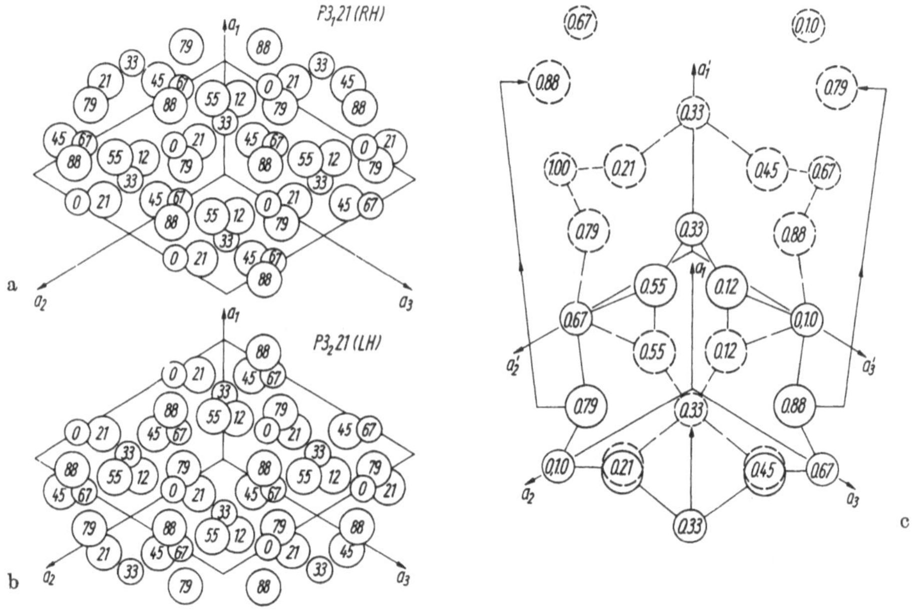

Fig. 9. Projections onto (0001) of the atomic positions in quartz.

a) Right-handed structure ($P3_121$).

b) Left-handed structure, mirrored about the normal to $\boldsymbol{a}_1$.

c) Transformation of left-handed structure (solid circles) to right-handed structure (broken circles) by basal shearing parallel to $\boldsymbol{a}_1$. For detailed discussion, see Section 4 of the text

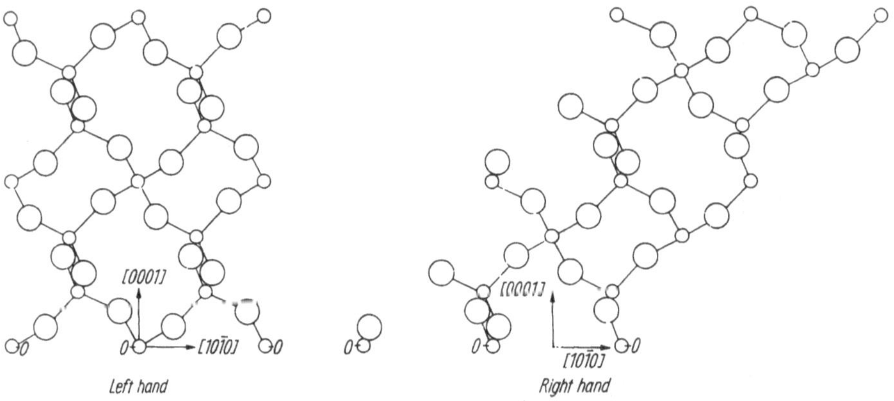

Fig. 10. Profile view of the left-handed structure looking toward $\boldsymbol{a}_2$ and the same atoms after shearing into a right-handed structure. The direction of translation is inclined 30° to the plane of the diagram, dipping in to the left. The centre of each structure is shown by the arrows at the bottom

Fig. 11. Schematic diagram of the sheared crystal

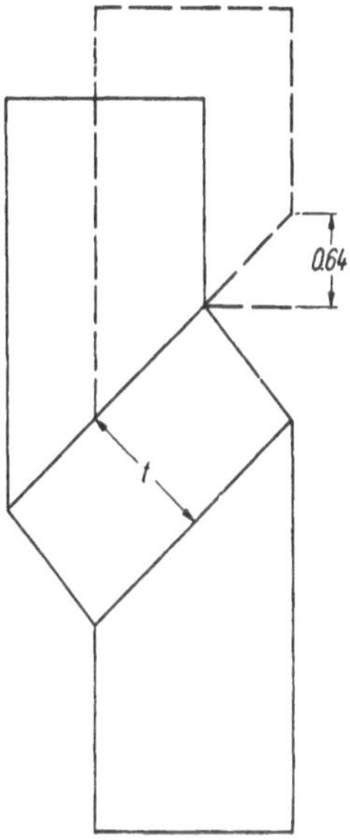

it is assumed that the latter of each pair is broken and reformed. The average shear produced in this transformation is $a/c = 0.91$.

Fig. 10 shows a profile view of this transformation looking towards $\boldsymbol{a}_2$, that is the atoms are projected onto $(1\bar{2}10)$. This is a much simpler projection than the projection onto $(0\bar{1}10)$ which contains the shear direction and the c-axis. As shown the shear is fore shortened, and the change of hand is not obvious in this projection, but the nature of the shear and the motion of the individual atoms is clear.

This shear causes the structure centre (screw axis) to move $0.465\,a$ along $[2\bar{1}\bar{1}0]$. Thus if a region many lattice dimensions thick were sheared in this way to produce a Brazil twin band the individual boundaries of the twin would have the same fault vector as the Brazil twin boundaries discussed by McLaren and Phakey [8]. The appearance of a single crystal specimen deformed by this mechanical twinning is shown in Fig. 11. It can be seen that each twin band of thickness t decreases the length of the cylinder by $0.64\,t$. From the electron micrographs of samples C-247 and C-253, which contain only planar defects, the average thickness of the twin bands is about 10^{-6} cm and there are about 10^5 per cm. The deformation calculated from these figures is about 6% which, in view of the limited accuracy of these figures, in is good agreement with the observed values (Table 1).

5. Nature of the Dislocations

Examination of the crystal structure of quartz suggests that the most probable slip vectors are $a\,\langle 11\bar{2}0\rangle$ since the smallest Burgers vector is any one of the three unit cell base vectors $\boldsymbol{a}_1$, $\boldsymbol{a}_2$, $\boldsymbol{a}_3$, each of lenght $a = 4.91$ Å. Another possible, but energetically less favourable, slip vector is $c\,[0001]$, where $c = 5.39$ Å.

In discussing the nature of the dislocations observed in samples of deformed natural quartz, we will consider firstly crystals in which the highest resolved shear stress in (0001) is along one of the a-axes, say $\boldsymbol{a}_1$. Therefore, the shear stress along $\boldsymbol{a}_2$ and $\boldsymbol{a}_3$ is only one-half that along $\boldsymbol{a}_1$, and a single slip vector $\boldsymbol{b} = \boldsymbol{a}_1$ is expected to operate. This is consistent with the macroscopic and optical microscopic observations of Christie, Griggs, and Carter [3].

The dislocations observed in the electron microscope are almost invariably straight (or consist of straight segments) and parallel to the crystallographic a-axes. Therefore, they cannot be pure edge dislocations with $\boldsymbol{b} = a\langle 11\bar{2}0\rangle$ as has been suggested [3].

In Fig. 5 it will be seen that $\boldsymbol{g} = (10\bar{1}1)$ and that the dislocations are predominantly parallel to $[\bar{1}2\bar{1}0]$. These dislocations cannot be pure screw with $\boldsymbol{b} = a\,[\bar{1}2\bar{1}0]$ for then $\boldsymbol{g}\cdot\boldsymbol{b} = 0$ and the dislocations would be out of contrast [17]. If there is only one Burgers vector of the type $a\,\langle 11\bar{2}0\rangle$, then $\boldsymbol{b}$ is either $a\,[\bar{1}\bar{1}20]$ or $a\,[2\bar{1}\bar{1}0]$. The dislocations parallel to $[\bar{1}2\bar{1}0]$ are, therefore, of mixed character, predominantly edge but with a 30° screw component. Therefore, the dislocations parallel to $[2\bar{1}\bar{1}0]$ or $[\bar{1}\bar{1}20]$ are either of similar mixed character or pure screw.

For crystals in which the highest resolved shear stress in (0001) is along $[0\bar{1}10]$, there is zero shear stress along $[2\bar{1}\bar{1}0]$, and equal shear stress along $[\bar{1}2\bar{1}0]$ and $[11\bar{2}0]$. In this case, two slip vectors $a\,[\bar{1}2\bar{1}0]$ and $a\,[11\bar{2}0]$ would operate in (0001), and we should expect each slip system to involve screw and mixed dislocations of the type already discussed. Our electron microscope observations are consistent with this.

However, for a crystal oriented for high shear stress on (0001), there will also be high shear stress on some or all of the $\{10\bar{1}0\}$ planes since (0001) and $\{10\bar{1}0\}$ are normal to each other. The slip system $\langle 0001 \rangle \{10\bar{1}0\}$ may be expected to operate, and evidence for this has been obtained by Christie, Griggs, and Carter [3]. Therefore, the straight dislocation lines parallel to $\langle 11\bar{2}0 \rangle$ may be edge dislocations with $\boldsymbol{b} = c\,[0001]$. For such dislocations $\boldsymbol{g} \cdot \boldsymbol{b} = 0$ for $\boldsymbol{g} = \{10\bar{1}0\}$ and the images would either be very weak or show the characteristic symmetrical contrast discussed by Howie and Whelan [17, 18]. We have found a few isolated examples of dislocations going out of contrast when $\boldsymbol{g} = \{10\bar{1}0\}$ but, in general, normal dislocation images are observed. Some examples can be seen in Fig. 8. We may conclude, therefore, that the contrast from the dislocations is consistent with their having a Burgers vector $\boldsymbol{b} = a\,\langle 11\bar{2}0 \rangle$.

Arrays of dislocations lying in or near basal Brazil twin boundaries can be seen in Fig. 3 and 4. Unfortunately, the diffraction vector $\boldsymbol{g}$ in both these micrographs was such that it is not possible to say whether the dislocations are of mixed character or pure screw. However, similar examples have been observed in which the dislocations are clearly of mixed character. The average spacing of the dislocations in the arrays is of the order of 10^{-5} cm. Due to their predominantly edge character, such an array of these dislocations could produce a stress field which is consistent with the optical effects observed at deformation lamellae.

6. Discussion and Conclusions

Macroscopic and optical microscopic observations have shown that natural quartz single crystals may be plastically deformed by basal slip under conditions of high pressure, elevated temperature, and high shear stress on the basal plane. Electron microscopy of replicas of etched surfaces showed lines of etch pits coincident with optically visible deformation lamellae. The stress-optical effects of arrays of pure edge dislocations with the spacing of the observed etch pits were in agreement with the measured optical effects. It was concluded that basal slip by the motion of basal edge dislocations was the easiest plastic deformation system in quartz.

The transmission electron microscope observations presented here show that the above conclusion needs to be modified. Sets of thin Brazil twins parallel to the base are the predominant structural defects found in specimens deformed at lower temperatures (500 °C). In specimens deformed at 700 °C mixed and screw dislocations of $\boldsymbol{b} = \boldsymbol{a}$ were also observed, and in one crystal (C-478) they were the predonimant feature.

The Brazil twin bands do not differ in crystal orientation from the host crystal, but the twin band is of opposite hand. No long-range stresses arise at the boundary, so the twin bands themselves will not give rise to the stress-optical effects observed in the optical microscope. Creation of Brazil twins will not *a priori* give rise to any deformation of the crystal, yet these were the only defects found in the low-temperature crystals. The basal shear model which has been proposed for the production of the thin twin bands was designed to transform the left-handed structure to the right-handed structure (or vice versa) with the minimum disturbance of tetrahedra and breaking the minimum number (2) of Si–O bonds. The shear involves a displacement $\boldsymbol{a}$ for every unit cell of height c. In crystals in which only the twin bands are observed (namely those deformed at 500 °C) the deformation calculated from the average thickness and spacing of the bands is in agreement with the measured deformation.

In general the twin bands traverse right across the crystal fragments studied in the electron microscope. However, occasionally a twin band terminates within the crystal. There is an example near the centre of Fig. 3. It can be seen that the twin band termination involves what appears to be a dislocation line parallel to an a-axis. From Fig. 10 it is clear that if a twin band one unit cell thick traversed only part of a crystal, then it would terminate along a line defect resembling a unit dislocation with Burgers vector $\boldsymbol{b} = \boldsymbol{a}$. Similarly if a twin band many unit cells thick changed its thickness by one unit cell, the step so formed would involve a similar dislocation. It is suggested that the mixed dislocations with $\boldsymbol{b} = \boldsymbol{a}$ which appear to lie in the twin bands (as in Fig. 3 and 4) arose in this way. Fig. 10 may be used to construct a model of such dislocations. Due to their predominantly edge character, an array of these dislocations (of the same sign) in a twin band would produce a stress field which is consistent with the optical effects observed at deformation lamellae [3]. Since the area of specimen examined at any one time in the electron microscope is only of the order of 5 μm^2, our failure to observe dislocation arrays in some crystals which contain deformation lamellae is almost certainly due to a low overall concentration of such lamellae.

The other slip mechanism involves dislocations only. Like the twin-terminating dislocations discussed above, these dislocations are almost invariably parallel to $\langle 11\bar{2}0 \rangle$ also. Their diffraction contrast is also consistent with their having a Burgers vecotr $\boldsymbol{b} = \boldsymbol{a}$. Both pure screw dislocations and dislocations of mixed character are observed.

These mixed dislocations are thought to be slip dislocations as distinct from the twin dislocations. The formation of such dislocations can readily be visualized in a crystal which slips by the breaking and reforming of two Si–O bonds e.g. Si(0)–O(0.12) or Si(0.33)–O(0.12) and Si(0)–O(0.21) or Si(0.33)–O(0.21). Dislocations of this sort, with the same Burgers vector and dislocation line as the twin dislocations, would be indistinguishable from the latter by their diffraction contrast.

As we have seen, the observed shear strength of these quartz crystals is somewhat more than half their theoretical strength. Therefore, the simple shear mechanism which has been proposed for the production of thin Brazil twin bands appears to be a reasonable mechanism. The high strength of these crystals also suggests that the observed slip dislocations were produced directly by the applied shear stress.

Synthetic quartz has been shown to have anomalously low strength above a critical temperature which depends on the H_2O content [18]. Transmission electron microscopy of such crystals deformed plastically above the critical temperature show 10^9 to 10^{10} dislocations per cm^2. These dislocations are characteristically curved and tangled. No Brazil twins are found. Synthetic quartz crystals deformed and annealed [19] exhibit dislocation networks characteristic of simple tilt and twist sub-boundaries. These observations will be published separately.

Acknowledgements

We wish to thank Mr. P. P. Phakey, Associate Professor W. A. Rachinger, and Dr. B. E. Hobbs for their critical comments during the course of this work. H. W. Green performed experiment C-478. J. D. Blacic permitted use of his unpublished data in Fig. 1. This work was supported by grants 678-A of the Petroleum Research Fund and 6P5575 of the National Science Foundation.

References

[1] C. Frondel, Dana's System of Mineralogy, Vol. 3, Wiley, New York 1962 (p. 109).
[2] J. M. Christie, H. C. Heard, and P. N. La Mori, Amer. J. Sci. **262**, 26 (1964).
[3] J. M. Christie, D. T. Griggs, and N. L. Carter, J. Geology **72**, 734 (1964).
[4] N. L. Carter, J. M. Christie, and D. T. Griggs, J. Geology **72**, 687 (1964).
[5] J. K. MacKenzie, Ph.D. Thesis, Bristol University, 1949 and private communication (1966).
[6] A. C. McLaren and P. P. Phakey, Austral. J. Phys. **18**, 135 (1965).
[7] A. C. McLaren and P. P. Phakey, Austral. J. Phys. **19**, 19 (1966).
[8] A. C. McLaren and P. P. Phakey, phys. stat. sol. **13**, 413 (1966).
[9] A. C. McLaren and P. P. Phakey, J. appl. Phys. **36**, 3244 (1965).
[10] H. Hashimoto, A. Howie, and M. J. Whelan, Proc. Roy. Soc. **A 269**, 80 (1962).
[11] R. Gevers, J. van Landuyt, and S. Amelinckx, phys. stat. sol. **11**, 689 (1965).
[12] J. van Landuyt, R. Gevers, and S. Amelinckx, phys. stat. sol. **7**, 519 (1964).
[13] H. Blank, P. Delavignette, R. Gevers, and S. Amelinckx, phys. stat. sol. **7**, 747 (1964).
[14] G. R. Booker and A. Howie, Appl. Phys. Letters **3**, 156 (1963).
[15] W. L. Bragg and G. F. Claringbull, Crystal Structures of Minerals, G. Bell & Sons, Ltd., London 1965.
[16] A. R. Lang, Acta cryst. **19**, 290 (1965).
[17] P. B. Hirsch, A. Howie, R. B. Nicholson, D. W. Pashley, and M. J. Whelan, Electron Microscopy of Thin Crystals, Butterworths, London 1965.
[18] D. T. Griggs and J. D. Blacic, Science **147**, 292 (1965).
[19] B. E. Hobbs, unpublished work.

(Received November 23, 1966)

phys. stat. sol. **19**, 645 (1967)

Subject classification: 10; 11; 22.6

General Electric Company, Nuclear Materials and Propulsion Operation, Cincinnati, Ohio

Clustering of Defects in Neutron-Irradiated Beryllium Oxide

By

R. C. Rau and G. A. Chase

Transmission electron microscopy and X-ray diffraction have been combined to study clustering of defects in BeO irradiated to neutron dosages ranging from 0.5 to 11.7×10^{20} nvt at pile ambient temperature (≈ 100 °C). Dot clusters which form in the microstructure at dosages greater than about 1.0×10^{20} nvt are interpreted as being small interstitial loops, oriented on basal planes. X-ray diffraction patterns contain duplex peaks consisting of sharp components, which reflect changes in the bulk lattice, and broad components which are related to the basal clusters. The observations are consistent with a model containing interstitial basal clusters and random point interstitials, but do not necessarily rule out the presence of point vacancies.

Elektronenmikroskopische Durchstrahlungsuntersuchungen und Beugungsmethoden mit Röntgenstrahlen werden kombiniert, um den Zusammenschluß von Defekten in neutronenbestrahltem BeO zu untersuchen. Die Neutronendosen lagen im Bereich 0,5 bis $11{,}7 \times 10^{20}$ nvt bei Reaktortemperaturen um ≈ 100 °C. Punktkomplexe, die sich in der Mikrostruktur bei Dosen größer als ungefähr $1{,}0 \times 10^{20}$ nvt bilden, werden als kleine Zwischengitterplatzschleifen gedeutet, die in der Basisebene orientiert sind. Die Beugungsbilder enthalten Doppelmaxima aus scharfen Komponenten, die Änderungen des Gitters im Kristallinneren anzeigen, und breite Komponenten, die den Komplexen der Basisebene zugeordnet werden. Die Beobachtungen sind konsistent mit einem Modell, das Zwischengitterkomplexe in der Basisebene und willkürlich verteilte Zwischengitterpunktdefekte enthält, aber nicht notwendig Punktlücken ausschließt.

1. Introduction

A fundamental effect of fast ($E_n \geqq 1$ MeV) neutron irradiation on beryllium oxide is the formation of defect clusters resulting from displacements produced by the neutron bombardment and ensuing collision cascades. Irradiation at pile ambient temperatures (≈ 60 to 150 °C) produces a very high density of small defects which appear as black dots in the electron microscope [1 to 3]. Annealing such specimens [2, 4] or irradiating at higher temperatures [2, 5, 6] causes these dot clusters to enlarge into resolvable prismatic dislocation loops lying primarily on the {0001} basal plane of the hexagonal BeO lattice. Recent analysis of such loops has established that they are interstitial in nature and have Burgers vectors of $c/2\ \langle 0001 \rangle$ [7].

The presence of defect clusters in irradiated BeO has also been deduced from X-ray diffraction studies. Patterns from irradiated single crystals show reciprocal lattice streaking in the c^*-direction, indicative of basal plane faulting, and duplex diffraction spots composed of sharp and broad components [3, 8 to 10]. Broadening [11, 12] and duplex structure [8, 13] have also been observed in diffraction line profiles obtained from irradiated polycrystalline BeO, and have been interpreted in terms of basal plane clusters [12, 14]. Recently, this interpretation was extended to include randomly distributed point defects in addition to the clusters [15].

In the present study, clustering of neutron-induced defects in irradiated BeO was observed directly by transmission electron microscopy, and was correlated with X-ray diffraction studies.

2. Specimen Preparation

The BeO specimens were polycrystalline rods formed by extrusion and sintering. They were irradiated in the Engineering Test Reactor (ETR) in Idaho at pile ambient temperatures (≈ 100 °C) to fast ($\geqq 1$ MeV) neutron dosages ranging from 0.5 to 11.7×10^{20} nvt. Neutron dosages were determined from dosimeters utilizing the $Fe^{54}(n, p)Mn^{54}$ transmutation.

Thin sections for transmission electron microscopy were obtained by crushing pieces of the irradiated rods and depositing the fine chips on carbon-filmed support grids. Thin edges of the chips were examined at 80 kV in the electron microscope, utilizing a tilting and rotating stage to orient the chips for the desired imaging conditions. Prominent prism plane cleavage in BeO caused most chips to be fragments perpendicular to the basal plane. Chips with orientations of $\{10\bar{1}0\}$ and $\{11\bar{2}0\}$ were especially common, and were found to be quite desirable from the standpoint of depicting the lattice defects. Comparison of the microstructures of chips obtained by crushing with chips clinging to replicas of freshly fractured surfaces indicated that crushing did not significantly alter the defect structure.

Specimens for X-ray diffraction analysis were prepared from powder obtained by crushing segments of the irradiated BeO rods. To minimize the effects of intergranular strains on the diffraction line profiles [8], the powder was crushed down to its approximate grain size in the sintered rods, i.e., $\approx 10\ \mu m$. The crushed powders were then mixed with a binder and spread in thin layers on glass microscope slides for use on the diffractometer. Strip chart recordings of the diffraction maxima were obtained with copper radiation, using scanning speeds of 1/4 or 1/8° per minute.

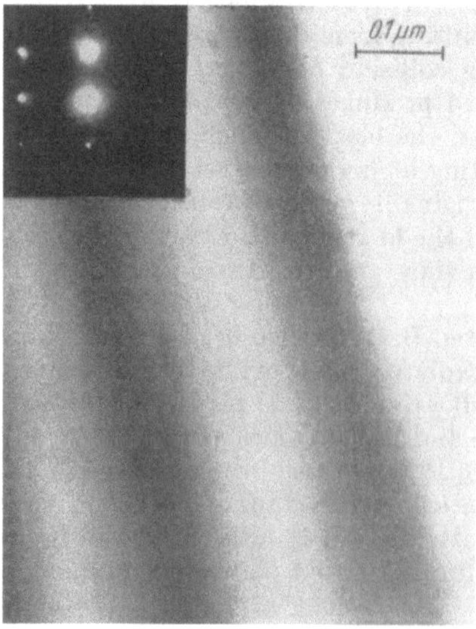

Fig. 1. Electron micrograph of unirradiated BeO showing defect-free structure. (0002) reflection operating

3. Results and Discussion

3.1 Electron microscopy

From a previous study [6] it was known that most of the visible defect clusters present in irradiated BeO were best imaged when the basal planes were oriented to produce diffraction contrast, and were at extinction when $\{hki0\}$ planes were operating. In the prism plane cleavage chips commonly produced by crushing, both $\{000l\}$ and $\{hki0\}$ diffraction conditions can be achieved by suitably orienting the chip with the goniometer stage. Therefore, in studying cluster formation, micrographs were usually taken with $\{0002\}$ planes operating to produce strong diffraction contrast images, and extinction conditions were checked with $\{10\bar{1}0\}$ or $\{11\bar{2}0\}$ reflections operating, as well as with various $\{hkil\}$ planes.

Examination of unirradiated BeO chips revealed relatively featureless microstructures with little or no visible detail. A typical micrograph of such a specimen is shown in Fig. 1. This micrograph shows the alternating dark and light extinction contours due to changing thickness of the wedge-shaped crystal, but

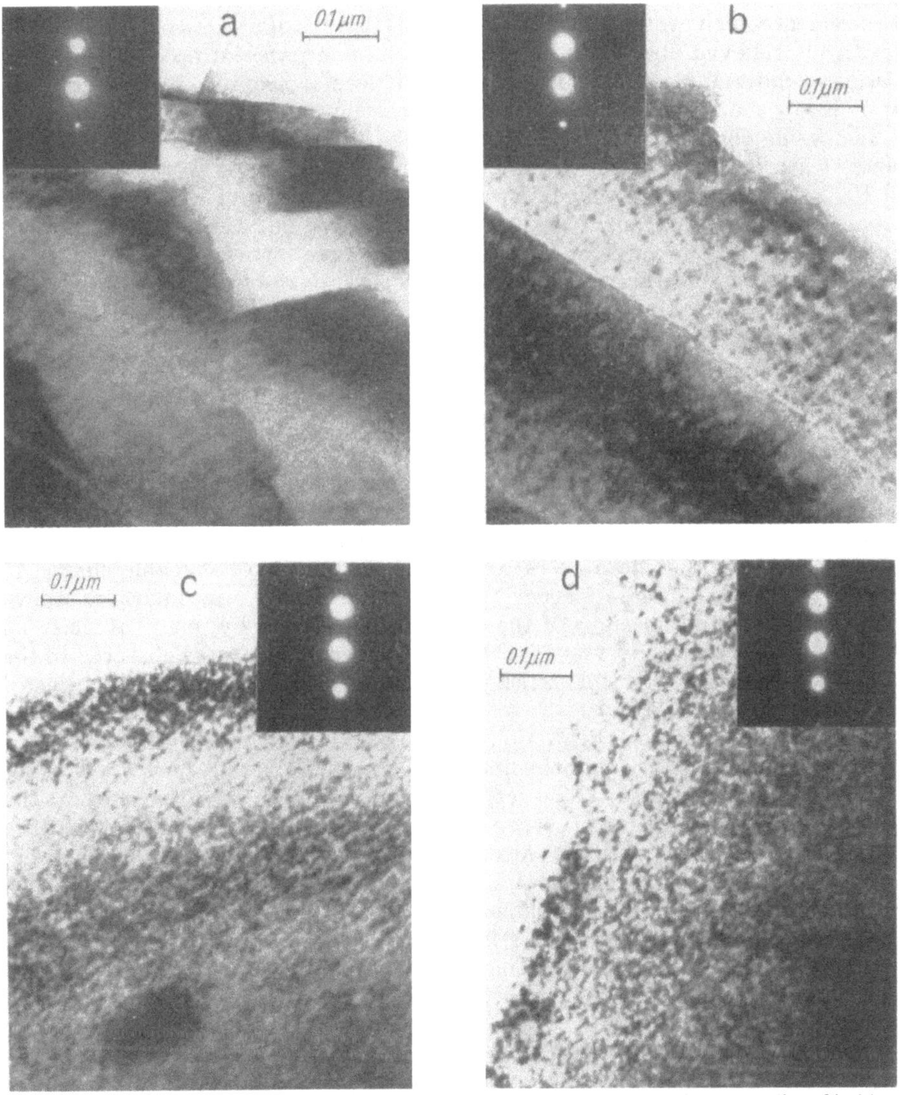

Fig. 2. Electron micrographs showing defect structure of BeO irradiated to various dosages at pile ambient temperature. All micrographs taken with (0002) reflection operating.
a) 0.5×10^{20} nvt. b) 1.8×10^{20} nvt.
c) 3.0×10^{20} nvt. d) 11.7×10^{20} nvt

no defects. Tilting such chips to other orientations also failed to bring any defects into contrast.

Clustering of neutron-induced defects was studied in a series of BeO specimens irradiated to various dosages ranging from 0.5 to 11.7×10^{20} nvt ($E_n \geqq 1$ MeV) at pile ambient temperature (≈ 100 °C). Fig. 2 shows typical micrographs of four of these specimens, illustrating the evolution of distinct dot clusters with increasing neutron dose.

Fig. 2a shows the microstructure of the lowest-dose sample, 0.5×10^{20} nvt. At this dose, no distinct clusters could be resolved, but a "textured" appearance was evident which contrasted with the microstructure of unirradiated BeO. This texture is believed due to the presence of a large number of point defects consisting of individual or multiple atomic or molecular groupings whose sizes are at or below the resolution limit of the electron microscope.

Increasing the irradiation dose caused resolvable dot clusters to form. Such clusters are seen superimposed on a "textured" background in the micrograph of Fig. 2b, showing a specimen irradiated to 1.8×10^{20} nvt. This specimen is representative of the transition range where the total number and density of the point defects become large enough that compressive forces within the strained lattice cause agglomeration of the point defects into resolvable clusters. These clusters were imaged under conditions such that $\{hkil\}$ reflections, with l nonzero, were operating, but were at extinction when $\{hki0\}$ reflections were operating. These diffraction conditions are the same as those required for imaging the enlarged basal loops in high-temperature irradiated BeO [6], indicating that the dot clusters are probably small basal loops.

Irradiation to dosages above 2.0×10^{20} nvt produced microstructures containing very large numbers of dot clusters estimated to be approximately 50 to 100 Å in diameter. Typical micrographs are shown in Fig. 2c (3.0×10^{20} nvt dose) and 2d (11.7×10^{20} nvt dose). Cluster densities could not be determined from the micrographs because of overlap of images. No noticeable change in size of the clusters was apparent as irradiation dosage was increased above 2.0×10^{20} nvt. Also, because of the limitations imposed by image overlap, no increase in cluster density was apparent. However, effects were observed in the electron diffraction patterns which could be attributed to increasing cluster concentration.

Selected area electron diffraction patterns of specimens irradiated to low dosages exhibited sharp Kikuchi lines, indicative of over-all long-range order and lattice homogeneity. However, with the onset of cluster formation the Kikuchi lines began changing, becoming progressively weaker and more diffuse with increasing neutron dose. Also, the orientation requirements for Bragg diffraction became considerably less stringent with increasing neutron dose. For the highest dosage samples examined (5.6×10^{20} nvt and above), no Kikuchi lines were seen and strict two-beam diffraction conditions could not be obtained, regardless of chip orientation. These diffraction effects are consistent with a continuing build-up of clusters with increasing dose, leading to severe distortions of the lattice.

Based on X-ray results, several authors [12, 14, 15] have concluded that the basal clusters in irradiated BeO are probably composed of interstitials. Recently, direct confirmation of this conclusion was achieved by electron microscope tilting experiments on large loops in BeO irradiated at 1000 and 1200 °C [7]. Because of the demonstrated increase in size of the clusters with increasing irradiation temperature [2, 6], it is probably fair to assume that the dot clusters shown in Fig. 2 are also composed of intersititials. Nevertheless, it was felt desirable to attempt direct identification in some of the specimens of the present study.

Standard tilting experiments for defect identification cannot be applied to clusters too small to be resolved into loops. Likewise, recently proposed dark-field techniques will not work on clusters in irradiated BeO for theoretical reasons (see [6] for a discussion). Therefore, the identification was carried out on a specimen annealed after irradiation. Pieces of a specimen which had been irradiated to 3.0×10^{20} nvt (Fig. 2c) were annealed for various lengths of time ranging from 15 min to 3 h at 1050 °C. The effect of this annealing was to reduce the cluster density and produce resolvable dislocation loops on the basal planes. The 3-hour annealed sample contained some loops approximately 300 to 500 Å in diameter which were suitable for analysis.

Fig. 3 shows micrographs illustrating typical results of tilting experiments in the electron microscope. Fig. 3a shows a $\{10\bar{1}0\}$ cleavage fragment before tilting, with the $(000\bar{2})$ reflection operating. In this orientation the loops are viewed edge-on, and appear linear perpendicular to the c-axis. After tilting approximately 15° about the $[2\bar{1}\bar{1}0]$ axis, the images became elliptical when diffraction positions were traversed. In Fig. 3b and 3c are shown the effects of tilting through the $(10\bar{1}3)$ diffraction position, changing the sign of the quantity $(\boldsymbol{g} \cdot \boldsymbol{b})\, s$ by changing s (i.e., by sweeping the $(10\bar{1}3)$ Kikuchi line through its Bragg diffraction spot). It can be seen that this small increase in tilt caused the loop images to widen considerably. This behavior is characteristic of interstitial loops [16]. This same result was repeated on several chips by similar tilting through various diffraction positions.

A final result from these tilting experiments was the observation of a line believed due to stacking fault contrast within most of the loops when $\{10\bar{1}3\}$

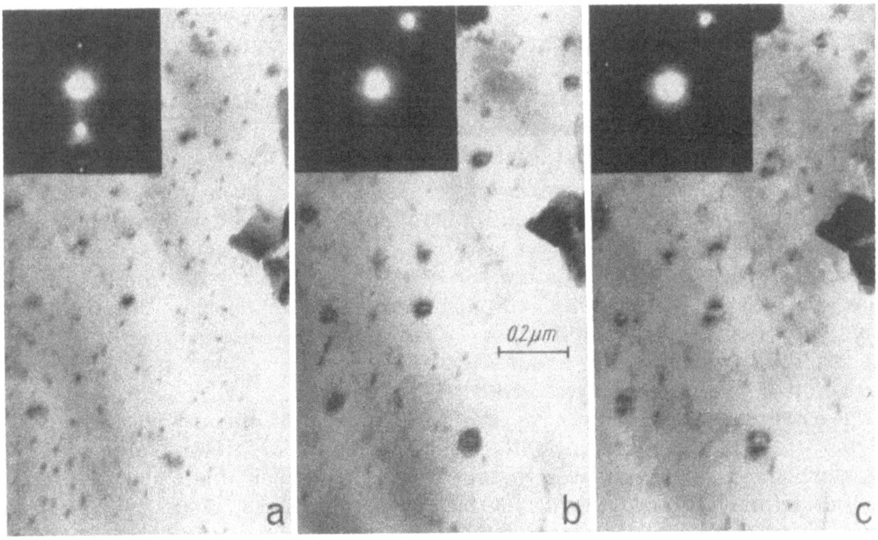

Fig. 3. Electron micrographs of irradiated and annealed BeO used for cluster identification.
a) Tilt angle ≈ 0°, $(000\bar{2})$ reflection operating.
b) Tilt angle ≈ 17°, $(\bar{1}0\bar{1}3)$ reflection operating, $s < 0$.
c) Tilt angle ≈ 17.5°, $(10\bar{1}3)$ reflection operating, $s > 0$

reflections were operating. The presence of faulted loops is supported by the reciprocal lattice streaking in the c^*-direction observed in single-crystal X-ray diffraction photographs [3, 14]. The appearance of stacking fault contrast for $\langle 10\bar{1}3\rangle$ $\boldsymbol{g}$ vectors leads to the conclusion that the Burgers vector $\boldsymbol{b}$ is $c/2\ \langle 0001\rangle$. Such a Burgers vector results when an interstitial layer is inserted into a C stacking position between A and B layers in a hexagonal lattice. In keeping with the BeO structure, these interstitial clusters are probably composed of BeO double layers, as Sabine [14] has suggested.

3.2 X-ray diffraction

Rather complex X-ray diffraction effects occur in neutron-irradiated BeO. In particular, diffraction peak positions and relative intensities change, duplex peaks consisting of sharp and broad components occur, and reciprocal lattice streaking is observed. For the present work, several of these effects were studied by slow-scan powder diffractometry of BeO samples irradiated to various dosages ranging from 0.5 to 11.7×10^{20} nvt at 100 °C.

Fig. 4 shows diffractometer traces of the $\{10\bar{1}0\}$, $\{0002\}$, and $\{10\bar{1}1\}$ BeO peaks which illustrate the progressive changes which occurred in the diffraction patterns with increasing neutron dose. Unirradiated BeO gave a diffraction pattern virtually identical with that of the lowest-dose irradiated sample, 0.5×10^{20} nvt, and hence is not included in the figure. The diffraction patterns are characterized by two sets of peaks, the sharp components and the broad components, respectively, of the duplex reflections. The sharp components remain sharp throughout irradiation, but shift toward smaller diffraction angles and decrease in relative intensity with increasing dose. The broad components increase in relative importance with increasing neutron dose, and also with decreasing angle of slope of the reflecting planes to the basal plane.

The fact that the sharp components remain sharp throughout irradiation indicates that over-all homogeneity of the BeO lattice is maintained. The position shifts of these peaks indicate that the lattice expands anisotropically with dose, with the increase of the c-parameter being greater than that of the a-parameter by approximately an order of magnitude. The lattice parameters plotted in Fig. 5 were computed from the sharp components of back reflection peaks ($2\,\theta > 113°$), using a weighted least-squares analytical extrapolation technique. Values of the a- and c-parameters were also deduced from the sharp $\{10\bar{1}0\}$ and $\{0002\}$ peaks, respectively, and were found to agree well with the computed values. As shown by the curves, the a-parameter approached saturation at higher dosages, while the c-parameter continued to increase. Similar results were obtained by Walker et al. [8, 17] in measurements made on irradiated polycrystalline and single crystal BeO. However, the sharp increase in c-axis expansion rate at dosages of about 2.0×10^{20} nvt reported by those authors was not as noticeable in the present study.

The coherent lattice expansion determined from the sharp components is attributed to the presence of point defects, randomly distributed on an atomic scale, which dilate the entire crystal lattice [15]. Such dilatation, occurring without appreciable broadening of the diffraction peaks, is analogous to the lattice changes of a solid solution. The dilating centers causing the expansion are probably single interstitials or small groups of interstitials dislodged from normal lattice sites by collision processes. However, the existence of isolated vacancies in the lattice is not precluded by this assumption. Tucker and Samp-

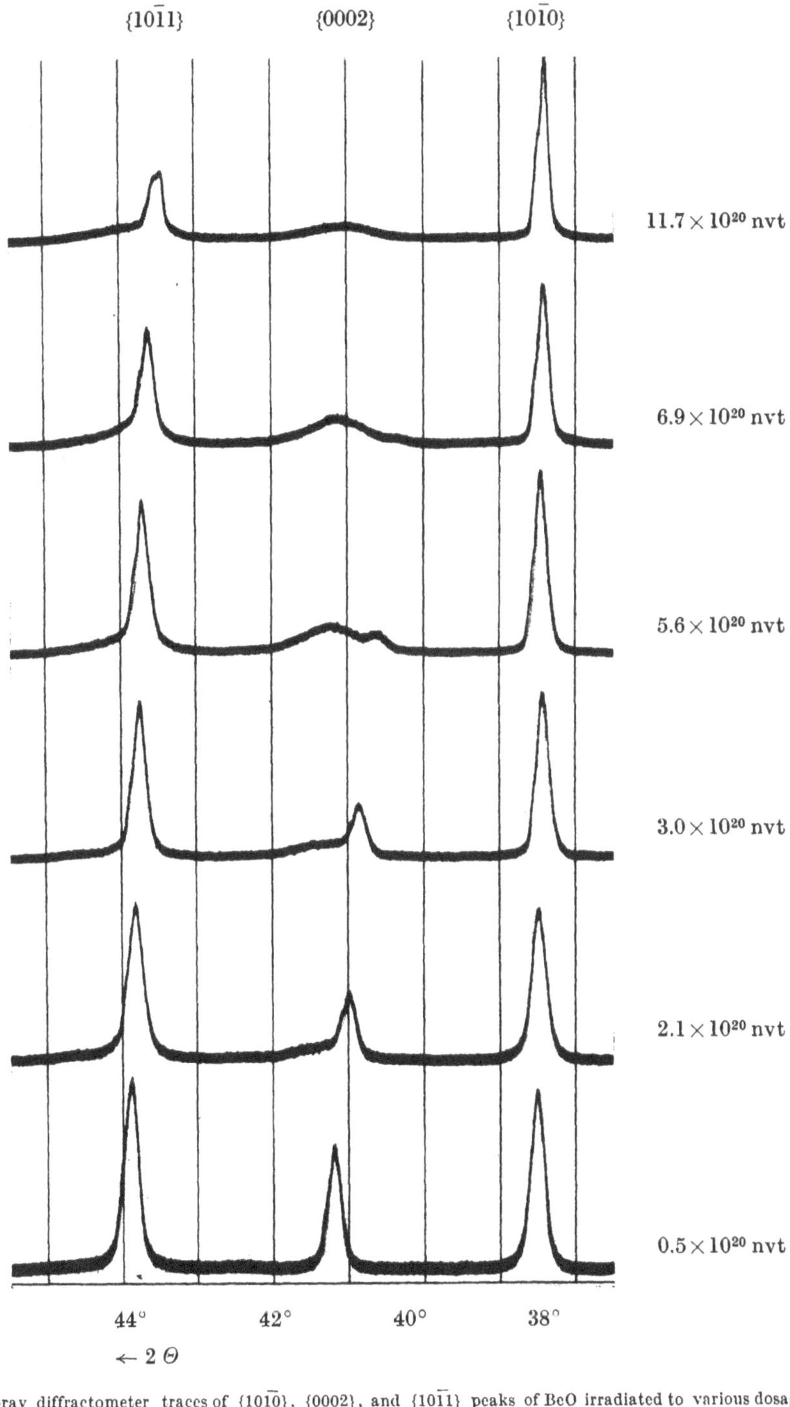

Fig. 4. X-ray diffractometer traces of {10$\bar{1}$0}, {0002}, and {10$\bar{1}$1} peaks of BeO irradiated to various dosages at pile ambient temperature

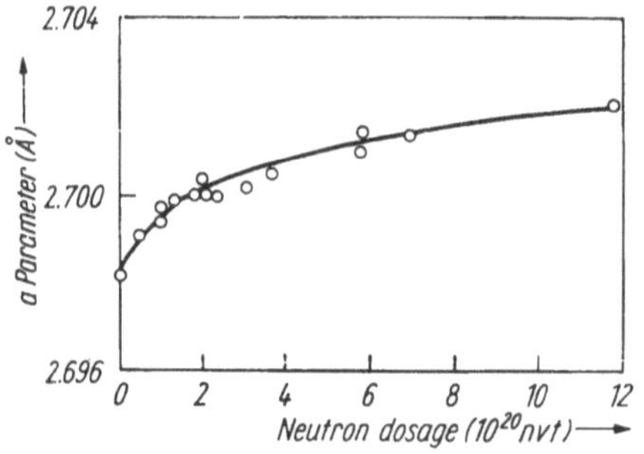

a

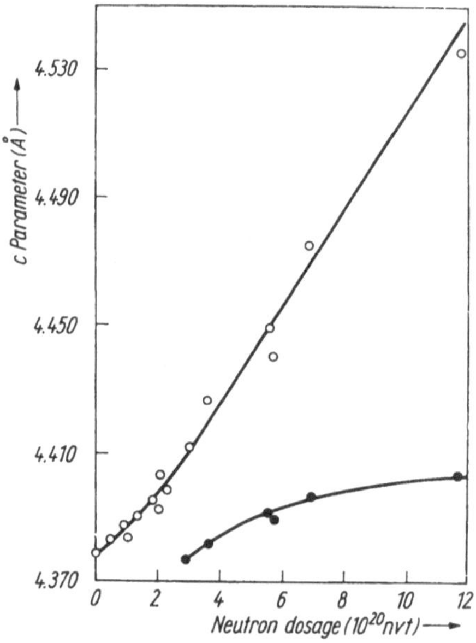

b

Fig. 5. Changes in lattice parameters of BeO with neutron dose. a) a-parameter. b) c-parameter; the lower curve represents values determined from the {0002} broad component

son [18] have shown that, although vacancies produce local relaxation of the lattice, the contraction is considerably less than the expansion caused by interstitials, and hence the lattice dimensions of crystals containing both types of defects are likely to be dominated by their interstitial content.

Changes in the relative intensities of the sharp components are evident in the three front reflection peaks of the BeO diffraction patterns of Fig. 4. These intensity changes are representative of those observed in the complete patterns, which may be summarized as follows: 1. $\{000l\}$ intensities were drastically reduced by irradiation, completely disappearing at the higher dosages. 2. $\{hki0\}$ intensities were practically unaffected by the dosages employed. 3. $\{hkil\}$ intensities were reduced in inverse proportion to the angle of slope of the reflecting planes to the basal plane. In other words, sharp reflections having large l index and small h and k indices suffered greater intensity reduction than those having l small and h and/or k large.

The reduction of intensities of the sharp components is similar to that caused by thermal motion, and can be attributed to displacement of atoms from the mean planar positions in the vicinity of the defects. From the observed behavior of the intensity reductions it is evident that these displacements must be highly anisotropic, being much greater in the c-direction than in the a-direction.

The diffraction patterns of Fig. 4 also illustrate the behavior representative of the broad components of the duplex reflections during irradiation. Evidence of the broad components first began to appear in the diffraction patterns after dosages in the range of 1.0 to 2.0×10^{20} nvt, and was more noticeable the more nearly parallel the diffracting planes were to the basal plane. These broad components appeared to increase in intensity at the expense of the sharp components, and, in contrast to the sharp components, shifted only slightly toward smaller diffraction angles with increasing neutron dose. This latter effect results in smaller lattice parameters being associated with the broad components than with the corresponding sharp components for a given dosage. This is illustrated

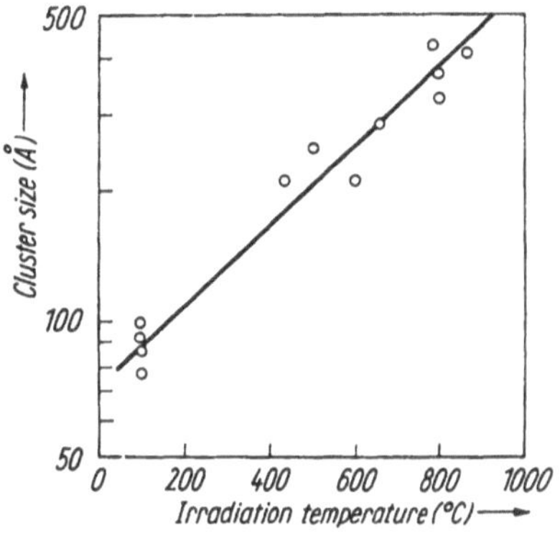

Fig. 6. Clusters sizes estimated from {0002} broad component peak widths for BeO irradiated at various temperatures

graphically by the lower curve of Fig. 5b, which shows values of the c lattice parameter deduced from the position of the maximum of the {0002} broad component.

It is particularly significant that the initial appearance of broad components in the X-ray diffraction patterns coincided with the onset of clustering visible in the electron microscope. The source of the broad components can therefore be attributed to the basal clusters. Crowding of these clusters between basal planes leads to localized reductions of spacings in the expanded lattice, thus accounting for the smaller position shifts of the broad components.

The correlation between the broad components and the clusters was investigated by X-ray line-broadening analyses of the {0002} diffraction peaks. This technique relates X-ray line widths, measured at half maximum intensity, to crystallite dimensions in the direction perpendicular to the diffracting planes [19]. For the case of irradiated BeO, the dimensions measured correspond to the strained region surrounding the clusters rather than true crystallite or cluster dimensions, but such a measurement is meaningful, since these strain regions are responsible for the cluster images seen in the electron microscope.

Clusters sizes obtained for pile ambient irradiated BeO ranged from 77 to 100 Å, in good agreement with the electron microscopy results, with the trend being toward larger values with increasing dose. Although this increase is slight, it may be significant. The analysis was extended by making similar measurements on a number of BeO specimens irradiated to various dosages in the 10^{20} nvt range at temperatures ranging up to 860 °C. Increasing irradiation temperature resulted in narrowing of the broad components, leading to increased estimated cluster sizes. These results are plotted as a function of irradiation temperature in Fig. 6, ignoring the possible dosage effect. The values indicated by the plot are in good agreement with cluster sizes which have been seen in the electron microscope [6]. Because of decreasing reliability as peak widths approach that due to instrumental effects, extension to even larger cluster sizes was not attempted.

4. Conclusions

The picture of the damage state in pile ambient irradiated BeO resulting from this study is one in which both point defects and clusters are involved, and is basically similar to that proposed by Austerman and Miller [15]. Irradiation first produces point defects at random sites throughout the lattice by direct neutron bombardment and secondary collision cascades. Although these point defects may be both vacancies and interstitials, only the interstitials are experimentally evident through their expansion of the lattice parameters. Continued irradiation adds point defects until their concentration reaches a level at which lattice forces begin to cause agglomeration into visible clusters. This occurs over the neutron dose range from 1.0 to 2.0×10^{20} nvt. The clusters are probably interstitial double layers of beryllium and oxygen atoms, inserted into rhombohedral stacking sequence in the otherwise hexagonal lattice. Within the limits of this study, the clusters increase in number and possibly also in size with increasing neutron dose.

From these studies, little can be concluded regarding the vacancy content of irradiated BeO. It seems certain that the clusters are all interstitial, and it likewise seems assured that the bulk lattice parameters are controlled by point interstitials. However, since isolated vacancies do not cause appreciable lattice distortion, considerable numbers of them may be present. The current feeling is that point vacancies are indeed generated in the lattice, but that some migrate to grain boundaries and disappear, some are annihilated by combination with interstitials, and some are left randomly distributed throughout the lattice. Many of the latter probably serve as sinks for helium and tritium formed by the (n, 2n) and (n, α) reactions with beryllium, and are effectively trapped in the lattice.

Acknowledgements

The authors are grateful to C. G. Collins for his encouragement and advice throughout the course of this work.

References

[1] D. H. Bowen, R. S. Wilks, and F. J. P. Clarke, J. nuclear Mater. **6**, 148 (1962).
[2] J. H. Chute and D. G. Walker, J. nuclear Mater. **14**, 187 (1964).
[3] R. C. Rau, S. F. Bartram, and K. Lacefield, J. Amer. Ceram. Soc. **48**, 233 (1965)
[4] A. Bisson, J. nuclear Mater. **10**, 321 (1963).
[5] R. S. Wilks and F. J. P. Clarke, J. nuclear Mater. **14**, 179 (1964).
[6] R. C. Rau, J. nuclear Mater. **20**, 141 (1966).
[7] R. C. Rau, Nature **211**, 71 (1966).
[8] D. G. Walker, R. M. Mayer, and B. S. Hickman, J. nuclear Mater. **14**, 147 (1964).
[9] D. G. Walker, J. nuclear Mater. **15**, 111 (1965).
[10] B. Belbeoch, J. Rodot, and M. Roulliay, J. nuclear Mater. **15**, 234 (1965).
[11] B. S. Hickman, T. M. Sabine, and R. A. Coyle, J. nuclear Mater. **6**, 190 (1962).
[12] H. L. Yakel and B. Borie, Acta cryst. **16**, 589 (1963).
[13] S. T. Konobeevskii, V. I. Klimenkov, and V. M. Kosenkov, Dokl. Akad. Nauk SSSR **165**, 524 (1965).
[14] T. M. Sabine, J. nuclear Mater. **14**, 159 (1964).
[15] S. B. Austerman and K. T. Miller, phys. stat. sol. **11**, 241 (1965).
[16] B. Edmondson and G. K. Williamson, Phil. Mag. **9**, 277 (1964).
[17] B. S. Hickman, D. G. Walker, and R. Hemphill, J. nuclear Mater. **14**, 167 (1964).
[18] C. W. Tucker and J. B. Sampson, Acta metall. **2**, 433 (1954).
[19] R. C. Rau, Adv. X-Ray Analysis **5**, 104 (1962); **6**, 191 (1963).

(Received November 23, 1966)

phys. stat. sol. **19**, 655 (1967)

Subject classification: 18.2; 10.1; 21.1

Institut für Physik am Max-Planck-Institut für Metallforschung, Stuttgart, und Institut für theoretische und angewandte Physik der Technischen Hochschule Stuttgart

Verformungs- und Temperaturabhängigkeit der Magnetisierungsprozesse in Nickeleinkristallen

II. Blochwandverschiebungen und Drehungen der spontanen Magnetisierung in ⟨001⟩-Nickeleinkristallen [1]

Von

E. KÖSTER

Die Magnetisierung und die magnetostriktive Längenänderung stabförmiger Nickeleinkristalle mit ⟨001⟩-Achsenrichtung wurden in Abhängigkeit von der plastischen Verformung und der Temperatur gemessen. In kleinen Feldern erfolgt die Magnetisierung gleichzeitig durch Wandverschiebungen und Drehungen; Magnetisierung und Längenänderung werden durch das Rayleigh-Gesetz beschrieben. Der Anteil der Wandverschiebungen an der Anfangssuszeptibilität ist proportional zu $(\tau - \tau_0)^{-1}$, der der Drehungen ist proportional zu $(\tau - \tau_0)^{-2}$ und macht etwa 15 bis 30% der Gesamtsuszeptibilität aus. Hierbei ist $(\tau - \tau_0)$ die Verfestigung der Kristalle. Aus den Messungen kann die Spannungsanisotropie bestimmt werden, die eine Aussage über die Versetzungsanordnung zuläßt. In der Arbeit werden weiter die Ergebnisse aus Untersuchungen irreversibler Magnetisierungsprozesse entlang der ganzen Magnetisierungskurve mitgeteilt.

Measurements are made of the magnetization and magnetostriction of rod-shaped ⟨001⟩-Ni single crystals at various temperatures and with various degrees of plastic deformation. The results show that the magnetization at low magnetic fields takes place simultaneously by domain wall displacements and rotational processes; magnetization and magnetostriction obey the Rayleigh law. The contribution of the wall displacements to the initial susceptibility is proportional to $(\tau - \tau_0)^{-1}$; the contribution of the rotational processes is proportional to $(\tau - \tau_0)^{-2}$ and amounts to about 15 to 30% of the total susceptibility. Here $(\tau - \tau_0)$ represents the hardening of the crystals. The stress anisotropy may be obtained from the measurements, which yields information concerning the dislocation arrangement. The results of an investigation of irreversible magnetization processes are also reported.

1. Einleitung

In der vorausgegangenen Veröffentlichung [1], im folgenden mit Teil I bezeichnet, wurden die Ergebnisse über Magnetisierungsprozesse durch Blochwandverschiebungen in stabförmigen ⟨110⟩- und ⟨111⟩-Nickeleinkristallen zusammen mit den Untersuchungsmethoden mitgeteilt. Die in der vorliegenden Arbeit untersuchten ⟨001⟩-Nickeleinkristalle geben die Möglichkeit, den von Kronmüller [2] theoretisch behandelten Fall von Wandverschiebungen mit gleichzeitig ablaufenden Drehungen der Magnetisierung zu untersuchen.

[1]) Dissertation, Teil II, Technische Hochschule Stuttgart, 1966.

2. Meßergebnisse

Die plastische Verformung der ⟨001⟩-Kristalle verlief in symmetrischer Doppelgleitung [3], wie aus lichtmikroskopischen Gleitlinienbeobachtungen und röntgenographischer Orientierungsbestimmung der Kristallachse abgeleitet wurde. Die Verfestigungskurven eines bei 77 °K (A) und eines bei 293 °K (B) verformten Kristalls sind in Fig. 1 dargestellt. Bei der Berechnung der Gesamtabgleitung a wurde die algebraische Summe der Abgleitungen in beiden Gleitsystemen gebildet.

Die Feldabhängigkeit der Magnetisierung, der Längenänderung und der reversiblen Suszeptibilität entlang der Neukurve bei 273 °K zeigt Fig. 2 für drei verschiedene Verformungsstufen 0, 3 und 8[2]) des Kristalls A. Dem steilen Teil der Magnetisierungskurven entspricht eine geringe Zunahme der reversiblen Suszeptibilität bis zur Knickfeldstärke H_K und der steile Abfall der reversiblen Suszeptibilität. Die Feldstärke am Wendepunkt der $J(H)$-Kurven ist gleich groß wie die Knickfeldstärke H_K. H_K ist durch die Extrapolation der geradlinigen Kurvenstücke bestimmt, wie dies am Beispiel der Kurve 8 angedeutet ist.

Ein Ausschnitt der Kurven von 0 bis 40 Oe ist in Fig. 3a bis c für die Meßtemperaturen 189, 273 und 363 °K der Verformungsstufen 0, 5 und 8 des Kristalls A bzw. 0, 6 und 16 des Kristalls B jeweils gleicher Verfestigung $(\tau - \tau_0)$ dargestellt. Die Magnetisierungskurven folgen bei allen Verformungen und Temperaturen bis zu einer Feldstärke $H_0 \approx H_K/2$ dem Rayleigh-Gesetz

$$J = \chi_A H_i + \alpha H_i^2 \quad (1)$$

mit der Anfangssuszeptibilität χ_A und der Rayleigh-Konstanten α. Die innere Feldstärke H_i ist die Differenz $H_i = H_a - N J$ aus der äußeren Feldstärke H_a und dem Streufeld $N J$ (N Entmagnetisierungsfaktor in Stabrichtung). Die Kurven der Längenänderung haben bei $H_a = 0$ eine Horizontaltangente und

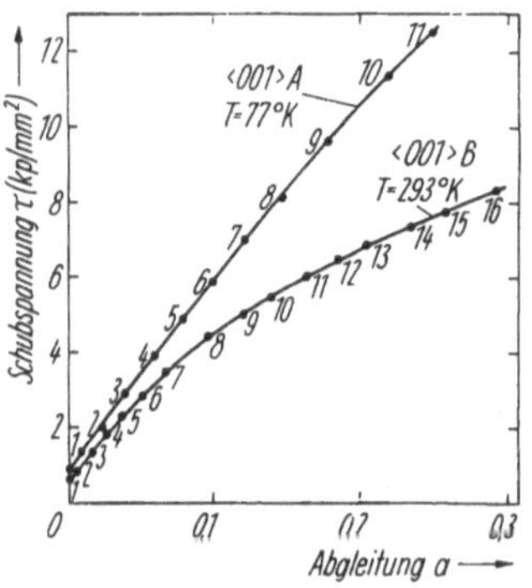

Fig. 1. Verfestigungskurven bei 77 und 293°K verformter ⟨001⟩-Nickeleinkristalle

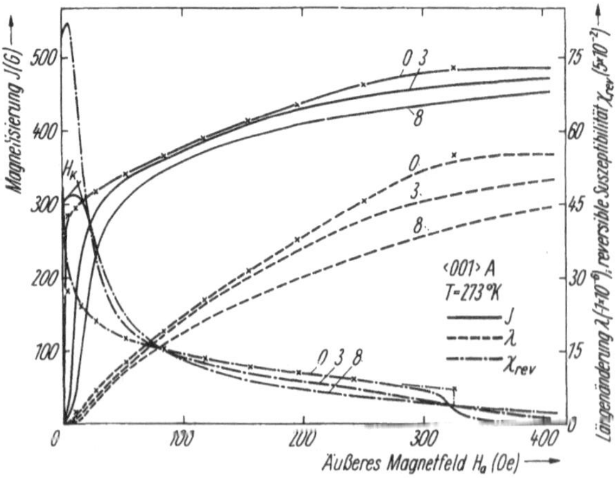

Fig. 2. Die Magnetisierung, die Längenänderung und die reversible Suszeptibilität verformter ⟨001⟩-Nickeleinkristalle bei 273 °K als Funktion der Feldstärke. Die durchgezogenen Linien stellen die Meßwerte, die Kreise berechnete Werte dar

---

[2]) Die Zahlen beziehen sich auf die Verfestigungskurven in Fig. 1.

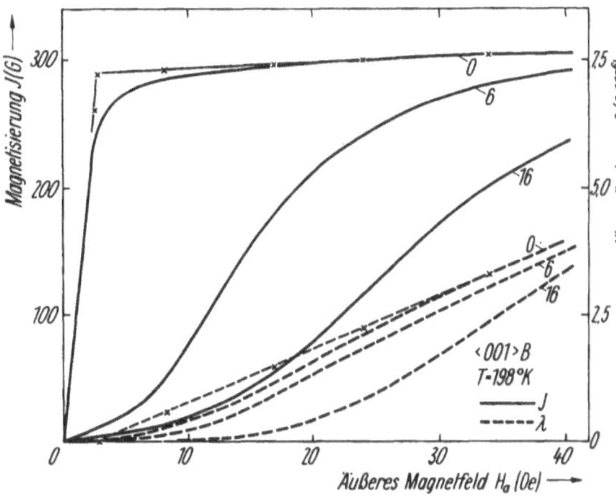

a

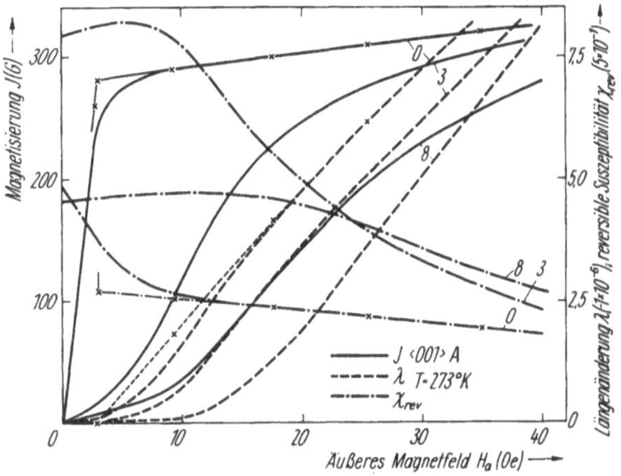

b

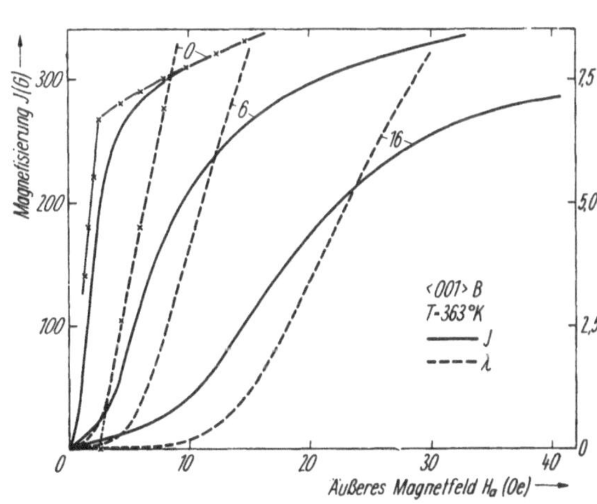

c

Fig. 3. a) Die Magnetisierung und Längenänderung verformter ⟨001⟩-Nickeleinkristalle bei 198 °K als Funktion der Feldstärke.
b) Die Magnetisierung, Längenänderung und reversible Suszeptibilität verformter ⟨001⟩-Nickeleinkristalle bei 273 °K als Funktion der Feldstärke.
c) Die Magnetisierung und Längenänderung verformter ⟨001⟩-Nickeleinkristalle bei 363 °K als Funktion der Feldstärke

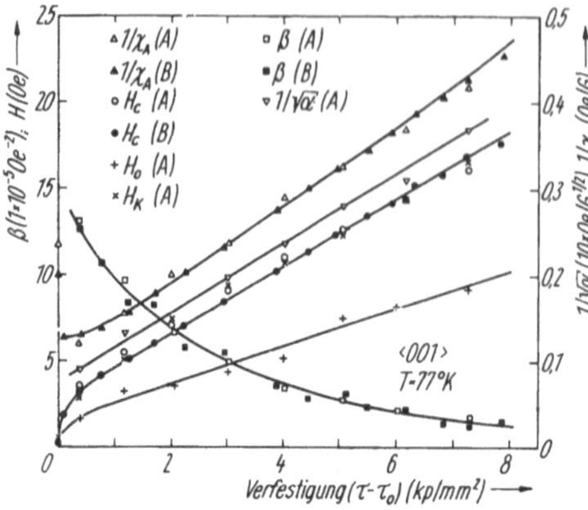

Fig. 4. Der Kehrwert der Anfangssuszeptibilität χ_A, der Kehrwert der Wurzel aus der Rayleigh-Konstanten α, der Koeffizient β, und die Koerzitivfeldstärke H_c eines ⟨100⟩-Nickeleinkristalls bei 273 °K als Funktion der Verfestigung $(\tau - \tau_0)$

steigen im gleichen Feldbereich bis H_0 gemäß der Beziehung

$$\lambda = \beta\, H_i^2 \tag{2}$$

proportional zu H_i^2 an. Der Koeffizient β wird wie χ_A und α mit zunehmender Verfestigung und abnehmender Temperatur kleiner. Bei 77 °K ist β so klein, daß unterhalb $J = 60$ G die Längenänderung nicht mehr gemessen werden konnte.

In Fig. 4 sind die Koerzitivfeldstärke H_c, der Gültigkeitsbereich des Rayleigh-Gesetzes H_0, der Kehrwert der Anfangssuszeptibilität χ_A, die Wurzel aus dem Kehrwert der Ralyeigh-Konstanten α und der Koeffizient β als Funktion der Verfestigung bei 273 °K zusammengestellt. Alle Parameter, außer β, nehmen in dieser Darstellung ab $(\tau - \tau_0) = 1$ kp/mm² linear mit der Verfestigung $(\tau - \tau_0)$ zu. Wie die gute Übereinstimmung der Werte der Kristalle A und B zeigt, ist das magnetische Verhalten beider Kristalle nur von der Verfestigung und nicht von der Verformungstemperatur abhängig.

Die Hysterese des Kristalls A bei 273 °K zeigt Fig. 5 für die Verformungsstufen 0 und 11. Sie ist bei dem unverformten Kristall verschwindend klein, bei dem verformten Kristall dagegen deutlich ausgeprägt.

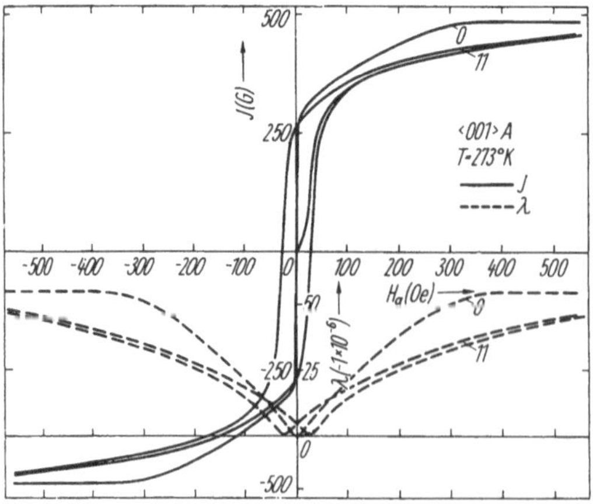

Fig. 5. Die Hysteresekurven der Magnetisierung und Längenänderung eines unverformten und eines verformten ⟨001⟩-Nickeleinkristalls bei 273 °K

3. Diskussion

3.1 Auswertung der Messungen

Da in einem $\langle 001 \rangle$-orientierten stabförmigen Nickeleinkristall alle $\langle 111 \rangle$-Richtungen symmetrisch zur Stabachse liegen, besteht die Bereichsstruktur in der Betrachtungsweise der Phasentheorie [4] im pauschal unmagnetischen Zustand aus acht gleich großen Phasenvolumina v. Beim Anlegen eines Magnetfeldes längs der Stabachse sind jeweils die vier günstig und die vier ungünstig orientierten Phasen energetisch gleichwertig und können zu den Superphasen v_+ und v_- zusammengefaßt werden, ihre Differenz sei

$$\Delta v = v_+ - v_- \,. \tag{3}$$

Der allgemeine Magnetisierungsprozeß ,,Wandverschiebung und Drehung" erfolgt durch 180°-Wandverschiebungen, die durch Δv beschrieben werden, und durch eine Drehung aller Magnetisierungsvektoren in den entsprechenden $\{110\}$-Ebenen um einen Winkel φ aus den $\langle 111 \rangle$-Richtungen heraus. Damit keine magnetischen Pole auf den Wänden entstehen, ist der Winkel φ in allen Phasen gleich groß. Aus Gleichung (1) und (2) in Teil I folgt für die pauschale Magnetisierung J

$$J = J_W + J_D \tag{4a}$$

mit

$$J_W = \sqrt{\frac{1}{3}}\, J_s\, \Delta v\,, \tag{4b}$$

$$J_D = J_s\left(\sqrt{\frac{2}{3}}\sin\varphi - \sqrt{\frac{1}{3}}\,\Delta v\,(1-\cos\varphi)\right). \tag{4c}$$

Für die Längenänderung gilt

$$\lambda = \lambda_{100}\left(\frac{1}{2}\sin^2\varphi + \Delta v\,\frac{1}{2}\sqrt{2}\sin 2\varphi\right). \tag{4d}$$

Hier bedeutet J_W den Anteil der Wandverschiebungen und J_D den Anteil der Drehungen an der Gesamtmagnetisierung. Bei der Ableitung der Gleichung (4) ist es nicht notwendig, spezielle Annahmen über die Bereichsstruktur des Kristalles zu machen; sie muß nur streufeldfreie Drehungen und die Unabhängigkeit der Wandverschiebungen und Drehungen voneinander zulassen.

Aus Gleichung (4) folgt

$$\sin\varphi = \sqrt{\frac{2}{3}}\left(\frac{J}{J_s} - \sqrt{\left(\frac{J}{J_s}\right)^2 - \frac{\lambda}{\lambda_{100}}}\right), \tag{5a}$$

$$\Delta v = \frac{\dfrac{J}{J_s} - \sqrt{\dfrac{2}{3}}\sin\varphi}{\sqrt{\dfrac{1}{3}}\cos\varphi} \tag{5b}$$

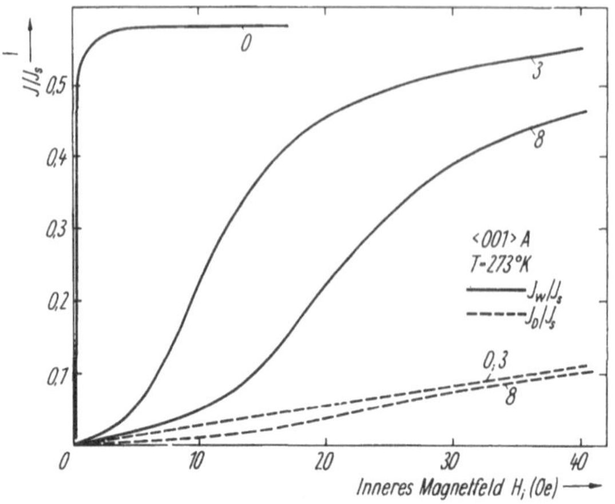

Fig. 6. Die Magnetisierung der Wandbewegungen J_W und die der Drehungen J_D verformter ⟨001⟩-Nickeleinkristalle bei 273 °K in kleinen Magnetfeldern

Damit können auch J_W und J_D aus Gleichung (4) berechnet werden, die in Fig. 6 für die gleichen Verformungsstufen wie in Fig. 3b mit auf J_s reduzierten Einheiten als Funktion der inneren Feldstärke H_i bei 273 °K aufgetragen sind. Die Auswertung der Messungen bei 333 °K ist in der Form sin φ (H) und $\Delta v(H)$, die für kleine Winkel proportional zu $J_D(H)$ und $J_W(H)$ ist, bei Köster und Kronmüller [5] angegeben.

3.2 Die Magnetisierung idealer ⟨001⟩-Nickeleinkristalle

Entsprechend der Phasentheorie von Néel [4], die von Rieger [6] sowie Birss und Hegarty [7] auf Nickeleinkristalle übertragen wurde, sind bis zur sogenannten idealen Remanenz $J_R = J_s/\sqrt{3}$ das innere Feld H_i sowie J_D Null. Für $J > J_R$ ist die Bereichsstruktur vierphasig, und die vier der Feldrichtung naheliegenden Magnetisierungsvektoren drehen sich in {110}-Ebenen mit zunehmendem innerem Feld zur Feldrichtung hin. Dabei bleibt J_W konstant, während J_D stetig zunimmt.

Durch die Minimalisierung der freien Enthalpie, die sich bei einem idealen Kristall aus dem Kristallenergieanteil, dem magnetostatischen Anteil und dem Streufeldanteil zusammensetzt, gewinnt man für den Bereich $J > J_R$ die folgenden Beziehungen für die äußere Feldstärke H_a, die Suszeptibilität χ und die Längenänderung λ als Funktion der Magnetisierung J:

$$H_a = \frac{K_1}{J_s}\left[\left(\frac{J}{J_s}\right) - 3\left(\frac{J}{J_s}\right)^3\right] + \frac{K_2}{J_s}\left[\frac{1}{2}\left(\frac{J}{J_s}\right) - 2\left(\frac{J}{J_s}\right)^3 + \frac{3}{2}\left(\frac{J}{J_s}\right)^5\right] + N\,J\,, \quad (6a)$$

$$\frac{1}{\chi} = \frac{K_1}{J_s^2}\left[1 - 9\left(\frac{J}{J_s}\right)^2\right] + \frac{K_2}{J_s^2}\left[\frac{1}{2} - 6\left(\frac{J}{J_s}\right)^2 + \frac{15}{2}\left(\frac{J}{J_s}\right)^4\right] + N\,, \quad (6b)$$

$$\frac{\lambda}{\lambda_{100}} = \frac{3}{2}\left(\frac{J}{J_s}\right)^2 - \frac{1}{2}\,. \quad (6c)$$

Mit den in Teil I verwendeten Kristallenergie- und Magnetostriktionskonstanten K_1, K_2, λ_{100} und λ_{111} wurden die mit Kreuzen in Fig. 2 und 3 eingezeichneten Werte berechnet, die unsere Messungen an unverformten Kristallen sehr gut wiedergeben. Auch der quadratische Zusammenhang zwischen λ und J entsprechend Gleichung (6c) stimmt mit unseren Messungen sehr gut überein.

3.3 Die Anfangssuszeptibilität

3.3.1 Der Magnetisierungsprozeß der Wandverschiebungen mit Drehungen

Aus Fig. 6 ist ersichtlich, daß in verformten ⟨001⟩-Einkristallen bereits in kleinen Feldern Drehungen an der Magnetisierung beteiligt sind. Die Anfangssuszeptibilität kann nach Kronmüller [2] für diesen Fall aus der Minimalisierung der gesamten freien Enthalpie des Kristalls berechnet werden, die aus den folgenden Beiträgen zusammengesetzt ist:

Die Kristallenergie beträgt für eine kleine Drehung um den Winkel φ aus den leichten Richtungen

$$\Phi_K = -\frac{2}{3}\left(K_1 + \frac{1}{3} K_2\right) \sin^2 \varphi \,. \tag{7a}$$

Die magnetostatische Energie mit J aus Gleichung (4) lautet

$$\Phi_H = -H_i J \,. \tag{7b}$$

Die magnetoelastische Wechselwirkungsenergie der Blochwände mit den Eigenspannungen der Versetzungen ist vom Ort abhängig. Im wechselfeldidealisierten Kristall liegen die Wände an Stellen, an denen die Wechselwirkungsenergie ein Minimum hat. Für eine kleine reversible Auslenkung kann man die Wechselwirkung durch eine quasielastische Bindung an diese Lage berücksichtigen. Die mittlere magnetoelastische Wechselwirkungsenergie der Blochwände pro Volumeneinheit ist dann

$$\Phi_M^W = \frac{1}{2} R \,\Delta v^2 \,, \tag{7c}$$

wobei R die Bedeutung einer Richtkraft pro Volumeneinheit hat.

Die magnetoelastische Wechselwirkung der spontanen Magnetisierung mit den Eigenspannungen der Versetzungen dreht die Magnetisierungsrichtungen etwas aus den ⟨111⟩-Richtungen heraus und führt zu Schwankungen des Magnetisierungsverlaufs um die leichten Richtungen. Für kleine Drehwinkel φ erhält man nach Kronmüller [5]

$$\Phi_M^D = K_\sigma \sin^2 \varphi \,. \tag{7d}$$

Die Spannungsanisotropiekonstante K_σ und K_1 mit K_2 können zu einer effektiven Kristallenergie

$$K_{\text{eff}} = K_\sigma - \frac{2}{3} K_1 - \frac{2}{9} K_2 \tag{8}$$

zusammengefaßt werden. Die Tatsache, daß der Anstieg zu Beginn der $J_D(H)$-Kurven in Fig. 6 mit zunehmender Verformung kleiner wird, läßt sofort auf eine Zunahme von K_{eff} und damit auf ein positives K_σ schließen.

Die Variation der Gesamtenthalpie bezüglich der Δv und φ liefert für kleine H_i und φ

$$\sin \varphi = \sqrt{\frac{2}{3}}\, H_i \frac{J_s}{2 K_{\text{eff}}} \,, \tag{9a}$$

$$\Delta v = H_i \frac{J_s}{\sqrt{3}\, R} \,. \tag{9b}$$

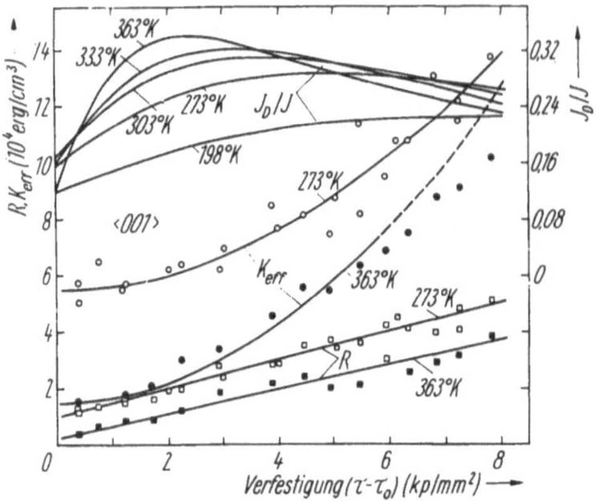

Fig. 7. Die Koeffizienten R, K_{eff} und J_D/J als Funktion der Verfestigung $(\tau - \tau_0)$ bei 273 und 363 °K

Aus den Gleichungen (5) und (9) ergibt sich für die Magnetisierung und die Längenänderung bei $H_i \to 0$

$$J = \frac{J_s^2}{3}\left(\frac{1}{R} + \frac{1}{K_{eff}}\right) H_i = \chi_A \, H_i\,, \tag{10a}$$

$$\lambda = \frac{J_s^2}{3}\lambda_{100}\left(\frac{1}{R\,K_{eff}} + \frac{1}{4\,K_{eff}^2}\right) H_i^2 = \beta\, H_i^2 \tag{10b}$$

in Übereinstimmung mit den experimentell gefundenen Beziehungen (1) und (2).

3.3.2 *Die Verformungs- und Temperaturabhängigkeit der Anfangssuszeptibilität*

Experimentell können R und K_{eff} entweder aus der Anfangssteigung der nach Gleichung (5) berechneten $\sin\varphi(H)$- und $\Delta v(H)$-Kurven oder aus den Koeffizienten β und χ_A nach Gleichung (10) berechnet werden. In Fig. 7 sind R und K_{eff} aus den Messungen bei 273 und 363 °K an den Kristallen A und B als Funktion der Verfestigung $(\tau - \tau_0)$ aufgetragen.[3]) Die Form der Gleichung (5a) bringt es mit sich, daß die Meßfehler der Magnetisierung und Längenänderung einen relativen Maximalfehler von 27% bei K_{eff} und von 15% bei R verursachen. Trotz der dadurch auftretenden großen Streuung der Werte in Fig. 7 können die Ergebnisse gut durch die Beziehungen

$$R = R_0 + R'\,(\tau - \tau_0)\,, \tag{11a}$$

$$K_{eff} = K_0 + K''\,(\tau - \tau_0)^2 \tag{11b}$$

wiedergegeben werden. Die Koeffizienten R_0, R', K_0 und K'' sind in Tabelle 1 zusammengefaßt.

Der Koeffizient R kann aus einer statistischen Theorie von Träuble [8] berechnet werden. Für die Bewegung von 180°-Wänden ist bei der Versetzungsdichte N und dem Schubmodul G nach Träuble

$$R \propto G\,\lambda_{111}\,K_1^{1/4}\,\sqrt{N}\,. \tag{12a}$$

[3]) Die Ergebnisse bei 333 °K sind in [5] aufgeführt.

Tabelle 1

Die experimentell bestimmten Koeffizienten der Gleichungen (11a) und (11b)

T (°K)	198	273	303	333	363
R_0 (10^4 erg/cm^3)	1,9	1,0	0,7	0,45	0,2
R' (10^{-5})	5,7	5,1	4,9	4,7	4,4
K_0 (10^4 erg/cm^3)	13,9	5,5	3,6	2,5	1,5
K'' (10^{-12} cm^3/erg)	0,12	0,13	0,15	0,16	0,18
$\frac{2}{3} K_1 + \frac{2}{9} K_2$ (10^4 erg/cm^3)	15,5	5,9	3,5	2,1	1,1

Die Verfestigungstheorie [3] liefert den Zusammenhang zwischen der Versetzungsdichte N und der Verfestigung $(\tau - \tau_0)$ in Bereich II der Verfestigungskurve:

$$(\tau - \tau_0) \propto \sqrt{N}\,, \tag{12b}$$

der mit Gleichung (12a) das experimentelle Ergebnis in Gleichung (11a) wiedergibt.

Die Temperaturabhängigkeit von R nach (12a) ist gut erfüllt, wie die Auftragung von R_0 und R' gegen $(K^{1/4}\, G\, \lambda_{111})$ in Fig. 8 zeigt.

Die Berechnung der Anisotropiekonstanten K_σ kann in ähnlicher Weise wie die der Suszeptibilität der Einmündung in die Sättigung erfolgen [9, 10]. Kronmüller [2] berechnete für nicht zu große Auslenkungen der Magnetisierungsvektoren aus der mittleren Magnetisierungsrichtung

$$K_\sigma = \frac{N\, G^2\, b^2\, c(\lambda_{100}, \lambda_{111})}{8\,\pi} \left(\frac{1}{\left| K_1 + \frac{1}{3} K_2 \right|} + \frac{C'}{J_s^2} \right), \tag{13}$$

wobei $c\ (\lambda_{100}, \lambda_{111})$ eine Funktion der magnetostriktiven Konstanten und der Versetzungsanordnung ist. Die Versetzungsordnung ist auch für die temperaturunabhängige Konstante C' maßgebend. Das erste Glied in der Klammer stellt den Beitrag der Kristallenergie, das zweite Glied den des Streufeldes der gegeneinander divergierenden Magnetisierungsvektoren zu K_σ dar [11].

Aus den Gleichungen (8) und (12b) erhält man das experimentelle Ergebnis in Gleichung (11b) mit $K_0 = |\frac{2}{3}\, K_1 + \frac{2}{9}\, K_2|$. In Tabelle 1 sind die aus K_1 und K_2 berechneten Werte für K_0 aufgenommen, die befriedigend mit den experimentellen Werten innerhalb der Fehlergrenze übereinstimmen.

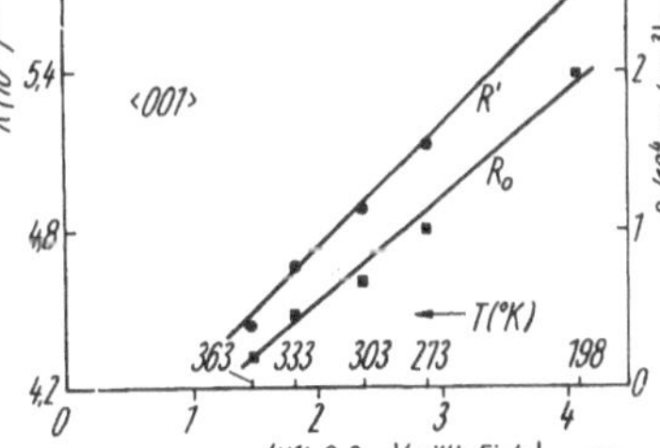

Fig. 8. R_0 und R' als Funktion von $(K^{1/4}\, G\, \lambda_{111})$

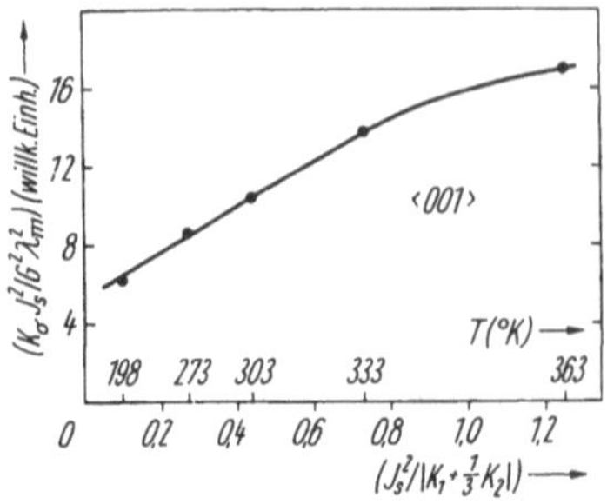

Fig. 9. $\left(K_\sigma J_s^2/G^2 \lambda_{111}^2\right)$ als Funktion von $\left(J_s^2 \Big/ \left|K_1 + \frac{1}{3} K_2\right|\right)$

Zur Prüfung der Temperaturabhängigkeit von K_σ nach (13) wurde K_σ für $(\tau - \tau_0) = 10$ kp/mm² mit K'' aus Tabelle 1 nach Gleichung (11b) berechnet, und $(K_\sigma J_s^2/G^2 \lambda_{111}^2)$ gegen $(J_s^2/|K_1 + \frac{1}{3} K_2|)$ in Fig. 9 aufgetragen, wobei berücksichtigt wurde, daß die Konstante $c\,(\lambda_{100}, \lambda_{111})$ proportional zum Quadrat der Magnetostriktionskonstanten ist, für die λ_{111}^2 eingesetzt wurde; mit λ_{100}^2 erhält man einen ähnlichen Verlauf der Temperaturabhängigkeit. Oberhalb 340 °K weicht die Kurve von der Geraden ab, d. h., die Temperaturabhängigkeit nach (13) ist nicht mehr erfüllt.

Der Ordinatenabschnitt der Darstellung in Fig. 9 entspricht der Konstanten C', die ohne die Kenntnis von N und $c\,(\lambda_{100}, \lambda_{111})$ aus dem Verhältnis von K_σ zweier verschiedener Temperaturen bestimmt werden kann. Für $c \sim \lambda_{111}^2$ und $c \sim \lambda_{100}^2$ erhält man $C' = 4$ bzw. $C' = 10$, d. h., der Streufeldbeitrag zur Spannungsanisotropie ist von der gleichen Größenordnung wie der Kristallanisotropiebeitrag.

Die Übereinstimmung der Beziehungen (11), (12) und (13) sowie die Tatsache, daß der quadratische Zusammenhang zwischen λ und J entsprechend Gleichung (6c) für alle Messungen ab etwa $J/J_s = 0{,}8$ bis zur Sättigung quantitativ erfüllt wird, zeigt, daß bei der in unseren Versuchen erreichten Verfestigung der Kristall bis etwa 340 °K eine ausgeprägte Bereichsstruktur besitzt. Dies ist im Einklang mit den Messungen der Anfangssuszeptibilität von Rieger [12]. Erst bei höheren Temperaturen überwiegt der Einfluß der Spannungsanisotropie gegenüber dem der Kristallanisotropie, und die Bereichsstruktur wird aufgelöst, wie die bei 363 °K auftretenden Abweichungen der Messungen von dem erwarteten Verlauf der Temperatur- und Verfestigungsabhängigkeit zeigen.

3.3.3 Der Beitrag der Drehungen zur Anfangssuszeptibilität

Der Anteil der Magnetisierung durch Drehungen J_D an der gesamten Magnetisierung J bei der Anfangssuszeptibilität kann aus K_{eff} und R wie folgt berechnet werden:

$$\frac{J_D}{J} = \frac{1}{\dfrac{K_{eff}}{R} + 1}\,. \tag{14}$$

Mit den in Tabelle 1 aufgeführten Werten erhält man den in Fig. 7 gezeigten Verlauf von J_D/J mit der Verfestigung. $J_D\,(\tau - \tau_0)$ nimmt mit steigender Temperatur zu und zeigt bei mittlerer Verformung ein Maximum, das mit zunehmender Temperatur zu kleineren Verfestigungswerten verschoben ist, wie es von Kronmüller [2] vorausgesagt wurde. Der Anteil der Drehungen zur

Anfangssuszeptibilität kann bis zu 30% ausmachen und sollte sich bei der Messung von χ_A bemerkbar machen. Wir berechneten χ_A mit Gleichung (10) für 273 °K in Abhängigkeit von der Verfestigung und haben die Werte als durchgezogene Kurve in Fig. 4 eingetragen, die sich mit den direkt ballistisch gemessenen Werten deckt. Die berechnete Kurve hat eine so geringe Krümmung, daß diese aus den experimentellen Werten nicht entnommen werden kann. Obwohl die Drehungen bis zu 30% zu χ_A beitragen, überwiegt doch der lineare Anteil der Wandverschiebungen, so daß bisher von verschiedenen Autoren (siehe hierzu Rieger [3] und die dort zitierten Arbeiten) eine lineare Abhängigkeit $1/\chi_A(\tau)$ gefunden wurde. Bei den Messungen von Mehrer, Kronmüller und Seeger [13] ist die Krümmung jedoch deutlich zu erkennen.

3.3.4 Der Einfluß einer Orientierungsabweichung der Stabachse von der $\langle 001\rangle$-Richtung auf die Meßergebnisse

Bei der symmetrischen Doppelgleitung wird die Stabachse um einen Winkel η aus der [001]-Richtung zur [111]-Richtung gedreht. Dadurch schließt die leichte Richtung in einem Teil der Phasen einen kleineren Winkel mit der Stabachse und in dem anderen Teil der Phasen einen größeren Winkel mit der Stabachse ein. Eine mathematische Berücksichtigung der höchstens $\eta = 6°$ betragenden Abweichung in den Gleichungen (4) bis (9) zeigt, daß für beide Teile der Phasen in diesen Gleichungen gleiche Zusatzglieder mit entgegengesetztem Vorzeichen auftreten, die sich bei Gleichverteilung der Phasen kompensieren. Die Bedingungen, daß keine Komponente der Magnetisierung senkrecht zur Stabachse auftreten darf, da sonst ein großes Streufeld entstehen würde, gibt eine Bevorzugung der Phasen, deren leichte Richtung näher zur Stabachse liegt. Aus dieser Bedingung kann die Abweichung aus der Gleichverteilung berechnet werden.

Die bei $\eta = 6°$ jeweils größte Abweichung von J beträgt etwa 1% nahe der Sättigung und die von λ etwa 0,2% für $J \to 0$. Beide Abweichungen können daher bei unseren Messungen vernachlässigt werden.

3.3.5 Die Bedeutung der Spannungsanisotropie bei der Untersuchung der Versetzungsstruktur in plastisch verformten Kristallen

In ähnlicher Weise wie aus der reversiblen Suszeptibilität im Gebiet der Einmündung in die Sättigung [10] kann aus der absoluten Größe der Spannungsanisotropie K_σ eine Aussage über die Versetzungsanordnung in plastisch verformten Einkristallen gewonnen werden, da die Größen c ($\lambda_{100}, \lambda_{111}$) und C' empfindlich von der Versetzungsanordnung abhängen. Beide Methoden geben im Detail aus den folgenden zwei Gründen etwas voneinander verschiedene Aussagen.

Einmal wird die Wechselwirkung der Magnetisierung mit den Versetzungen bei ein und demselben Kristall in zwei verschiedenen Richtungen gemessen; dies sind in kleinen Feldern die $\langle 111\rangle$-Richtungen, und bei der Einmündung in die Sättigung ist es die Feldrichtung. Zum anderen ist für beide Fälle eine verschieden große Austauschlänge $\varkappa^{-1}$ maßgebend, nämlich für kleine Feldstärken [2]

$$\varkappa_K^{-1} = \left(\frac{3\,A}{2\,|K_1|}\right)^{1/2} \tag{15a}$$

und für die Einmündungssuszeptibilität [10]

$$\varkappa_H^{-1} = \left(\frac{2\,A}{H\,J_s}\right)^{1/2}, \tag{15b}$$

die angibt, ab welchem gegenseitigen Abstand Versetzungen als Einzelversetzungen gemessen werden. Durch eine Veränderung von $\varkappa^{-1}$ kann die Versetzungsstruktur sozusagen abgetastet und die Zahl der über $l < \varkappa^{-1}$ angeordneten Versetzungen, d. h. die in der Verfestigungstheorie [3] wichtige Zahl der aufgestauten Versetzungen pro Versetzungsgruppe, angegeben werden. In Nickel erhält man bei 273 °K und $H = 2700$ Oe mit der Austauschkonstanten $A = 8{,}6 \times 10^{-7}$ erg/cm [14] und der Anisotropiekonstanten $K_1 = 7{,}7 \times 10^4$ erg/cm³ die beiden Austauschlängen

$$\varkappa_K^{-1} = 410\ \text{Å} \tag{16a}$$

und

$$\varkappa_H^{-1} = 114\ \text{Å}\,. \tag{16b}$$

Die $\varkappa_K$ entsprechende Feldstärke bei Einmündungsmessungen aus $\varkappa_K = \varkappa_H$ ist

$$H = 210\ \text{Oe}\,, \tag{17}$$

bei der der Kristall jedoch bereits in Weißsche Bezirke aufgeteilt ist. Beide Arten von Messungen zusammen gestatten im Prinzip eine Untersuchung der Versetzungsstruktur über einen weiten Bereich der Austauschlänge $\varkappa^{-1}$, der durch Temperaturerhöhung, d. h. Verkleinerung von K_1, nach Gleichung (15a) zu noch größeren Werten ausgedehnt werden kann.

Ein quantitativer Vergleich der Ergebnisse beider Untersuchungsmethoden erfordert eine eingehende theoretische Untersuchung, die zur Zeit von Holz [15] in Stuttgart ausgeführt wird. Die bisher vorliegenden Ergebnisse zeigen, daß sowohl die in Abschnitt 3.6 wiedergegebenen Messungen der reversiblen Suszeptibilität im Gebiet der Einmündung in die Sättigung als auch der Betrag von K_σ nur durch weitreichende Spannungsfelder aufgestauter Versetzungsgruppen im Bereich II der Verfestigungskurve erklärt werden kann.

3.4 Das Rayleigh-Gebiet

Für das Rayleigh-Gesetz eines einachsigen Polykristalls leitete Néel [16] die Beziehung

$$J = \chi_A\,H + \alpha\,H^2 = a\left(0{,}27\,\frac{J_s^2}{P_0}\,H + 0{,}16\,\frac{J_s^3}{P_0^2}\,H^2\right) \tag{18}$$

ab. Hierbei ist χ_A die Anfangssuszeptibilität, α die Rayleigh-Konstante, a die mittlere reziproke Anzahl der Minima der Energiefunktion der Blochwand $E(x)$ und P_0 der mittlere Gradient dE/dx der Energiefunktion. Übernimmt man, wie Néel vorgeschlagen hat, die qualitative Aussage der Gleichung (18) für kubische Kristalle, so folgt aus einem Vergleich der Gleichung (18) mit den Gleichungen (10) und (12)

$$\alpha \propto \frac{\chi_A^2}{J_s} \propto \frac{J_s^3}{R^2} \propto \frac{1}{N} \propto \frac{1}{(\tau - \tau_0)^2} \tag{19}$$

in guter Übereinstimmung mit unseren in Fig. 4 eingetragenen Resultaten. Nielsen [17] und Kneller [18] fanden für verschieden verformtes polykristallines Nickel ebenfalls $\alpha = z\,\chi_A^2/J_s$ mit dem Proportionalitätsfaktor $z = 5{,}2$, der überraschend gut mit dem unserer Messungen ($z = 5{,}6$) übereinstimmt.

Die für reversible Prozesse abgeleitete Beziehung (10b) $\lambda \sim H^2$ ist im ganzen Rayleigh-Gebiet erfüllt; dies bedeutet, daß der Anteil irreversibler 180°-Wandverschiebungen in diesem Bereich sehr klein ist, wie man auch an der geringen Krümmung der Magnetisierungskurve im Rayleigh-Gebiet (Fig. 3) erkennt. Der geringe irreversible Magnetisierungsanteil wird aus der folgenden Überlegung verständlich: Nach Oberflächenbeobachtungen von Träuble [8] beträgt der Blochwandabstand in verformten Nickeleinkristallen etwa 30 μm. An der Grenze des Gültigkeitsbereichs des Rayleigh-Gesetzes ist die Magnetisierung durch Wandverschiebungen $J_W/J_s = 0{,}03$ und somit jede Wand im Mittel um die Strecke $x = 0{,}8\,\mu$m verschoben. Die Wechselwirkungsfunktion einer Blochwand mit den Gitterstörungen hat nach Träuble [8] eine Wellenlänge von der vierfachen Blochwanddicke δ_W. Dies bedeutet, daß die Blochwand bei der Bewegung aus einem Energieminimum einen Weg von etwa δ_W reversibel zurücklegt und dann eine Strecke von etwa $3\,\delta_W$ irreversibel springt. Nach Lilley [19] beträgt die Dicke einer 180°-Blochwand etwa $\delta_W = 0{,}2\,\mu$m. Diese Abschätzung zeigt, daß bis zum Ende des Rayleigh-Gebiets die einzelnen Blochwände höchstens einen irreversiblen Barkhausen-Sprung ausgeführt haben und somit der irreversible Magnetisierungsanteil klein ist.

3.5 *Die irreversiblen Wandverschiebungen*

Mit zunehmender Feldstärke H nimmt der Anteil der irreversiblen Wandverschiebungen stark zu. Man kann die differentielle Suszeptibilität χ_{diff} in einen reversiblen Anteil χ_{rev} und einen irreversiblen Anteil χ_{irr} aufspalten entsprechend

$$\chi_{\text{diff}} = \chi_{\text{rev}} + \chi_{\text{irr}}\,. \tag{20}$$

Aus der Steigung der Magnetisierungskurven in Fig. 3b wurde χ_{diff} entnommen und daraus mit den gemessenen χ_{rev}-Werten χ_{irr} berechnet. In Fig. 10 sind die Ergebnisse zusammen mit der aus Fig. 6 entnommenen Suszeptibilität der Drehungen χ_D aufgetragen. Bis zum Maximum von χ_{irr} bleibt die gesamte Blochwandfläche konstant, wie aus dem nahezu konstanten χ_{rev} folgt, das proportional zur gesamten Blochwandfläche ist. Das auch von Beller [20] gefundene schwache Maximum der $\chi_{\text{rev}}(H)$-Kurven kann damit erklärt werden,

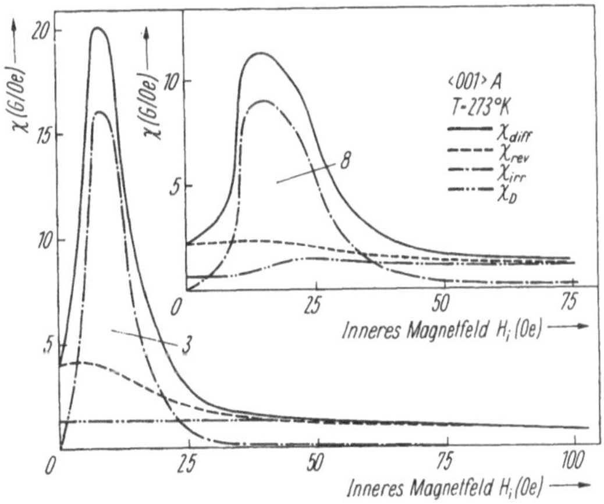

Fig. 10. χ_{diff}, χ_{rev}, χ_{irr} und χ_D verformter ⟨001⟩-Nickeleinkristalle bei 273 °K als Funktion der Feldstärke

daß die Wände nach irreversiblen Sprüngen an Stellen mit einem geringeren Gradienten der Wechselwirkungskraft der Blochwände mit den Gitterstörungen gelangen.

Die Zunahme von χ_{irr} folgt aus der mit der Feldstärke zunehmenden Zahl der einen Barkhausen-Sprung ausführenden Wände. Wie Stierstadt und Boeckh [21] an polykristallinem Nickel gezeigt haben, durchläuft die Gesamtzahl aller Barkhausen-Sprünge pro Feldstärkenintervall als Funktion der Feldstärke ebenso wie χ_{irr} unserer Messungen bei der Feldstärke $H_K = H_c$ ein Maximum. Bei eben dieser Feldstärke beginnt der Abfall der $\chi_{rev}(H)$-Kurven, d. h., die Blochwände beginnen sich zu annihilieren, die Zahl der Barkhausen-Sprünge und damit χ_{irr} nehmen ab.

Die Übereinstimmung der Koerzitivfeldstärke H_c mit der Knickfeldstärke H_K hängt damit zusammen, daß beide durch eine Bewegung der 180°-Wände über große Strecken charakterisiert sind, die in beiden Fällen zu einem Wendepunkt in der Hystereseschleife (Fig. 5) führt. Wegen der Symmetrie aller acht leichten Richtungen bezüglich der Stabachse ist in beiden Fällen H_i parallel zu H_a.

Nach Rechnungen von Träuble [8] ist für Bereich II der Verfestigungskurve [3]

$$H_c \sim \sqrt{N} \sim (\tau - \tau_0) \tag{21}$$

in Übereinstimmung mit unseren Messungen (Fig. 4).

Die Wandbewegungen sind bei allen Verformungsstufen bei einer Sättigungsfeldstärke von $H_{SW} \approx 10\, H_c$ abgelaufen. Es zeigt sich also, daß alle von irreversiblen Wandbewegungen abhängenden Feldgrößen, wie die Koerzitivfeldstärke H_c, der Gültigkeitsbereich H_0 des Rayleigh-Gesetzes, die Feldstärke H_K des größten irreversiblen Magnetisierungsanteils und die Sättigungsfeldstärke H_{SW} der Wandverschiebungen in gleicher Weise von der Verfestigung abhängen. Sie sind alle proportional zur Wurzel aus der Versetzungsdichte N.

Die Fläche unter der $\chi_{irr}(H)$-Kurve ist für alle Verformungsstufen konstant; der Beitrag der genannten irreversiblen zur gesamten durch Wandverschiebungen erreichbaren Magnetisierung ist für alle Verformungsstufen $J_{irr}/J_W = 0{,}7$. Die Wände legen also unabhängig von der Verformung etwa 30% ihres Weges bis zur Annihilation reversibel zurück. Dieses Ergebnis stützt wesentlich die theoretische Vorstellung [8], daß die plastische Verformung nur den Betrag, jedoch nicht die mittlere Wellenlänge der Wechselwirkungskraft der Versetzungen mit den Blochwänden beeinflußt.

Der Anteil der Drehungen χ_D zu χ_{diff} ist bei kleinen Verformungen über einen großen Feldbereich nahezu konstant und bei $H > H_{SW} \approx 10\, H_c$ im Rahmen der Meßgenauigkeit identisch mit χ_{rev}. Bei größeren Verformungen dagegen nimmt χ_D für $H > H_0$ zu und durchläuft ein Maximum, das etwa einem mittleren Drehwinkel von $\varphi = 3°$ entspricht, und wird bei H_{SW} wieder gleich χ_{rev}. Dieses Verhalten wird mit zunehmender Temperatur ausgeprägter, wie man an der stark gekrümmten Zunahme von $\sin\varphi(H)$ bei 333 °K in [5] sieht. Bei höheren Feldstärken gelten für reversible Prozesse die gleichen Beziehungen (9) wie bei der Anfangssuszeptibilität, so daß eine Zunahme von χ_D bei reversiblen Drehungen eine Abnahme von K_{eff} mit zunehmender Feldstärke bedeuten würde. Dies ist aber wegen der damit auftretenden Instabilität nicht möglich. Diese Tatsache führt zu der Annahme sehr kleiner irreversibler Drehungen, die dadurch zustande kommen, daß die lokale magnetoelastische Kopplungsenergie

Φ_M^D in der Nähe des Minimums der Kristallenergie Φ_K ein Maximum hat und die Summe beider einen Höcker aufweist, der in Analogie zu den irreversiblen Wandverschiebungen zu irreversiblen Drehungen führt. Die Annahme irreversibler Drehungen erklärt qualitativ die Temperatur- und Verformungsabhängigkeit dieses Effektes. Mit zunehmender Temperatur ist der Anstieg von Φ_K bei etwa gleichem Φ_M^D geringer, die Höcker werden ausgeprägter und damit der irreversible Drehanteil größer; in gleicher Richtung wirkt eine Vergrößerung von Φ_M^D durch Verformung. Die stark hervortretende Hysterese der Magnetisierungs- und Längenänderung im Bereich reiner Drehungen $H > H_{SW}$ in Fig. 5 und bei dem $\langle 110 \rangle$-Kristall in Teil I deutet ebenfalls auf irreversible Drehprozesse hin. Auch die Remanenz der Längenänderung des $\langle 001 \rangle$-Kristalls (Fig. 4) kann nur so erklärt werden. In diesem Zusammenhang sind noch genauere theoretische und experimentelle Untersuchungen notwendig, um alle Einzelheiten solcher Prozesse zu verstehen. Experimentell erscheinen uns hierfür quantitative Messungen der reversiblen Suszeptibilität und des Barkhausen-Effektes zusammen mit der Magnetisierung und Längenänderung im Bereich reiner Drehprozesse oberhalb H_{SW} besonders geeignet.

3.6 Der Bereich makroskopischer Drehungen und die Einmündung in die Sättigung

In Feldern, die größer als die Sättigungsfeldstärke H_{SW} der Wandverschiebungen sind, ist die reversible Suszeptibilität χ_{rev} im Rahmen der Meßgenauigkeit gleich der differentiellen Suszeptibilität χ_{diff}. Dies ist bei allen Verformungsgraden und Temperaturen oberhalb etwa $J/J_s = 0{,}8$ der Fall. Auf Fig. 2 ist zu erkennen, daß χ_{rev} in diesem Teil der Magnetisierungskurve mit zunehmender Verformung abnimmt. Bei verformten Kristallen ist Gleichung (6b) für die reversible Suszeptibilität durch Zusatzglieder mit der Spannungsanisotropie K_σ zu ergänzen. Deren Betrag $(1/\chi_{rev} - 1/\chi_{rev}^0)$ ist bei konstanter Magnetisierung J in Fig. 11 als Funktion der Verfestigung $(\tau - \tau_0)$ aufgetragen. χ_{rev}^0 ist die Suszeptibilität des unverformten Kristalls. Bei der Berechnung der Spannungsanisotropie ist in den auftretenden hohen Magnetfeldern außer der Kristallenergie auch die magnetostatische Energie des Feldes zu berücksichtigen. Die Auswertung der Messungen sollte folglich mit den beiden sich gegenseitig ausschließenden Bedingungen $J = \text{const}$ und $H = \text{const}$ erfolgen. Die Vernachlässigung der Feldenergie in der Auswertung mit $J = \text{const}$ verursacht das Flacherwerden der Kurven in Fig. 11 mit zunehmender Magnetisierung, so daß zu erwarten ist, daß eine korrekte Auswertung zu dem Ergebnis in kleinen Feldstärken $K_\sigma \sim (\tau - \tau_0)^2$ führen wird.

Im Gebiet der Einmündung in die Sättigung nimmt die reversible Suszeptibilität mit der plastischen Verformung zu. Die Messungen der Einmündungssuszeptibilität des $\langle 001 \rangle$-Einkristalls A bei 273 °K sind in Fig. 12 dargestellt.

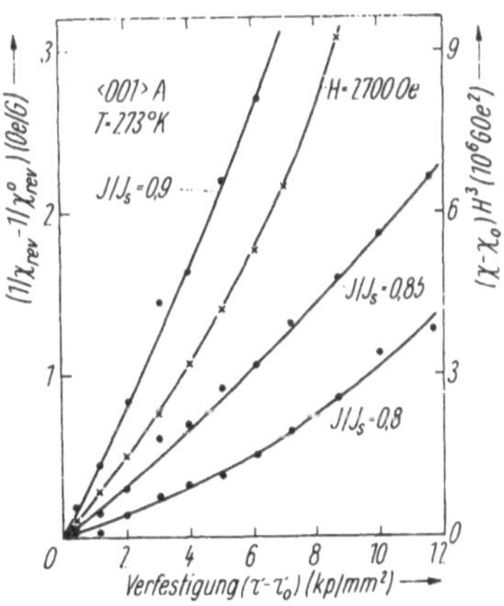

Fig. 11. $(1/\chi_{rev} - 1/\chi_{rev}^0)$ bei drei verschiedenen Magnetisierungen und $(\chi - \chi_0)$ bei $H = 2700$ Oe als Funktion der Verfestigung $(\tau - \tau_0)$

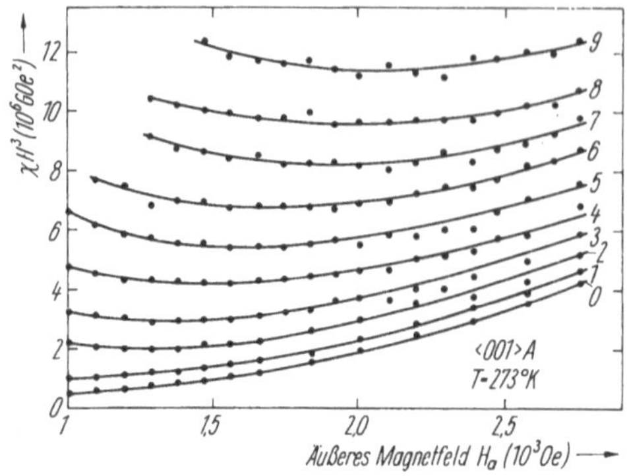

Fig. 12. χH^3 verformter ⟨001⟩-Nickeleinkristalle in Abhängigkeit von der Feldstärke

Dabei wurde wie üblich $\chi\, H^3$ aufgetragen. Da nach Becker und Döring [21] die Einmündungssuszeptibilität eines idealen eckorientierten Einkristalls Null ist, stellt die Kurve für $(\tau - \tau_0) = 0$ die Suszeptibilität des Paraprozesses dar. Die quantenmechanische Rechnung von Holstein und Primakoff [23] ergibt für den Paraprozeß die Feldabhängigkeit

$$\chi_0 = \alpha(T)\left[\left(\frac{4\,\pi\,J_0}{H}\right)^{1/2} + \arcsin\left(\frac{4\,\pi\,J_0}{H + 4\,\pi\,J_0}\right)^{1/2}\right]. \tag{22}$$

Es bedeutet J_0 die spontane Magnetisierung am absoluten Nullpunkt. Unsere Messung bei $(\tau - \tau_0) = 0$ erfüllt diese Beziehung mit α (273 °K) $= 0{,}72 \times 10^{-4}$. Von Ehlermann [24] wurde in wesentlich größeren Feldern α (273 °K) $= 0{,}5 \times 10^{-4}$ gemessen.

Die Verformungsabhängigkeit der Einmündungssuszeptibilität ist für $H = 2700$ Oe als $(\tau - \tau_0)\, H^3$ in Abhängigkeit von $(\tau - \tau_0)$ in Fig. 11 dargestellt, wobei χ_0 die Suszeptibilität des unverformten Kristalls ist. Die Krümmung der Kurve ist etwas geringer, wie sie von Kronmüller [25] an einem mittelorientierten Nickeleinkristall gemessen wurde. Die Abhängigkeit der reversiblen Suszeptibilität von der Verformung und der Feldstärke wurde von Seeger und Kronmüller [8] quantitativ für einen mittelorientierten Nickeleinkristall berechnet. Eingehende theoretische Untersuchungen für beliebig orientierte Kristalle und über den Vergleich mit den Messungen der Anfangssuszeptibilität werden, wie bereits in Abschnitt 3.3 erwähnt, zur Zeit von Holz [15] ausgeführt.

Herrn Professor Dr. A. Seeger und Herrn Dr. H. Kronmüller bin ich für die wertvollen Ratschläge und anregenden Diskussionen, die zum Gelingen dieser Arbeit wesentlich beigetragen haben, sehr dankbar. Der Deutschen Forschungsgemeinschaft danke ich für die Bereitstellung von Sachmitteln.

Literatur

[1] E. Köster, phys. stat. sol. **19**, 153 (1967).

[2] H. Kronmüller, in Moderne Probleme der Metallphysik, Bd. II, Springer-Verlag, Berlin/Heidelberg/New York 1966.

[3] A. Seeger, Kristallplastizität, in Hdb. Phys., Bd. VII/2, Springer-Verlag, Berlin/Göttingen/Heidelberg 1958.

[4] L. Néel, J. Phys. Radium **5**, 241 (1944).

[5] E. Köster und H. Kronmüller, Phys. Letters (Netherlands) **20**, 476 (1966).

[6] H. Rieger, Z. angew. Phys. **15**, 232 (1963).
[7] R. R. Birss und B. C. Hegarty, eingereicht bei Brit. J. appl. Phys.
[8] H. Träuble, in Moderne Probleme der Metallphysik, Bd. II, Springer-Verlag, Berlin/Heidelberg/New York 1966.
[9] W. Brown, Phys. Rev. **58**, 736 (1940); **60**, 132 (1941).
[10] A. Seeger und H. Kronmüller, J. Phys. Chem. Solids **12**, 298 (1960); **18**, 93 (1961).
[11] H. Kronmüller, private Mitteilung.
[12] H. Rieger, phys. stat. sol. **8**, 283 (1965).
[13] H. Mehrer, H. Kronmüller und A. Seeger, phys. stat. sol. **11**, 313 (1965).
[14] B. E. Argyle, S. H. Charap und E. W. Pugh, Phys. Rev. **132**, 2051 (1963).
[15] A. Holz, unveröffentlicht.
[16] L. Néel, Cahiers Phys. **12**, 1 (1942).
[17] H. Nielsen, Diplomarbeit, Stuttgart 1954.
[18] E. Kneller, in Beiträge zur Theorie des Ferromagnetismus, Springer-Verlag, Berlin/Göttingen/Heidelberg 1956.
[19] B. A. Lilley, Phil. Mag. **41**, 792 (1950).
[20] M. Beller, Diplomarbeit, Stuttgart 1965.
[21] K. Stierstadt und W. Boeckh, Z. Phys. **186**, 154 (1965).
[22] R. Becker und W. Döring, Ferromagnetismus, Springer-Verlag, Berlin 1939.
[23] T. Holstein und H. Primakoff, Phys. Rev. **58**, 1089 (1940).
[24] D. Ehlermann, Diplomarbeit, Stuttgart 1965.
[25] H. Kronmüller, Z. Phys. **154**, 574 (1959).

(Received November 23, 1966)

phys. stat. sol. **19**, 673 (1967)

Subject classification: 20; 13.1; 14.4.1

Institut für Theoretische Physik der Humboldt-Universität zu Berlin

On the Theory of Electro-Absorption and Electro-Reflectance in Semiconductors

By

R. ENDERLEIN and R. KEIPER

The dielectric response of a semiconductor in a static electric field is investigated. The dielectric function is found to be an integral operator. It should be possible to confirm experimentally the predicted generation of "harmonics" displaced with respect to the frequency of the incident light by an integer times the Zener oscillation frequency.

Es wird die dielektrische „response" eines Halbleiters bei Anwesenheit eines statischen elektrischen Feldes untersucht. Die Dielektrizitätsfunktion erweist sich dabei als Integraloperator. Als experimentell beobachtbare Konsequenz folgt daraus die Erzeugung von „Harmonischen", die um ganzzahlige Vielfache der Frequenz der Zenerschwingung gegenüber der Lichtfrequenz verschoben sind.

1. Introduction

Previous experimental [1–4] and theoretical works [5–9] show that there exists a similarity in semiconductors between electro-absorption and electro-reflectance and absorption and reflectance in the absence of an electric field in that there is a direct relation to the energy spectrum of the crystal electrons which is not significantly disturbed by collision processes. From measurements of electro-reflectance it was in fact possible to draw conclusions regarding the position and type of the critical points of interband energy [2, 10], which are more comprehensive than what can be implied from an analysis of reflectivity in the field-free case. An essential factor in these results is the characteristic change in the dispersion of absorption in the vicinity of critical points induced by a static electric field (Franz-Keldysh effect).

The cited investigations are based on the assumption that the propagation of a light wave in a semiconductor in a quasi-static electric field can be described completely by a complex refractive index which, as is well known, can be done in the field-free case. The purpose of the present work is to show that this is not the case. By studying the dielectric response of a semiconductor in an electric field, it is shown that then the dielectric function must be regarded as an integral operator rather than as a tensorial factor. An analysis of the propagation of light in such a medium using this property of the dielectric function yields results which can be confirmed by experiment. Agreement with the results of previous authors is established with respect to the variation of the absorption and reflectivity in a static electric field.

2. Kubo's Formula for the Field-Modified Dielectric Function

Our aim is to derive a general expression for the dielectric function $\hat{\varepsilon}(\omega)$ of a crystal in the presence of a static electric field $\boldsymbol{E}_0$. We begin with the equations

$$\boldsymbol{D}(\omega) = \hat{\varepsilon}(\omega)\, \boldsymbol{E}(\omega)\,, \tag{1}$$

$$\boldsymbol{D}(\omega) = \boldsymbol{E}(\omega) + 4\,\pi\, \boldsymbol{P}(\omega)\,, \tag{2}$$

connecting the Fourier transforms of $\boldsymbol{E}(t)$ (electric field of the light wave), $\boldsymbol{D}(t)$ (dielectric displacement) and $\boldsymbol{P}(t)$ (polarization). $\boldsymbol{P}(\omega)$ follows from the Fourier transform $\boldsymbol{J}(\omega)$ of the average current density $\boldsymbol{J}(t)$ generated by the light field $\boldsymbol{E}(t)$:

$$\boldsymbol{P}(\omega) = \frac{i}{\omega}\, \boldsymbol{J}(\omega)\,. \tag{3}$$

The total current density $\boldsymbol{J}'(\omega)$ differs from $\boldsymbol{J}(\omega)$ by an amount $\boldsymbol{J}_0$ equal to the direct current set up by the static electric field $\boldsymbol{E}_0$ and which does not depend on $\boldsymbol{E}(t)$:

$$\boldsymbol{J}'(\omega) = \boldsymbol{J}_0 + \boldsymbol{J}(\omega)\,. \tag{4}$$

As we are dealing with the electro-optical effects which exhibit a non-linear dependence on $\boldsymbol{E}_0$, it is necessary to give an accurate description of the interaction of the crystal with $\boldsymbol{E}_0$. We therefore incorporate the static electric field into the unperturbed system. The Hamiltonian of the whole system is

$$H'(t) = H + \Delta H(t)\,, \tag{5}$$

where

$$\Delta H(t) = -\sum_i e\, \boldsymbol{E}(t)\, \boldsymbol{r}_i \tag{6}$$

describes the interaction of the crystal electrons with the light wave $\boldsymbol{E}(t)$ in the self-consistent-field approximation. ($\boldsymbol{r}_i$ is the radius vector of the i-th electron.) The Hamiltonian H of the unperturbed system can be written, using the adiabatic approximation, in the form

$$H = H_0 - \sum_i e\, \boldsymbol{E}_0\, \boldsymbol{r}_i\,. \tag{7}$$

Here H_0 denotes the Hamiltonian of the crystal electrons in the absence of external fields. The second term in H describes the interaction of the electrons with the static field $\boldsymbol{E}_0$.

The density operator $\varrho'(t)$ of the system satisfies the equation of motion

$$i\,\hbar\, \dot{\varrho}'(t) = [H'(t), \varrho'(t)]\,. \tag{8}$$

We assume that both external fields, $\boldsymbol{E}_0$ and $\boldsymbol{E}(t)$, are switched on at the time t_0 and that the system was previously in thermodynamical equilibrium. The initial condition for $\varrho'(t)$ is therefore

$$\varrho'(t_0) = \varrho_0\,, \quad \varrho_0 = [\mathrm{Sp}\; \mathrm{e}^{-\beta(H_0-\mu N)}]^{-1} \cdot \mathrm{e}^{-\beta(H_0-\mu N)}\,, \tag{9}$$

where μ and N are the chemical potential and the total particle-number operator, respectively.

Denoting by $\varrho(t)$ that part of $\varrho'(t)$ which is independent of $\boldsymbol{E}(t)$ and by $\Delta\varrho(t)$ the part linear in $\boldsymbol{E}(t)$ we have, neglecting terms of higher order in $\boldsymbol{E}(t)$,

$$\varrho'(t) = \varrho(t) + \Delta\varrho(t) . \tag{10}$$

We then obtain, using (8) and (9),

$$\varrho(t) = \mathrm{e}^{-\frac{i}{\hbar} H(t-t_0)} \varrho_0 \, \mathrm{e}^{\frac{i}{\hbar} H(t-t_0)} , \qquad t > t_0 , \tag{11}$$

$$\Delta\varrho(t) = -\frac{i}{\hbar} \int_{t_0}^{t} \mathrm{d}t' \, \mathrm{e}^{-\frac{i}{\hbar} H(t-t')} [\Delta H(t'), \varrho(t')] \, \mathrm{e}^{\frac{i}{\hbar} H(t-t')} , \qquad t > t_0 . \tag{12}$$

The splitting of $\varrho'(t)$ into $\varrho(t)$ and $\Delta\varrho(t)$ corresponds to the separation of $\boldsymbol{J}'(\omega)$ into $\boldsymbol{J}_0$ and $\boldsymbol{J}(\omega)$ in equation (4). $\boldsymbol{J}(\omega)$ is therefore given by

$$\boldsymbol{J}(\omega) = \frac{1}{2\pi} \int_{-\infty}^{\infty} \mathrm{d}t \, \mathrm{e}^{i\omega t} \, \mathrm{Sp} \, \{\Delta\varrho(t) \, \boldsymbol{j}\} , \tag{13}$$

where $\boldsymbol{j}$ is the current operator for 1 cm³ of the crystal. After a short calculation using (2), (3), and (12) the dielectric displacement $\boldsymbol{D}(\omega)$ becomes

$$\boldsymbol{D}(\omega) = \int_{-\infty}^{\infty} \mathrm{d}\omega' \, \hat{\varepsilon} \, (\omega', \omega - \omega') \, \boldsymbol{E}(\omega') , \tag{14}$$

where

$$\hat{\varepsilon} \, (\omega', \omega - \omega') = \left[1 - \frac{\omega_\mathrm{p}^2}{\omega^2}\right] \delta \, (\omega - \omega') +$$
$$+ \frac{4\pi}{\omega\omega'} \frac{1}{2\pi} \int_{-\infty}^{\infty} \mathrm{d}t \, \mathrm{e}^{i(\omega-\omega')t} \int_{t_0}^{t} \mathrm{d}t' \, \mathrm{e}^{i\omega'(t-t')} \left\langle \frac{i}{\hbar} [\boldsymbol{j} \, (t - t_0), \boldsymbol{j} \, (t' - t_0)] \right\rangle_0 . \tag{15}$$

Moreover, we have made use of

$$A(t) = \mathrm{e}^{\frac{i}{\hbar} H t} A \, \mathrm{e}^{-\frac{i}{\hbar} H t} , \tag{16}$$

$$\langle B \rangle_0 = \mathrm{Sp} \, \{\varrho_0 \, B\} . \tag{17}$$

(ω_p is the plasma frequency.)

It can be seen from (14) that the dielectric function $\hat{\varepsilon}(\omega)$ defined by (1) is not a simple tensorial factor, as it is in the absence of the static field $\boldsymbol{E}_0$, but is an integral operator acting on the field $\boldsymbol{E}(\omega')$. One can easily confirm that in the case $\boldsymbol{E}_0 = 0$ the relation (14) is equivalent to the usual local relation between $\boldsymbol{D}(\omega)$ and $\boldsymbol{E}(\omega)$. One finds in this case

$$\hat{\varepsilon} \, (\omega', \omega - \omega') = \delta \, (\omega - \omega') \left[1 - \frac{\omega_\mathrm{p}^2}{\omega^2} + \frac{4\pi}{\omega^2} \int_0^{\infty} \mathrm{d}t \, \mathrm{e}^{i\omega t} \left\langle \frac{i}{\hbar} [\boldsymbol{j}(t), \boldsymbol{j}(0)] \right\rangle_0 \right] . \tag{18}$$

In the special case of harmonic fields

$$\boldsymbol{E}(\omega') = \boldsymbol{E}(\omega) \, \delta \, (\omega - \omega') \tag{19}$$

equation (14) can be written, after performing the inverse Fourier transformation in the form

$$D(t) = \hat{\varepsilon}(\omega, t)\, E(t)\,, \tag{20}$$

where

$$\hat{\varepsilon}(\omega, t) = \int_{-\infty}^{\infty} \mathrm{d}\omega'\, \mathrm{e}^{-i\omega' t}\, \hat{\varepsilon}(\omega, \omega')\,. \tag{21}$$

We have here a local relation between $D(t)$ and $E(t)$, the dielectric function being time-dependent (in contrast to the case $E_0 = 0$).

3. Explicit Expression for $\hat{\varepsilon}\,(\omega', \omega - \omega')$

In order to evaluate the expression (15) for $\hat{\varepsilon}\,(\omega', \omega - \omega')$ we simplify the situation as follows. The crystal is assumed to possess cubic symmetry. Neglecting the non-diagonal elements $\hat{\varepsilon}_{ij}\,(\omega', \omega - \omega')\; i \neq j$ of the tensor $\hat{\varepsilon}\,(\omega', \omega - \omega')$ we have

$$\hat{\varepsilon}_{ij}\,(\omega', \omega - \omega') = \delta_{ij}\, \varepsilon\,(\omega', \omega - \omega')\,. \tag{22}$$

The static field

$$E_0 = (F, 0, 0) \tag{23}$$

is directed along the x-axis of the coordinate system defined by the cubic axes. Neglecting collisions we can approximate the Hamiltonian H_0 by a sum of one-electron Bloch Hamiltonians:

$$H_0 = \sum_{\nu \boldsymbol{k}} E_\nu(\boldsymbol{k})\, c^+_{\nu \boldsymbol{k}}\, c_{\nu \boldsymbol{k}}\,. \tag{24}$$

Neglecting interband transitions induced by E_0 we obtain[1]) the current operator

$$\boldsymbol{j}(t) = \frac{e}{m} \sum_{\mu\nu} \sum_{\boldsymbol{k}} \mathrm{e}^{i \int_0^t \mathrm{d}\tau\, \omega_{\mu\nu}(\boldsymbol{k}_\tau)}\, (\mu\, \boldsymbol{k}_t\, |\boldsymbol{p}|\, \nu\, \boldsymbol{k}_t)\, c^+_{\mu \boldsymbol{k}}\, c_{\nu \boldsymbol{k}}\,, \tag{25}$$

where

$$\boldsymbol{k}_t = \boldsymbol{k}_\perp + \boldsymbol{n}_x\, k_{xt}\,, \tag{26}$$

$$k_{xt} = k_x + \frac{e\,F}{\hbar} t \mod \frac{2\pi}{a}\,, \tag{27}$$

$$\hbar\, \omega_{\mu\nu}(\boldsymbol{k}_t) = E_\mu(\boldsymbol{k}_t) - E_\nu(\boldsymbol{k}_t)\,. \tag{28}$$

Here $\boldsymbol{p}$, $\boldsymbol{k}_\perp$, $\boldsymbol{n}_x$, and a denote the momentum operator of an electron, the component of $\boldsymbol{k}$ perpendicular to the field E_0, a unit vector directed along the x-axis, and the lattice constant, respectively.

In the following we choose t_0 to have the form $t_0 = M\,T$, where M is an integer. Using (25) the current correlator in (15) can be written

$$\langle [\boldsymbol{j}\,(t - t_0), \boldsymbol{j}\,(t' - t_0)] \rangle_0 =$$

$$= \left(\frac{e}{m}\right)^2 \sum_{\mu\nu} \sum_{\boldsymbol{k}} |(\mu\; 0\; |p_x|\; \nu\; 0)|^2\, [\langle N_{\nu \boldsymbol{k}} \rangle - \langle N_{\mu \boldsymbol{k}} \rangle]\, \mathrm{e}^{-i \int_{t'}^{t} \mathrm{d}\tau\, \omega_{\mu\nu}(\boldsymbol{k}_\tau)}\,, \tag{29}$$

where

$$\langle N_{\nu \boldsymbol{k}} \rangle = [\mathrm{e}^{\beta\, [E_\nu(\boldsymbol{k}) - \mu]} + 1]^{-1} \tag{30}$$

is the average number of particles in the state $|\nu\, \boldsymbol{k})$. Here the periodicity of

---

[1]) See the appendix.

$\omega_{\mu\nu}(\boldsymbol{k}_t)$ with respect to t with the period $T = \frac{2\pi\hbar}{eFa}$ has been used. After substituting (29) in (15) we have

$$\varepsilon(\omega', \omega - \omega') = \left[1 - \frac{\omega_p^2}{\omega^2}\right] \delta(\omega - \omega') + + \frac{4\pi i}{\omega} \sum_{\mu\nu} \sum_{\boldsymbol{k}} \left(\frac{e}{m}\right)^2 \frac{|(\mu\,0\,|p_x|\,\nu\,0)|^2}{\hbar\,\omega'} [\langle N_{\nu\boldsymbol{k}}\rangle - \langle N_{\mu\boldsymbol{k}}\rangle]\, C_{\mu\nu}^{\boldsymbol{k}\omega'}(\omega - \omega') . \quad (31)$$

Here $C_{\mu\nu}^{\boldsymbol{k}\omega'}(\omega - \omega')$ denotes the Fourier transform of

$$C_{\mu\nu}^{\boldsymbol{k}\omega'}(t) = e^{if(t)} \int_{t_0}^{t} dt'\, e^{-if(t')}, \quad (32)$$

where

$$f(t) = \int_0^t d\tau\, [\omega' - \omega_{\mu\nu}(\boldsymbol{k}_\tau)] . \quad (33)$$

Now, as t_0 approaches $-\infty$ we obtain, using the relation $f(t + T) = f(t) + f(T)$,

$$C_{\mu\nu}^{\boldsymbol{k}\omega'}(t) = e^{if(t)} \left[\frac{1}{e^{-if(T)} - 1} \int_{-T/2}^{T/2} dt'\, e^{-if(t')} + \int_{-T/2}^{t} dt'\, e^{-if(t')}\right], \quad -\frac{T}{2} \leqq t \leqq \frac{T}{2}, \quad (34)$$

and

$$C_{\mu\nu}^{\boldsymbol{k}\omega'}(t + T) = C_{\mu\nu}^{\boldsymbol{k}\omega'}(t) . \quad (35)$$

Because of the periodicity (35) we can expand $C_{\mu\nu}^{\boldsymbol{k}\omega'}(t)$ into a Fourier series

$$C_{\mu\nu}^{\boldsymbol{k}\omega'}(t) = \sum_l C_{\mu\nu l}^{\boldsymbol{k}\omega'}\, e^{-il\Omega t}, \qquad \Omega = \frac{2\pi}{T}, \quad (36)$$

where

$$C_{\mu\nu l}^{\boldsymbol{k}\omega'} = \frac{1}{T} \int_{-T/2}^{T/2} dt\, e^{il\Omega t}\, C_{\mu\nu}^{\boldsymbol{k}\omega'}(t) . \quad (37)$$

Introducing this into (31) we obtain finally

$$\varepsilon(\omega', \omega - \omega') = \sum_l \varepsilon^{(l)}(\omega')\, \delta(\omega - \omega' - l\,\Omega), \quad (38)$$

where

$$\varepsilon^{(l)}(\omega') = \left[1 - \frac{\omega_p^2}{\omega^2}\right] \delta_{l0} + + \frac{4\pi i}{(\omega' + l\,\Omega)} \sum_{\mu\nu} \sum_{\boldsymbol{k}} \left(\frac{e}{m}\right)^2 \frac{|(\mu\,0\,|p_x|\,\nu\,0)|^2}{\hbar\,\omega'} [\langle N_{\nu\boldsymbol{k}}\rangle - \langle N_{\mu\boldsymbol{k}}\rangle]\, C_{\mu\nu l}^{\boldsymbol{k}\omega'} . \quad (39)$$

4. Propagation of Light in the Presence of a Static Field

In order to investigate the physical consequences of the coefficients $\varepsilon^{(l)}(\omega')$ in (14) we follow the propagation of an incident light wave:

$$\boldsymbol{E}_e = \boldsymbol{E}^{(0)}\, e^{i(qz - \omega t)}, \qquad \boldsymbol{H}_e = \boldsymbol{H}^{(0)}\, e^{i(qz - \omega t)} \quad (40)$$

in a medium with dielectric properties as defined by (14), (22), and (38). We

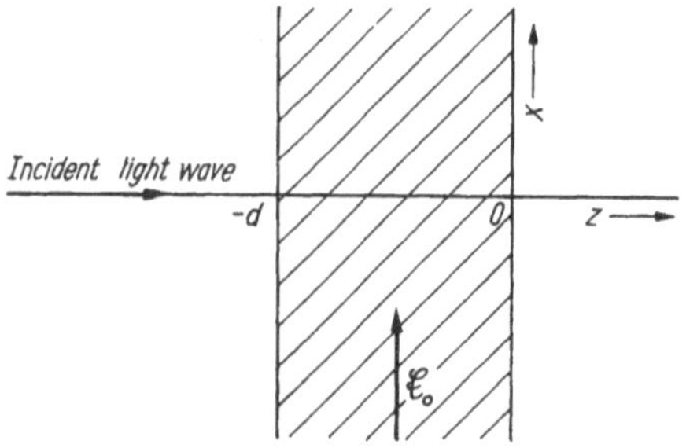

Fig. 1. Propagation of a light wave in a region with a dielectric function which depends on time and one coordinate (the shaded section represents the crystal)

assume

$$\varepsilon^{(l)}(\omega') = 0 \quad \text{for} \quad l \neq -1, 0, 1 \tag{41}$$

so that results published in a previous paper on a similar problem [11] can be utilized. An inspection of (38) suggests for $\boldsymbol{E}(\omega')$ an expression of the form

$$\boldsymbol{E}(\omega') = \sum_{l=-\infty}^{\infty} \boldsymbol{V}^{(l)} \, \delta\,(\omega - \omega' - l\,\Omega)\,. \tag{42}$$

As Ω, the frequency of the Zener oscillation, is much smaller under experimental conditions than the frequency ω of the light wave, we can use as a good approximation

$$\varepsilon\,(\omega \pm \Omega\,,\ \omega - \omega') = \varepsilon\,(\omega, \omega - \omega')\,. \tag{43}$$

In this approximation we then have

$$\boldsymbol{D}(t) = \varepsilon(\omega, t)\,\boldsymbol{E}(t)\,, \tag{44}$$

where $\varepsilon(\omega, t)$ is the time-dependent dielectric function (21) in the static field $\boldsymbol{E}_0$.

We assume the dielectric function without external field to, be equal to unity outside the crystal and equal to $1 + \Delta\varepsilon$ within the crystal. In the presence of $\boldsymbol{E}_0$ we have, expressing the dependence on the spatial coordinates explicitly (see Fig. 1),

$$\varepsilon'(\omega, t, z) = 1 + \varepsilon_1(\omega, t, z)\,, \tag{45}$$

where

$$\varepsilon_1(\omega, t, z) = [\theta\,(z - d) - \theta(z)]\,[\varepsilon^{(0)} - 1 + \varepsilon^{(-1)}\,e^{i\,\Omega t} + \varepsilon^{(1)}\,e^{-i\,\Omega t}]\,. \tag{46}$$

In [11] the propagation of the light wave (40) in a medium defined by (45) and (46) was examined under the assumption $\varepsilon_1 \ll 1$ in the lowest non-vanishing approximation with respect to ε_1. One then obtains for the relative intensities S_T and S_R of the energy current of the l-th partial wave in (42)

$$S_\mathrm{T}^{(0)} = \left|1 + \frac{i}{2}\,q\,d\,(\varepsilon^{(0)} - 1)\right|^2, \tag{47}$$

$$S_\mathrm{T}^{(l)} = |\varepsilon^{(l)}|^2\,\varkappa_l^2\,\frac{\sin^2\frac{1}{2}\,(\varkappa_l - q)\,d}{(\varkappa_l - q)^2} \qquad (l = 1,\ -1) \tag{48}$$

for transmission and

$$S_R^{(0)} = \frac{1}{16} |\varepsilon^{(0)} - 1|^2 , \tag{49}$$

$$S_R^{(l)} = \frac{1}{4} |\varepsilon^{(l)}|^2 \frac{\varkappa_l^2}{(\varkappa_l + q)^2} \quad (l = 1, -1) \tag{50}$$

for reflection, where

$$\varkappa_l^2 = \frac{1}{c^2} (\omega + l\,\Omega)^2 . \tag{51}$$

From (47) one obtains approximately

$$S_T^{(0)} = 1 - \alpha\, d , \tag{52}$$

where

$$\alpha = q \operatorname{Im} \varepsilon^{(0)} \tag{53}$$

is the absorption coefficient in the static field $\boldsymbol{E}_0$. The reflectivity R in the presence of $\boldsymbol{E}_0$ follows from (49), assuming that $\Delta\varepsilon$ is real, and is given approximately by

$$R = \frac{1}{8} \Delta\varepsilon \operatorname{Re} \{\varepsilon^{(0)} - 1\} . \tag{54}$$

Besides the variation in absorption and reflectivity the static field $\boldsymbol{E}_0$ produces another effect according to (48) and (50), namely the crystal becomes a source of light radiation with the frequency $\omega \pm \Omega$ transmitted in both positive and negative directions along the z-axis. The generation of "harmonics" $\omega \pm \Omega$ is quadratic in the field-induced quantity $\varepsilon^{(\pm 1)}$ in contrast to electro-absorption and electro-reflectance which are linear in the field contribution $\varepsilon^{(0)} - (1 + \Delta\varepsilon)$. The relative intensity of the "harmonics" is of the order of magnitude of $|\varepsilon^{(0)} - (1 + \Delta\varepsilon)|^2$ and depends on the band structure.

5. Franz-Keldysh Effect and Electro-Reflectance

As was shown in the previous section, the absorption and reflectivity of a crystal in a static field $\boldsymbol{E}_0$ is determined by the Fourier coefficient $\varepsilon^{(0)}(\omega)$. At the absolute zero and for values of ω in the vicinity of the threshold we obtain from (39) approximately

$$\varepsilon^{(0)}(\omega) = 1 - \frac{\omega_p^2}{\omega^2} + \frac{4\pi i}{\omega} \left(\frac{e}{m}\right)^2 \frac{|(20\ |p_x|\ 10)|^2}{\hbar\,\omega} \cdot \frac{1}{4\pi^3} \int \mathrm{d}^3\boldsymbol{k}\ C_{210}^{\boldsymbol{k}\omega} , \tag{55}$$

where $\nu = 1$ and $\mu = 2$ denote the valence band and the conduction band, respectively.

In order to determine the absorption coefficient α we consider according to (53) and (55) the real part of $C_{210}^{\boldsymbol{k}\omega}$. We have

$$\operatorname{Re} \left\{\frac{1}{\mathrm{e}^{-if(T)} - 1}\right\} = \lim_{\gamma \to 0} \frac{1}{2} \left[\frac{\cosh\gamma - \mathrm{e}^{-\gamma}}{\cosh\gamma - \cos f(T)} - 1\right] \tag{56}$$

or

$$\operatorname{Re} \left\{\frac{1}{\mathrm{e}^{-if(T)} - 1}\right\} = \pi \sum_{s=-\infty}^{\infty} \delta\left(f(T) - 2\pi s\right) - \frac{1}{2} . \tag{57}$$

Taking (37) and (34) into account it follows that

$$\operatorname{Re} C_{210}^{\boldsymbol{k}\,\omega} = \frac{1}{T}\left|\int_{-T/2}^{T/2} \mathrm{d}t\ \mathrm{e}^{-i f(t)}\right|^2 \cdot \sum_s \pi\,\delta\big(f(T) - 2\,\pi\,s\big)\,. \tag{58}$$

The second factor in (58) describes the Stark ladder in the absorption spectrum [7, 12]. If we replace it according to

$$\sum_s \delta\big(f(T) - 2\,\pi\,s\big) \to \frac{1}{2\,\pi} \tag{59}$$

by 1/2 we find the standard expression

$$\alpha = K\,\theta^{1/2} \int_{\frac{\omega_0-\omega}{\theta}}^{\infty} \mathrm{d}x\ |\mathrm{Ai}(x)|^2\,, \tag{60}$$

where

$$\theta = \left(\frac{e^2\,F^2}{2\,\hbar\,m^*}\right)^{1/3}, \qquad K = \frac{2}{c}\left(\frac{2\,m^*}{\hbar}\right)^{3/2}\left(\frac{e}{m}\right)^2 \frac{|(20\,|p_x|\,10)|^2}{\hbar\,\omega}\,, \tag{61}$$

and m^* is the reduced effective mass. (60) was derived using the effective-mass approximation and the assumption $\theta\,T \gg 1$.

One can easily show that the formal substitution (59) is equivalent to making allowance for collisions. We assume that the frequency $\omega_{21}(\boldsymbol{k})$ of an interband oscillator becomes a complex number if collission processes are taken into account

$$\omega_{21}(\boldsymbol{k}) \to \omega_{21}(\boldsymbol{k}) - \frac{i}{\tau_{21}}\,. \tag{62}$$

The relaxation time τ_{21} must satisfy

$$\Omega\,\tau_{21} \ll 1 \ll \omega\,\tau_{21}\,. \tag{63}$$

(56) is not changed by (62) provided T/τ_{21} is substituted for γ. Because of (63) T/τ_{21} is a very large number and we have approximately

$$\operatorname{Re}\left\{\frac{1}{\mathrm{e}^{-i f(T)} - 1}\right\} = 0 \tag{64}$$

and therefore

$$\operatorname{Re} C_{210}^{\boldsymbol{k}\,\omega} = \frac{1}{2\,T}\left|\int_{-T/2}^{T/2} \mathrm{d}t\ \mathrm{e}^{-i f(t)}\right|^2. \tag{65}$$

Collisions have been neglected in the integrals in (37). A comparison of (65) and (58) shows in fact that taking collisions into account in our approximation is equivalent to the substitution (59) (see also [12]).

The imaginary part of $C_{210}^{\boldsymbol{k}\,\omega}$ which determines the reflectivity in the presence of a static electric field can be found from (65) with the help of a Kramers-Kronig relation.

Acknowledgement

The authors wish to express their thanks to Dr. M. Porsch for numerous discussions and valuable suggestions.

Appendix

Let φ_n denote the elements of a complete set of orthonormal functions in the Hilbert space of a single electron and c_n^+ and c_n the creation and annihilation operators for an electron in the state φ_n. An arbitrary one-particle operator $\boldsymbol{A}$ is then in second quantization

$$A = \sum_{n' n} (\varphi_{n'} |\boldsymbol{A}| \varphi_n)\, c_{n'}^+\, c_n \,. \tag{A1}$$

The functions $\varphi_n(t)$ are defined as solutions of the time-dependent Schrödinger equation

$$i\,\hbar\,\dot{\varphi}_n(t) = H\,\varphi_n(t) \tag{A 2}$$

with the initial condition

$$\varphi_n(0) = \varphi_n \,. \tag{A 3}$$

We then have the following operator identity

$$A(t) = \sum_{n' n} (\varphi_{n'} |\boldsymbol{A}| \varphi_n)\, c_{n'}^+(t)\, c_n(t) = \sum_{n' n} \big(\varphi_{n'}(t) |\boldsymbol{A}| \varphi_n(t)\big)\, c_{n'}^+\, c_n \,. \tag{A 4}$$

This identity can easily be proved by differentiation.

In the case in which we are interested, i.e. the Hamiltonian H is given by (7) and the initial condition (A 3) by $\varphi_n = |\nu\,\boldsymbol{k})$, the Schrödinger equation is solved in good approximation by Houston functions [13]

$$\varphi_{\nu\boldsymbol{k}_0}(t) = \mathrm{e}^{-\frac{i}{\hbar}\int\limits_0^t \mathrm{d}\tau\, E_\nu(\boldsymbol{k}_\tau)}\, |\nu\,\boldsymbol{k}_t)\,, \tag{A 5}$$

where

$$\boldsymbol{k}_t = \boldsymbol{k}_0 + \frac{e}{\hbar}\boldsymbol{E}_0\, t\,. \tag{A 6}$$

The approximation consists of neglecting interband transitions induced by $\boldsymbol{E}_0$.

If in (A 4) $\boldsymbol{A}$ is the single-electron current operator $\frac{e}{m}\boldsymbol{p}$ and $\varphi_n(t)$ the Houston functions, then expression (25) follows.

References

[1] A. Frova, P. Handler, F. A. Germano, and D. E. Aspnes, Phys. Rev. **145**, 575 (1966).
[2] B. O. Seraphin, Phys. Rev. **140**, A 1716 (1965).
[3] K. L. Shaklee, F. H. Pollak, and M. Cardona, Phys. Rev. Letters **15**, 883 (1965).
[4] V. S. Vavilov, V. B. Stopachinskii, and V. Sh. Chanbarisov, Fiz. tverd Tela 8, 2660 (1966).
[5] W. Franz, Z. Naturf. **13a**, 484 (1958).
[6] V. L. Keldysh, Zh. eksper. teor. Fiz. **34**, 1138 (1958).
[7] J. Callaway, Phys. Rev. **130**, 549 (1963); **134**, A 998 (1964).
[8] K. Tharmalingam, Phys. Rev. **130**, 2204 (1963).
[9] K. S. Viswanathan and J. Callaway, Phys. Rev. **143**, 564 (1966).
[10] B. O. Seraphin and N. Bottka, Phys. Rev. **145**, 628 (1966).
[11] R. Enderlein, phys. stat. sol. **14**, 97 (1966).
[12] L. Fritsche, phys. stat. sol. **11**, 381 (1965); **13**, 487 (1966).
[13] W. V. Houston, Phys. Rev. **57**, 184 (1940).

(Received November 3, 1966)

phys. stat. sol. 19, 683 (1967)
Subject classification: 1; 4; 10; 22.6

S.C.K.-C.E.N., Mol

On the Superstructures of Ta_2O_5 and Nb_2O_5

By

J. SPYRIDELIS[1]), P. DELAVIGNETTE, and S. AMELINCKX[2])

The superstructures of tantalum and niobium pentoxide are studied. The observations made support the assumption that these, and related crystals, have a "shear structure" and that this shear structure can be observed directly. It is shown that the size of the unit cell depends on the exact stoichiometry of the compound. A model is proposed which explains both the details of the diffraction pattern and the geometry of the line system.

Es werden die Superstrukturen von Tantal- und Niobiumpentoxyd untersucht. Die gemachten Beobachtungen führen zu der Annahme, daß diese und ähnliche Kristalle eine „Scherungsstruktur" haben, und daß diese Scherungsstruktur direkt beobachtet werden kann. Es wird gezeigt, daß die Größe der Einheitszelle von der exakten Stöchiometrie der Verbindung abhängt. Ein Modell wird vorgeschlagen, das die Einzelheiten des Beugungsbildes und die Geometrie des Liniensystems erklärt.

1. Introduction

The structures of the oxides of tantalum and niobium pentoxide have been puzzling the crystallographers for a long time.

As well for Ta_2O_5 as for Nb_2O_5 a large number of different phases with different unit cells is known. Some of these phases appear not to be well defined; there is usually no sharp transition between them; moreover it is not always clear whether the compound has the simple stoichiometric composition Nb_2O_5 or Ta_2O_5 or is in fact a compound with a more complicated stoichiometry. The fact that different authors have given different symbols to the same compound only adds to the confusion.

A survey of the characteristics of all the known Nb_2O_5 phases was recently published by Schäfer et al. [1], who also tried to correlate the names given by the different authors.

A number of these oxides have unit cells which are large in one or two dimensions; they are therefore interesting objects for study by means of direct lattice resolution and by electron diffraction. Moreover, it was hoped to obtain some information by direct methods concerning the so-called "shear structures" which are common to a number of metal oxides.

A large number of different features of interest was observed, especially in Ta_2O_5, the appearance of which depends on the exact preparation method and heat treatment. This suggests that different phases have been observed. This paper gives a preliminary account and a tentative interpretation of some of the observations made on foils treated at around 700 to 800 °C and at higher temperatures than 1000 °C. Most of the observations refer to Ta_2O_5 but some observations on Nb_2O_5 will be discussed as well.

---

[1]) Permanent address: Department of Physics, Aristoteles University of Thessaloniki.
[2]) Also at Solid State Physics Laboratory, Rijksuniversitair Centrum Antwerpen, Middelheimlaan, Antwerpen.

2. The Crystalline Forms of Ta_2O_5 and Nb_2O_5

2.1 Niobium pentoxide

According to the review by Schäfer et al. [1] Nb_2O_5 occurs in 13 known crystallographic forms of which the characteristics are summarized in Table 1, excerpted from this review paper.

The form which has been observed in this work is possibly one of the M- or H-forms for which the unit cells have dimensions of the same order of magnitude as the one observed in this paper.

The most detailed structure determination of one of these H-phases is due to Gatehouse and Wadsley [2]. These authors propose a structure for Nb_2O_5 consisting of large blocks having a defective rhenium oxide structure, the

Table 1

Lattice parameters of some forms of Nb_2O_5

Form of Nb_2O_5	Lattice constants				Symbol given by the author	Author
	a	*b*	*c*	*β*		
TT	3.607		3.925		δ-pseudo-hexagonal	L. K. Frevel and H. W. Rinn [17]
T	7.31	15.72	10.75	120°42′	γ-monoclinic	L. K. Frevel and H. W. Rinn [17]
H	21.34	3.82	19.47	120°20′	α-monoclinic	L. K. Frevel and H. W. Rinn [17]
T	6.19	3.65	3.94		orthorhombic	W. T. Holser [18]
TT	3.62		3.93		δ-hexagonal	N. Terao [19]
T	6.17	3.66	3.93		γ-orthorhombic	N. Terao [19]
B	7.348	5.962	13.646	115°30′	γ′-monoclinic	N. Terao [19]
M	22.10	7.638	19.52	118°15′	β-monoclinic	N. Terao [19]
H	21.20	3.82	19.39	120°10′	α-monoclinic	N. Terao [19]
T	6.24	43.79	3.92		γ-orthorhombic	R.A. Zvinchuk [20]
M	20.28	3.83	20.28	120°	β-monoclinic	R.A. Zvinchuk [20]
H	20.39	3.82	19.47	115°39′	α-monoclinic	R.A. Zvinchuk [20]
H	21.50	3.825	20.60	121°45′	α-monoclinic	A. Magnéli and S. Lagergren [21]
H	21.34	3.816	19.47	120°20′	α-monoclinic	F. Holtzberg et al. [22]
H	21.16	3.822	19.35	119°50′	α-monoclinic	B. M. Gatehouse and A. D. Wadsley [23]
B	12.73	5.56	4.88	105°05′	ζ-monoclinic	F. Laves et al. [24]
H	19.63	3.835	20.93	120°	I-monoclinic	M. W. Schafer and R. Roy [25]
TT	3.98	3.83	12.80	90°45′	R monoclinic	R. Gruehn [26]

TT: very low temperature (< 600 °C);
T: low temperature (600 to 800 °C);
B: plates (750 to 850 °C);
M: medium temperature (900 to 1000 °C);
H: high temperature (> 1100 °C)

Table 2
Lattice parameters of some forms of Ta_2O_5

Form of Ta_2O_5	Lattice constants			Symbol given by the author	Author
	a	*b*	*c*		
T	6.20	3.66	3.89	β-orthorhombic	K. Lehovec [10]
T	7.77	7.66	12.67	β-orthorhombic	Y. P. Simonov et al. [27]
T	7.76	12.38	12.57	β-orthorhombic	Y. P. Simonov et al. [27]
T	7.24	6.47	7.47	β-orthorhombic	A. V. Lapitskii et al. [28]
T	6.18	43.93	3.89	β-orthorhombic	A. I. Zaslavskii et al. [29]
T	12.38	7.32	7.79	β-orthorhombic	A. I. Zaslavskii et al. [29]
H	3.80		35.67	α-tetragonal	A. I. Zaslavskii et al. [29]
T	6.18	3.66	3.88	β-orthorhombic	B. W. King et al. [30]
T	6.20	69.60	3.90	β-orthorhombic	J. Harvey and H. Wilman [6]
H	6.17		11.70	α-hexagonal	J. Harvey and H. Wilman [6]

T: low temperature phase;
H: high temperature phase

blocks being joined together along lines of discontinuity materialized by rows of metal ions in tetrahedral interstices. Within the rhenium oxide blocks the niobium occupies octahedral interstices in the oxygen packing.. The size of these blocks would be either $3 \times 5 \times \infty$ or $3 \times 4 \times \infty$ oxygen octahedra depending on the considered layer. The unit cell proposed by these authors is large in *two* dimensions, as is evident from Table 1.

We shall interpret our results on the basis of a structural scheme related to that of Gatehouse and Wadsley except that our observations suggest a unit cell which is large in *one* dimension only. However, we shall suggest a possible explanation for the fact that X-diffraction may lead to a unit cell with two large dimensions whereas in electron diffraction one finds a unit cell which is only large in one dimension.

2.2 Tantalum pentoxide

Tantalum pentoxide occurs in at least two phases: the low-temperature β-phase and the high-temperature α-phase. According to Kofstad [3] the transformation temperature is around 1300 °C. The crystallographic parameters of these two forms, according to the different authors are summarized in Table 2, excerpted from the review paper by Steeb and Renner [14]. According to various authors β-Ta_2O_5 is isomorphous with γ-Nb_2O_5. The phase studied here is presumably a superstructure of the β-phase (terminology used by Kofstad).

3. Experimental

3.1 Preparation of specimens

Thin foils of tantalum oxide were obtained by anodizing a tantalum sheet of a purity of 99.999%. The sheet was first polished chemically in a fresh solution of 5 parts by volume sulfuric acid (95%), 2 parts nitric acid (70%), and 2 parts hydrofluoric (40%) [4]. Subsequently the sheet was anodized to about 30 V at a constant current density of 1 mA/cm² in a solution of 2% by volume nitric acid during 30 minutes. Taking into account the thickness–voltage relationship in these circumstances, which was determined to be 16 Å V^{-1} [5], the thickness of the tantalum pentoxide film should be about 500 Å. Using other electrolytes, e.g. 0.5% sodium sulfate or 0.2% potassium fluoride, gave essentially the same results.

The oxide film was removed from the metal substrate either mechanically (by stripping) or chemically (i.e. by reversing the anodizing current). For the first procedure use was made of adhesive tape which was dissolved in benzene to liberate the oxide film. With the second procedure it is the hydrogen formation at the interface metal oxide which causes the oxide to become free from the metal. With this procedure clean large-area specimens can more easily be obtained (up to 5 mm²) [6]. In both cases the oxide films were carefully washed in distilled water before being transferred on nickel, copper, or platinum grids.

The thin foils of Nb_2O_5 were prepared in a completely similar way. A niobium sheet of a purity of 99.999% was anodized at 20 V, which leads to a thickness of about 500 Å in view of the fact that the thickness–voltage relationship for Nb_2O_5 is 26 Å/V [7, 8, 9].

3.2 Observation method

The observations as well as the selected area diffraction patterns were made partly by means of the Siemens Elmiskop I and partly by using the Philips EM 200 electron microscope operated at 100 kV. The microscope was also used to perform the final specimen preparation, which was a recrystallization treatment. This was done either by using a Philips hot stage at a temperature of 700 to 800 °C or by means of beam heating. By carefully increasing the beam current the crystallization takes place in a controlled way; it starts already at a beam current of about 10 μA (condensor lens focused, condensor aperture out).

During this treatment the grain boundaries were found to sharpen and sometimes grain boundary melting was observed. From these observations it was concluded that during this second heat treatment the specimen temperature must have been close to the melting point, i.e. between 1000 and 1300 °C. Fig. 1 shows how the crystallized area which exhibits extinction contours, expands from the heated spots. Recrystallization was also performed by heating the oxide film on its grid in air at different temperatures between 700 and 800 °C.

4. Observations

We shall first discuss the observations on Ta_2O_5. Afterwards we shall compare these observations with those on Nb_2O_5.

4.1 Images

The oxide films formed by anodizing are apparently amorphous; the electron diffraction patterns consist of diffuse halos. After heat treatment the oxide foils are recrystallized. Especially beam heating in the microscope allows to produce relatively large grains. We shall discuss mainly specimens prepared in this way. Such foils are found to consist of large grains each grain being further fragmented by a line pattern. Some grains contain only one family of

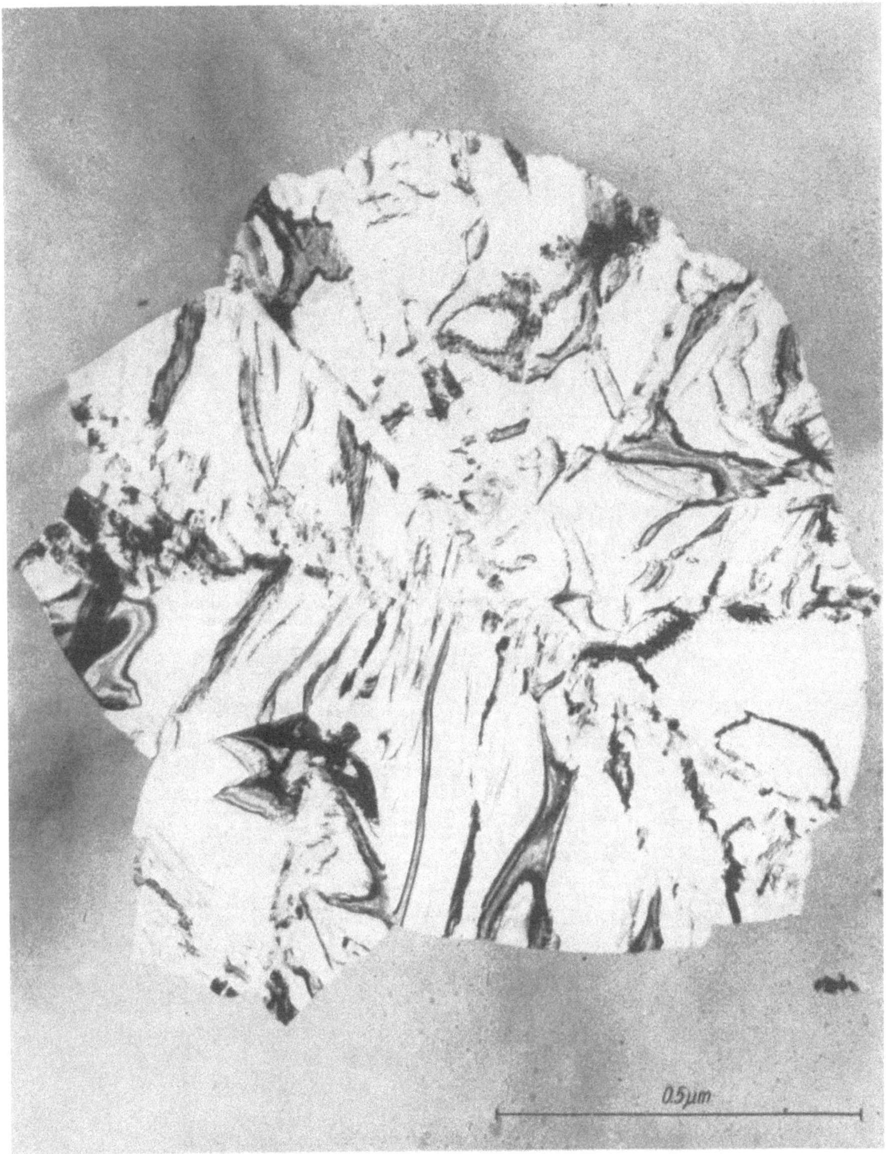

Fig. 1. Recrystallization of amorphous tantalum pentoxide by beam heating in the electron microscope. The recrystallized area exhibits extinctions contours

striations; others contain up to a maximum of three different families of lines enclosing angles which are approximately 60° (or 120°) (Fig. 2). Usually the striations do not interpenetrate but occasionally in thick crystals they do so and then a two-dimensional line system results locally (Fig. 3). As a consequence of the striations the grains are further subdivided into subgrains, each subgrain containing only one family of lines. The orientation of the striations in the different subgrains of a given grain are clearly parallel to one out of three pos-

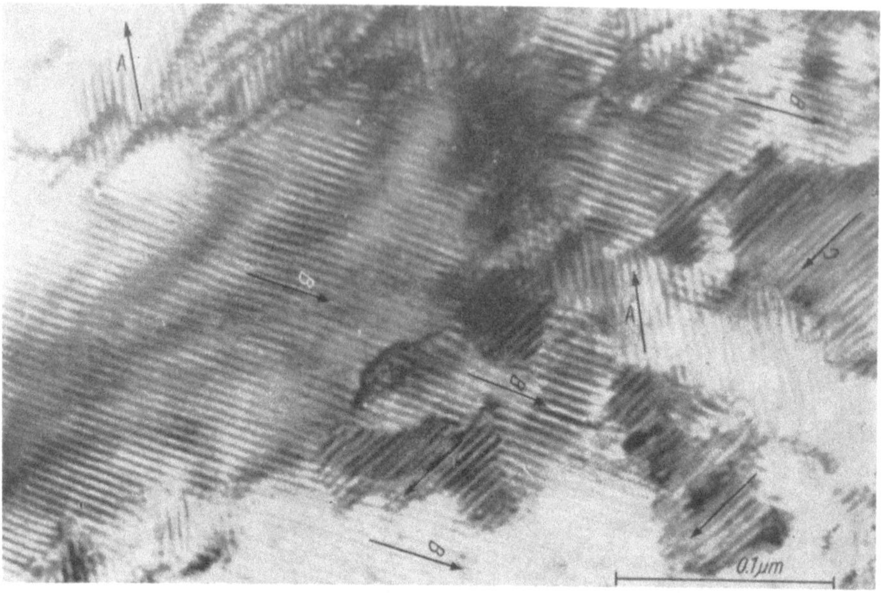

Fig. 2. Fragmentation of the crystal into regions exhibiting striations in three different directions, labelled A, B, and C, enclosing approximately angles of 60° (tantalum pentoxide)

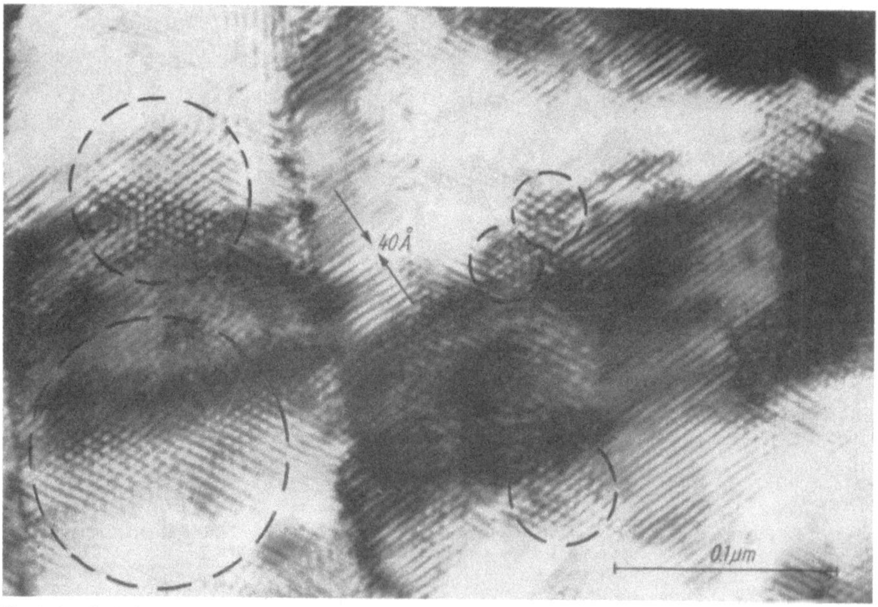

Fig. 3. In thick foils regions exhibiting striations in different directions sometimes overlap, producing pseudo-hexagonal meshes in the encircled areas (tantalum pentoxide)

sible crystallographic directions of some common lattice, presumably the lattice of the original grain, before the striations were formed. We shall call this *hypothetical structure* the "*basic structure*". The angles between the directions of these striations and also the geometry of the intense spots (subsequently

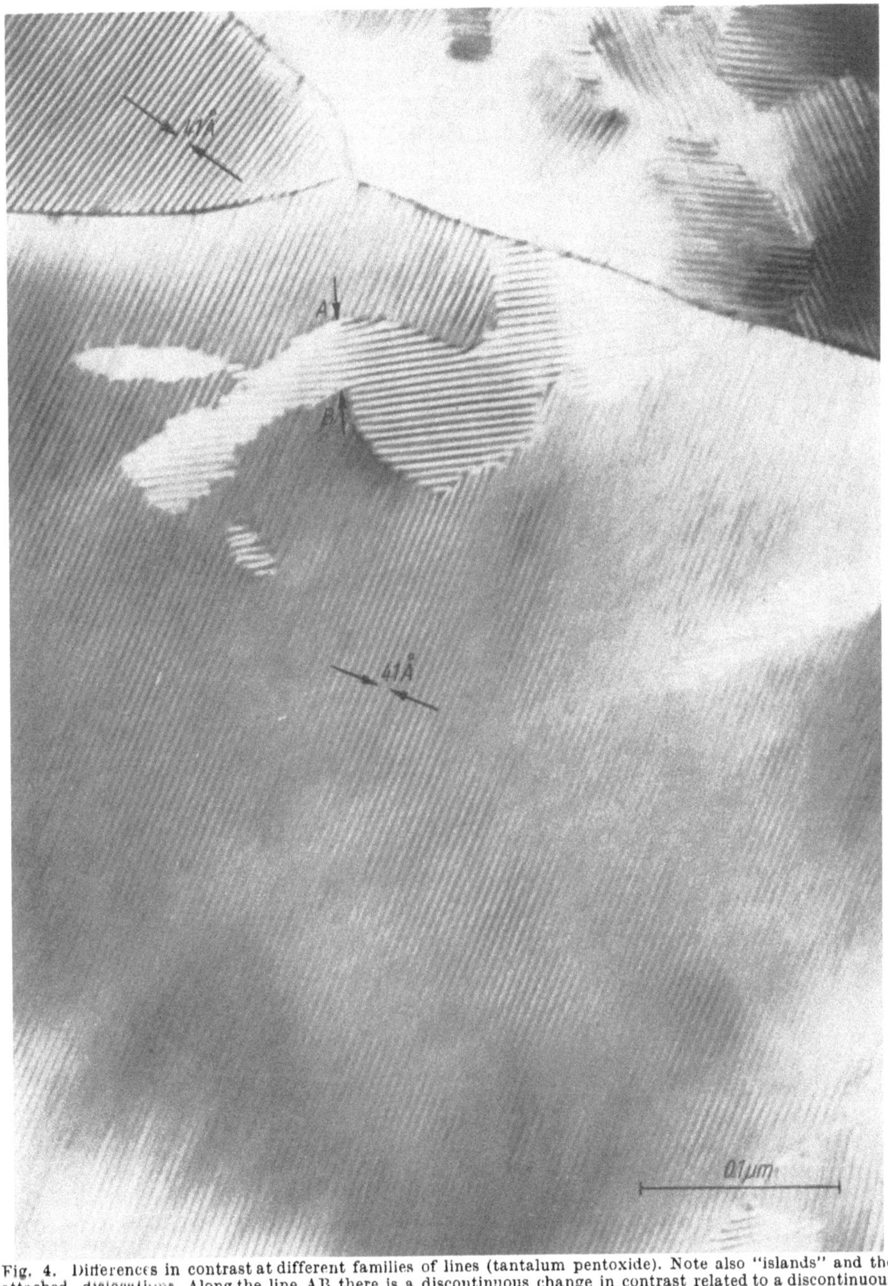

Fig. 4. Differences in contrast at different families of lines (tantalum pentoxide). Note also "islands" and the attached dislocations. Along the line AB there is a discontinuous change in contrast related to a discontinuous change in orientation of the fringes

called main spot) of the diffraction pattern suggest that the basic crystal should presumably be hexagonal or pseudo-hexagonal (see Fig. 13a).

Apparently the line pattern starts forming simultaneously and independently in different parts of a given grain having this basic structure. Where

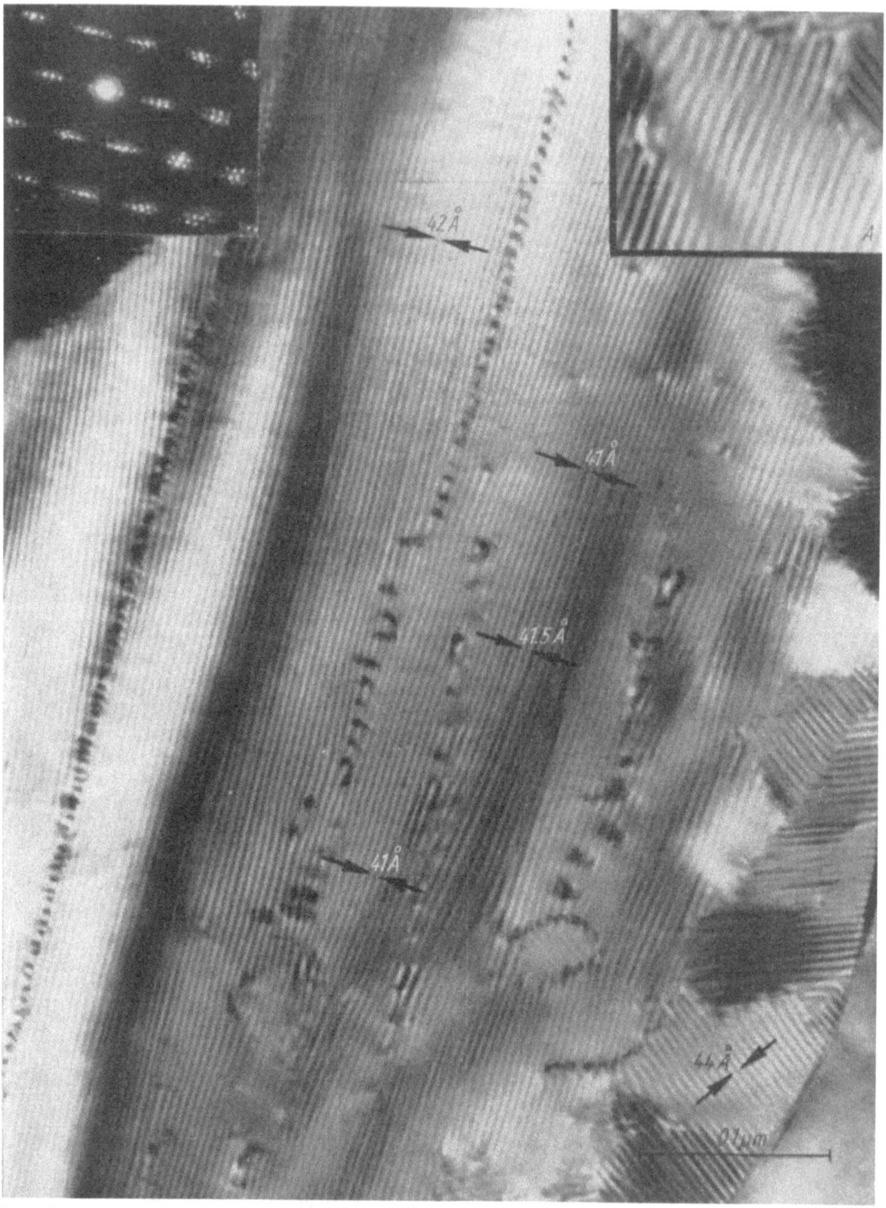

Fig. 5. Well recrystallized area of tantalum pentoxide specimens containing isolated dislocations and subgrain boundaries. Note also the "out of step" boundary in the right bottom corner and which is also reproduced as an inset A

lines with different orientations meet, subboundaries are formed (Fig. 2). Since the line systems do not usually interpenetrate well defined subgrains are formed. It is found that the directions of the line system do not depend on the diffraction conditions, however their contrast may vary with the orientation of the specimen. Also the contrast may be quite different for the different families of lines (Fig. 4). This suggests that the lines represent a feature which is directly related to the lattice. Apart from the small regions where two line systems overlap, one does not find two line systems in the same subgrain which are visible either simultaneously or alternatively. Occasionally parallel families of lines meet "out of step". This phenomenon may occur at inclination extinction contours; in such cases it is a pure contrast phenomenon [15]. In a few exceptional cases (Fig. 5) we believe that the "out of step" boundary is *not* due to contrast effects. The generation of such boundaries is easy to visualize; if the formation of the linear domains starts independently in different parts of the grain they will generally not match on meeting, even if they are parallel.

The observations on niobium pentoxide are quite similar to those on tantalum pentoxide. However, as a result of the smaller spacing of the lines Nb_2O_5 does not show images as detailed as those seen in Ta_2O_5; an example of the line pattern is shown in Fig. 8.

4.2 Dislocations

In view of the large repeat distance of the "superstructure" it is of course unlikely that the dislocations would have a Burgers vector related to a "lattice vector" of the superlattice. A more likely interpretation relates the Burgers vector to the lattice of the "basic" structure. Let us assume that we have a "shear" structure and that the vector which describes the relative displacement of one block with respect to the next block is $\boldsymbol{R}$. The shear planes can then be considered as fault planes with a displacement vector $\boldsymbol{R}$. The vector $\boldsymbol{R}$ will not be a lattice vector with respect to the "basic" structure (i.e. the structure within the blocks). If such a shear plane now terminates within the crystal a dislocation with Burgers vector $\boldsymbol{R}$ must be present at the end point. This dislocation is a partial with respect to the basic structure and a fortiori with respect to the superstructure. It is now clear that the Burgers vector of the dislocation is quite unrelated to the fringe spacing.

If two adjacent shear planes terminate they may join and end at the same dislocation with a Burgers vector which is now equal to the sum of the displacement vectors of the two ending shear planes. This dislocation may be a "perfect" dislocation with respect to the basic structure.

One could imagine dislocations within the blocks; such dislocations would not be revealed as ending fringes.

The systems of lines have a direct relationship with the presence of dislocations; at the point where a dislocation is present a line usually terminates there (Fig. 5). Contrast due to the strain field as well as the supplementary fringes are simultaneously visible. At some dislocations there is only *one* terminating fringe; at others there are *two* terminating fringes.

Fig. 6 shows the detailed way in which the lines terminate in a subboundary. Some of the lines have a pronounced black dot at their endpoint; others just fade out. By comparison with the black dots in the interior of the grains and which also form the endpoint of lines we conclude that also the black dots in the subboundaries represent dislocations. There is a dislocation every two fringes in the boundary AB of Fig. 6 whereas there is a dot for every fringe in the subboundary CD.

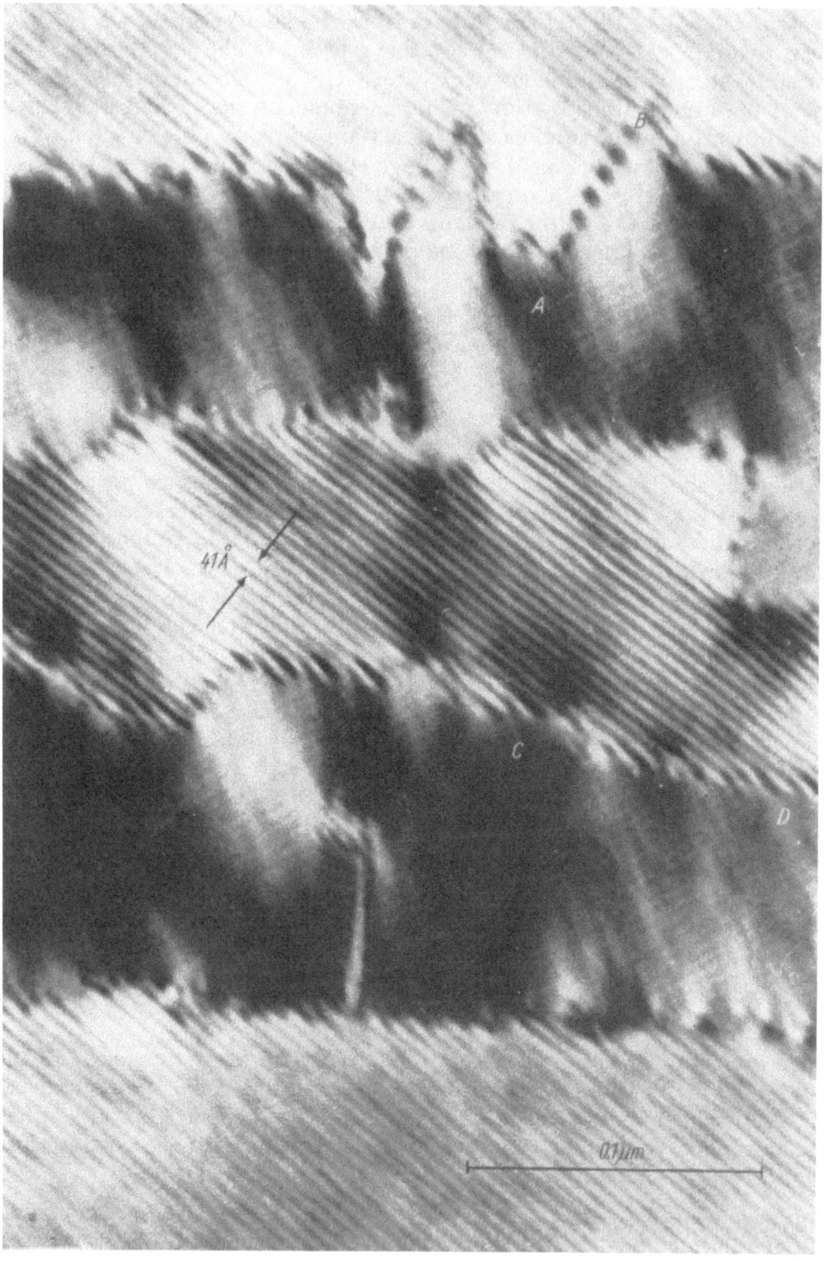

Fig. 6. Detailed structure of some of the subgrain boundaries. The black dots at terminating fringes mark dislocations in the subboundary. For the boundary AB there is one dot every two fringes, whereas in the boundary CD there is one dot per fringe

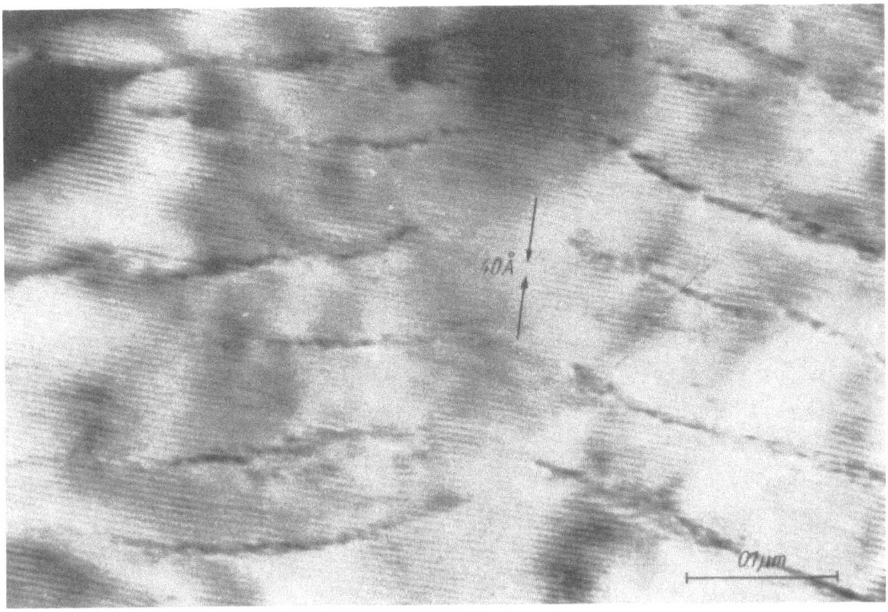

Fig. 7. Dislocations parallel to the foil plane and their effect on the line pattern

Fig. 4 shows a small island of one orientation enclosed within a region of another orientation. Note the presence of dislocations associated with this small island.

Fig. 7 shows the effect of dislocations which are almost parallel to the foil plane on the line pattern; although they terminate in the foil surface no extra half line is attached to their emergence point. The reason is presumably that these are glide dislocations in glide planes parallel to the foil plane. As a result their Burgers vectors are roughly parallel to the foil plane and presumably very much shorter than the long spacing which produces the line pattern. This observation would suggest that the glide plane is the plane c in the description of Lehovec [10].

4.3 Diffraction pattern

We shall again first discuss the diffraction pattern of tantalum pentoxide, and afterwards we shall point out the characteristic differences for niobium pentoxide.

The diffraction pattern depends on the selected area. The simplest type of pattern is obtained if the selected area is confined to a single subgrain containing therefore only one line system within the selected area. The type of pattern obtained in these circumstances is shown in Fig. 9; it consists of *main spots* forming a hexagonal or pseudo-hexagonal grid. These spots are indicated by means of small circles in Fig. 9, and in Fig. 10 the pattern is drawn schematically. There are further dense rows of spots passing through the main spots; we shall call these *superlattice spots*. The intensity of the superlattice spots varies in a periodic way; the intense spots are grouped around the positions of the main spots and also around the centres of all the triangles (which will

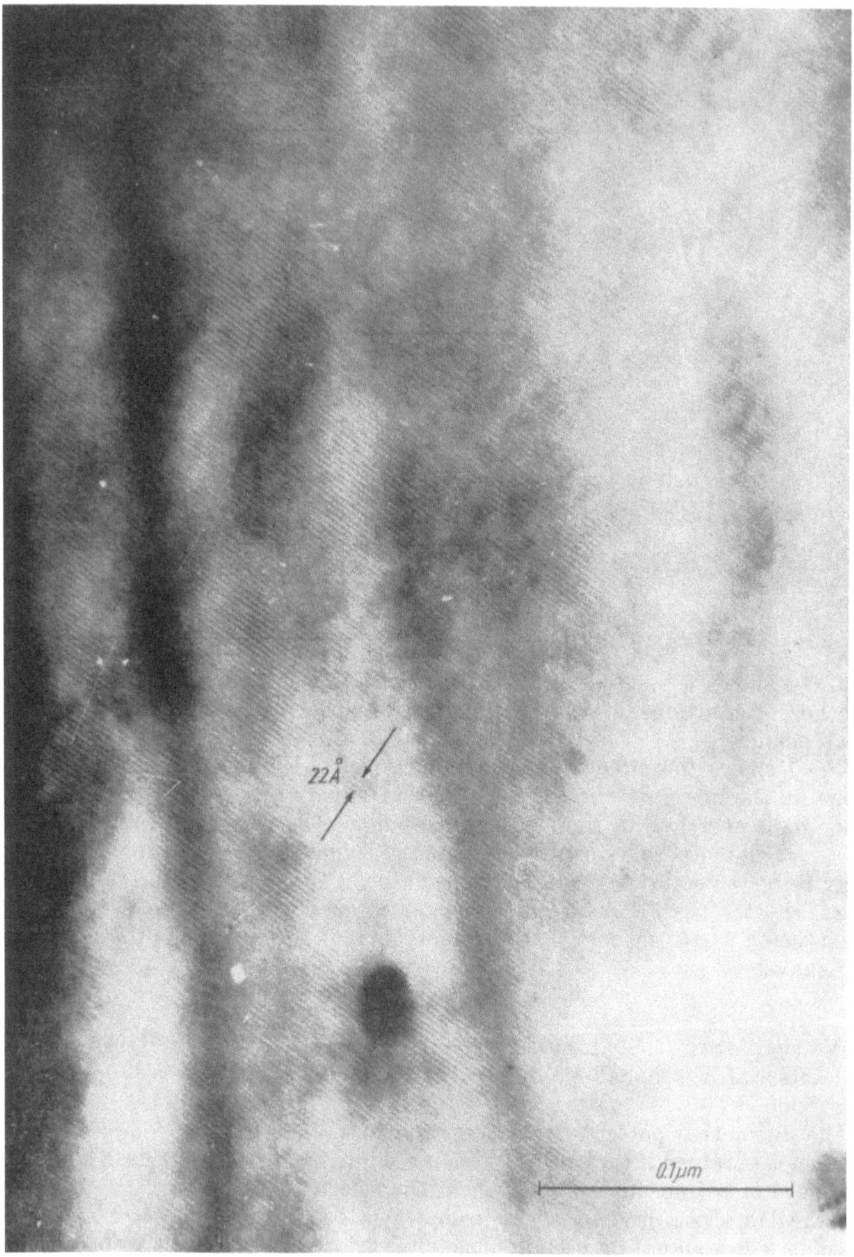

Fig. 8. Fringe system in niobium pentoxide

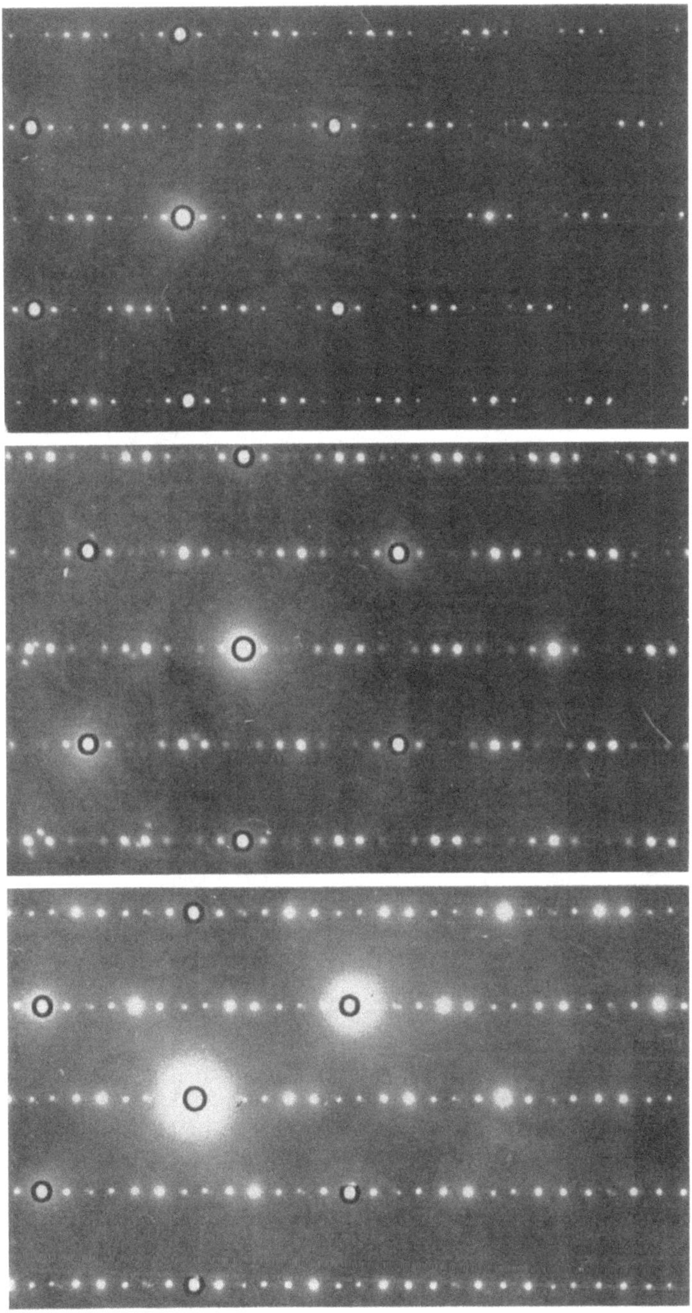

a

b

c

Fig. 9. Diffraction pattern of niobium pentoxide due to a region containing a) only one line system and b) showing "spacing anomalies"; c) spacing and orientation anomaly

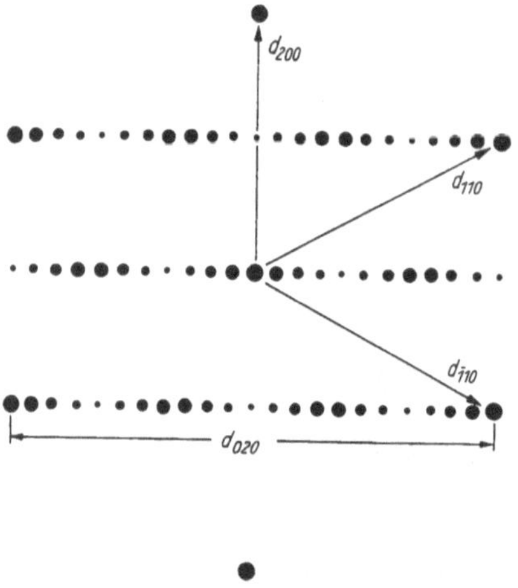

Fig. 10. Schematic drawing of a diffraction pattern without anomalies. The "main" spots are represented as the black dots of the largest size

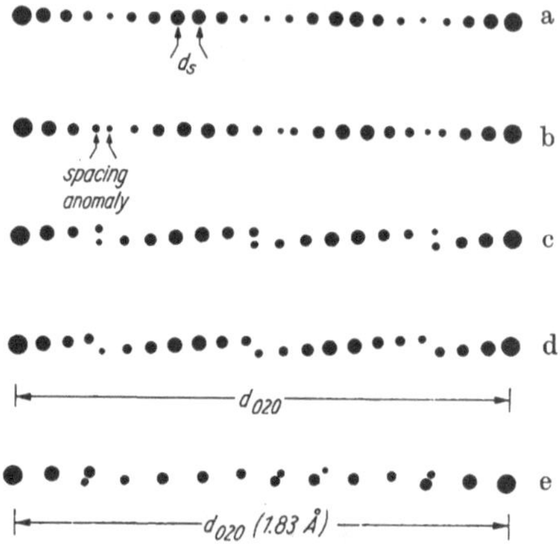

Fig. 11. Schematic drawing of the sequences of diffraction spots due to the "superlattice".
a) Normal sequence without anomaly;
b) spacing anomaly;
c) orientation anomaly;
d) orientation and spacing anomaly in Ta_2O_5;
e) orientation and spacing anomaly in Nb_2O_5 (see also Fig. 9c)

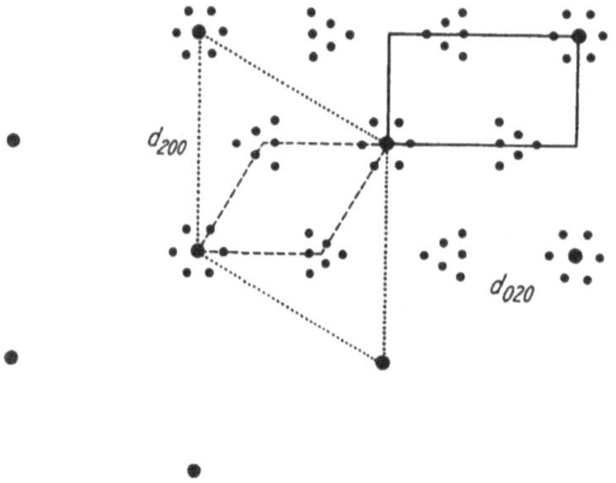

Fig. 12. Comparison of different reciprocal unit cells used to describe the reciprocal lattice as they appeared in the literature. In reality the crystal is built in a superlattice. Full line: Lehovec; dashed line: Wilman; dotted line: unit cell corresponding to the pattern of Fig. 13a

be called subsidiary spots) formed by the pseudo-hexagonal pattern of the main spots; in those centres the most intense superlattice spots are arranged in a triangular configuration as schematized in Fig. 12 and observed in Fig. 13b, c, and d. Main and subsidiary spots form together another hexagonal pattern. The groups of superlattice spots form linear arrangements of the type shown in Fig. 11a, b, c, d, or e. The intensity of the spots in a given group decreases with increasing distance from the main or subsidiary spot. The spots within a given group are equally spaced. The directions of the linear groups often enclose a very small angle with the principal directions of the pattern of main spots; this angle is somewhat variable (Fig. 9). Fig. 11c, d is a schematic and an exaggerated representation of the effect. In some cases the angular mis-orientation cannot be detected but even then a spacing anomaly is usually present at the point where two linear groups join (Fig. 9b). The "gap" between two of these linear arrangements is not an integer number of spacings between superlattice spots.

The simplest assumption that could explain the anomalies would of course be to accept that we are not observing one simple plane of the reciprocal lattice, but that in reality the reciprocal lattice mesh is also small in the direction perpendicular to the foil plane. We would then see a slightly oblique section of the reciprocal lattice, the reinforcement occuring at the intersection of Ewald's sphere with rows in successive levels of the reciprocal lattice. This hypothesis could explain the spacing anomalies as well as the angular misorientation effect. Repeated attempts were made to find the exact orientation which would eliminate all anomalies. Such attempts lead to reproducible patterns with the anomalies still present. It is therefore suggested that anomalies are genuine. Accepting them at their face value and assuming the patterns of Fig. 9 and 13 to represent single reciprocal lattice planes we shall attempt to explain them.

The presence of the "gap" between two linear arrangements would imply that either the mesh size of the superlattice is larger than one would conclude simply from the spacing of the superlattice spots, or that the superlattice spots are not part of a reciprocal lattice strictly speaking. This last hypothesis seems to be favoured for the following reasons:

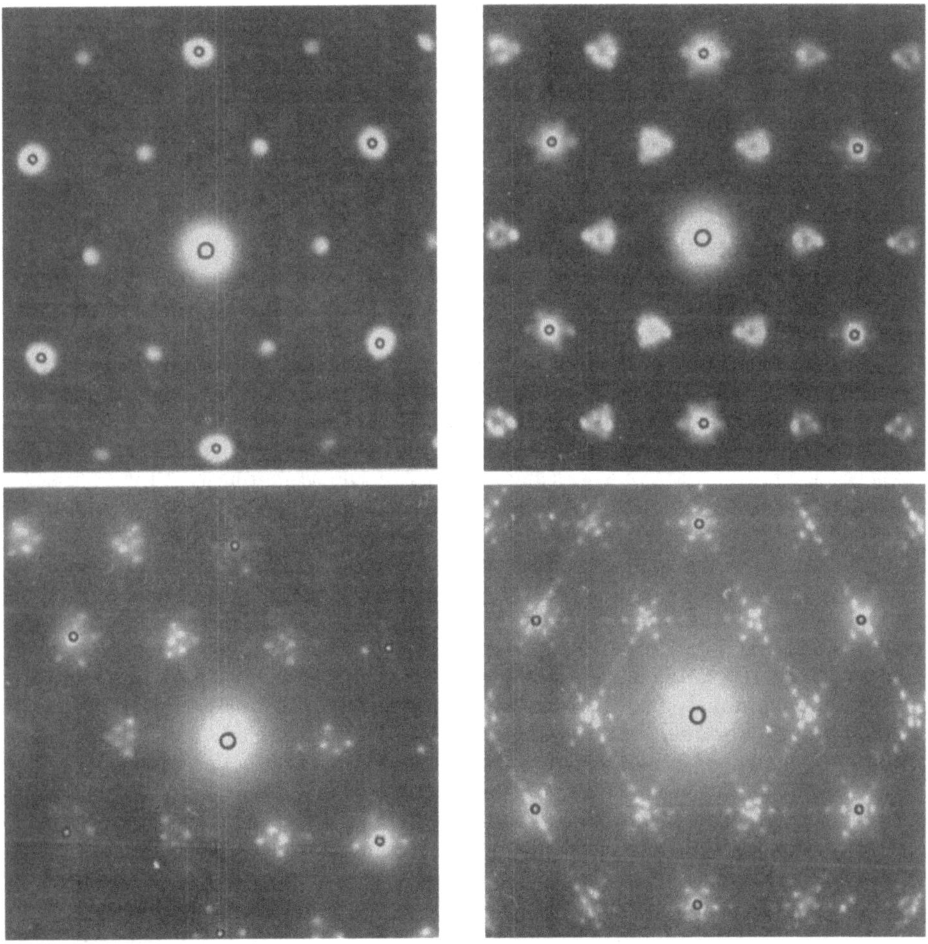

Fig. 13. Diffraction patterns from areas containing line systems of the three orientations in foils in successive stages of recrystallization

a) The spacing of the superlattice spots and hence the corresponding average-long spacing in direct space is somewhat variable among different specimens. This is *not* the case for the spacings corresponding with the main spots.

b) As a result the spacing anomaly associated with the "gap" between different linear groups of spots is variable as well. Sometimes even no measurable spacing anomaly is found.

c) In poorly recrystallized specimens the number of superlattice spots associated with each main or subsidiary spot (i.e. in a given linear group) decreases. For such specimens also the line system is poorly developed and variable in spacing within the same area (Fig. 13).

d) The fact that the direction of linear groups of spots does not coincide with the principle direction of the pattern of main spots also points in that direction.

The pattern of main spots can be indexed on the orthorhombic unit cell proposed by Lehovec [10] and which has the dimensions

$$a = 6.18\ \text{Å}\,, \quad b = 3.66\ \text{Å}\,, \quad \text{and} \quad c = 3.88\ \text{Å}\,.$$

The corresponding reciprocal cell is shown in Fig. 12 (full line) and it is clear that the "main" spots occupy node points at the reciprocal lattice generated by this cell.

One can also describe, within the experimental error, the lattice of "main" spots with respect to an hexagonal cell with $a = 3.56$ Å; the reciprocal of this is shown as well (dotted line).

The pattern of main *and* subsidiary spots is also hexagonal, and can with good approximation be indexed on a hexagonal cell with $a = 6.17$ Å, outlined by the dashed line in Fig. 12, and which is identical with the cell proposed by Harvey and Wilman [6].

The lattice spacing corresponding to the spacing between superlattice spots varies from 20 to 40 Å for different specimens depending on the heat treatment. The pattern of main spots is independent of heat treatment.

Although our unit cell is different from that proposed by Wadsley et al. [2], it is not excluded that we are in fact looking at the same phase. Since the specimen temperature is difficult to estimate on beam heating, we may well look at the high-temperature phase in the beam-heated specimens.

If the selected area covers different subgrains of a single grain one obtains in general pseudo-hexagonal patterns. Depending on whether the number of line systems present is one, two or three, one obtains a diffraction pattern containing one, two, or three directions in which there are rows of closely spaced spots. Since the angle between the rows is about 60° (or 120°) one obtains a pseudo-hexagonal pattern of the type shown in Fig. 13d.

In niobium pentoxide the spacing between the superlattice spots in the linear group is somewhat larger than in tantalum pentoxide corresponding to a smaller long spacing in direct space (which varies from 20 to 30 Å).

The angular difference between the directions connecting the main spots and the direction of the linear arrangements of superlattice spots is larger. This gives rise in the "gap" to apparent spot doubling of the type shown in Fig. 11e and in Fig. 14. This phenomenon is due to the combination of angular misorientation and spacing anomaly. Apart from the difference in long spacing the pattern of main spots is the same as that for Ta_2O_5.

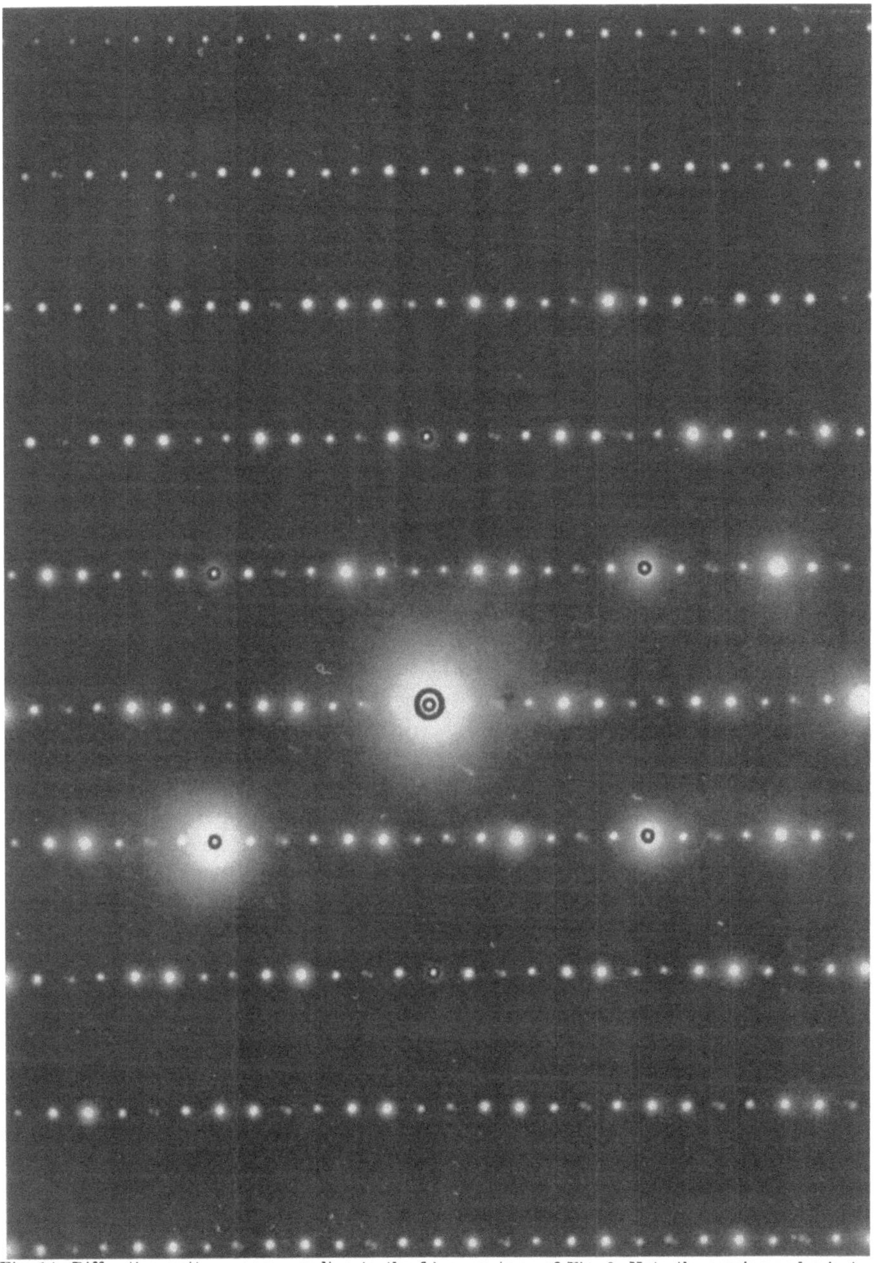

Fig. 14. Diffraction patterns corresponding to the fringe systems of Fig. 8. Note the spacing and orientation anomaly represented schematically in Fig. 11e

4.4 Relation between the diffraction pattern and the image

It is quite obvious that a close relationship must exist between the observed line systems and the long spacing revealed by the diffraction pattern. The line

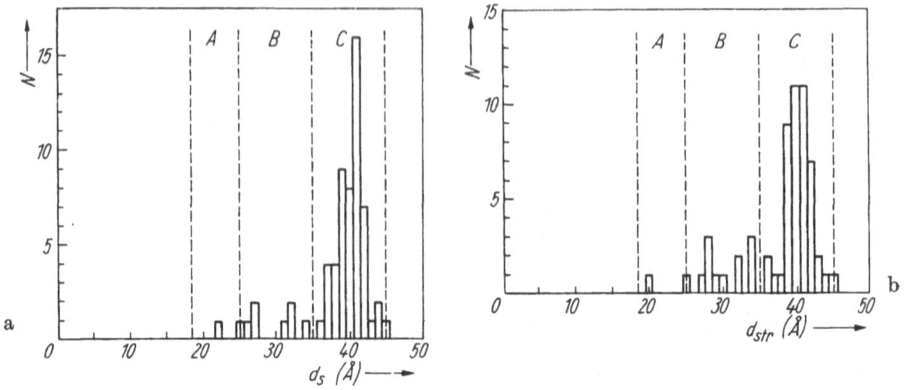

Fig. 15. a) Histogram of interplanar spacings as deduced from the diffraction spots.
b) Histogram of line spacings as deduced from the image

pattern is formed if the aperture admits as well the direct beam as the superlattice reflection on each side of the 000 spot [16]. It was verified that in all cases the rows of superlattice spots are perpendicular to the striations in the image. We have evaluated for a number of somewhat differently treated specimens the long spacing as deduced from the diffraction pattern and we have on the other hand measured the average spacing for the line patterns.

The histograms of Fig. 15a and b summarize the results of such measurements on Ta_2O_5. The frequency of observation is plotted as a function of the spacings deduced from measurements of the fringe spacing (Fig. 15a) and from measurements of spot separation in the diffraction pattern (Fig. 15b). In both histograms one can distinguish three groups A, B, and C of long spacings. The values around 40 Å refer to specimens crystallized from the amorphous phase either by beam heating or by heating in the hot stage in the microscope vacuum. Long spacings of around 30 Å (26 to 34 Å) are obtained from specimens crystallized in sealed quartz tubes containing a reduced pressure of air ($\approx 10^{-2}$ Torr) at a temperature of about 700 to 800 °C.

Spacings of only 20 Å are obtained in specimens crystallized in air at 700 to 800 °C. If such specimens are subsequently treated by beam heating in the microscope the spacing of 40 Å appears.

The same frequency distribution is found in both histograms proving the direct relationship between fringe spacing and spot separation.

In badly recrystallized specimens with poorly developed line systems only the hexagonal pattern of main and subsidiary spots is visible (Fig. 13a). However, these spots are then broadened and they have a somewhat hexagonal outline (Fig. 12 and 13b) suggesting that perhaps superlattice spots are in fact present at a small distance from the main or subsidiary spot, but that they are submerged in the latter spot due to its broadening (Fig. 13c).

4.5 Fringe contrast effects

In some regions of the images, especially in extinction contours, it is found that the fringes are doubled giving the impression that the fringe spacing is half of the repeat distance deduced from the diffraction pattern. A similar phenomenon was noted and explained by Labaw [11] as being the result of the dynamical interaction between spots of the first-layer line and spots of the zero-layer line. The conditions on which Labaw's explanation is based do not seem to be fulfilled in our case however. We believe that this phenomenon is due to the fact that the image results from the interference between the direct beam and the superlattice spots *on either side* of the direct beam.

The total wave function at the exit face of the crystal is then of the form

$$\Psi = \Psi_0 + \Psi_h \, e^{2\pi i h x} + \Psi_{-h} \, e^{-2\pi i h x},$$

when Ψ_0, Ψ_h, and Ψ_{-h} are the amplitudes of the transmitted beam and of the scattered beams $+\boldsymbol{h}$ and $-\boldsymbol{h}$, respectively.

The intensity distribution in the fringes is then of the general form

$$I = I_0 + I_1 \cos(2\pi h x + \varphi_1) + I_2 \cos(4\pi h x + \varphi_2),$$

where x is measured along the exit face in the direction of the diffraction vector $\boldsymbol{h}$ (and $-\boldsymbol{h}$). φ_1 and φ_2 are phase angles which determine the fringe position and which depend on the diffraction conditions. The first term gives rise to a continuous background. The second term has a period $1/h$ and the third has a period of $1/2\,h$. The value of I_1 depends on the amplitudes of the three reflections whereas the magnitude of I_2 depends only on the amplitudes of the superlattice reflections. In a dark extinction contour for the strong (main) reflection the intensity of the 000 beam may become very weak. The coefficient I_2 may now become larger than I_1; the period $1/2\,h$ will then show up prominently. The presence of the second term, however, will usually cause the two fringes in one period to have unequal intensities and unequal spacing. The half-period harmonic of the fringes in fact results from the interference between the two superlattice reflections.

In the bright contours of the main reflection on the other hand I_1 is always much larger than I_2 and therefore only the spacing $1/h$ shows up normally.

This explains qualitatively the fringe doubling in extinction contours. A detailed discussion of the lattice fringes found in the circumstances described here will be the subject of another paper.

5. Discussion and Interpretation of the Observations

5.1 Model

We shall present a model which seems to be consistent with all the observed features and which also accounts for the relationship between the image and the diffraction pattern. According to this model the crystal would consist of narrow linear strips forming domains within which for instance a defective ReO_3 or some other simple structure would be present. The boundary between the strips would be formed either as a result of ordering following a diffusion type mechanism or as a result of shear motions on a set of closely spaced parallel planes. The ordering or the shear motions would transform a single crystal of some other phase into the fragmented structure observed within a single grain. The fact that the line system can develop along three equally favoured directions under angles of 120° seems to suggest that the basic structure should have hexagonal or pseudo-hexagonal symmetry. On transforming from this basic structure into the observed phase, it is quite possible that due to a change in symmetry transformation stresses are created which can be relieved by shear on equally spaced planes. An alternative common way of releaving stress would be to fragment into fine twin lamella; there is no evidence for twinning however.

Strain would be releaved even further by shearing or twinning on three systems of planes.

However, it is also possible that due to the particular stoichiometry the metal ions cannot all be accomodated in equivalent octahedral sites of the oxygen framework and that as a result of this some other sites, e.g. tetrahedral sites, are occupied. This was in fact proposed for Nb_2O_5 by Wadsley [2] and for molybdenum and tungsten oxides by Magnéli [12]. Ordering of the metal ions in tetrahedral interstices into linear arrangements may also explain the fragmentation of the crystal into narrow strips. If the ordering starts simultaneously along different directions, out of the three possible ones, in different parts of a grain of the basic structure the observed line structure would result. The final result would not be very different from that due to a shear transformation.

5.2 Relation between model and diffraction pattern

The main features of the diffraction patterns can qualitatively be accounted for on the basis of the model proposed above. According to this model the crystal is in fact a "double grating" shown schematically in Fig. 16. The narrow strips each form a grating with a small well defined repeat distance containing a somewhat variable number of unit cells of the basic structure. These gratings form a grating themselves because they are regularly repeated. The spacing of this "super-grating" is large but somewhat variable; the total number of "lines" of the super-grating is equal to the number of lattice fringes observed within one subgrain.

It has been shown by Heidenreich [13] on a linear model that a double grating of this type produces a diffraction pattern of the kind observed, i.e. consisting of short linear sequences of superlattice spots grouped around the positions of the diffraction spots due to the basic structure. A similar feature was observed experimentally by Magnéli in β-tungsten oxide [12].

In poorly recrystallized specimens the large spacing is not precisely defined and therefore the number of superlattice spots in a given group is small because the intensity of the superlattice spots decreases rapidly with the distance from the positions of the main spots. The better the long spacing is defined, the larger will be the number of superlattice spots in a sequence.

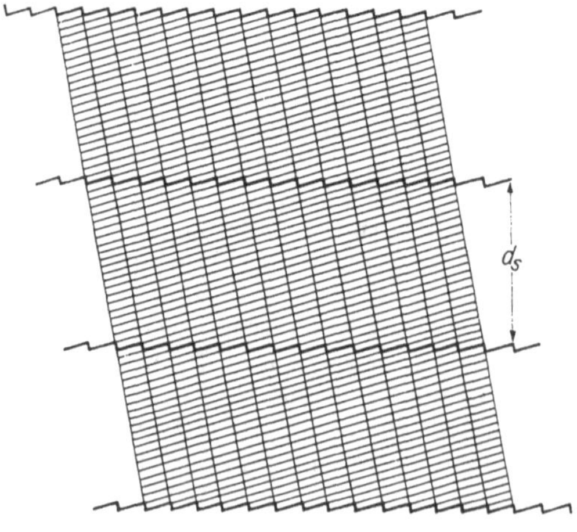

Fig. 16. Schematic model for the structure of Nb_2O_5 and Ta_2O_5

We finally have to account for the orientation difference between the rows of main spots and the short rows of superlattice spots. The interface between domains need not coincide precisely with the main direction of the fundamental structure and therefore the linear group of superlattice spots need not have directions which are strictly parallel to the main directions of the fundamental structure. Expressed in terms of the lattice this could be reformulated as follows: The low index planes of the superlattice need not be parallel to those of the basic lattice.

The small angular deviation might for instance be due to small rotations of the lattice (with respect to the lattice of the basic structure) accompanying the shear motion on equally spaced planes.

The spacing between interfaces of lines of discontinuity is furthermore not necessarily an integer number of unit cells of the fundamental structure. This explains the spacing anomaly. By chance the average spacing might be an integer number of repeat distances of the fundamental structure; in this case no spacing anomaly would result.

The variation of the line spacing with the oxygen pressure and with the heat treatment strongly suggests that the stoichiometry should play an important role in determining the superlattice spacing. The fact that a superlattice is formed seems to suggest further that even compounds with deviations from stoichiometry or expressed otherwise: with a complicated stoichiometry, tend to adopt regular structures.

6. Relation with Previous Work

Specimens of tantalum oxide of the same type have been examined previously by means of electron diffraction by Harvey and Wilman [6]; their diffraction patterns do not seem to be the same as the ones observed here. In particular the large spacing was found to be ≈ 35 Å [3]), which is comparable to some of the long spacings observed in this work. Also Harvey and Wilman did not find the spacing anomaly, neither did they observe the line patterns.

The model for the structure of tantalum pentoxide and niobium pentoxide proposed here is a variation of the "block scheme" proposed for the latter oxides by Wadsley and coworkers. In particular we found only *one long spacing* within each subgrain although a number of differently oriented foils were examined.

The discrepancy in unit cell size presumably results from the fact that X-ray diffraction was used to determine the unit cell and the structure. Therefore inevitably use was made of crystals which were fragmented into domains along the three directions observed in our specimens. This of course leads to a diffraction pattern which suggests a unit cell large in *two* dimensions. This would also be the case in electron diffraction if a pattern of the type shown in Fig. 13d were used.

7. Conclusions

The observations presented here show that the method of direct "lattice resolution" is a fruitful approach for the study of structural details of the so-called "shear structure" with large unit cells. The observed variability in spacing of the line structure and the corresponding variations in superlattice spacing

[3]) The unit cell he proposes has dimensions $a = 6.20$ Å, $b = 69.6$ Å, $c = 11.7$ Å.

seem to suggest that some of the observed "phases" of Nb_2O_5 might in fact have very closely related structures, only the block sizes being different. There is presumably a close relationship between deviations from stoichiometry and the block size.

The shear compounds presumably form from a phase of higher symmetry either by fine scale periodic shear or perhaps by periodic twinning on a fine scale, generating in this way a superlattice. Superlattice unit cells determined by means of X-rays have to be considered with caution; they may in fact be too large due to the fact that "single crystals" have in fact a domain structure.

There is also evidence that isolated "lines" are visible by a contrast mechanism which is independent of the "lattice fringe" contrast.

References

[1] H. SCHÄFER, R. GRUEHN, and F. SCHULTE, Z. angew. Chem. **78**, 28 (1966).
[2] B. M. GATEHOUSE and A. D. WADSLEY, Acta cryst. **19**, 26 (1965).
[3] P. KOFSTAD, J. less-common Metals **5**, 158 (1963).
[4] M. HARTL, Z. Naturf. **20 b**, 392 (1965).
[5] D. A. VERMILYEA, Acta metall. **1**, 282 (1953).
[6] J. HARVEY and H. WILMAN, Acta cryst. **14**, 1278 (1961).
[7] R. E. POWEL and J. J. CAMPBELL, J. Electrochem. Soc. **111**, 1230 (1964).
[8] L. YOUNG, Proc. Roy. Soc. **A 244**, 41 (1958).
[9] L. YOUNG, Canad. J. Chem. **38**, 1141 (1960).
[10] K. LEHOVEC, J. less-common Metals **7**, 397 (1964).
[11] L. W. LABAW, Acta cryst. **17**, 46 (1964).
[12] A. MAGNÉLI, Ark. Kemi (Sweden) **1**, 513 (1950).
[13] R. D. HEIDENREICH, Fundamentals of Transmission Electron Microscopy, Interscience Publishers, New York 1964 (p. 300).
[14] S. STEEB and J. RENNER, J. less-common Metals **9**, 181 (1965).
[15] H. HASHIMOTO, M. MANNAMI, and T. NAIKI, Phil. Trans. Roy. Soc. (London) **A 253**, 459 (1961).
[16] J. W. MENTER, Proc. Roy. Soc. **A 236**, 119 (1956).
[17] L. K. FREVEL and H. W. RINN, Analytic. Chem. **27**, 1329 (1955).
[18] W. T. HOLSER, Acta cryst. **9**, 196 (1956).
[19] T. TERAO, Japan. J. appl. Phys. **2**, 156 (1963); **4**, 8 (1965).
[20] R. A. ZVINCHUK, Soviet Phys.-Cryst. **3**, 750 (1959).
[21] A. MAGNÉLI and S. LAGERGREN, A.S.T.M. Index Card 5-0379/80.
[22] F. HOLTZBERG, A. REISMAN, M. BERRY, and M. BERKENBLIT, J. Amer. Chem. Soc. **81**, 3182 (1959); **79**, 2039 (1957).
[23] B. M. GATEHOUSE and A. D. WADSLEY, Acta cryst. **17**, 1545 (1964).
[24] F. LAVES, R. MOSER, and W. PETTER, Naturwissenschaften **51**, 356 (1964); **51**, 633 (1964).
[25] M. W. SCHAFER and R. ROY, Z. Kristallogr., **110**, 241 (1958).
[26] R. GRUEHN, J. less-common Metals **11**, 119 (1966).
[27] Y. P. SIMANOV, A. V. LAPITSKII, and E. P. ARTEMONOVA, Vestn. Mosk. Univ. **9** (9), Ser. Fiz. Mat. i Estestven. Nauk **6**, 109 (1954).
[28] A. V. LAPITSKII, Y. P. SIMANOV, K. N. SEMENEN, and E. I. YAREMBASH, Vestn. Mosk. Univ. **9** (3), Ser. Fiz. Mat. i Estestven. Nauk **2**, 85 (1959).
[29] A. I. ZASLAVSKII, R. A. ZVINCHUK, and A. G. TUTOW, Dokl. Akad. Nauk SSSR **104**, 409 (1955).
[30] B. W. KING, J. SCHULTZ, E. A. DURBIN, and W. DUCKWORTH, A.S.T.M. Index Card 8-255.

(Received Ncvember 25, 1966)

phys. stat. sol. **19**, 705 (1967)

Subject classification: 9; 22.2.1

Institute of Semiconductors, Academy of Sciences of the USSR, Leningrad

The Effect of Applied Electric Field on Diffusion of Impurities in Gallium Arsenide

By

B. I. BOLTAKS and T. D. DZHAFAROV

Electrical migration of lithium and copper in gallium arsenide is studied. It is shown that lithium (between 740 and 980 °C) and copper (between 820 and 1000 °C) migrate in n-type gallium arsenide as positive ions with a charge of less than +1. The decrease of the ionic charge is due to ion drag by the electrons. In heavily doped p-type gallium arsenide, copper again migrates in the form of positive ions but in this case with a charge close to +1. This is due to the ion drag by the electrons being almost balanced by the drag from the holes which acts in the opposite direction. Experimental evidence suggests that thermal conversion in gallium arsenide is caused by copper diffusing from outside. Copper can be successfully removed from specimens by heating them in an applied dc electric field. Electrical migration of zinc and cadmium is also studied. Cadmium is shown to migrate in the form of positive ions. In the temperature range 830 to 1130 °C, zinc also migrates as positive ions, the effective mobility decreasing with temperature. This temperature dependence of the mobility is in agreement with a theory of the ion drag effect.

Es wird die Elektrodiffusion von Lithium und Kupfer in Galliumarsenid untersucht. Es wird gezeigt, daß Lithium (zwischen 740 und 980 °C) und Kupfer (zwischen 820 und 1000 °C) in n-leitendem Galliumarsenid als positive Ionen mit einer Ladung kleiner als +1 wandern. Die Verkleinerung der Ionenladung wird durch den Ionendrag durch Elektronen verursacht. In stark dotiertem p-leitendem Galliumarsenid wandert Kupfer auch als positives Ion, aber in diesem Falle mit einer Ladung nahe +1. Das wird hervorgerufen durch den Ionendrag von Elektronen, der fast durch den in entgegengesetzter Richtung wirkenden der Löcher ausgeglichen wird. Die experimentellen Fakten weisen darauf hin, daß die thermische Konversion in Galliumarsenid durch Kupfer verursacht wird, das von außen eindiffundiert. Kupfer läßt sich gut aus den Proben entfernen, wenn diese bei angelegtem elektrischen Gleichfeld aufgeheizt werden. Auch die Elektrodiffusion von Zink und Kadmium wurde untersucht. Es wird gezeigt, daß Kadmium als positives Ion wandert. Im Temperaturbereich 830 bis 1130 °C wandert auch Zink als positives Ion. Die effektive Beweglichkeit nimmt mit der Temperatur ab. Diese Temperaturabhängigkeit befindet sich mit der Theorie des Ionendrageffekts in Übereinstimmung.

In the presence of an applied electric field, besides the drift of impurity particles in this field there arises also an interaction between diffusing particles and free carriers of a semiconductor. In a number of cases this interaction affects considerably the migration of these particles.

There have appeared up to now only a limited number of papers dealing with the effect of an applied electric field on the diffusion of impurities in semiconductors. The present paper supplements the available experimental data on this subject with data on electric migration of lithium, copper, zinc, and cadmium in gallium arsenide.

1. Theory

1.1 Diffusion and drift of impurity ions in a dc electric field

As is well known [1], the application of dc electric field to a crystal changes the form of the crystal lattice potential in such a way that the neighboring potential "humps" shift relative to one another by $\pm \frac{1}{2} q E \delta$ (Fig. 1), where q is the charge of the diffusing ion, δ the separation between "humps" or "wells", and E is the applied electric field. This shift results, in its turn, in a change of the probability of ion transition from one equilibrium position to another. In this case the probabilities of ion transition in the direction of the field, W_+, and against the field, W_-, will be

$$W_+ = \frac{1}{2} \exp\left(-\frac{\Delta U - q E \delta/2}{kT}\right), \tag{1}$$

$$W_- = \frac{1}{2} \exp\left(-\frac{\Delta U + q E \delta/2}{kT}\right). \tag{2}$$

Now since $W_+ > W_-$, a drift flow of charged particles $J = \bar{v} N$ will appear in the direction of the field which will be superimposed on the diffusion flow of the same particles $J = -D \frac{\partial N}{\partial x}$ (where N is the concentration of diffusing particles, D the diffusion coefficient, and $\bar{v}$ the drift velocity of particles).

The drift velocity is determined by the following relation:

$$\bar{v} = \frac{\nu_0 \delta^2}{2} \frac{q}{kT} \exp\left(-\frac{\Delta U}{kT}\right) E = \mu E, \tag{3}$$

where μ is the mobility and ν_0 the frequency of ion oscillations in the equilibrium position.

The concentration distribution of diffusing ions in a crystal in the presence of an applied electric field can be found by solving the differential equation

$$\frac{\partial N}{\partial t} = D \frac{\partial^2 N}{\partial x^2} + \bar{v} \frac{\partial N}{\partial x} \tag{4}$$

at the corresponding initial and boundary conditions.

The study of the free diffusion and diffusion in the presence of applied electric field permits to determine the diffusion coefficient and mobility of diffusing ions. Besides, by using Einstein's relation

$$\frac{\mu}{D} = \frac{q}{kT}, \tag{5}$$

one can find the charge state of the diffusing ions.

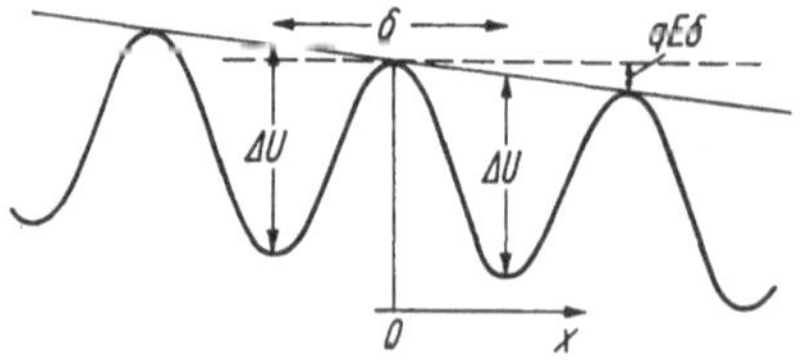

Fig. 1. Potential distribution in a crystal with an applied electric field

1.2 The effect of ion drag by carriers

If there is a sufficiently high concentration of moving current carriers (electrons and holes) in a crystal to which a dc electric field is applied, an effect of drag of impurity particles by current carriers appears [2 to 4]. The force excited by electrons on an impurity particle is

$$F_{el} = n\, l_n\, \sigma_{in}\, q_0\, E\ , \tag{6}$$

where n is the electron concentration, l_n the mean free path of an electron, σ_{in} the effective scattering cross section of electrons by ions, q_0 the ion charge.

The resultant force acting on a positively charged ion from an applied electric field and from electrons (due to the drag effect) will be

$$F = q_0\,(1 - n\, l_n\, \sigma_{in})\, E = q_{eff}\, E\ , \tag{7}$$

where

$$q_{eff} = q_0\,(1 - n\, l_n\, \sigma_{in})\ . \tag{7'}$$

Here q_{eff} is the experimentally measured effective ion charge. By substituting q_{eff} into the expression for the mobility it is easy to demonstate that the effective mobility μ_{eff} measured experimentally will be related to the true mobility of ions μ_0 through the following expression:

$$\mu_{eff} = \mu_0\,(1 - n\, l_n\, \sigma_{in}) \tag{8}$$

when $n\, l_n\, \sigma_{in} \ll 1$, $\mu_{eff} \approx \mu_0$. As $n\, l_n\, \sigma_{in}$ approaches unity, the effect of drag becomes noticeable, and μ_{eff} begins to decrease and vanishes at $n\, l_n\, \sigma_{in} = 1$.

In a more general case of mixed conduction where ions are dragged not only by electrons but by holes as well,

$$q_{eff} = q_0\,(1 - n\, l_n\, \sigma_{in} + p\, l_p\, \sigma_{ip})\ , \tag{9}$$

$$\mu_{eff} = \mu_0\,(1 - n\, l_n\, \sigma_{in} + p\, l_p\, \sigma_{ip})\ . \tag{9'}$$

Here p, l_p are the concentration and mean free path of holes, respectively, and σ_{ip} the scattering cross section of holes on ions.

Thus both the magnitude and sign of the ion mobility measured experimentally are determined completely by the formula (9′). In the intrinsic conduction region where $n = p$ the effect of drag by holes can be neglected since $l_p \ll l_n$ and $\sigma_{in} \approx \sigma_{ip}$. In semiconductors the concentration of current carriers in the intrinsic conduction region increases sharply with temperature, and therefore the effect of ion drag by electrons will increase whereas the effective charge of ions should at first decrease.

2. Experimental

2.1 Techniques

The electric migration of impurities was studied on single crystal specimens of n-type gallium arsenide ($n = 3 \times 10^{17}$ cm^{-3}, $\mu_n = 3400$ cm^2 V^{-1} s^{-1}). Specimens were cut from an ingot and represented rectangular parallelepipeds with dimensions $\approx 4 \times 4 \times 5$ mm^3. A thin layer of the impurity to be studied was deposited electrolytically on the end faces of two identical specimens which were then put together with these faces in contact and clamped between the electrodes of the apparatus. In another arrangement three specimens were fixed in the apparatus, viz. a specimen with the impurity already introduced into it and two pure ones. Dc electric current was then passed through them which served both for heating and as a source of electric field. The electric field strength was measured with two tungsten probes pressed by means of special springs to the lateral surface of the specimens. The specimens were heated in argon vapor. The temperature of specimens was measured with a chromel-alumel thermocouple led in through a hole in the specimen.

By studying the impurity concentration distribution in specimens after the electric migration had taken place one could determine the mobility of ions from the formula of reference [5]:

$$\mu = \frac{Q_+ - Q_-}{N_0 E t}, \tag{10}$$

where Q_+ and Q_- are the amounts of impurity which had diffused into the anode and cathode specimens in a dc electric field E, N_0 is the impurity concentration in the plane $x = 0$, t is the duration of electric migration. On the other hand, if one knows the concentration distribution in the specimens being studied, it is possible to determine the ratio of mobility to the diffusion coefficient:

$$\frac{\mu}{D} = \frac{1}{E} \frac{\mathrm{d}}{\mathrm{d}x} \ln \frac{N_-(x, t)}{N_+(x, t)}. \tag{11}$$

Here $N_-(x, t)$ and $N_+(x, t)$ are the impurity concentration distributions in the cathode and anode specimens, respectively. Knowing μ/D and using Einstein's relation (5) one can calculate the charge of the impurity ions.

Since all the impurities (Li, Cu, Zn, Cd) investigated by us in gallium arsenide are acceptors, the analysis of the impurity concentration profile in the study of electric migration was supplemented with the p–n junction technique. In this case the ion mobility was determined by the formula

$$\mu = \frac{x_- - x_+}{2 E t}, \tag{12}$$

where x_- and x_+ are the depths at which the p–n junctions are located in the cathode and anode specimens, respectively. The lateral surfaces of specimens were freed of the oxide layer, and the p–n junctions were revealed by etching or by the method of electrolytic deposition of copper. The distance between the specimen welding boundaries and p–n junctions in the cathode and anode specimens was measured on a type ИЗА-2 comparator to ± 1 μm.

The electric migration of zinc and cadmium was studied by a somewhat different technique. Gallium arsenide specimens shaped as parallelepipeds ($\approx 4 \times 4 \times 2$ mm^3) with thoroughly ground-in end faces were subjected to diffusion annealing in the vapor of the impurity being investigated (Zn or Cd). After the annealing p^+–p and p–n junctions were revealed on both sides by etching the lateral surfaces, and the depths at which the junctions were located were measured. Next such a specimen was clamped between two pure specimens, and the three specimens were mounted in the apparatus for the study of electric migration. After the migration the junctions were again revealed on the lateral surfaces, and their depths were measured. The measured depths of the p–n junctions were then used to determine the ion mobility from formula (12).

In addition to the study of electric migration of some impurities (Cu, Zn), free diffusion was also investigated. This permitted to obtain independent data on the concentration distribution and coefficient of diffusion of these impurities in gallium arsenide.

Free diffusion of impurities was studied with radioactive isotopes. The radioactive impurity was electrolytically deposited on two opposite specimen surfaces. The diffusion annealing was carried out in evacuated (to 10^{-2} mm Hg) and sealed quartz ampoules. After the diffusion annealing the lateral surfaces of the specimens were cleaned. Next plane-parallel layers were successively removed from the operating surfaces, and their γ-activity was measured. The concentration distribution of the radioactive impurity was studied by the standard layer removal technique or by the contact radiography method [1].

2.2 Diffusion and electric migration of copper

The diffusion of copper in gallium arsenide was investigated by a number of authors [7, 8]. They have established that the diffusion coefficient of copper in this material is as large by the order of magnitude as that for germanium. The analysis of concentration curves obtained by means of autoradiography has shown that the distribution of copper in gallium arsenide does not obey Fick's law.

In order to get additional information on the character of copper concentration distribution and the parameters of diffusion, we studied the diffusion of Cu-64 into GaAs. Diffusion of Cu into GaAs results in a p^+–p–n structure. The obtained curve for copper, as seen from Fig. 2, can be divided into two parts. The first part characterized by a high copper solubility ($\approx 1 \times 10^{19}$ cm^{-3}) and a steep fall-off of the concentration distribution corresponds to the surface layer of the specimen. The second part illustrating copper distribution in the bulk of the specimen is characterized by a lower solubility and a less steep concentration decrease.

As the diffusion heating becomes longer, the concentration distribution of copper in the bulk rapidly attains saturation. At the same time the copper concentration distribution in the surface layer changes only slightly with the increase of heating time. A similar pattern of concentration distribution was observed earlier in the diffusion of other group I elements in GaAs and other $A^{III}B^{V}$ compounds [16, 17]. As noted in references [16, 17], the first part of the concentration distribution is due to the presence in the specimen surface layer of an enhanced concentration of vacancies or other defects occupied by the diffusing impurity. Quite naturally this part of the concentration distribution characterizes the diffusion of vacancies or other defects rather than of the impurity. Therefore for the determination of the impurity diffusion coefficient one should use the second part of the concentration distribution curve. Our estimates yield for the coefficient of copper diffusion in GaAs at 940 °C the value $\approx 10^{-6}$ cm^2 s^{-1} and for the limiting value of copper solubility $N_0 \approx 10^{17}$ cm^{-3}.

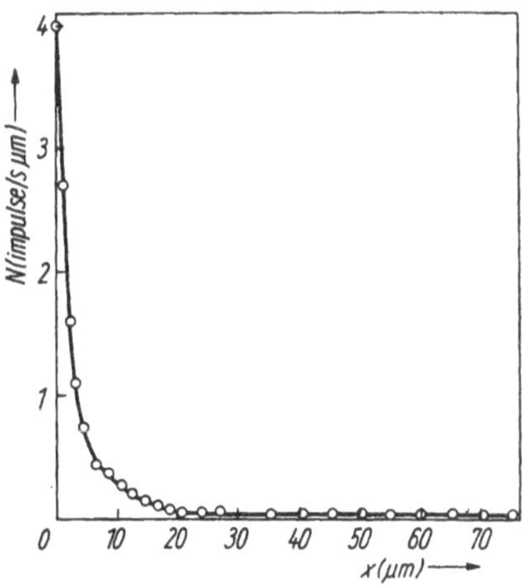

Fig. 2. Copper distribution in gallium arsenide

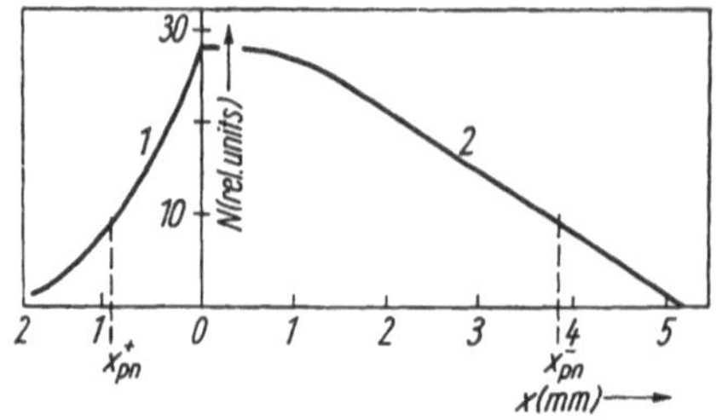

Fig. 3. Copper distributions in n-type gallium arsenide. $T = 1000$ °C, $t = 11$ min; 1 — anode specimen, 2 — cathode specimen. x^+_{pn} and x^-_{pn} are locations of p–n junctions in the anode and cathode specimens, respectively

The electric migration of radioactive copper in n-type GaAs was studied in the 820 to 1000 °C range. The current density in these experiments was varied from 55 to 97 Acm^{-2} and the time of migration was varied from 3 to 17 min. Fig. 3 shows the distribution curves for Cu-64 obtained by the method of contact radiography for a number of gallium arsenide specimens studied. It is seen that the depth of copper penetration is larger in the specimen adjoining the cathode than in the anode specimen. Such asymmetry of the curves indicates a predominant migration of copper in gallium arsenide to cathode which means that it acquires a positive charge. Using formulas (10) and (11), one can derive from the curves of copper distribution in the cathode and anode specimens μ and μ/D and from Einstein's relation one can determine the charge of copper ions. The values of μ and q thus obtained are listed in Table 1.

Table 1

T (°C)	t (s)	μ_{eff} ($cm^2 V^{-1} s^{-1}$)	q_{eff}	x^-_{pn} (mm)	x^+_{pn} (mm)	σ_{in} (cm^2)
820	720	9.8×10^{-6}	0.6	3.04	1.00	8.9×10^{-13}
850	1020	4.9×10^{-5}	—	4.05	0.40	—
930	135	7.4×10^{-5}	0.4	1.60	0.70	1.4×10^{-13}
950	480	9.9×10^{-5}	0.4	3.66	1.33	1.3×10^{-13}
1000	660	8.5×10^{-5}	0.3	3.85	1.00	1.1×10^{-13}

The measured depths of locations of p–n junctions in the cathode and anode specimens are also presented in Table 1. These experiments on the electric migration of copper show that the copper ion charge as measured experimentally is always less than +1, indeed, within the 820 to 1000 °C range it decreases from +0.6 to +0.3.

We have also carried out a number of experiments on the electric migration of copper in heavily doped (with copper up to concentrations of 1×10^{18} and 5×10^{18} cm^{-3}) specimens of p-type gallium arsenide. The electric migration of radioactive copper in such specimens was studied at two temperatures: 830 and 870 °C. These experiments have shown that radioactive copper migrates in heavily doped p-type GaAs specimens in the form of positive ions just as in the case of n-type GaAs. However, in contrast to n-type gallium arsenide, in p-type samples the experimentally measured charge of copper ions is close to +1 (at 830 and 870 °C the experimental values of the copper ion charge are +1.07 and +0.94, respectively). The reason for such difference will be discussed below.

2.3 Electric migration of thermal acceptors

It was reported in a number of articles [9, 10] that if a specimen of n-type gallium arsenide with electron concentration $\leqq 5\times10^{17}$ cm^{-3} is quenched after having been heated to a high temperature, it will acquire hole conduction. This effect of thermal conversion is supposed to be a result of diffusion of a rapidly diffusing acceptor impurity from outside into the specimen.

In order to determine the diffusion coefficient, as well as the charge state of particles responsible for thermal conversion (or thermal acceptors), we have carried out a number of experiments. A specimen of gallium arsenide subjected to thermal conversion at 1090 °C for $t = 77$ h and having after conversion a hole concentration $p = 2\times10^{16}$ cm^{-3} and electric conductivity $\sigma = 0.48\ \Omega^{-1}$ cm^{-1}, was fixed between two gallium arsenide specimens (with electron concentration $n = 1.2\times10^{16}$ cm^{-3} and $\sigma = 7.7\ \Omega^{-1}$ cm^{-1}) and electric migration was investigated in the system. After the migration, measurements of the type of conduction, Hall effect, and electric conductivity were performed on each specimen. These measurements have shown that the specimen adjoining the anode remained n-type but its resistance became high; the middle specimen subjected preliminarily to thermal conversion became again n-type after migration ($n = 5.4\times10^{16}$ cm^{-3}, $\sigma = 320\ \Omega^{-1}$ cm^{-1}); the cathode specimen became p-type ($p = 1.7\times10^{16}$ cm^{-3}, $\sigma = 0.38\ \Omega^{-1}$ cm^{-1}). It follows from this experiment that the larger part of thermal acceptors have passed from the middle into the cathode specimen, i.e. thermal acceptors migrate in gallium arsenide as positive ions. Similar experiments performed at various durations and temperatures yielded the same result on the sign of thermal acceptors. The values for the mobility and diffusion coefficient of thermal acceptors obtained from these experiments lie close to those for copper in gallium arsenide.

Thus our data on electric migration of thermal acceptors support the current hypothesis on the thermal conversion in gallium arsenide being due to the diffusion of copper. It has also been established that by heating thermally converted gallium arsenide specimens in a dc electric field, copper can be removed from the specimens. A similar effect was revealed earlier in the study of the electric migration of thermal acceptors in germanium [11].

2.4 Electric migration of lithium

The effect of an applied electric field on the diffusion of lithium in gallium arsenide was investigated within the range 740 to 980 °C. The electric migration was carried out on a lithium-saturated specimen fixed between two pure specimens of n-type GaAs. After the migration there appeared in the specimens adjoining the cathode and anode, p–n junctions from whose location one could determine the mobility and charge of lithium ions. These experiments have shown that lithium migrates in the form of positive ions just as copper does. In the temperature range specified the mobility turned out to increase from 1.6×10^{-5} to 9.4×10^{-5} cm^2 V^{-1} s^{-1}. The obtained values of mobility and the data on the diffusion coefficient taken from reference [12] and extrapolated to high temperatures were used to calculate the values of the lithium ion charge in gallium arsenide. The effective charge of lithium ions in the 740 to 980 °C range has been found to lie within +0.3 to +0.2.

It was found from electric conductivity and Hall effect measurements at room temperature that at a diffusion saturation of n-type gallium arsenide specimens ($n \approx 3\times10^{17}$ cm^{-3}) they convert to p-type; whereas doping p-type specimens results in an increase of hole concentration. This supports the available data on the acceptor properties of lithium in gallium arsenide.

2.5 Electric migration of zinc and cadmium

As is well known, the diffusion of zinc and cadmium in gallium arsenide exhibits a complicated pattern. Various explanations were suggested by a number of authors to account for the anomalous behavior of these impurites. In particular, various conjectures were made concerning the charge state of zinc during diffusion in gallium arsenide [13, 14]. Experiments on diffusion in an applied electric field permit to determine the charge state of an impurity and thus shed light on the mechanism of diffusion. Preliminary data on the electric migration of zinc in gallium arsenide have already been published [4].

Fig. 4 shows experimental data for the mobility of zinc ions in gallium arsenide. The full line 1 given this same figure represents the temperature dependence of mobility calculated from Einstein's relation using experimental values of the diffusion coefficients and assuming the charge of the zinc ions to be +2. As seen from the figure, up to 980 °C there is a good agreement between the experimental and calculated values of mobility. Above 980 °C the experimental values of mobility begin to decrease with increasing temperature. Such a decrease of the effective mobility of zinc ions can be due to the effect of ion drag by free electrons moving to the anode.

We have also conducted a number of experiments on the electric migration of cadmium in gallium arsenide. The corresponding experimental technique was similar to that used in the study of the electric migration of zinc. It was found that at 950 °C and 980 °C cadmium migrates in the form of positive ions, the experimentally determined mobility of ions being 2.6×10^{-10} and 8×10^{-10} cm^2 V^{-1} s^{-1}, respectively.

2.6 Discussion

It follows from our experiments on the electric migration of lithium, copper, zinc, and cadmium that these impurities migrate in gallium arsenide as positive ions. The positive sign of the ion charge indicates that these impurities diffuse

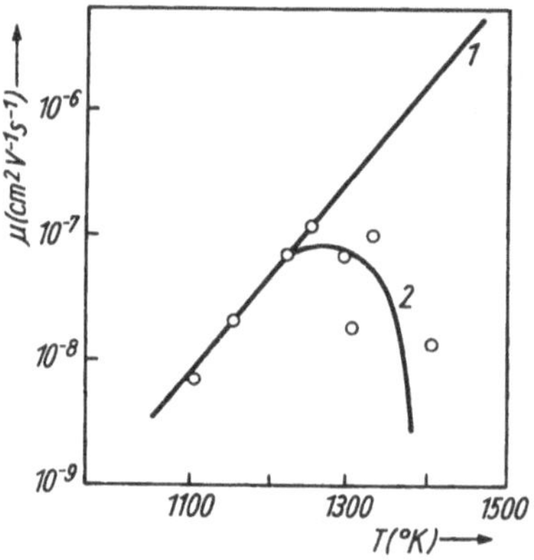

Fig. 4. Temperature dependence of zinc ion mobility in gallium arsenide;
1 – curve calculated from Einstein's relation,
2 – curve calculated taking into account ion drag by electrons

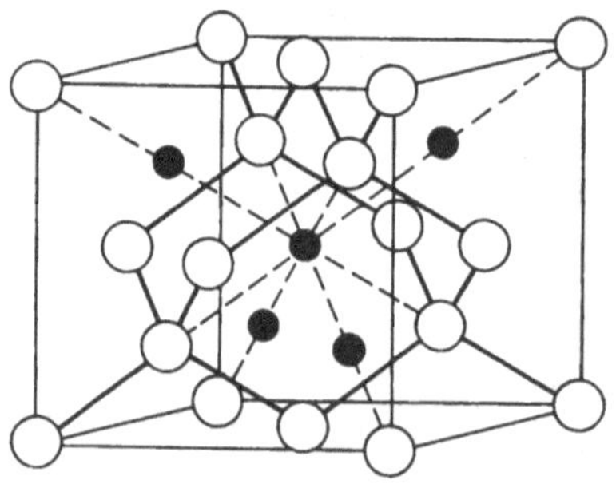

Fig. 5. Lattice sites and interstitials in a diamond-type lattice;
○ – lattice sites, ● – interstitials

mainly interstitially. The magnitudes of the ionic radii of the impurities studied are favourable for the movement through interstitial positions in gallium arsenide. Indeed, it is easy to demonstrate that the most narrow passage for a particle moving from one interstitial position to another in a diamond-type lattice is $0.87\,a$, where a is the lattice constant (Fig. 5). Hence ions with a diameter $d < 0.87\,a$ can move freely from one interstitial position to another. The diameters of the ions Li^+, Cu^+, Zn^{++}, and Cd^{++} which are 1.56 Å, 1.92 Å, 1.48 Å, and 1.94 Å, respectively [15], are smaller than the most narrow passage on the way from one interstitial position to another which is equal to 2.12 Å in gallium arsenide, and therefore for them the interstitial mechanism of diffusion is possible.

On ionization, elements of the groups I and II can transfer in gallium arsenide to positive ions with the charges $q = +1$ and $q = +2$, respectively. However, as follows from our experimentally measured positive charge of lithium, copper, and zinc ions, (i) it is fractional, and (ii) it decreases with increasing temperature. Such a decrease of ionic charge is due to the effect of ion drag by electrons. As seen from formula (7′), in the case of positive ions being dragged by electrons the magnitude and sign of the effective ion charge are determined by the product $n\, l_n\, \sigma_{in}$, which in semiconductors depends strongly on temperature in the intrinsic conduction region. Curve 2 in Fig. 4 corresponding to the electric migration of zinc in n-type gallium arsenide was calculated using formula (8). As seen from this figure, with increasing temperature μ_{eff} first increases, and then, because of the increase of $n\, l_n\, \sigma_{in}$ passes through a maximum and begins to decrease vanishing at the inversion temperature. Although there is a spread in experimental data, it is easily seen that they are in agreement with theory.

Experiments on the electric migration of copper have yielded for the copper ion charge a value less than $+1$. Let us evaluate the effective charge of copper ions being dragged by electrons at 1000 °C. Using the data on the electron concentration at 1000 °C given in reference [8], and the values for l_n, extrapolated to 1000 °C, taken from reference [18] and calculated values of the effective cross section for the scattering of electrons by ions σ_{in} by formula (6) of reference [3], one can show, that $n\, l_n\, \sigma_{in} = 0.5$. Then formula (7′) will yield for the effective charge of copper ions $q_{eff} \approx +0.5$. This value of q_{eff} lies close to the experimentally determined value of q_{eff} at 1000 °C which is equal to $+0.3$. Since in our calculation of the effective copper ion charge by formula (7′) we used a crude estimate for the effective cross section for electron scattering by ions σ_{in} one can hardly expect a better agreement with experiment. However, one can approach the problem in the opposite way, viz. use the experimental values of q_{eff} to estimate σ_{in} from formula (7′). Such values of σ_{in} derived from experiments on the electric migration of copper are presented in Table 1.

When calculating the effective ionic charge in the electric migration in n-type gallium arsenide ($n \leq 3 \times 10^{17}$ cm^{-3}) one may neglect the drag by holes to the cathode since the mean free path of holes l_p is much smaller than that of electrons, and at the temperatures of diffusion investigated by us $n = p$ and $\sigma_{in} \approx \sigma_{ip}$. However, in the case of a heavily doped p-type gallium arsenide (in which the concentration of holes exceeds that of free current carriers at the temperature of diffusion) one should take into account the drag of impurity ions not only by electrons but by holes as well. Indeed, our experiments on the electric migration of copper in heavily doped p-type gallium arsenide yielded for $q_{eff} \approx +1$. Calculations show that in this case $n\, l_n\, \sigma_{in} \approx p\, l_p\, \sigma_{ip}$.

It follows from our experiments that all the impurities studied by us and exhibiting in gallium arsenide at room temperature acceptor properties migrate as positive ions at elevated temperatures. This can be attributed to the fact that impurities can be present in a semiconductor in two states — in substitutional and in interstitial positions. The solubility of the substitutional species is much higher than that of the interstitial one and therefore electric measurements at room temperature reveal an acceptor behavior. However, the rate of migration through interstitial positions exceeds by far that of substitutional migration and therefore the electric migration yields in the first place information on the impurity particles in interstitials.

The results obtained by us support the current hypothesis on the mechanism of diffusion of group I and II elements in gallium arsenide which consists in a rapid interstitial migration in the form of positive ions with a subsequent trapping by vacancies where these impurities exhibit acceptor properties.

Second, our experimental data provide evidence for the effect of drag of diffusing ions by current carriers in a semiconductor which was predicted by the theory.

The data obtained in electric migration experiments can be used to evaluate the effective cross section for the scattering of electrons on impurity ions in a semiconductor.

In conclusion we should like to express our gratitude to G. S. Kulikov and R. Sh. Malkovich for their assistance in carrying out the present work.

References

[1] B. I. Boltaks, Diffuziya v poluprovodnikakh (Diffusion in Semiconductors), Fizmatgiz, Moscow 1961.
[2] V. B. Fiks, Fiz. tverd. Tela **1**, 16 (1959).
[3] V. B. Fiks, Fiz. tverd. Tela **1**, 1321 (1959).
[4] V. B. Fiks, Fiz. tverd. Tela **6**, 1589 (1964).
[5] R. Sh. Malkovich, Fiz. tverd. Tela **2**, 2784 (1960).
[6] B. I. Boltaks, T. D. Dzhafarov, V. I. Sokolov, and F. S. Shishiyanu, Fiz. tverd. Tela **6**, 1511 (1964).
[7] C. S. Fuller and J. M. Whelan, J. Phys. Chem. Solids **6**, 173 (1958).
[8] R. Hall and J. Raccete, J. appl. Phys. **35**, 379 (1964).
[9] J. J. Wysocki, J. appl. Phys. **31**, 1696 (1960).
[10] B. I. Boltaks and F. S. Shishiyanu, Fiz. tverd. Tela **7**, 1021 (1965).
[11] V. A. Zhidkov and V. E. Lashkarev, Zh. tekh. Fiz. **27**, 877 (1957).
[12] C. S. Fuller and K. B. Wolfstirn, J. appl. Phys. **33**, 745 (1962); **33**, 2507 (1962).
[13] L. R. Weisberg and J. Blanc, Phys. Rev. **131**, 1548 (1963).
[14] R. L. Longini, Solid State Electronics **5**, 127 (1962).
[15] L. Pauling, Priroda khimicheskai svyazi, Goskhimizdat, Moscow-Leningrad 1947.
[16] B. I. Boltaks and F. S. Shishiyanu, Fiz. tverd. Tela **5**, 2310 (1963).
[17] V. I. Sokolov and F. S. Shishiyanu, Fiz. tverd. Tela **6**, 328 (1964).
[18] F. Williams, J. Electrochem. Soc. **112**, 876 (1965).

(Received December 7, 1966)

phys. stat. sol. **19**, 715 (1967)

Subject classification: 14.2; 12

Institut für Metallphysik der Universität Göttingen

Elastische Konstanten des Flußfadengitters in Supraleitern zweiter Art

Von

R. Labusch

Die elastischen Konstanten des Flußfadengitters in Supraleitern zweiter Art werden erneut aus dem Wechselwirkungspotential zwischen den Flußlinien berechnet. Hierbei werden Irrtümer früherer Autoren vermieden. Unter der vernünftigen Annahme von Dreieckssymmetrie lassen sich alle Konstanten durch die Parameter der gemessenen reversiblen Magnetisierungskurve ausdrücken. Als Beispiel werden die berechneten elastischen Konstanten des Flußfadengitters in Niob bei 4,15 °K als Funktion des angelegten Magnetfeldes aufgetragen.

The elastic constants of the flux line lattice in type II superconductors are recalculated from the interaction potential between the flux lines, avoiding an error made in previous calculations. Assuming triangular symmetry, it is shown that the constants can be expressed in terms of the measured reversible magnetization curve. The calculated elastic constants of the flux line lattice in niobium at 4.15 °K are plotted as functions of the applied magnetic field.

Die elastischen Konstanten des Flußfadengitters in Supraleitern zweiter Art wurden von Matricon [1] und von Fetter et al. [2] berechnet. In beiden zitierten Arbeiten wurde die Änderung der freien Energie des Supraleiters im Magnetfeld $H > H_{c1}$, die durch Einführung von homogenen Verzerrungen u_{ij} entsteht, aus dem Wechselwirkungspotential der Flußfäden berechnet, und die Entwicklungskoeffizienten zweiter Ordnung in u_{ij} den elastischen Konstanten zugeordnet. Dieses Verfahren liefert aber im allgemeinen nicht die richtigen elastischen Konstanten [3]. Die Zuordnung ist überdies nicht eindeutig. Es ist daher verständlich, daß die obengenannten Autoren zu verschiedenen Ergebnissen kommen.

Wegen dieser Unsicherheit leiten wir zunächst den Tensor der elastischen Konstanten aus dem Wechselwirkungspotential $V(r^2)$ der Flußfäden her. $V(0)$ wird als doppelte Selbstenergie eines Flußfadens definiert, so daß die Gesamtenergie gleich $\frac{1}{2} n \sum_{\nu} V(r_\nu^2)$ ist. $n = \frac{B}{\Phi_0}$ ist die Dichte der in x_3-Richtung liegenden Flußfäden, r_ν durchläuft alle Schnittpunkte mit einer x_1–x_2-Ebene.

Durch Anbringen kleiner Verschiebungen $u_j = u_{j,k} x_k + \frac{1}{2} u_{j,kl} x_k x_l$ entsteht eine Volumenkraft

$$K_i = \frac{1}{2} \tilde{C}_{ij,kl} u_{j,kl}$$

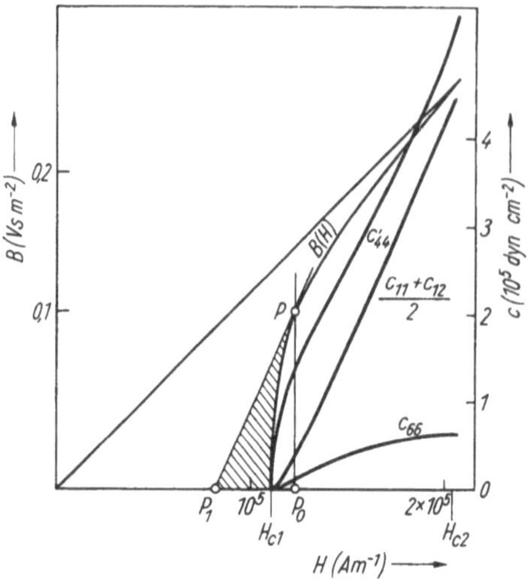

Fig. 1. Die elastischen Konstanten des Flußfadengitters in Nb bei 4,15 °K. Für die miteingezeichnete $B(H)$-Kurve gilt die linke Skala. $\overline{PP_1}$ ist die Tangente an $B(H)$. Der Wert von c_{66} für den Berührungspunkt ist gleich der schraffierten Fläche; außerdem gilt $\frac{c_{11}+c_{12}}{2} = \overline{P_1 P_0} \times \overline{P_0 P}$ und $c'_{44} = \overline{OP_0} \times \overline{P_0 P}$

Die Gleichungen (7) gelten ohne Einschränkung. Eine Messung der Magnetisierungskurve $B(H)$ ist daher zugleich eine Messung der elastischen Konstanten. Alle elastischen Konstanten sind positiv, wenn $\frac{\partial^2 H}{\partial B^2}$ überall positiv ist.

Als Anwendungsbeispiel wurden die Moduln des Flußfadengitters in reinem Niob bei 4,14 °K als Funktion von H aus der von Stromberg und Swenson [7] gemessenen Magnetisierungskurve berechnet. Die Ergebnisse sind in Fig. 1 zusammen mit der benutzten $B(H)$-Kurve aufgetragen. In der Abbildung wird zugleich ein graphisches Verfahren demonstriert, mit dessen Hilfe man leicht die Größe der verschiedenen Moduln aus der $B(H)$-Kurve ablesen kann.

Der reine Schubmodul c_{66} ist viel kleiner als die übrigen Moduln. Das ergab sich qualitativ schon in den früheren Arbeiten [1, 2]. Die Grenzwerte, die bei H_{c2} erreicht werden, lassen sich durch das thermodynamische kritische Feld H_c bzw. B_c und $\varkappa$ ausdrücken. Bei H_{c2} gilt

$$\int_0^{B_{c2}} H(\xi)\,d\xi = \frac{1}{2} H_c\, B_c\,(1 + 2\,\varkappa_1^2) \quad \text{und} \quad \frac{\partial H}{\partial B} = \frac{1{,}16\,(2\,\varkappa_2^2 - 1)}{1{,}16\,(2\,\varkappa_2^2 - 1) + 1} \cdot \frac{H_c}{B_c}\,.$$

Daraus erhält man den zweidimensionalen Kompressionsmodul

$$K = \frac{c_{11} + c_{12}}{2} = \frac{H_c\, B_c}{2} \cdot 4\,\varkappa_1^2\, \frac{1{,}16\,(2\,\varkappa_2^2 - 1)}{1{,}16\,(2\,\varkappa_2^2 - 1) + 1}\,,$$

den Schubmodul

$$c_{66} = \frac{H_c\, B_c}{2} \cdot \frac{1{,}16\,(2\varkappa_2^2 - 1) - (2\,\varkappa_1^2 - 1)}{1{,}16\,(2\,\varkappa_2^2 - 1) + 1}$$

und das Verhältnis

$$\mu_{H_{c2}} = \frac{2\,c_{66}}{c_{11} + c_{12}} = \frac{1}{4\,\varkappa_1^2} \cdot \frac{1{,}16 - \left(\dfrac{2\,\varkappa_1^2 - 1}{2\,\varkappa_2^2 - 1}\right)}{1{,}16}\,.$$

In schmutzigen Supraleitern ist $\varkappa_1 = \varkappa_2$. Dann ist $\mu_{H_{c2}} = \frac{1}{29\,\varkappa^2}$. In Supraleitern mit großer freier Weglänge, wie in reinem Niob, ist $\varkappa_2 > \varkappa_1$ [8, 9, 10] und daher $\mu_{H_{c2}}$ größer. Für Nb bei 4,15 °K ergibt sich $\mu_{H_{c2}} \approx 7{,}4$.

Literatur

[1] J. Matricon, Phys. Letters **9**, 289 (1964).
[2] A. L. Fetter, P. C. Hohenberg und P. Pincus, Phys. Rev. **147**, 140 (1966).
[3] G. Leibfried, Handb. Phys. VII, 1, Springer Verlag, 1955 (S. 185 ff).
[4] A. L. Fetter, Phys. Rev. **147**, 153 (1966).
[5] D. Cribier, B. Jacrot, B. Farnoux und L. Madhav Rao, J. appl. Phys. **37**, 952 (1966).
[6] W. Fite und A. G. Redfield, Phys. Rev. Letters **17**, 381 (1966).
[7] T. F. Stromberg und C. A. Swenson, Phys. Rev. Letters **9**, 370 (1962).
[8] K. Maki und T. Tsuzuki, Phys. Rev. **139**, A 868 (1965).
[9] D. K. Finnemore, T. F. Stromberg und C. A. Swenson, Phys. Rev. **149**, 231 (1966).
[10] L. C. Skinner, Techn. Report No. 1, Massachusetts Inst. of Techn., Januar 1966.

(Received November 29, 1966)

phys. stat. sol. **19**, 721 (1967)

Subject classification: 6; 22.1.3

Institut für Neutronenphysik der Kernforschungsanlage Jülich

Streuung langsamer Neutronen an polykristallinem Selen und Tellur

Von

A. Axmann und W. Gissler

Mit Hilfe eines Drehkristall-Flugzeitspektrometers wurde die inelastische Streuung langsamer Neutronen an polykristallinem Selen und Tellur untersucht. Es wurden verschiedene Bandenschwerpunkte gefunden, die in ihrer Lage gut mit den durch Infrarotmessungen [1] ermittelten übereinstimmen, jedoch ist ihre Anzahl kleiner. Dieser Umstand zeigt, daß die neutronenspektroskopisch nicht beobachteten Übergänge von Kombinationen herrühren. Bezüglich des Tellurspektrums ergibt sich eine gute Übereinstimmung mit der Theorie von Hulin [4]. Die Neutronenspektren der beiden isomorphen Kristalle sind einander ähnlich.

The inelastic scattering of slow neutrons by polycrystalline selenium and tellurium is studied using a rotating crystal spectrometer. The frequencies of the observed transitions are consistent with the results of infrared spectroscopy [1], but the number of transitions is lower, which indicates that many of the infrared transitions are combinations. The tellurium spectrum is in good agreement with Hulin's theory [4]. The neutron spectra of the two isomorphous lattices are similar.

1. Gitterschwingungen in Selen und Tellur

Obgleich Selen und Tellur bezüglich ihrer Halbleitereigenschaften eingehend untersucht wurden, ist der Erforschung ihres gitterdynamischen Verhaltens vergleichsweise nur wenig Aufmerksamkeit geschenkt worden.

Selen und Tellur sind isomorph und kristallisieren in einem hexagonalen Dreipunktgitter entsprechend Fig. 1. Die Atome sind auf Spiralen angeordnet, deren Achsen parallel zur hexagonalen C-Achse verlaufen. Die Bindung der auf einer Spirale liegenden Atome ist kovalent und stärker als die teils van der Waals- teils metallische Bindung von Atomen verschiedener Spiralen.

Von Caldwell und Fan [1] liegen für Selen und Tellur Infrarot-Transmissionsmessungen vor, die Aufschlüsse über die optischen Schwingungsbranchen im Zentrum der Brillouin-Zone ergeben. Die Interpretation dieser Messungen durch Boitsov [2 und 3] und durch Hulin [4] ist jedoch teilweise widersprechend.

Wegen der schwachen Wechselwirkung der Atome auf verschiedenen Spiralen legte Boitsov seinen Rechnungen das Modell einer Spirale ohne Wechselwirkung mit den Nachbarspiralen zugrunde. Die Kraftkonstanten des Modells wurden den optisch bestimmten Übergangsfrequenzen angepaßt. Bei diesem Verfahren interpretiert Boitsov die experimentell bestimmten Übergänge als direkte Übergänge, obwohl Auswahlregeln einige dieser Übergänge — zumindest in sehr reinen Kristallen — verbieten müßten. Für Selen konnte eine gute Kon-

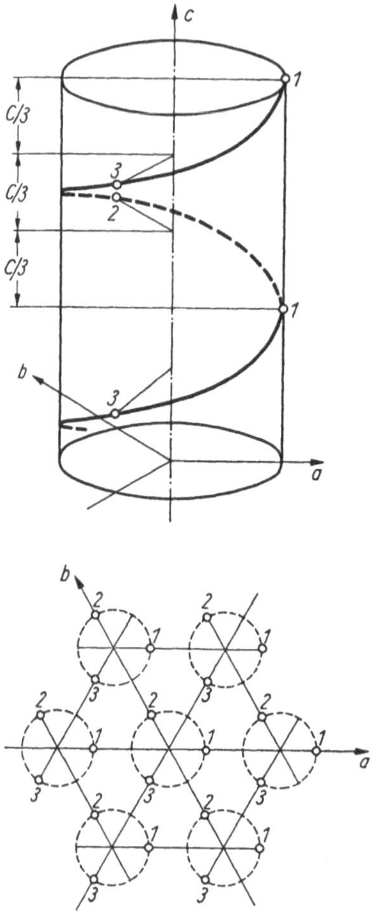

Fig. 1. Anordnung der Selen- bzw. Telluratome in der hexagonalen Gitterzelle

sistenz des errechneten mit dem experimentellen Spektrum erzielt werden; bei Tellur ergaben sich besonders bei niedrigen Frequenzen größere Abweichungen.

Hulin schlug für Tellur ein verfeinertes dreidimensionales Modell vor, welches die Wechselwirkung verschiedener Spiralen berücksichtigt. Im Gegensatz zu Boitsov paßt er die Kraftkonstanten seines Modells den elastischen Konstanten an. Dadurch werden diese nicht von einer a priori-Interpretation der optisch bestimmten Übergänge abhängig. Hulin findet nur eine Übereinstimmung mit den ersten beiden optisch bestimmten Übergängen, dagegen erklärt er höherenergetische Übergänge nicht als direkte Übergänge, sondern unter Berücksichtigung von Auswahlregeln als Kombinationsübergänge.

Die inelastische Neutronenstreuung sollte weiteren Aufschluß über die Gitterdynamik liefern; insbesondere sollte sie in der Lage sein, die Natur der optischen Übergänge aufzuklären: 1. Hier gelten nicht die für Infrarotmessungen gültigen Auswahlregeln. 2. Der Anteil von Multiphononprozessen ist weitaus kleiner als bei Infrarotmessungen.

2. Inelastische Neutronenstreuung an Polykristallen

Da Einkristalle der vorzugsweise kohärenten Streuer Selen und Tellur nur schwer erhältlich sind, haben wir Pulverproben benutzt. Die hierbei durch die Streuung gelieferten Informationen sind geeignet gebildete Mittelwerte über die verschiedenen Richtungen der Mikrokristalle im Pulver. Der kohärent inelastische Streuquerschnitt $(\partial^2\sigma/\partial\Omega\,\partial\omega)$ für Polykristalle wurde von Egelstaff [5] angegeben:

$$\frac{\partial^2\sigma}{\partial\Omega\,\partial\omega} = \frac{\sigma_{\text{coh}}}{4\pi} I Z\,. \tag{1}$$

I stellt im wesentlichen die inkohärente Streufunktion dar:

$$I = \frac{\hbar}{2m} \mathrm{e}^{-2W} \frac{k'}{k_0} \sum_i T(\omega_i) \frac{\varkappa^2}{\omega_i} \cos^2 \vartheta_i\, q_i^2 \left(\frac{\mathrm{d}q_i}{\mathrm{d}\omega_i}\right). \tag{2}$$

Der Faktor Z berücksichtigt, daß bei der kohärenten Streuung die Impulserhaltungsbedingung $\delta\,(\boldsymbol{\varkappa} + \boldsymbol{q} - 2\pi\,\boldsymbol{\tau})$ erfüllt werden muß; beim Polykristall muß diese Bedingung über alle Orientierungen von $\boldsymbol{\varkappa}$ und $\boldsymbol{q}$ bezüglich der Richtung von $2\pi\boldsymbol{\tau}$ gemittelt werden. Dies ergibt

$$Z = \begin{cases} \dfrac{\pi F_\tau}{2 B \varkappa q \tau} & \text{für} \quad 2\pi|\boldsymbol{\tau}| - |\boldsymbol{q}| \leqq |\boldsymbol{\varkappa}| \leqq 2\pi|\boldsymbol{\tau}| + |\boldsymbol{q}|, \\ 0 & \text{im übrigen Bereich.} \end{cases} \tag{3}$$

Gleichung (2) und (3) sind unter der Voraussetzung isotroper Dispersionsbeziehungen $\omega = \omega(|\boldsymbol{q}|)$ gültig. Hier verwenden wir jedoch Gleichung (3) näherungsweise auch für nichtisotrope Dispersionsbeziehungen.

Die in (1), (2) und (3) verwendeten Symbole haben folgende Bedeutung: σ_{coh} ist der kohärente Streuquerschnitt, m die Masse des Streukerns, $2W$ der Debye-Waller-Faktor, k' bzw. k_0 die Wellenzahl des gestreuten bzw. des einfallenden Neutrons, $\hbar\boldsymbol{\varkappa}$ ist der Impuls- und $\hbar\omega$ der Energieübertrag der Neutronen bei der Streuung. Der Index i bezeichnet eine der drei Polarisationsrichtungen. F_τ ist die Summe der Strukturfaktoren aller Netzebenen mit dem Netzebenenabstand $d = 1/|\boldsymbol{\tau}|$. $T(\omega) = 1/[\exp(\hbar\omega/kT) - 1]$ ist die Besetzungszahl und ϑ_i der Winkel zwischen Polarisationsvektor und $\boldsymbol{\varkappa}$; es gilt $\cos\vartheta_i = \{(2\pi\boldsymbol{\tau})^2 - \boldsymbol{\varkappa}^2 - \boldsymbol{q}^2\}/(2|\boldsymbol{\varkappa}||\boldsymbol{q}|)$. B ist das Volumen der Gitterzelle. Die Bereiche, in denen Streuung möglich ist, sind nach (3) durch die von $|\boldsymbol{\varkappa}| = 2\pi|\boldsymbol{\tau}|$ (Lagen der Debye-Scherrer-Reflexe) ausgehenden Dispersionskurven $\pm q(\omega)$ gegeben. In Fig. 2 sind diese Bereiche für einzelne isotrope akustische und optische Dispersionskurven qualitativ dargestellt. Die sich für Messungen mit konstantem Streuwinkel δ auf Grund von Energie- und Impulssatz ergebenden Wege sind in der $\varkappa$, ω-Ebene für verschiedene δ gestrichelt eingezeichnet.

Da die Streuintensität nach (2) und (3) proportional zu $q\,(\mathrm{d}q/\mathrm{d}\omega)$ ist, erhalten wir Intensitätsmaxima bei den Energien der optischen Zweige und in der Nähe der Grenzschwingungen der akustischen Zweige. Die Intensität für Übergänge an den optischen Zweigen ist am größten, wenn der betreffende Streubereich durch Meß„wege" in der Nähe von $|\boldsymbol{\varkappa}| \approx 2\pi|\boldsymbol{\tau}|$ untersucht wird, da hier Energie- und Impulssatz für die meisten Phononen zu erfüllen ist.

Auf Grund von (3) erwarten wir an den Grenzen der Bereiche möglicher Streuung scharfe Intensitätskanten für die longitudinalen Moden ($\cos\vartheta = 1$) der akustischen Zweige. Cocking und Guner [6] sowie Larsson et al. [7] haben diese Intensitätskanten an polykristallinem Zinn bzw. Aluminium angedeutet gefunden. Im besonders günstig gelagerten Fall des Argons konnten Kroó et al. [8] aus der Lage dieser Intensitätskanten eine „mittlere" Dispersionskurve gewinnen.

In vielen Fällen, wie hier bei Selen und Tellur, wird dies jedoch nur schlecht

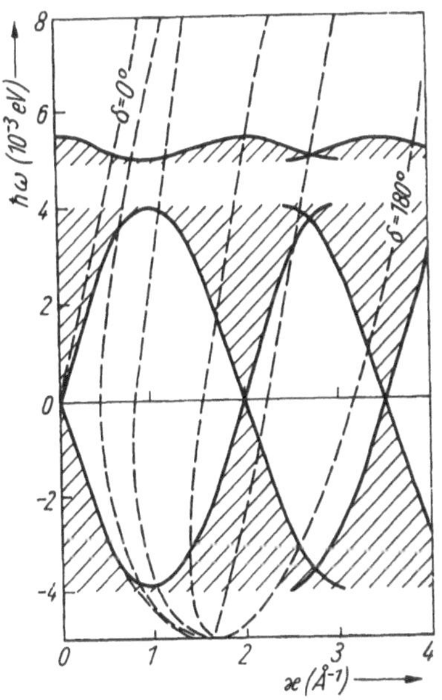

Fig. 2. Bereiche möglicher Streuung in der $\varkappa$, ω-Ebene. Die gestrichelten Kurven stellen die für konstante Streuwinkel möglichen „Wege" in der $\varkappa$, ω-Ebene dar. Schnittpunkte dieser Linien mit den Grenzen der Streubereiche ergeben für longitudinale Moden Intensitätsmaxima

möglich sein. Bei großer Anisotropie der Kristalle bilden die den verschiedenen Kristallrichtungen zugehörigen Dispersionskurven ein breites Band, was zu einer Verschmierung der Intensitätskanten führt. Außerdem wirkt sich erschwerend aus, daß bei weniger symmetrischen Gittern als dem kubisch flächenzentrierten mangels Auslöschungen und kleinerer Multiplizitäten die Debye-Scherrer-Reflexe dichter liegen und ihre Intensitäten kleiner sind. Dieser Umstand bewirkt eine Überlappung der Bereiche möglicher Streuung, so daß im Spektrum Intensitätskanten sich so überlagern können, daß deren Auflösung nicht möglich ist.

3. Messungen und Resultate

Die Messungen wurden mit einem Drehkristall-Flugzeitspektrometer am FRJ-2 Reaktor in Jülich durchgeführt. Fig. 3 zeigt die schematische Meßanordnung. Der durch einen totalreflektierenden Neutronenleiter geführte subthermische Neutronenstrahl (gefiltert mit einem gekühlten Be-Kristall) wird von einem sich schnell drehenden Aluminium-Einkristall (ca. 10000 U/min) monochromatisiert und in kurze Pulse zerhackt. Die an der Probe gestreuten Neutronen werden auf zwei 3,20 m langen Strecken einer Flugzeitanalyse unterzogen. Die Detektorbänke bestehen aus jeweils zwei $\times$ acht BF_3-Zählern (Durchmesser 5 cm, aktive Länge 40 cm), die in zwei hintereinanderliegenden Schichten angeordnet sind. Die Energieauflösung der einfallenden Neutronen betrug $\Delta E = 1{,}5 \times 10^{-4}$ eV. Die Energieauflösung der gestreuten Neutronen betrug für $E = 5 \times 10^{-3}$ eV (elastische Streuung) $\Delta E = 3 \times 10^{-4}$ eV und für $E = 25 \times 10^{-3}$ eV $\Delta E = 2 \times 10^{-3}$ eV.

Durch eine Pulver-Röntgenaufnahme von Selen bzw. Tellur haben wir uns davon überzeugt, daß keine andere als die hexagonale Phase in merkbarer Konzentration vorlag. Die Reinheit der Proben betrug nach Angaben der Lieferfirma 99.9%. Die Pulverproben

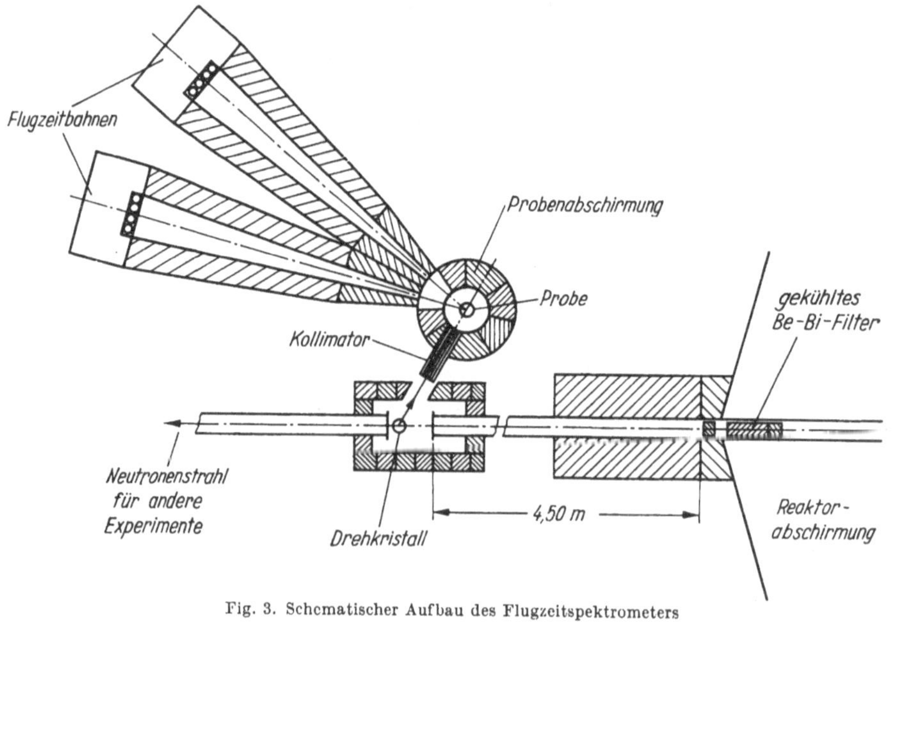

Fig. 3. Schematischer Aufbau des Flugzeitspektrometers

Fig. 4. Flugzeitspektren von Selen und Tellur für Streuwinkel von 80 und 105°. Die gestrichelten Kurven stellen den errechneten Beitrag der Mehrphononenstreuung dar

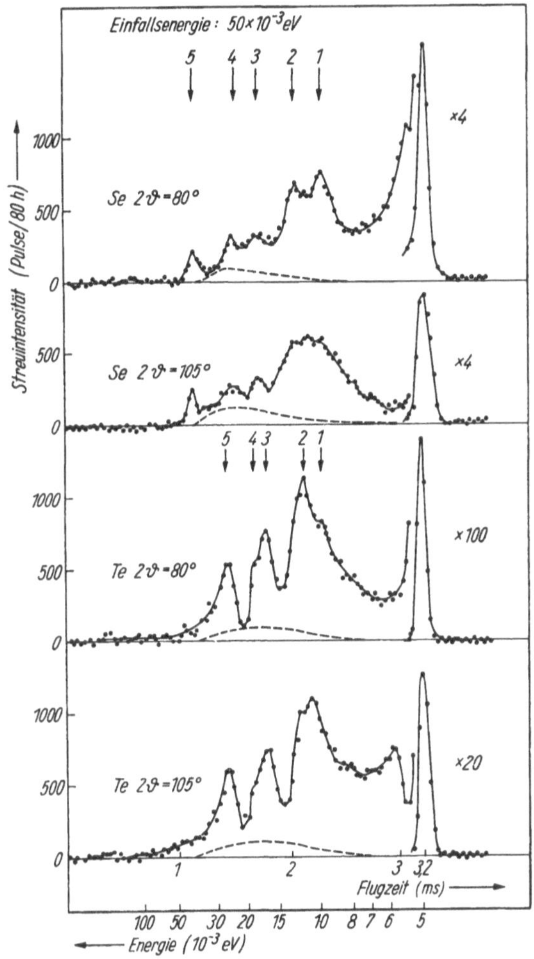

wurden in eine 1 bzw. 2 cm dicke plattenförmige Aluminiumküvette (Wandstärke 1 mm) eingefüllt. Aus der gemessenen Transmission, dem bekannten Absorptionsquerschnitt bei 5×10^{-3} eV für Selen von 23,2 *b* bzw. für Tellur von 10 *b* [9] und der jeweils experimentell bestimmten Pulverdichte ergab sich für die Selenprobe eine 12- für die Tellurprobe ein 13prozentige Streuwahrscheinlichkeit. Zweifachstreuprozesse sind damit, wie eine einfache Abschätzung zeigt, zu vernachlässigen.

Fig. 4 zeigt die gemessenen Flugzeitspektren für Streuwinkel von 80 und 105° für Selen bzw. Tellur. Die Messungen sind bereits bezüglich Untergrund korrigiert und auf die gleiche Monitorrate bezogen. Es wurden jeweils zwei der 16-μs-Meßkanäle zu einem Punkt zusammengefaßt. Die Zeitauflösung entspricht für Neutronenenergien $E > 10 \times 10^{-3}$ eV etwa dem Abstand dreier Meßpunkte. Das Signal–Untergrund-Verhältnis beträgt im Maximum Nr. 1 ca. 4:1. In diesem Maximum wurden ca. 800 Signalimpulse pro Meßpunkt in vier Tagen ausgezählt.

Neben den Messungen für Streuwinkel von 80 und 105° wurden an Selen noch weitere für Streuwinkel von 40, 50, 60 und 70°, allerdings bei einer um den Faktor 2 schlechteren Auflösung, durchgeführt. Diese Messungen zeigten im wesentlichen die gleiche Struktur wie die der Fig. 3; insbesondere ließen diese Meßergebnisse keine Veränderungen der Lage der in Fig. 3 bezifferten Maxima erkennen.

Energieverlustprozesse der Neutronen könnten infolge von „Zyklusüberlappung“ als Maxima auf der hochenergetischen Seite des Energiegewinn-Neutronenspektrums erscheinen. Die Lage dieser Maxima ist jedoch von der Pulsfolgefrequenz der einfallenden Neutronen abhängig. Durch Messung bei zwei verschiedenen Drehzahlen (120 und 90 U/s) haben wir uns davon überzeugt, daß keines der von uns beobachteten Maxima der Fig. 4 von der Pulsfolgefrequenz abhängig ist.

Mehrphononenprozesse werden pauschal durch eine effektive Debye-Temperatur Θ_{eff} in inkohärenter Näherung nach einem Ausdruck von Sjölander [10] abgeschätzt. Wir setzten dabei für Tellur $\Theta_{\text{eff}} = 130$ °K [11] und für Selen den $\sqrt{m_{\text{Te}}/m_{\text{Se}}} \approx 1{,}3$fachen Wert ein ($m_{\text{Te}}$ bzw. m_{Se} ist die Masse des Tellur- bzw. des Selenatoms). Die Resultate sind als gestrichelte Linien in Fig. 4 eingezeichnet.

4. Diskussion

Die Tabellen 1 und 2 zeigen die Lagen der von uns beobachteten Maxima in 10^{-3} eV abzüglich der Neutroneneinfallsenergie für Selen und Tellur. Zum Vergleich sind die durch optische Messungen [1] bestimmten Übergangsfrequenzen und die theoretischen Werte für Normalschwingungen von Tellur von Hulin [4] mit angeführt (eine Entzerrung des Flugzeitmaßstabs in den Energiemaßstab unter Berücksichtigung des Temperaturfaktors hatte keinen Einfluß auf die Lage der Maxima).

Tabelle 1
(Selen)

Maximum Nr.	1	2	3	4	5	6	7	8	9
Neutronen-Mess.	5,1	7,8	13	18	35	—	—	—	—
Infrarot-Mess.	—	—	—	17,5	34	44	56	59	82

Tabelle 2
(Tellur)

Maximum Nr.	1	2	3	4	5	6	7	8	9	10	11
Neutronen-Mess.	5,3	6,6	11,1	13,6	20,5	—	—	—	—	—	—
Infrarot-Mess.	—	—	11,0	13,1	—	22,5	26,2	39,7	44,3	51,0	*)
Theorie Hulin**)	—	—	9	15	(20)	—	—	—	—	—	—

*) Die Infrarotmessungen für Tellur erstrecken sich nur bis etwa 6×10^{-2} eV.
**) Werte für die Grenzschwingung $q = 0$.

Phononenübergänge akustischer Zweige: Unsere Meßergebnisse zeigen bei niedrigen Energien zwei je nach Meßwinkel mehr oder weniger gut aufgelöste Maxima (Nr. 1 und 2); wir nehmen an, daß sie von den akustischen Zweigen herrühren. Ihre Lagen stimmen im Falle des Selen gut mit der bekannten Debye-Temperatur Θ_D überein ($K\,\Theta_D = 6{,}5 \times 10^{-3}$ eV). Messungen bei verschiedenen Streuwinkeln ergaben keine Änderung der Lagen dieser Maxima, obwohl dies nach (2) und (3) zu erwarten wäre. Eine Erklärung hierfür ist wahrscheinlich in dem in Abschnitt 2 beschriebenen Umstand zu suchen, der auf Grund einer starken Anisotropie und einer Überlappung der Streubereiche die Intensitätskanten verschmiert und die Übergänge in der Nähe der Zonengrenze begünstigt.

Phononenübergänge optischer Zweige: Die Energien der übrigen von uns beobachteten Maxima zeigen innerhalb der Meßfehlergrenzen ebenfalls keine Abhängigkeit vom Streuwinkel. Wir deuten sie als Übergänge optischer Zweige.

Auf Grund der Isomorphie und ähnlicher Bindungsverhältnisse von Selen und Tellur erwarten wir eine Ähnlichkeit der Spektren beider Substanzen. Tatsächlich konnte diese auch durch die Infrarotmessungen nachgewiesen werden. Sämtliche hierbei beobachteten Übergangsfrequenzen stehen ungefähr im Wurzelverhältnis der Atommassen $\sqrt{\frac{m_{Te}}{m_{Se}}} \approx 1{,}3$. Diese Ähnlichkeit wird auch in dem von uns gemessenen Spektrum hinsichtlich der Maxima 1 bis 5 beobachtet.

Ein Vergleich mit den Infrarotwerten zeigt, daß die von uns gefundenen Übergangsfrequenzen zwar geringer an Zahl sind, jedoch hinsichtlich ihrer Lage sehr gut mit einzelnen Infrarotwerten übereinstimmen. Da bei den Neutronenmessungen Mehrphononenprozesse unter den hier betrachteten optischen Energiebanden vernachlässigbar sind, folgt daraus, daß einige der Infrarotwerte Übergänge sind, die von Mehrphononenprozessen herrühren, was den Annahmen von Boitsov widerspricht.

Nach Hulin entsprechen nur die beiden ersten Infrarotfrequenzen direkten Übergängen, während ein weiterer direkter Übergang zu einem optischen Band bei 20×10^{-3} eV für Infrarotmessungen verboten ist und alle übrigen gemessenen Infrarotfrequenzen durch Mehrfachprozesse erklärt werden. Unsere Messungen zeigen eine gute Übereinstimmung mit diesen Annahmen[1]); tatsächlich beobachten wir neben den bereits optisch bestimmten Übergängen 3 und 4 den von Hulin vorausgesagten Übergang bei 20×10^{-3} eV. Weitere Übergänge sind nicht festzustellen.[2])

Herrn Dr. T. Springer verdanken wir die Anregung der Arbeit und zahlreiche Diskussionen. Herrn Dr. H. Stiller danken wir für seine Unterstützung.

Literatur

[1] R. C. Caldwell und H. J. Fan, Phys. Rev. **114**, 664 (1959).
[2] V. G. Boitsov, Fiz. tverd. Tela **5**, 1050 (1963).
[3] V. G. Boitsov, Fiz. tverd. Tela **5**, 1822 (1963).
[4] M. Hulin, Ann. Phys. (France) **8**, 647 (1963).
[5] P. A. Egelstaff, AERE Rep. N/R 1164 (1953).
[6] S. J. Cocking und Z. Guner, Inelastic Scattering of Neutrons in Solids and Liquids, IAEA, Vienna 1963 (S. 237).
[7] K. E. Larsson, U. Dahlborg und D. Jovic, Inelastic Scattering of Neutrons in Solids and Liquids, IAEA, Vienna 1965 (S. 117).
[8] N. Kroó, G. Borgonovi, K. Sköld und K. E. Larsson, Inelastic Scattering of Neutrons in Solids and Liquids, IAEA, Vienna 1965 (S. 115).
[9] D. J. Hughes und R. B. Schwartz, Neutron Cross Sections, U.S. Atomic Energy Commission BNL 325 (1955).
[10] A. Sjölander, Ark. Fys. (Sweden) **17**, 315 (1958).
[11] L. Gmelin, Handbuch der anorganischen Chemie, Teil A (Selen), Verlag Chemie, Weinheim /Bergstr. 1953.
[12] R. Geick, persönliche Mitteilung.

(Received November 29, 1966)

1) Auch neuere Messungen mit der Infrarot Reststrahlenmethode von Geick [12] scheinen diese Annahmen zu bestätigen.

2) Bei der Untersuchung von Tellurproben, die Verunreinigungen von Sauerstoff und Schwefel von einigen Gewichtsprozenten enthielten, konnten im Neutronenspektrum bei 40×10^{-2} eV und 80×10^{-2} eV Maxima beobachtet werden. Ihre Natur soll in einer anderen Arbeit geklärt werden.

phys. stat. sol. **19**, 729 (1967)

Subject classification: 6; 21.1

Department of Physics, Banaras Hindu University, Varanasi

The Lattice Vibrations of Nickel in Krebs's Model

By

M. M. SHUKLA and B. DAYAL

The vibration spectrum of nickel is studied using Krebs's model. The phonon dispersion curves for the three symmetry directions [100], [110], and [111] are presented and are found to be in very good agreement with neutron scattering measurements. Good agreement is also obtained with the curve of θ against T. The computed $G(\nu)$ curve is compared with the experimental curves and some agreement is found.

Das Schwingungsspektrum von Nickel wird mit dem Krebsschen Modell untersucht. Die Dispersionskurven der Phononen für die drei Symmetrierichtungen [100], [110] und [111] werden angegeben und befinden sich in sehr guter Übereinstimmung mit Neutronenstreuungsmessungen. Gute Übereinstimmung wird auch für die θ, T-Kurve gefunden. Die berechnete $G(\nu)$-Abhängigkeit wird mit experimentellen Kurven verglichen und Übereinstimmungen werden gefunden.

1. Introduction

The lattice dynamics of nickel presents several points of intrinsic interest. It is a ferromagnetic transition metal with electronic structure $3d^8 4s^2$. A striking feature presented by this metal is that it displays similarity in its properties with some of the metals which belong to two different groups of the periodic table. The incomplete shell and the ferromagnetic character show its strong resemblance with iron and cobalt. Its electronic structure and Fermi surface, on the other hand, are found to be similar to those of copper in certain respects. A number of workers have investigated the experimental dispersion curves of this metal by means of cold neutron scattering. Amongst such measurements the work of Birgeneau et al. [1] is the most recent and detailed. They found that the shapes of the phonon distribution curves of this metal are similar to those of copper to a large extent. As a matter of fact the normal mode frequencies in nickel were found to be roughly 1.24 times the corresponding frequencies in copper. In spite of these similarities with copper, nickel presents some peculiarities of its own in its specific heat data. Its ferromagnetic properties give rise to a spin-wave contribution of the form $AT^{3/2}$ to the lattice specific heats in addition to the usual electronic contribution γT. The spin-wave contribution introduces complication in the analysis of the heat capacity data.

An important reason for the choice of nickel for the present study is that nickel is one of the few metals whose frequency distribution curve has been observed experimentally by means of inelastic scattering of neutrons. This makes the comparison of the theoretical frequency distribution curve with the experimental one possible and thus provides an independent check to the validity of the force models.

Recent studies of Krebs [2, 3], Mahesh and Dayal [4, 5], Shukla [6, 7], and Shukla and Dayal [8, 9] have shown that Krebs's model yields very good results in the case of some body-centered and face-centered cubic metals, the agreement with the experimental data being almost exact in a few cases. Although this model is phenomenological in its nature like those of de Launay [10], Bhatia [11], and Sharma and Joshi [12], it is more satisfactory inasmuch as it satisfies the requirements of the lattice periodicity in the reciprocal space. We have, therefore, thought it proper to study the lattice dynamics of nickel also with the help of this model. The results are reported in the present paper.

2. Numerical Computation

The secular determinant for calculating the frequencies and the method of evaluating the frequency distribution have been described earlier by us [9] in a paper dealing with the lattice vibrations in aluminium. The same procedure is followed in this paper. As in that paper the force constants have been calculated from the experimental values of the elastic constants and one observed phonon frequency. The elastic constants have been taken from the work of Alers et al. [13] and refer to a temperature of 296 °K. The experimental phonon frequency is taken to be the one corresponding to the zone boundary for the longitudinal branch of [100]. This is taken from the work of Birgeneau et al. [1]. These data together with the other input data are given below:

$$C_{11} = 2.46 \times 10^{12}\ \text{dyn/cm}^2,$$
$$C_{12} = 1.50 \times 10^{12}\ \text{dyn/cm}^2,$$
$$C_{44} = 1.22 \times 10^{12}\ \text{dyn/cm}^2,$$
$$\nu = 7.30 \times 10^{12}\ \text{Hz},$$
$$a = 3.5239\ \text{Å}.$$

The force constants calculated from the above data are found to have the following values:

$$\alpha_1 = 36.347 \times 10^3\ \text{dyn/cm},$$
$$\alpha_2 = -2.291 \times 10^3\ \text{dyn/cm},$$
$$\alpha_3 = 1.007 \times 10^3\ \text{dyn/cm}.$$

In the evaluation of force constants we have to take into account the screening parameter (λ) and the number of free electrons per atom (n_{eff}). In the case of transition metals n_{eff} is rather uncertain. In our present study we have used $n_{\text{eff}} = 2$ along with the Bohm and Pines value of the screening parameter λ. The specific heat data allow us to choose the value of $(n_{\text{eff}}/\lambda)^{1/6}$. Any uncertainty in n_{eff} is thus balanced by the corresponding uncertainty in λ.

The computed dispersion curves along the three symmetry directions [100], [110], and [111] are shown in Fig. 1, 2, and 3 respectively. The experimental data of Birgenau et al. [1] have also been plotted there to facilitate comparisons.

The specific heats have been calculated by the same method as reported in the previous work from this laboratory (see, e.g., reference [7]) by solving the secular determinant for only 48 non-equivalent points distributed uniformly within the 1/48th part of the first Brillouin zone. The whole spectrum was

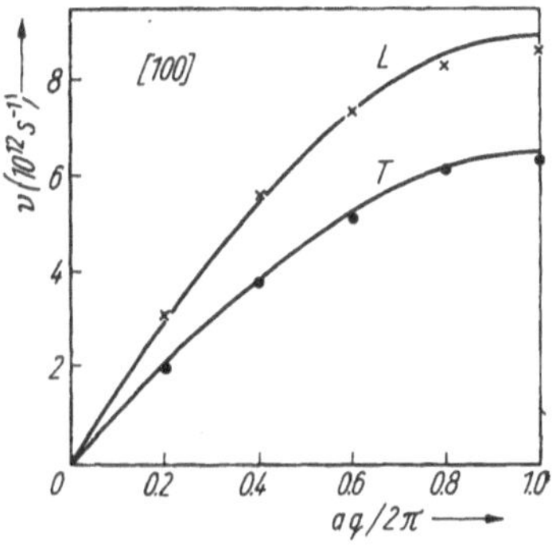

Fig. 1. Dispersion curve for nickel in [100] direction
—— calculated by us,
× × × ● ● ● } experimental points of Birgeneau et al. [1]

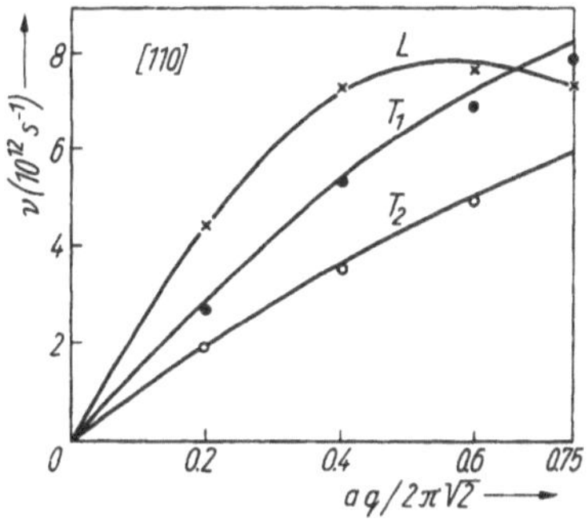

Fig. 2. Dispersion curve for nickel in [110] direction
—— calculated by us,
× × × ● ● ● ○ ○ ○ } experimental points of Birgeneau et al. [1]

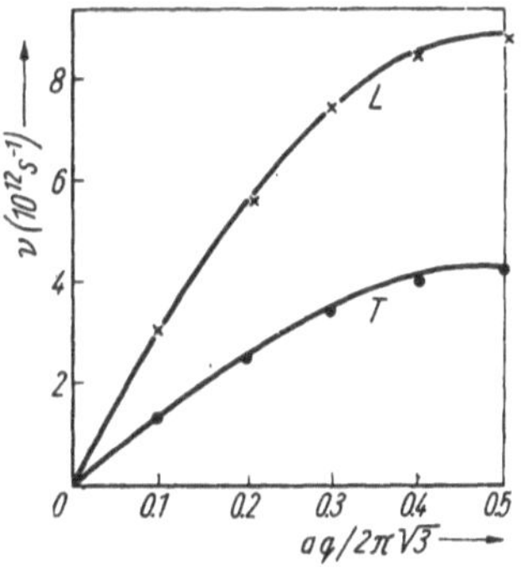

Fig. 3. Dispersion curve for nickel in [111] direction
—— calculated by us,
× × × ● ● ● } experimental points of Birgeneau et al. [1]

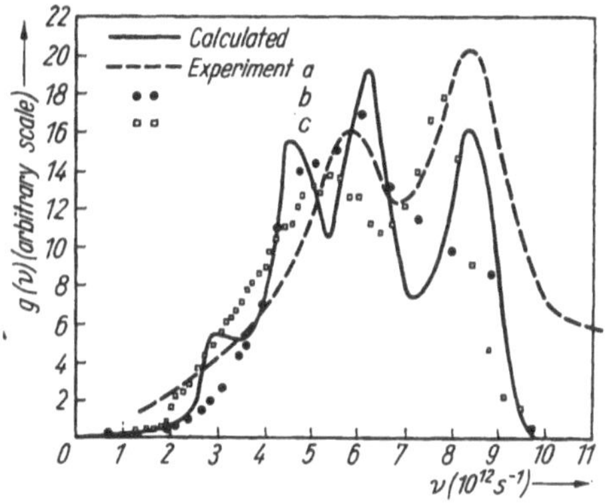

Fig. 4. Frequency distribution curve for nickel
— — — calculated by us,
- - - - - ● ● ● □ □ □ } experimental of Chernoplekov et al. [14] Brugger [15] Mozer et al. [16]

divided into intervals of $\Delta\nu = 4 \times 10^{12}$. The frequency distribution curve obtained in this way is shown in Fig. 4. The experimental frequency distribution curves of Chernoplekov et al. [14], Brugger [15], and Mozer et al. [16] are also shown there. The results of Brugger [15] and Mozer et al. [16] have not yet been published. Birgeneau et al. [1] have, however, plotted them in their paper. The present authors have, therefore, reproduced these results from the paper of Birgeneau et al. [1]. The specific heats C_v were not calculated below 15 °K because the mesh becomes too coarse to give accurate results.

The experimental data available on the heat capacities of nickel are those of Keesom and Clark [17], Busey and Giauque [18], and Rayne and Kemp [19]. Keesom and Clark [17] have carried out their measurements from 1 to 19 °K and Busy and Giauque [18] from 13 to 300 °K. Rayne and Kemp [19] confined themselves to liquid helium temperatures. Since nickel is ferromagnetic below 631 °K, there is also a spin-wave contribution to the lattice specific heat in addition to the usual electronic contribution. These two contributions have to be subtracted from the experimental heat capacities in order to obtain the

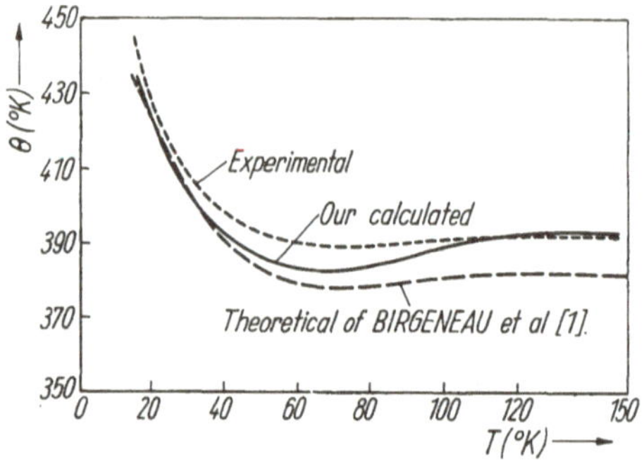

Fig. 5. ($\theta - T$) curve for nickel
——— calculated by us,
- - - - - experimental of Rayne and Kemp [19],
– – – calculated by Birgeneau et al. [1]

lattice contribution. Rayne and Kemp [19] have analysed the data of Busey and Giauque [18] and have given a ($\theta - T$) curve after subtracting the electronic and spin-wave contributions from the experimental heat capacities. In this way they have found the value of electronic heat capacity coefficient γ to be 7.05×10^{-3} J/mol deg^2. The value of C_m, the spin-wave contribution to the heat capacities, was found to be $C_m = 8.8 \times 10^{-5}\ T^{3/2}$ J/mol deg$^{3/2}$. The experimental ($\theta - T$) curve presented by Rayne and Kemp [19] has been reproduced in Fig. 5 together with the theoretical curve obtained by us using Krebs's model. We have also given the theoretical curve calculated by Birgeneau et al. [1] from the general Born-Karman force model including fourth neighbours.

3. Discussion

A study of Fig. 1 to 3 shows that Krebs's model yields a very good description of the experimental dispersion curves in the three symmetry directions. At low vectors the calculated frequencies almost coincide with the experimental results. The maximum divergence of less than 3% is found only at the zone boundaries. Since the experimental phonon frequencies were determined with an accuracy of about 2% the agreement must be considered excellent. Birgeneau et al. [1] have also attempted to fit their experimental results by the use of general interatomic forces up to fourth nearest neighbours, and also with an axially symmetric model in which the interactions are extended up to fifth nearest neighbours. They have also reported an excellent agreement in both cases. This is, however, not surprising in view of the fact that their general force model and the atomic-shell model both employ a considerably large number of parameters (twelve and ten respectively), all of which have been determined entirely from neutron scattering results by the method of least square fit. The agreement with the experimental results, therefore, depicts nothing but the self-consistency of the experimental results. In spite of this self-consistency the elastic constants and the low-frequency region of the disperison curves are not described correctly as these autors have pointed out themselves. This appears to be due to the fact that they did not use the elastic constants to determine their force constants. The great superiority of Krebs's model over the models employed by Birgeneau et al. [1] lies in the fact that it gives excellent agreements with elastic data as well as with the dispersion curves with a very small number (four only) of parameters. These are the central force constants with the three neighbours and the electronic force constant k_e.

A study of Fig. 5 reveals that our calculated ($\theta - T$) curves show an excellent agreement with the experimental one. We have also compared our results with the theoretical curve of Birgeneau et al. [1] calculated by them by means of the general interatomic force model extended up to fourth neighbours. In the low-temperature region our curve almost resembles to the curve of Birgeneau et al. [1], but at higher temperatures it reproduces the experimental results much better than is given by the latter.

Fig. 4 gives the frequency distribution of nickel calculated on the basis of Krebs's model. Since there are three different sets of experimental $g(\nu)$ curves available we have compared our theoretical curve with all of them. The real difficulties about these experimental curves are that they differ from each other considerably. Our calculated curve is also not found to describe fully the course of anyone of these experimental curves but, on the whole, it makes a good compromise amongst them in a wide range of spectrum.

Acknowledgement

The authors are thankful to Dr. R. P. Gupta for useful discussions.

References

[1] R. J. BIRGENEAU, J. CORDES, G. DOLLING, and A. D. B. WOODS, Phys. Rev. **136**, 1359 (1964).
[2] K. KREBS, Phys. Letters (Netherlands) **10**, 12 (1964).
[3] K. KREBS, Phys. Rev. **138**, 143 (1965).
[4] P. S. MAHESH and B. DAYAL, phys. stat. sol. **9**, 351 (1965).
[5] P. S. MAHESH and B. DAYAL, Phys. Rev. **143**, 443 (1966).
[6] M. M. SHUKLA, phys. stat. sol. **7**, K11 (1964).
[7] M. M. SHUKLA, phys. stat. sol. **8**, 475 (1965).
[8] M. M. SHUKLA and B. DAYAL, J. Phys. Chem. Solids **26**, 1343 (1965).
[9] M. M. SHUKLA and B. DAYAL, phys. stat. sol. **16**, 513 (1966).
[10] J. DE LAUNAY, J. chem. Phys. **21**, 1979 (1953).
[11] A. B. BHATIA, Phys. Rev. **97**, 363 (1955).
[12] P. K. SHARMA and S. K. JOSHI, J. chem. Phys. **39**, 2633 (1963).
[13] G. A. ALERS, J. R. NEIGHBOURS, and H. SATO, Bull. Amer. Phys. Soc. **4**, 131 (1959).
[14] N. A. CHERNOPLEKOV, M. G. ZEMLYANOV, A. G. CHETSERIN, and B. G. LYASHCHENKO, Inelastic Scattering of Neutrons in Solids and Liquids, Vol. 2, International Atomic Energy Agency, Vienna 1962 (p. 159).
[15] R. M. BRUGGER, A.E.R.E. (Harwell) Rep. R4562 (1964) (unpublished).
[16] B. MOZER, K. OTNES, and H. PALEVSKY, J. Phys. Chem. Solids, to be published.
[17] W. H. KEESOM and C. W. CLARK, Physica **2**, 513 (1935).
[18] R. H. BUSEY and W. F. GIAUQUE, J. Amer. Chem. Soc. **74**, 3157 (1952).
[19] J. A. RAYNE and W. R. G. KEMP, Phil. Mag. **1**, 918 (1956).

(Received November 29, 1966)

phys. stat. sol. 19, 735 (1967)
Subject classification: 18.2; 10.1; 21.1; 21.1.1

Institut für Physik am Max-Planck-Institut für Metallforschung, Stuttgart, und Institut für theoretische und angewandte Physik der Technischen Hochschule Stuttgart

Wechselwirkung zwischen Versetzungen und ebenen Blochwänden mit starrem Magnetisierungsverlauf[1])

Von
K.-H. PFEFFER

Mit Hilfe der Peach-Koehlerschen Formel werden die Wechselwirkungskräfte zwischen 180°-Blochwänden, deren Magnetisierungsverlauf als starr angenommen wird, und verschiedenen Versetzungskonfigurationen berechnet. Numerische Ergebnisse werden für die (110)- und $(\bar{1}\bar{1}2)$-180°-Blochwände in Nickel und für die (001)- und (110)-180°-Blochwände in Eisen angegeben. Die Temperaturabhängigkeit der für die verschiedenen Versetzungskonfigurationen charakteristischen effektiven Wechselwirkungslängen und die in der statistischen Analyse der Blochwandbewegung zu beachtenden geometrisch-statistischen Parameter werden diskutiert.

The interaction forces between 180°-Bloch walls, the magnetization of which is assumed to be fixed and different dislocation configurations are calculated by the Peach-Koehler method. Numerical results are given for the (110)- and $(\bar{1}\bar{1}2)$-180°-Bloch walls in nickel and for the (001)- and (110)-180°-Bloch walls in iron. The temperaturedependence of the effective interaction lengths of the different dislocation configurations and the geometrical parameters relating to a statistical analysis of the Blochwall movement are discussed.

1. Einleitung

Die ausgeprägte Abhängigkeit der Kenngrößen der Magnetisierungskurven ferromagnetischer Kristalle von deren Verformungszustand beruht vorwiegend auf der magnetoelastischen Wechselwirkung zwischen der Magnetisierung und den durch Versetzungen hervorgerufenen inneren Spannungen.

Durch Arbeiten von Dietrich und Kneller [1], Krause [2], Rieger [3], Träuble und Bilger [4] und Träuble [5] ist experimentell gesichert, daß in den magnetisch mehrachsigen Kristallen Eisen und Nickel die Koerzitivfeldstärke H_c und die Anfangssuszeptibilität χ_a bei hinreichend tiefen Temperaturen (bei Fe unterhalb 500 °C, bei Ni unterhalb Raumtemperatur) durch Verschiebungen von 180°-Blochwänden bestimmt werden.

Die Wechselwirkung der 180°-Blochwände mit inneren Spannungen ist bisher in verschiedenen Näherungen und unter verschiedenen Annahmen untersucht worden. In der von Kondorskii [6] und Kersten [7] entwickelten Spannungstheorie der Koerzitivfeldstärke wird ein mittlerer Eigenspannungsbetrag $\bar{\sigma}_i$ eingeführt, ohne die Art und die Anordnung der Spannungsquellen zu spezifizieren. Vicena [8], Rieder [9] und Träuble [5] berücksichtigten die spezielle Natur der Eigenspannungsquellen und berechneten die Wechselwirkungskraft zwischen einer Blochwand und einer Versetzung unter folgenden Annahmen:

1. Der Magnetisierungsverlauf innerhalb und außerhalb der Blochwand wird durch die inneren Spannungen der Versetzung nicht beeinflußt.
2. Die Versetzungen verlaufen parallel zur Blochwand.

[1]) Dissertation, Teil I, Technische Hochschule Stuttgart, 1966.

Zur Begründung der Annahme 1. schätzte Vicena [8] den Einfluß der durch die inhomogene Magnetisierungsverteilung in der Umgebung einer Versetzung [10] bedingten Streufelder auf die Wechselwirkung zwischen Blochwand und Versetzung ab und fand, daß dieser Streufeldeffekt [5] eine um zwei bis drei Größenordnungen kleinere Wechselwirkungskraft liefert, als die Näherung der starren Blochwand ergibt. Bei dieser Abschätzung bleibt jedoch die entscheidende Tatsache unberücksichtigt, daß sich die Magnetisierungsverteilungen in der Umgebung einer Versetzung und im Innern einer Blochwand gegenseitig beeinflussen, wodurch eine zusätzliche Energieänderung im System Blochwand–Versetzung auftritt. Bevor wir eine mikromagnetische Behandlung dieses Sachverhaltes durchführen, untersuchen wir in der vorliegenden Arbeit unter Beibehaltung der Annahme 1. die Wechselwirkung von Blochwänden mit gekrümmten Versetzungen und schaffen damit die Grundlage für eine genauere Berücksichtigung der in einem Kristall vorliegenden Versetzungsanordnung bei der Berechnung von H_c und χ_a.

2. Methoden zur Berechnung der Wechselwirkungskraft

In allen Verfahren zur Berechnung der Wechselwirkungskraft zwischen Versetzungen und Blochwänden wird der Magnetisierungsverlauf innerhalb einer isolierten Blochwand und damit der Tensor $\boldsymbol{\epsilon}^{\mathrm{M}}$ der magnetostriktiven Extradehnungen derselben als gegeben betrachtet. Bei Kenntnis des Spannungstensors $\boldsymbol{\sigma}^{\mathrm{V}}$ der Versetzung läßt sich die magnetoelastische Kopplungsenergie $-\boldsymbol{\epsilon}^{\mathrm{M}} \cdot \boldsymbol{\sigma}^{\mathrm{V}}$ berechnen und hieraus durch Differentiation die Wechselwirkungskraft bestimmen (Methode von Vicena [8]). Rieder [11] dagegen ermittelt den Tensor $\boldsymbol{\sigma}^{\mathrm{M}}$ der mit dem inhomogenen Magnetisierungsverlauf in der Blochwand verknüpften Extraspannungen und berechnet mit Hilfe der Peach–Koehlerschen Formel [12] die Kraft, die die Blochwand auf die Versetzung ausübt. Durch Anwendung des Gesetzes „Actio = Reactio" ergibt sich hieraus die Kraft, die die Versetzung auf die Blochwand ausübt. Letztere Methode ist die bequemere, da nur der Verlauf und der Burgersvektor, nicht aber der Spannungshof der Versetzung in die Rechnung eingeht. Wir werden im folgenden diese Methode benutzen.

Die Kraft, die eine geschlossene Versetzungsschleife auf eine Blochwand ausübt, ist als Flächenintegral [13] zu berechnen:

$$\boldsymbol{P}_{\mathrm{W}}^{(0)} = -\int_{S} (\mathrm{d}\boldsymbol{S} \times \nabla) \times \boldsymbol{\sigma}^{\mathrm{M}} \cdot \boldsymbol{b} \,. \tag{1}$$

$\boldsymbol{b}$ ist der Burgersvektor der Versetzungslinie, S eine beliebige, von der Versetzungslinie berandete Fläche. $\boldsymbol{\sigma}^{\mathrm{M}}$ ist der Extraspannungstensor der Blochwand. Weist die Blochwandnormale in z-Richtung, so sind nur die Spannungskomponenten σ_{ij}^{M} mit $(i, j) = (1, 2)$ von Null verschieden. Diese berechnen sich für eine isolierte 180°-Blochwand²) nach Rieder [11] zu

$$\sigma_{ij}^{\mathrm{M}} = A_{ij} \sin^2 \Phi_0 + B_{ij} \sin 2\,\Phi_0 \,; \qquad 0 \leqq \Phi_0 \leqq \pi \,. \tag{2}$$

Die Konstanten A_{ij} und B_{ij} enthalten die magnetostriktiven und elastischen Konstanten des betreffenden Kristalls. Für die (110)-180°- und die $(\bar{1}\bar{1}2)$-180°-Blochwand in Ni bzw. die (001)-180°- und die (110)-180°-Blochwand in Fe sind die A_{ij} und B_{ij} aus Tabelle 1 bzw. 2 zu entnehmen.

²) Die y-Richtung fällt mit einer leichten Magnetisierungsrichtung zusammen; bei Ni ist dies eine ⟨111⟩-Richtung, bei Fe eine ⟨100⟩-Richtung.

$\Phi_0(z)$ ist der Verlauf des Azimutwinkels des Magnetisierungsvektors in der isolierten Blochwand und durch die Gleichung

$$\operatorname{ctg} \Phi_0 = - a \sinh b\, \varkappa_K\, z + c \tag{3}$$

gegeben. Dabei ist $\varkappa_K^2 = |K_1|/A$, wobei K_1 die Anisotropie- und A die Austauschkonstante ist. Die Konstanten a, b, c sind für die obengenannten Wände in Ni bzw. Fe in Abhängigkeit von der Temperatur in Tabelle 3 bzw. 4 angegeben. Hierbei wurde die mit den Extraspannungen verbundene Extraenergie berücksichtigt, wobei elastische und magnetostriktive Isotropie angenommen wurde. Weiterhin wurde nur die Anisotropiekonstante K_1 als temperaturabhängig betrachtet. Für die (110)-180°- und $(\bar{1}\bar{1}2)$-180°-Blochwand in Ni ist der Verlauf von Φ_0, $\sin^2 \Phi_0$ und $\sin 2\, \Phi_0$ in Abhängigkeit von $b\, \varkappa_K\, z$ für verschiedene Temperaturen in Fig. 1a, b und 2a, b dargestellt. In Fig. 3a, b und 4a, b sind

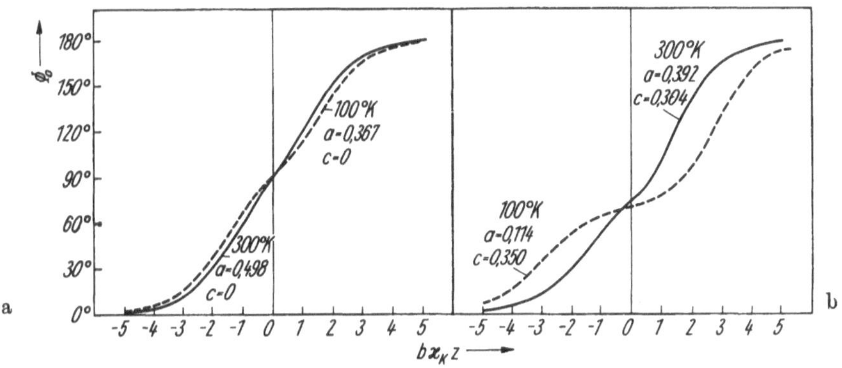

Fig. 1. Verlauf des Azimutwinkels in 180°-Blochwänden in Nickel.
a) $(\bar{1}\bar{1}2)$-180°-Blochwand, b) (110)-180°-Blochwand

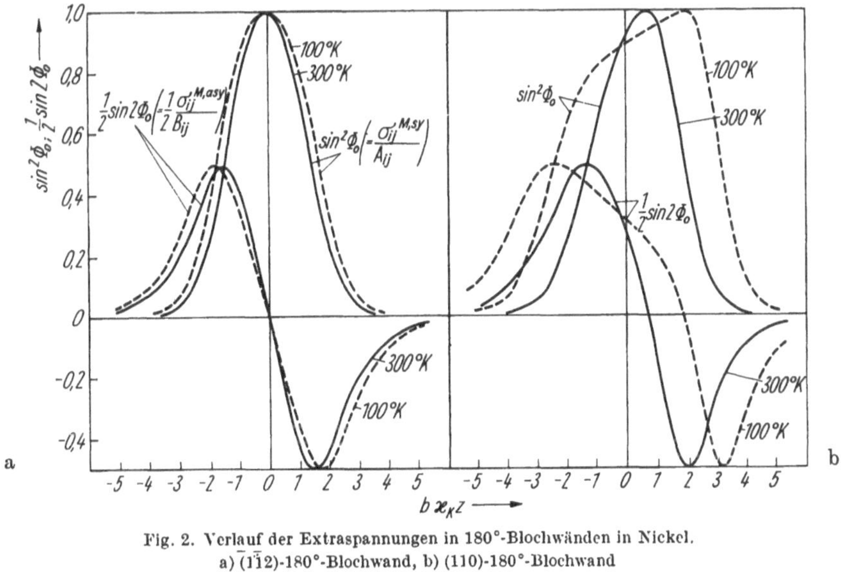

Fig. 2. Verlauf der Extraspannungen in 180°-Blochwänden in Nickel.
a) $(\bar{1}\bar{1}2)$-180°-Blochwand, b) (110)-180°-Blochwand

Tabelle 1

Magnetoelastische Konstanten A_{ij} und B_{ij} für Ni (elastisch isotrop)

	(110)-180°-Blochwand		$(\bar{1}\bar{1}2)$-180°-Blochwand	
ij	A_{ij}	B_{ij}	A_{ij}	B_{ij}
11	$\frac{2G\lambda_{100}}{1-\nu}\left(\frac{3}{2}\nu\frac{\lambda_{111}}{\lambda_{100}}-\frac{1}{4}\left(1+5\frac{\lambda_{111}}{\lambda_{100}}\right)\right)$	$\frac{G\lambda_{100}}{1-\nu}\frac{\sqrt{2}}{2}\left(1-\frac{\lambda_{111}}{\lambda_{100}}\right)$	$\frac{2G\lambda_{100}}{1-\nu}\left(\frac{3}{2}\nu\frac{\lambda_{111}}{\lambda_{100}}-\frac{1}{4}\left(1+5\frac{\lambda_{111}}{\lambda_{100}}\right)\right)$	0
22	$\frac{2G\lambda_{100}}{1-\nu}\left(\frac{3}{2}\frac{\lambda_{111}}{\lambda_{100}}-\frac{1}{4}\nu\left(1+5\frac{\lambda_{111}}{\lambda_{100}}\right)\right)$	$\frac{G\lambda_{100}}{1-\nu}\frac{\sqrt{2}}{2}\nu\left(1-\frac{\lambda_{111}}{\lambda_{100}}\right)$	$\frac{2G\lambda_{100}}{1-\nu}\left(\frac{3}{2}\frac{\lambda_{111}}{\lambda_{100}}-\frac{1}{4}\nu\left(1+5\frac{\lambda_{111}}{\lambda_{100}}\right)\right)$	0
12	$-G\lambda_{100}\frac{\sqrt{2}}{2}\left(1-\frac{\lambda_{111}}{\lambda_{100}}\right)$	$G\lambda_{100}\left(1+\frac{1}{2}\frac{\lambda_{111}}{\lambda_{100}}\right)$	0	$G\lambda_{100}\left(1+\frac{1}{2}\frac{\lambda_{111}}{\lambda_{100}}\right)$

Tabelle 2

Magnetoelastische Konstanten A_{ij} und B_{ij} für Fe (elastisch isotrop)

	(110)-180°-Blochwand		(001)-180°-Blochwand	
ij	A_{ij}	B_{ij}	A_{ij}	B_{ij}
11	$\frac{2G\lambda_{100}}{1-\nu}\left(\frac{3}{2}\nu-\frac{3}{4}\left(1+\frac{\lambda_{111}}{\lambda_{100}}\right)\right)$	0	$-3\lambda_{100}G$	0
22	$\frac{2G\lambda_{100}}{1-\nu}\left(\frac{3}{2}-\frac{3}{4}\nu\left(1+\frac{\lambda_{111}}{\lambda_{100}}\right)\right)$	0	$3\lambda_{100}G$	0
12	0	$\frac{3}{2}\lambda_{111}G$	0	$\frac{3}{2}\lambda_{111}G$

Tabelle 3

Blochwandkonstanten a, b, c für Ni (elastisch und magnetostriktiv isotrop)

	(110)-180°-Blochwand			$(\bar{1}\bar{1}2)$-180°-Blochwand		
T (°K)	$a = \left(1 - \frac{7}{12}\frac{1}{b^2} - \frac{1}{18}\frac{1}{b^4}\right)^{1/2}$	$b = \left(\frac{2}{3} + \frac{9}{2}\frac{\lambda^2 G}{\lvert K_1\rvert}\right)^{1/2}$	$c = \sqrt{2}\,\Big/\left(4 + 27\frac{\lambda^2 G}{\lvert K_1\rvert}\right)$	$a = \left(1 - \frac{7}{12}\frac{1}{b^2}\right)^{1/2}$	$b = \left(\frac{2}{3} + \frac{9}{2}\frac{\lambda^2 G}{\lvert K_1\rvert}\right)^{1/2}$	c
100	0,114	0,821	0,350	0,367	0,821	0
140	0,138	0,823	0,348	0,376	0,823	0
180	0,182	0,828	0,343	0,387	0,828	0
220	0,226	0,835	0,338	0,406	0,835	0
260	0,292	0,850	0,327	0,439	0,850	0
300	0,392	0,880	0,304	0,498	0,880	0

Tabelle 4

Blochwandkonstanten a, b, c für Fe (elastisch und magnetostriktiv isotrop)

	(110)-180°-Blochwand			(001)-180°-Blochwand		
T (°K)	$a = \left(1 - \frac{3}{4b^2}\right)^{1/2}$	$b = \left(1 + \frac{9}{2}\frac{\lambda^2 G}{K_1}\right)^{1/2}$	c	$a = \left(1 - \frac{1}{b^2}\right)^{1/2}$	$b = \left(1 + \frac{9}{2}\frac{\lambda^2 G}{K_1}\right)^{1/2}$	c
100	0,501	1,001	0	0,050	1,001	0
200	0,501	1,001	0	0,052	1,001	0
300	0,503	1,002	0	0,056	1,002	0
400	0,503	1,002	0	0,062	1,002	0
500	0,505	1,003	0	0,072	1,003	0
600	0,506	1,004	0	0,084	1,004	0

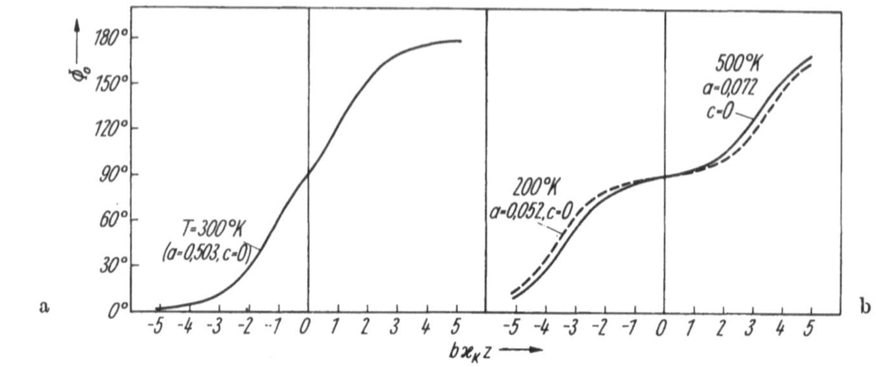

Fig. 3. Verlauf des Azimutwinkels in 180°-Blochwänden in Eisen.
a) (110)-180°-Blochwand, b) (001)-180°-Blochwand

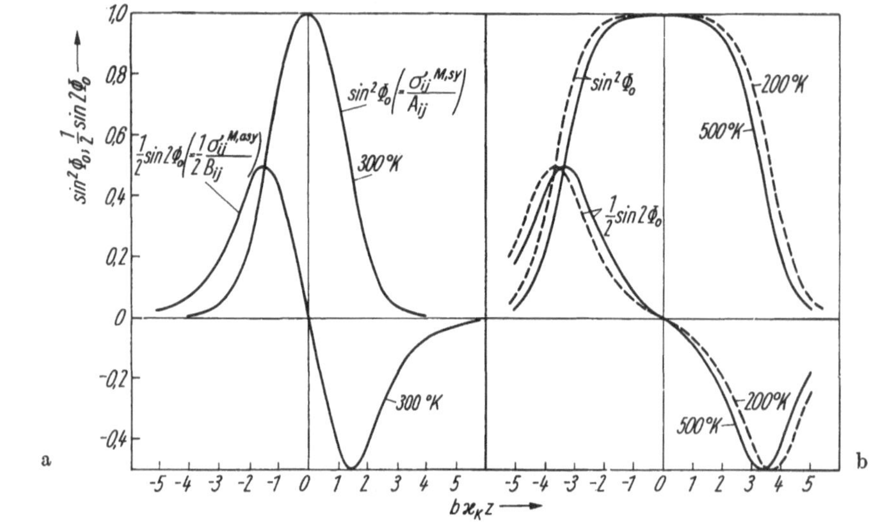

Fig. 4. Verlauf der Extraspannungen in 180°-Blochwänden in Eisen.
a) (110)-180°-Blochwand, b) (001)-180°-Blochwand

die entsprechenden Größen für die (110)-180°- und die (100)-180-°Blochwand in Fe dargestellt.

Die in Gleichung (2) angegebenen Extraspannungskomponenten sind nur für eine 180°-Blochwand gültig, die sich in einem sonst idealen, unendlich ausgedehnten Kristall befindet. In diesem Fall verschwinden die Extraspannungen in den angrenzenden Weißschen Bezirken. Betrachtet man dagegen eine äquidistante Anordnung von parallelen 180°-Blochwänden, so findet man, daß bei hinreichend großem Blochwandabstand D ($D > 20$mal Blochwanddicke) der Verlauf des Azimutwinkels Φ_0 und der Extraspannungen σ_{ij}^{M} innerhalb einer Blochwand durch die Anwesenheit anderer Blochwände praktisch nicht beeinflußt wird, die Extraspannungen in den Weißschen Bezirken jedoch einen von Null verschiedenen Betrag c_{ij} annehmen. Näherungsweise ist

$$c_{ij} = -\frac{1}{D}\int_0^D (A_{ij}\sin^2\Phi_0 + B_{ij}\sin 2\,\Phi_0)\,\mathrm{d}z\,. \tag{4}$$

Da für die Wechselwirkungskraft zwischen einer geschlossenen Versetzungsschleife und einer Blochwand nach Gleichung (1) nur der Gradient des Extraspannungstensors maßgebend ist, spielen die konstanten Spannungskomponenten c_{ij} in den Weißschen Bezirken für die Kraftwirkung keine Rolle. Mit anderen Worten: Die Wechselwirkungskraft zwischen Blochwand und geschlossener Versetzungsschleife ist unabhängig von der Anwesenheit anderer Blochwände und wird nur durch den Verlauf der Versetzungsschleife innerhalb der Blochwand bestimmt. Diese Tatsache ist für die statistische Behandlung der Blochwandbewegung von großer Bedeutung.

3. Anwendung auf verschiedene Versetzungskonfigurationen

Zur expliziten Berechnung der Wechselwirkungskräfte zwischen einer Blochwand und verschiedenen Versetzungskonfigurationen schreiben wir Gleichung (1) in kartesischen Koordinaten:

$$P^{(0)}_{\mathrm{W},x} = P^{(0)}_{\mathrm{W},y} = 0\,, \tag{5a}$$

$$P^{(0)}_{\mathrm{W},z} = -\int_{S_1}\mathrm{d}S_1\frac{\mathrm{d}}{\mathrm{d}z}\left(\sigma^{\mathrm{M}}_{11}\,b_1 + \sigma^{\mathrm{M}}_{12}\,b_2\right) - \int_{S_2}\mathrm{d}S_2\frac{\mathrm{d}}{\mathrm{d}z}\left(\sigma^{\mathrm{M}}_{12}\,b_1 + \sigma^{\mathrm{M}}_{22}\,b_2\right). \tag{5b}$$

S_1 bzw. S_2 ist der Bereich, der durch die Projektion der Versetzungsschleife auf die Ebene $x = 0$ bzw. $y = 0$ berandet wird. Bezeichnen wir die Berandungslinie von S_2 mit $x_{1,2}(z)$, so geht Gleichung (5b) mit Hilfe von Gleichung (2) unter der Annahme, daß die Versetzungsschleife in der Gleitebene liegt, deren Normalenvektor die Komponenten n_i besitzt, über in

$$P^{(0)}_{\mathrm{W},z} = -\frac{n_i\,b_j}{|n_2|}\left(A_{ij}\int_{z_1}^{z_2}(x_1(z) - x_2(z))\frac{\mathrm{d}}{\mathrm{d}z}\sin^2\Phi_0\,\mathrm{d}z + \right.$$
$$\left. + B_{ij}\int_{z_1}^{z_2}(x_1(z) - x_2(z))\frac{\mathrm{d}}{\mathrm{d}z}\sin 2\,\Phi_0\,\mathrm{d}z\right).^{3)} \tag{6}$$

$z_2 - z_1$ ist der Durchmesser der Versetzungsschleife in z-Richtung.

Anhand von Gleichung (6) diskutieren wir nun die Kraftwirkung zwischen einer Blochwand und Versetzungssegmenten, die parallel zur Blochwand verlaufen oder diese durchstoßen.

3.1 Zur Blochwand parallele Versetzungssegmente

Zur Behandlung dieses Falles nehmen wir eine parallelogrammförmige Versetzungsschleife derart an, daß zwei Seiten parallel zur Blochwand verlaufen. Die Projektion der parallelen Segmente auf die Ebene $y = 0$ habe die Länge

[3] Ist $n_2 = 0$, $n_1 \neq 0$, so sind n_2 durch n_1 und $x_{1,2}$ durch $y_{1,2}$ zu ersetzen. Für $n_1 = n_2 = 0$ wird $P^{(0)}_{\mathrm{W},z}$ Null.

$x_1 - x_2 = l_p$. Dann ergibt sich aus Gleichung (6)

$$P^{(0)}_{W,z} = - l_p \frac{n_i b_j}{|n_2|} (A_{ij} \sin^2 \Phi_0(z) + B_{ij} \sin 2 \Phi_0(z)) \Big|_{z=z_1}^{z=z_2} . \tag{7a}$$

Hieraus erhält man für die Kraft, die ein einzelnes paralleles Versetzungssegment im Abstand z_0 von der Blochwand auf diese ausübt,

$$P^{(0)\,P}_{W,z} (z_0) = \pm l_p \frac{n_i b_j}{|n_2|} (A_{ij} \sin^2 \Phi_0(z_0) + B_{ij} \sin 2 \Phi_0(z_0)) . \tag{7b}$$

Das Vorzeichen wird durch die Orientierung des Versetzungssegmentes festgelegt.

Den Ausdruck (7b) werden wir zur Definition von effektiven Wechselwirkungslängen [5] im Falle durchstoßender Versetzungssegmente benutzen (siehe Abschnitt 3.2).

3.2 Die Blochwand durchstoßende Versetzungssegmente

Wir müssen zwischen Versetzungssegmenten unterscheiden, die die Blochwand vollständig durchstoßen (siehe Fig. 5b), und solchen, die innerhalb der Blochwand umkehren (siehe Fig. 5c).

Zur Behandlung des ersten Falles stellen wir den Verlauf $x_1 - x_2$ der Versetzungssegmente innerhalb der Blochwand durch eine Taylorreihe dar:

$$x_1(z) - x_2(z) = \sum_{n=0} \frac{1}{n!} \Delta f^{(n)}(0)\, z^n . \tag{8}$$

Nach Einsetzen von (8) in (6) und anschließender partieller Integration erhält man für die Wechselwirkungskraft

$$P^{(0)\,D}_{W,z} = \frac{n_i b_j}{|n_2|} \Bigg(A_{ij} \sum_{n=1} \frac{1}{(n-1)!} \Delta f^{(n)}(0) \int_{-\infty}^{+\infty} z^{n-1} \sin^2 \Phi_0(z)\, dz + $$
$$+ B_{ij} \sum_{n=1} \frac{1}{(n-1)!} \Delta f^{(n)}(0) \int_{-\infty}^{+\infty} z^{n-1} \sin 2 \Phi_0(z)\, dz \Bigg) . \tag{9}$$

Der Index „D" soll an „durchstoßend" erinnern.

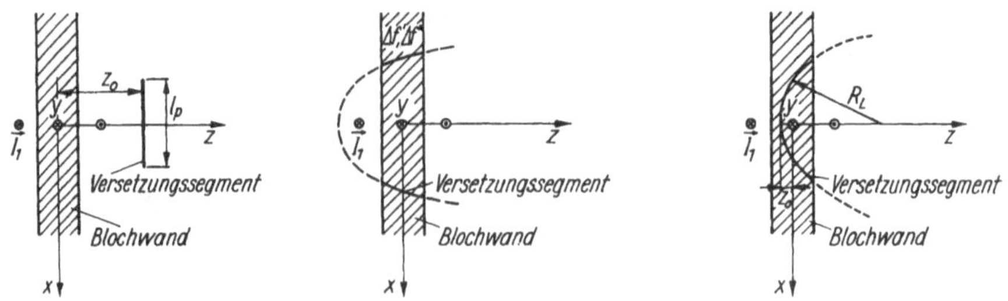

a b c

Fig. 5. Versetzungskonfigurationen. a) Zur Blochwand paralleles, b) die Blochwand durchstoßendes, c) in der Blochwand umkehrendes Versetzungssegment. I_1 leichte Richtung

Durch gliedweisen Vergleich von Gleichung (9) mit Gleichung (7b) können wir zwei effektive Wechselwirkungslängen definieren:

$$l^{\mathrm{s}}_{\mathrm{eff,D}} = \sum_{n=1} \frac{1}{(n-1)!} \Delta f^{(n)}(0) \frac{1}{(b\,\varkappa_K)^n} \underbrace{\int_{-\infty}^{+\infty} x^{n-1} \sin^2 \Phi_0(x)\,\mathrm{d}x}_{I_n^{\mathrm{s}}}\,, \tag{10a}$$

$$l^{\mathrm{a}}_{\mathrm{eff,D}} = \sum_{n=1}' \frac{1}{(n-1)!} \Delta f^{(n)}(0) \frac{1}{(b\,\varkappa_K)^n} \underbrace{\int_{-\infty}^{+\infty} x^{n-1} \sin 2\,\Phi_0(x)\,\mathrm{d}x}_{I_n^{\mathrm{a}}}\,. \tag{10b}$$

Dabei ist $x = b\,\varkappa_K\,z$. Der Index „s" bzw. „a" rührt davon her, daß $l^{\mathrm{s}}_{\mathrm{eff}}$ bzw. $l^{\mathrm{a}}_{\mathrm{eff}}$ mit der bezüglich der Blochwandmitte ($\Phi_0 = \pi/2$) geraden Funktion $\sin^2 \Phi_0$ bzw. ungeraden Funktion $\sin 2\,\Phi_0$ verknüpft ist. Die Integrale I_n^{s} und I_n^{a} in Gleichung (10a, b) sind für $n = 1$ und $n = 2$ sowohl für die (110)-180°- als auch die $(\bar{1}\bar{1}2)$-180°-Blochwand in Ni bei verschiedenen Temperaturen in Tabelle 5 zahlenmäßig angegeben. Tabelle 6 enthält die entsprechenden Werte für die (110)-180°- und die (001)-180°-Blochwand in Fe.

Tabelle 5

I_1^{s}, I_1^{a}, I_2^{s}, I_2^{a} für Ni

	(110)-180°-Blochwand				$(\bar{1}\bar{1}2)$-180°-Blochwand			
T (°K)	I_1^{s}	I_2^{s}	I_1^{a}	I_2^{a}	I_1^{s}	I_2^{s}	I_1^{a}	I_2^{a}
100	5,43	1,95	2,49	−17,06	3,57	0	0	−11,20
140	5,10	1,77	2,25	−16,00	3,51	0	0	−11,07
180	4,63	1,50	1,91	−14,52	3,48	0	0	−10,91
220	4,26	1,30	1,66	−13,37	3,39	0	0	−10,65
260	3,84	1,06	1,35	−12,04	3,26	0	0	−10,23
300	3,37	0,79	1,01	−10,56	3,05	0	0	− 9,55

Tabelle 6

I_1^{s}, I_1^{a}, I_2^{s}, I_2^{a} für Fe

	(110)-180°Blochwand				(001)-180°-Blochwand			
T(°K)	I_1^{s}	I_2^{s}	I_1^{a}	I_2^{a}	I_1^{s}	I_2^{s}	I_1^{a}	I_2^{a}
100	3,04	0	0	−9,54	7,39	0	0	−23,12
200	3,04	0	0	−9,54	7,31	0	0	−22,88
300	3,03	0	0	−9,52	7,16	0	0	−22,42
400	3,03	0	0	−9,52	6,96	0	0	−21,80
500	3,03	0	0	−9,50	6,66	0	0	−20,88
600	3,02	0	0	−9,50	6,36	0	0	−19,93

Zur Untersuchung der Wechselwirkung mit Versetzungen, die in der Blochwand umkehren, nehmen wir einen elliptischen Verlauf der Versetzungsschleife gemäß

$$\frac{(z - z_0 - R_1)^2}{R_1^2} + \frac{x^2}{R_2^2} = 1 \tag{11}$$

an. R_1 und R_2 sind die Halbachsen der Ellipse, z_0 ist die Koordinate des linken Hauptscheitels, der innerhalb der Blochwand liegen möge. Berechnet man aus Gleichung (11) $x_1 - x_2$ und setzt diesen Wert in Gleichung (6) ein, so gewinnt man folgenden Ausdruck für die Wechselwirkungskraft:

$$\left. \begin{aligned} P^{(0)\,\mathrm{UL}}_{\mathrm{w},z}(z_0, R_1, R_\mathrm{L}) &= -\frac{n_i b_j}{|n_2|} 2 \sqrt{\frac{2 R_\mathrm{L}}{b \varkappa_K}} \times \\ &\times \Bigg(A_{ij} \underbrace{\int\limits_0^{2 b \varkappa_K R_1} \sqrt{x - \frac{1}{2} \frac{x^2}{b \varkappa_K R_1}} \frac{\mathrm{d}}{\mathrm{d}x} \sin^2 \Phi_0 (x + b \varkappa_K z_0) \, \mathrm{d}x}_{I^\mathrm{s}_\mathrm{L}(z_0, R_1)} + \\ &+ B_{ij} \underbrace{\int\limits_0^{2 b \varkappa_K R_1} \sqrt{x - \frac{1}{2} \frac{x^2}{b \varkappa_K R_1}} \frac{\mathrm{d}}{\mathrm{d}x} \sin 2 \Phi_0 (x + b \varkappa_K z_0) \, \mathrm{d}x}_{I^\mathrm{a}_\mathrm{L}(z_0, R_1)} \Bigg). \end{aligned} \right\} \quad (12)$$

Dabei sind $x = b \varkappa_K z$ und $R_\mathrm{L} = R_2^2 / R_1$ der Krümmungsradius im linken Hauptscheitel. Der Index „U" soll an „umkehrend" erinnern. Die beiden von R_1 und z_0 abhängigen Integrale I^s_L und I^a_L in Gleichung (12) sind in Fig. 6a, b für die $(\bar{1}\bar{1}2)$-180°- und die (110)-180°-Wand in Ni bei 300 °K dargestellt. Die Fig. 7a, b beziehen sich auf Fe.

Führen wir dieselbe Rechnung für den rechten Hauptscheitel durch, so ergibt sich für die $(\bar{1}\bar{1}2)$-180°-Wand in Ni und die (001)-180°- und (110)-180°-Blochwand in Fe

$$I^\mathrm{s}_\mathrm{R}(z_0) = -I^\mathrm{s}_\mathrm{L}(-z_0)\,, \quad I^\mathrm{a}_\mathrm{R}(z_0) = I^\mathrm{a}_\mathrm{L}(-z_0)\,,$$

wobei z_0 jetzt die Koordinate des rechten Hauptscheitels bedeutet. Die entsprechenden Integrale I^s_R und I^a_R für die (110)-180°-Wand in Ni sind in Fig. 8 dargestellt.

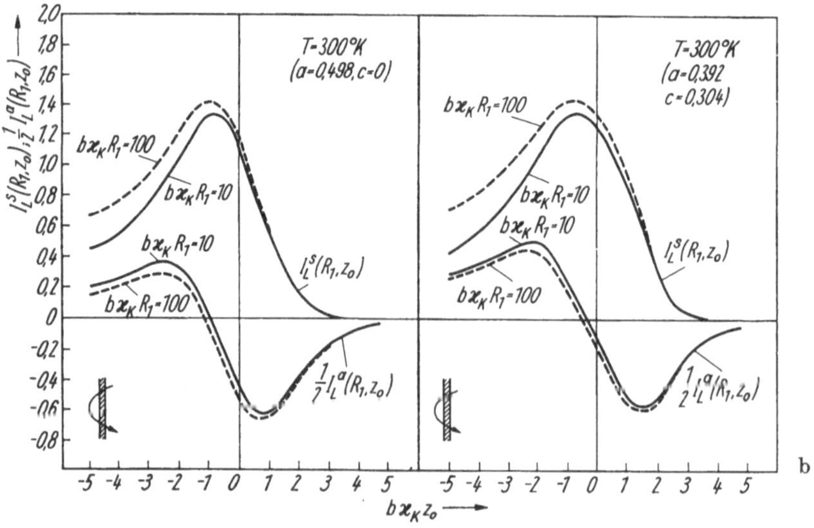

Fig. 6. Funktionen $I^\mathrm{s}_\mathrm{L}(z_0, R_1)$ und $I^\mathrm{a}_\mathrm{L}(z_0, R_1)$ für Nickel.
a) $(\bar{1}\bar{1}2)$-180°-Blochwand, b) (110)-180°-Blochwand

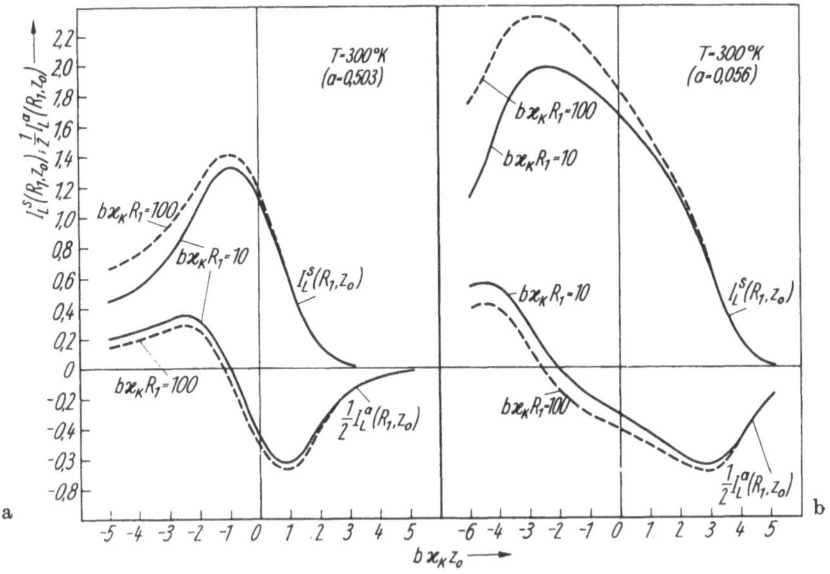

Fig. 7. Funktionen I_L^s (z_0, R_1) und I_L^a (z_0, R_1) für Eisen.
a) (110)-180°-Blochwand, b) (001)-180°-Blochwand

Die Integrale $I^s_{L,R}$ und $I^a_{L,R}$ zeigen eine nur geringe Veränderung bei Variation von $b\,\varkappa_K\,R_1$, so daß wir $I^s_{L,R}$ und $I^a_{L,R}$ in erster Näherung als von $b\,\varkappa_K\,R_1$ unabhängig betrachten können. Durch Vergleich von (12) mit (6) erhalten wir folgende effektiven Wechselwirkungslängen:

$$l^s_{\text{eff, U L}} = 2\sqrt{\frac{2\,R_L}{b\,\varkappa_K}}\;I^s_{L,\max}\,, \tag{13a}$$

$$l^a_{\text{eff, U L}} = 2\sqrt{\frac{2\,R_L}{b\,\varkappa_K}}\;\frac{I^a_{L,\max} - I^a_{L,\min}}{2}\,. \tag{13b}$$

Entsprechende Ausdrücke erhält man für den rechten Hauptscheitel.

4. Bestimmung des Strukturfaktors S_{ij}

Nach Gleichung (6) benötigt man zur vollständigen Berechnung der Wechselwirkungskraft neben den effektiven Wechselwirkungslängen noch die sogenannten Strukturfaktoren $S_{ij} = n_i\,b_j$. In kubisch-flächenzentrierten Kristallen, zu denen Ni zählt, sind die Gleitebenen {111}-Ebenen, die Burgersvek-

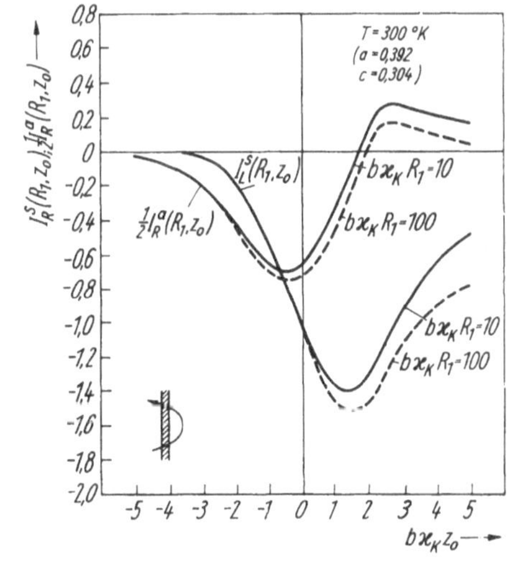

Fig. 8. Funktionen I_R^s (z_0, R_1) und I_R^a (z_0, R_1) für die (110)-180°-Blochwand in Nickel

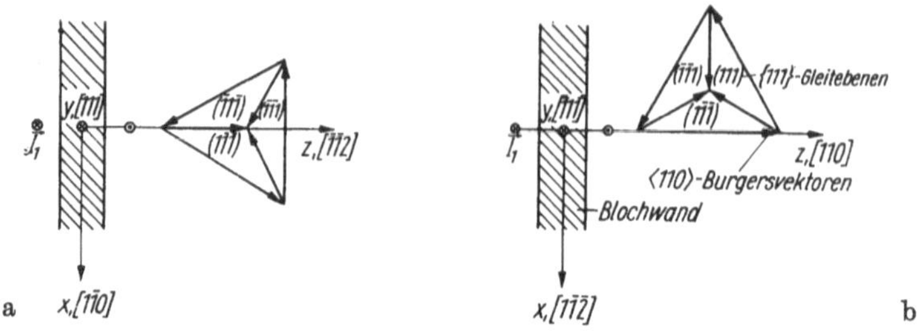

Fig. 9. Blochwandtypen, Gleitebenen und Burgersvektoren in Nickel.
a) $(\bar{1}\bar{1}2)$-180°-Blochwand, b) (110)-180°-Blochwand. I_1 leichte Richtung

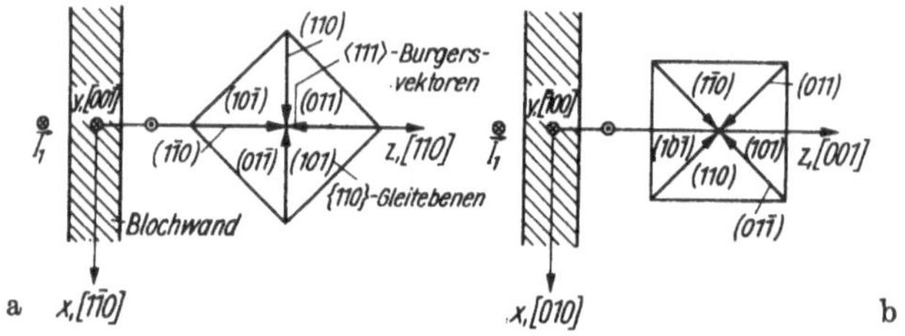

Fig. 10. Blochwandtypen, Gleitebenen und Burgersvektoren in Eisen.
a) (110)-180°-Blochwand, b) (001)-180°-Blochwand. I_1 leichte Richtung

toren $\boldsymbol{b}$ sind vom Typ $a/2\ \langle 110\rangle$ (a Kantenlänge des Elementarwürfels). Die Lage der Gleitebenen in den in Ni auftretenden Blochwandtypen $(\bar{1}\bar{1}2)$ und (110) angepaßten Koordinatensystemen ist in Fig. 9a, b dargestellt. Die Komponenten der Gleitebenenormalen und der Burgersvektoren in diesen Koordinatensystemen sind in den Tabellen 7 und 8 angegeben. Tabelle 9 gibt die Linienrichtungen derjenigen Versetzungen an, die in einer {111}-Ebene liegen und parallel zur Blochwand verlaufen.

In Eisen, einem kubisch-raumzentrierten Metall, sind die Verhältnisse komplizierter, da als Gleitebenen mehrere Ebenen, vorwiegend die {110}-, {112}- und {321}-Ebenen, in Frage kommen. Wir beschränken uns auf die {110}-Gleitebenen. Ihre Lagen in den den in Fe auftretenden Blochwandtypen angepaßten Koordinatensystemen sind in Fig. 10a, b dargestellt. Die Komponenten der Normalenvektoren dieser Gleitebenen sowie die Komponenten der Burgersvektoren, die fast ausschließlich vom $a/2\ \langle 111\rangle$-Typ sind, sind in den Tabellen 10 und 11 zusammengefaßt. Die Linienrichtungen der parallel zu den Blochwänden verlaufenden Versetzungen, die dem {110}-Gleitsystem angehören, sind in Tabelle 12 enthalten.

5. Diskussion

Die Bedeutung der in Abschnitt 3 gewonnenen Ergebnisse ist erst im Rahmen einer allgemeinen Theorie der Koerzitivfeldstärke und Anfangssuszeptibilität zu beurteilen. Da wir diese Theorie erst in einer späteren Arbeit entwickeln werden, beschränken wir uns an dieser Stelle allein auf die Diskussion der oben abgeleiteten Ausdrücke für die Wechselwirkungskräfte, insbesondere der Ausdrücke für die effektiven Wechselwirkungslängen.

Tabelle 7
Komponenten der Normalenvektoren der {111}-Gleitebenen in Ni

Gleitebene	(110)-180°-Blochwand	($\bar{1}\bar{1}2$)-180°-Blochwand
(111)	$\frac{1}{3}[-\sqrt{2}, -1, \sqrt{6}]$	[0, 1, 0]
($\bar{1}\bar{1}1$)	$\frac{1}{3}[-\sqrt{2}, -1, -\sqrt{6}]$	$\frac{1}{3}[0, -1, 2\sqrt{2}]$
($\bar{1}1\bar{1}$)	[0, 1, 0]	$\frac{1}{3}[-\sqrt{6}, -1, -\sqrt{2}]$
($1\bar{1}\bar{1}$)	$\frac{1}{3}[2\sqrt{2}, -1, 0]$	$\frac{1}{3}[\sqrt{6}, -1, -\sqrt{2}]$

Tabelle 8
Komponenten der ⟨110⟩-Burgersvektoren in Ni

Gleitebene	(110)-180°-Blochwand	($\bar{1}\bar{1}2$)-180°-Blochwand
(111)	$\frac{a}{2}\left[\frac{2}{\sqrt{6}}, -\frac{2}{\sqrt{3}}, 0\right]$, $\frac{a}{2}\left[\frac{3}{\sqrt{6}}, 0, \frac{1}{\sqrt{2}}\right]$, $\frac{a}{2}\left[\frac{1}{\sqrt{6}}, \frac{2}{\sqrt{3}}, \frac{1}{\sqrt{2}}\right]$	$\frac{a}{2}[\sqrt{2}, 0, 0]$, $\frac{a}{2}\left[\frac{1}{\sqrt{2}}, 0, -\frac{3}{\sqrt{6}}\right]$, $\frac{a}{2}\left[\frac{1}{\sqrt{2}}, 0, \frac{3}{\sqrt{6}}\right]$
($\bar{1}\bar{1}1$)	$\frac{a}{2}\left[\frac{2}{\sqrt{6}}, -\frac{2}{\sqrt{3}}, 0\right]$, $\frac{a}{2}\left[-\frac{1}{\sqrt{6}}, -\frac{2}{\sqrt{3}}, \frac{1}{\sqrt{2}}\right]$, $\frac{a}{2}\left[-\frac{3}{\sqrt{6}}, 0, \frac{1}{\sqrt{2}}\right]$	$\frac{a}{2}[\sqrt{2}, 0, 0]$, $\frac{a}{2}\left[\frac{1}{\sqrt{2}}, \frac{2}{\sqrt{3}}, \frac{1}{\sqrt{6}}\right]$, $\frac{a}{2}\left[-\frac{1}{\sqrt{2}}, \frac{2}{\sqrt{3}}, \frac{1}{\sqrt{6}}\right]$
($\bar{1}1\bar{1}$)	$\frac{a}{2}\left[0, 0, \frac{2}{\sqrt{2}}\right]$, $\frac{a}{2}\left[\frac{3}{\sqrt{6}}, 0, \frac{1}{\sqrt{2}}\right]$, $\frac{a}{2}\left[-\frac{3}{\sqrt{6}}, 0, \frac{1}{\sqrt{2}}\right]$	$\frac{a}{2}\left[0, \frac{2}{\sqrt{3}}, -\frac{2}{\sqrt{6}}\right]$, $\frac{a}{2}\left[\frac{1}{\sqrt{2}}, 0, -\frac{3}{\sqrt{6}}\right]$, $\frac{a}{2}\left[-\frac{1}{\sqrt{2}}, \frac{2}{\sqrt{3}}, \frac{1}{\sqrt{6}}\right]$
($1\bar{1}\bar{1}$)	$\frac{a}{2}\left[0, 0, \frac{2}{\sqrt{2}}\right]$, $\frac{a}{2}\left[-\frac{1}{\sqrt{6}}, -\frac{2}{\sqrt{3}}, \frac{1}{\sqrt{2}}\right]$, $\frac{a}{2}\left[\frac{1}{\sqrt{6}}, \frac{2}{\sqrt{3}}, \frac{1}{\sqrt{2}}\right]$	$\frac{a}{2}\left[0, \frac{2}{\sqrt{3}}, -\frac{2}{\sqrt{6}}\right]$, $\frac{a}{2}\left[\frac{1}{\sqrt{2}}, \frac{2}{\sqrt{3}}, \frac{1}{\sqrt{6}}\right]$, $\frac{a}{2}\left[\frac{1}{\sqrt{2}}, 0, \frac{3}{\sqrt{6}}\right]$

Tabelle 9

Linienrichtungen der zur Blochwand parallelen Versetzungen in Ni (Komponenten bezüglich der kubischen Achsen)

Gleitebene	(110)-180°-Blochwand		$(\bar{1}\bar{1}2)$-180°-Blochwand	
	$\boldsymbol{t}$	$\boldsymbol{I}_1 \cdot \boldsymbol{t}$	$\boldsymbol{t}$	$\boldsymbol{I}_1 \cdot \boldsymbol{t}$
(111)	$[1\bar{1}0]$	$-\sqrt{\frac{2}{3}}$	$[1\bar{1}0]$	0
$(\bar{1}\bar{1}1)$	$[1\bar{1}0]$	$-\sqrt{\frac{2}{3}}$	$[1\bar{1}0]$	0
$(\bar{1}1\bar{1})$	$[1\bar{1}\bar{2}]$	0	[132]	—
$(1\bar{1}\bar{1})$	$[1\bar{1}2]$	$-\frac{2}{3}\sqrt{2}$	[312]	—

Tabelle 10

Komponenten der Normalenvektoren der {110}-Gleitebenen in Fe

Gleitebene	(110)-180°-Blochwand	(001)-180°-Blochwand
(110)	[0, 0, 1]	$\frac{1}{\sqrt{2}}\,[1\bar{1}0]$
$(1\bar{1}0)$	[1, 0, 0]	$\frac{1}{\sqrt{2}}\,[\bar{1}\bar{1}0]$
(101)	$\frac{1}{2}\,[1, -\sqrt{2}, 1]$	$\frac{1}{\sqrt{2}}\,[0\bar{1}1]$
$(10\bar{1})$	$\frac{1}{2}\,[1, \sqrt{2}, 1]$	$\frac{1}{\sqrt{2}}\,[0\bar{1}\bar{1}]$
(011)	$\frac{1}{2}\,[-1, -\sqrt{2}, 1]$	$\frac{1}{\sqrt{2}}\,[101]$
$(01\bar{1})$	$\frac{1}{2}\,[\;1, \sqrt{2}, 1]$	$\frac{1}{\sqrt{2}}\,[10\bar{1}]$

Tabelle 11
Komponenten der ⟨111⟩-Burgersvektoren in Fe

Gleitebene	(110)-180°-Blochwand	(001)-180°-Blochwand
(110)	$\frac{a}{2}[\sqrt{2}, -1, 0]$, $\frac{a}{2}[\sqrt{2}, 1, 0]$	$\frac{a}{2}[\bar{1}\bar{1}1]$, $\frac{a}{2}[111]$
$(1\bar{1}0)$	$\frac{a}{2}[0, -1, \sqrt{2}]$, $\frac{a}{2}[0, 1, \sqrt{2}]$	$\frac{a}{2}[1\bar{1}1]$, $\frac{a}{2}[\bar{1}11]$
(101)	$\frac{a}{2}[0, 1, \sqrt{2}]$, $\frac{a}{2}[\sqrt{2}, 1, 0]$	$\frac{a}{2}[1\bar{1}\bar{1}]$, $\frac{a}{2}[111]$
$(10\bar{1})$	$\frac{a}{2}[0, -1, \sqrt{2}]$, $\frac{a}{2}[-\sqrt{2}, 1, 0]$	$\frac{a}{2}[1\bar{1}1]$, $\frac{a}{2}[11\bar{1}]$
(011)	$\frac{a}{2}[0, 1, \sqrt{2}]$, $\frac{a}{2}[\sqrt{2}, -1, 0]$	$\frac{a}{2}[1\bar{1}\bar{1}]$, $\frac{a}{2}[\bar{1}\bar{1}1]$
$(01\bar{1})$	$\frac{a}{2}[0, -1, \sqrt{2}]$, $\frac{a}{2}[\sqrt{2}, 1, 0]$	$\frac{a}{2}[1\bar{1}1]$, $\frac{a}{2}[\bar{1}\bar{1}\bar{1}]$

Tabelle 12
Linienrichtungen der parallel zur Blochwand verlaufenden Versetzungen in Fe (Komponenten bezüglich der kubischen Achsen)

Gleitebene	(110)-180°-Blochwand		(001)-180°-Blochwand	
	$\boldsymbol{t}$	$\boldsymbol{I}_1 \cdot \boldsymbol{t}$	$\boldsymbol{t}$	$\boldsymbol{I}_1 \cdot \boldsymbol{t}$
(110)	beliebig	beliebig	$[1\bar{1}0]$	$\frac{1}{\sqrt{2}}$
$(1\bar{1}0)$	[001]	1	[110]	$\frac{1}{\sqrt{2}}$
(101)	$[1\bar{1}\bar{1}]$	$\frac{1}{\sqrt{3}}$	[010]	0
$(10\bar{1})$	$[1\bar{1}1]$	$\frac{1}{\sqrt{3}}$	[010]	0
(011)	$[1\bar{1}1]$	$\frac{1}{\sqrt{3}}$	[100]	1
$(01\bar{1})$	$[\bar{1}11]$	$\frac{1}{\sqrt{3}}$	[100]	1

Tabelle 13
Temperaturabhängigkeit der effektiven Wechselwirkungslängen; statistische Parameter

Versetzungs-konfiguration	$l_{eff} \sim \delta_B^n \sim K_1^{-n/2}$	statistische Parameter
parallel	$n = 0$	z_0, l_p
durchstoßend	$n \geqq 1$	$\Delta f'(0)$, $\Delta f''(0)$, . . .
umkehrend	$n \leqq 1$	z_0, R_1, R_2

Für jede oben besprochene Konfiguration tritt eine spezifische Verknüpfung von geometrischen Parametern, die den Verlauf der Versetzungslinie bezüglich der Blochwand charakterisieren, mit der zu $\varkappa_K^{-1}$ proportionalen Blochwanddicke δ_B auf. Die Verhältnisse sind in Tabelle 13 zusammenfassend dargestellt.

Da die Blochwanddicke über die Anisotropiekonstante K_1 empfindlich von der Temperatur abhängt, zeigt die Tabelle 13 die Möglichkeit an, aus einer genauen Analyse der Temperaturabhängigkeit von H_c und χ_a Aussagen über die Versetzungsanordnung zu gewinnen. Die Temperaturabhängigkeit von H_c und χ_a, die die verschiedenen Versetzungskonfigurationen verursachen, kann jedoch erst nach einer statistischen Untersuchung der Blochwandbewegung angegeben werden. Hierbei sind die oben aufgezählten geometrischen Parameter als statistisch unabhängige Zufallsgrößen zu behandeln. Diese Zufallsgrößen lassen sich in zwei Gruppen aufteilen. Die zur ersten Gruppe gehörigen Zufallsgrößen (z_0) beschreiben die räumliche Lage der Versetzungen im Kristall; die Zufallsgrößen der zweiten Gruppe (l_p, R_1, $\Delta f'$, $\Delta f''$, R) charakterisieren die Form der Versetzungslinie. Entsprechend dieser Aufteilung wird die statistisch zu ermittelnde Kraftwirkung der Versetzungen auf die Blochwand außer durch die Anzahlschwankung sowohl durch die räumliche Schwankung als auch durch die Formschwankung der Versetzungen bestimmt. Die Formschwankung umfaßt Schwankungen in der Ausdehnung der Versetzungsschleifen (l_p, R_1), Richtungsschwankungen ($\Delta f'$) und Krümmungsschwankungen ($\Delta f''$, R).

Literatur

[1] H. DIETRICH und E. KNELLER, Z. Metallk. **47**, 716 (1956).
[2] D. KRAUSE, Z. Phys. **168**, 239 (1962).
[3] H. RIEGER, phys. stat. sol. **8**, 283 (1965).
[4] H. BILGER und H. TRÄUBLE, phys. stat. sol. **10**, 755 (1965).
[5] H. TRÄUBLE, in Moderne Probleme der Metallphysik, Bd. 2, Hrsg. A. SEEGER, Springer-Verlag, Berlin/Göttingen/Heidelberg 1966.
[6] E. KONDORSKII, Sowjet Phys. **11**, 597 (1937).
[7] H. KERSTEN, in Probleme der technischen Magnetisierungskurve, Hrsg. R. BECKER, Springer-Verlag, Berlin 1938.
[8] F. VICENA, Czech. J. Phys. **5**, 480 (1955).
[9] G. RIEDER, Z. angew. Phys. **9**, 187 (1957).
[10] W. F. BROWN, Phys. Rev. **58**, 736 (1940).
[11] G. RIEDER, Abh. Braunschweig. Wiss. Ges. **11**, 20 (1959).
[12] M. PEACH und J. S. KOEHLER, Phys. Rev. **80**, 436 (1950).
[13] A. SEEGER, H. KRONMÜLLER, H. RIEGER und H. TRÄUBLE, J. appl. Phys. **35**, 740 (1964).

(Received November 30, 1966)

phys. stat. sol. 19, 751 (1967)
Subject classification: 6; 22.6

Department of Physics, Banaras Hindu University, Varanasi

Lattice Dynamics of MgO

By
M. P. VERMA and B. DAYAL

A phenomenological model developed by the author in an earlier paper [1] is applied to the case of MgO. The two-phonon combined density of states (C.D.S.) is computed from the calculated frequencies. The C.D.S. peaks show good agreement with the subsidiary maxima in the infrared absorption spectrum of the solid. The calculated dispersion curves agree with Pekham's measurements except for the large wave-vector region, and the specific heats agree with the experimental observations of Barron et al.

Ein vom Autor in einer früheren Arbeit entwickeltes phänomenologisches Modell [1] wird auf MgO angewendet. Die kombinierte Zwei-Phononen-Zustandsdichte (C.D.S.) wird aus den berechneten Frequenzen ermittelt. Die C.D.S.-Maxima zeigen gute Übereinstimmung mit den Nebenmaxima im Infrarotspektrum des Festkörpers. Die berechneten Dispersionskurven stimmen bis auf den Bereich großer Wellenvektoren mit den Messungen von Pekham überein, die spezifischen Wärmen mit den experimentellen Beobachtungen von Barron et al.

1. Introduction

Calculations of the cohesive energy of an ionic solid in the Heitler London approximation lead to a many-body component which is supposed to be responsible for the failure of the Cauchy relation in such solids. In an earlier paper [1] (hereafter referred to as I) the present authors have shown that the character of the three-body forces in ionic solids with cubic symmetry is such that the elastic constants must satisfy Fuchs's relations. This suggests that these forces could be approximated by a volume dependent interaction. This idea was used in I to develop a phenomenological model. The lattice potential in this model is assumed to consist of a volume dependent part in addition to the conventional two-body central potential. The Coulomb term in the two-body potential is modified through an effective charge parameter and the exchange repulsion term is supposed to be significant between the nearest as well as the next-nearest neighbours. The next-nearest neighbour interaction is, however, assumed to be appreciable amongst the negative ions only.

The model was applied to LiF in I and gave an excellent fit with the observed heat capacity. One of the reasons why LiF was chosen is that this alkali halide has a relatively large Cauchy discrepancy and hence should have a significant volume dependent interaction. Another ionic solid with a large Cauchy discrepancy is MgO. In this paper we report the results obtained by applying the model described in I to a study of the lattice dynamics of this ionic solid. The advantage of choosing this solid is that its phonon dispersion curves in certain directions of the wave vector space have been obtained by Pekham [2] from neutron scattering measurements.

The solid has an additional interest because of some confusion regarding the principal infra-red absorption maximum. The infra-red absorption spectra of MgO as observed by Willmott [3], Burstein et al. [4], Fock [5], and other workers[1]) contain a number of maxima. All these authors have recognised a strong peak near 17.3 μm as the principal absorption peak and the others as subsidiaries. Saksena and Vishwanathan [6] have critically examined the absorption spectrum in relation to refractive index data and have shown that the main absorption peak should occur at 25.26 μm. These latter authors therefore consider the peaks at shorter wavelengths including the peak at 17.3 μm as subsidiary maxima.

The absorption maxima at shorter wavelengths can be interpreted as being due to multiphonon (two-phonon) absorption by the lattice. These multiphonon processes in ionic solids can be the result of two distinct effects: one due to the second and higher order terms in the expansion of dipole moments as a power series in ionic displacements and the other due to the presence of third and higher order anharmonic terms in the lattice potential energy. Szigeti [7, 8] has studied the effect of the second-order dipole moment term and the third-order anharmonic term on the infra-red absorption spectra of solids. He has shown that these terms give rise to sum and difference modes of lattice vibration frequencies. The difference modes are distributed over a large frequency interval and are not expected to give rise to any absorption peaks. Therefore, the observed peaks can be interpreted as being due to the sum modes or in other words due to consecutive absorption of two phonons. Such two-phonon processes are subjected to the following selection rules:

1. If one of the phonons corresponds to a wave vector $\boldsymbol{q}_\alpha$ the other must correspond to $-\boldsymbol{q}_\alpha$.

2. In solids where every atom (or ion) is at a centre of inversion symmetry the two phonons must come from different branches of the spectrum.

The density of states $N(\omega_1 + \omega_2)$ of the sum frequencies $(\omega_1 + \omega_2)$ can be computed from the lattice vibration frequencies in accordance with the two selection rules as has been done by Smart et al. [9]. In view of Szigeti's analysis the subsidiary maxima in the infra-red absorption spectrum should coincide with the peaks of this so-called combined density of states (C.D.S.) curve. This provides a test of the validity of the theoretical model which is sensitive towards the higher frequency side of the vibrational spectrum. Since the heat capacity calculations are sensitive towards the low frequency side of the spectrum, measurements of both quantities should provide an effective test of any theoretical model. This is particularly so where neutron scattering data are not extensive or not very accurate. Since Pekham's [2] observed dispersion curves have a rather large scatter and have very few points in certain branches we have preferred to apply the heat capacity and the C.D.S. tests also to our model.

2. Results and Discussions

In the present study the model described in I has been used. This model has overall six disposable parameters A_1, B_1, A_2, B_2, α_v, and s. The first four belong to the nearest and next-nearest neighbour exchange repulsions, α_v belongs to the volume dependent forces, and s is the effective charge parameter. These

[1]) Other references are given in [6].

parameters are determined from equations expressing the three elastic constants, the two optical frequencies ω_t and ω_l corresponding to zero wave vector and the equilibrium condition. The longitudinal optical frequency ω_l is itself determined from ω_0 and the high- and low-frequency dielectric constants ε_∞ and ε_0 in accordance with the Lyddane-Sachs and Teller relation

$$\omega_l = \omega_0 \sqrt{\frac{\varepsilon_0}{\varepsilon_\infty}}\,. \tag{1}$$

The six equations used for determining the six parameters are best satisfied at 0 °K. The values of the elastic constants of MgO at 0 °K have been taken from measurements of Durand [10] and are $c_{11} = 29.89$, $c_{12} = 8.57$, and $c_{44} = 15.679$ in units of 10^{12} dyn cm^{-2}. Because of lack of measurements of ε_0 and ε_∞ at 0 °K we have used the values $\varepsilon_0 = 9.8$ and $\varepsilon_\infty = 2.95$ as given by Born and Huang [11] (Table 17, p. 85). Calculations were made for two different values 10.9 and 7.462×10^{13} s^{-1} of the infra-red absorption frequency ω_0 corresponding to $\lambda_0 = 17.3$ and 25.26 μm respectively. We are, however, reporting the results only for $\omega_0 = 7.462 \times 10^{13}$ s^{-1} because the other value does not give even an approximate agreement with the observed heat capacity except at very low temperatures. The values of the six parameters obtained by solving equations (10), (12), and (13) of I are as follows:

$$A_1 = 9.364,\; B_1 = -0.697,\; A_2 = 5.104,\; B_2 = -0.468,\; \alpha_v = -1.264,$$

and $s = 1.377$.

The lattice vibration frequencies have been calculated for a 1000 point division of the first Brillouin zone on an I.B.M. 1620 computer at I.I.T., Kanpur. The Debye temperatures θ_D in the range 0 to 300 °K have then been determined in the usual way. The plot of θ_D versus T together with the corresponding experimental graph obtained from the measurements of Barron et al. [12] has been shown in Fig. 1. The agreement is good and proves that Saksena and Vishwanathan's value of the principal infra-red absorption frequency is correct.

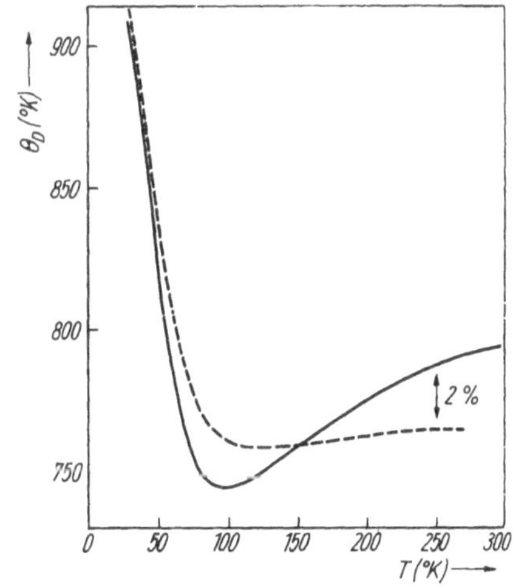

Fig. 1. θ_D-T curve for MgO
—— theoretical,
- - - - - experimental

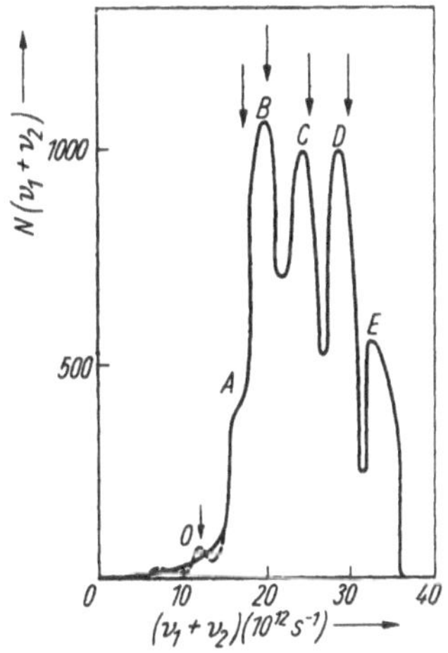

Fig. 2. The combined density-of-states curve for MgO. Arrows indicate positions of infra-red absorption peaks observed by Willmott [3]

The combined density of states can be obtained from the lattice vibration frequencies in accordance with the two selection rules. Since the vibrational frequencies at wave vector $-\boldsymbol{q}_a$ must be the same as those at $+\boldsymbol{q}_a$ the sum frequencies have to be obtained from the six frequencies (in general) belonging to every wave vector. In view of the second selection rule a particular frequency will not combine with itself. Thus there will in general be ${}^6C_2 = 15$ possible sum modes corresponding to each wave vector. Taking the statistical weight of a particular sum mode to be the same as that of the corresponding wave vector, a histogram can be prepared. When smoothed out this histrogram gives the combined density of states curve $N(\nu_1 + \nu_2)$ versus $\nu = (\nu_1 + \nu_2)$. This curve has been shown in Fig. 2. The arrows indicate positions of infra-red absorption peaks as observed by Willmott [3].

The curve has five distinct peaks at 0, B, C, D, and E and a kink at A. The dotted curve near 0 obtained from the histogram and showing the peak 0 has been smoothed out into the continuous curve because in this part of the curve the number of frequencies is not very large and one may not be very definite about the existence of the peak. The wavelengths corresponding to the six points in the C.D.S. curve are given in Table 1 together with the wavelengths of the infra-red absorption maxima observed by Willmott. The peak 0 has been included though, as discussed above, its existence must be taken with caution.

Table 1

The C.D.S. and infra-red absorption peaks

Sample No.	C.D.S. peaks		Infra-red absorption peaks (Willmott), λ (μm)
	Designation	λ (μm)	
1	0	25.0	24.3
2	A	18.1	17.5
3	B	15.2	14.8
4	C	12.3	11.7
5	D	10.5	10.07
6	E	9.1	–

Willmott's infra-red studies cover a region between 7 and 25 μm. The absorption peak at 24.3 μm is close to the position of the main absorption maximum of 25.26 μm and may not be a subsidiary maximum. The other peaks observed by him agree fairly well with those observed by Burstein et al. [4] and by Fock [5]. None of these authors have, however, observed an absorption peak that can correspond to the C.D.S. peak at E (9.1 μm).

The infra-red absorption spectrum of MgO as observed by Barnes et al. [13] presents a fine structure giving all together about 40 maxima. This has not been confirmed by the observations of other workers. Apart from the fine structure, however, the more pronounced absorption peaks appear approximately at 8.8, 10.2, 11.8, 14.4, and 15.4 μm. The peak at 8.8 μm may be taken to correspond to the C.D.S. peak E at 9.1 μm.

The C.D.S. curve has been obtained from vibrational frequencies calculated in the harmonic approximation. As pointed out by Smart et al. [9] the anharmonic terms in the lattice potential are not expected to change the positions

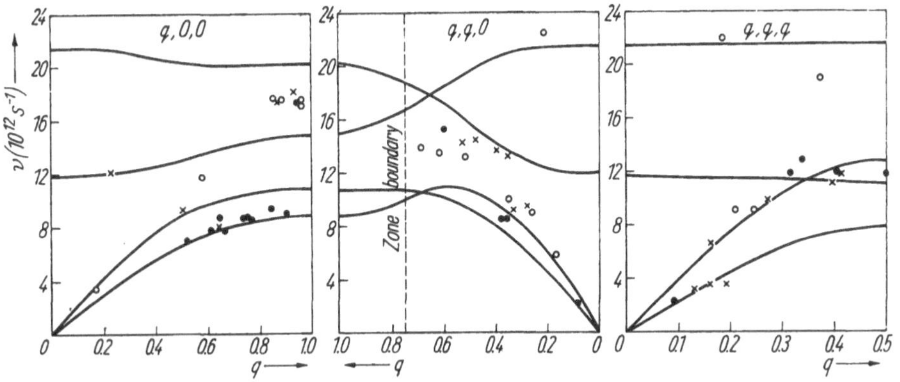

Fig. 3. Dispersion relations for MgO.
Experimental points are from Pekham [2]
● transversal,
○ longitudinal,
× undefined

of the peaks although they could modify their relative magnitudes. As discussed above the C.D.S. peaks present a general agreement with the subsidiary maxima in the infra-red absorption spectrum. Considering the spread in the positions of the peaks as observed by different authors the agreement appears to be quite good.

The dispersion curves ν versus q in the symmetry directions have been shown in Fig. 3 together with Pekham's [2] results. The solid circles indicate points on a transverse branch and the empty circles those on a longitudinal one. There are other points for which it was not possible for Pekham to decide whether they belong to a longitudinal or a transverse branch. Such points have been shown by a cross.

The theoretical curves agree with the experimental observations at low wave vectors but the differences become rather large for high wave vectors. The maximum difference appears near the zone boundary. This is perhaps due to the fact that the volume dependent forces have been introduced in our model in such a way that the model ceases to be symmetric in the reciprocal space. The general agreement obtained, however, shows that our model gives a good description of this class of solids.

One other point with respect to the C.D.S. peak E corresponding to the shortest wavelength (9.1 μm) or the highest frequency deserves to be mentioned. This peak must arise from the absorption of two phonons of the optical branches. The neutron scattering measurements show that even if our calculated frequencies for the optical branches are somewhat in error the longitudinal frequencies are higher and the transverse frequencies are lower, by approximately the same amount, than the experimental values. Therefore the peak E in the sum mode is not expected to be very much in error. It is possible that this peak is weak and has been missed by Willmott in the infra-red absorption spectrum.

Major part of this paper forms a part of a thesis submitted by one of us (M.P.V.) and accepted for the award of a Ph. D. degree of the Banaras Hindu University.

We are grateful to the authorities of the computer centre at the I.I.T., Kanpur, for extending free computation facilities to us.

References

[1] M. P. VERMA and B. DAYAL, phys. stat. sol. **13**, 251 (1966).
[2] G. PEKHAM, Lattice Dynamics, Proc. Internat. Conf., Copenhagen, 1963, Ed. R. F. WALLIS (p. 49).
[3] J. C. WILLMOTT, Proc. Phys. Soc. **A 63**, 389 (1950).
[4] E. BURSTEIN, J. J. OBERLEY, and E. K. PLYLER, Proc. Indian Acad. Sci. **28 A**, 398 (1948).
[5] J. FOCK, Z. Phys. **90**, 44 (1934).
[6] B. D. SAKSENA and S. VISHWANATHAN, Proc. Phys. Soc. **B 69**, 129 (1956).
[7] B. SZIGETI, Proc. Roy. Soc. **A 252**, 217 (1959).
[8] B. SZIGETI, Proc. Roy. Soc. **A 258**, 337 (1960).
[9] C. SMART, G. R. WILKINSON, A. M. KARO, and J. R. HARDY, Lattice Dynamics, Proc. Internat. Conf., Copenhagen 1963, Ed. R. F. WALLIS (p. 387).
[10] M. A. DURAND, Phys. Rev. **50**, 449 (1936).
[11] M. BORN and K. HUANG, Dynamical Theory of Crystal Lattices, Clarendon Press, Oxford 1954.
[12] T. H. K. BARRON, W. T. BERG, and J. A. MORRISON, Proc. Roy. Soc. **A 250**, 70 (1959).
[13] R. B. BARNES, R. BRATTAIN, and F. SEITZ, Phys. Rev. **48**, 582 (1935).

(Received December 2, 1966)

phys. stat. sol. **19**, 757 (1967)

Subject classification: 20

*II. Physikalisch-Technisches Institut
der Deutschen Akademie der Wissenschaften zu Berlin*

Zur Theorie der Erzeugung der zweiten Harmonischen an der Oberfläche eines unmagnetischen, quadratischen, optisch einachsigen Kristalls[1])

Von

R. FISCHER

Fällt eine intensive monochromatische, elektromagnetische Welle auf die Grenzfläche eines optisch nichtlinearen Kristalls, so erfolgt eine Wechselwirkung, die zur Erzeugung einer zweiten Harmonischen führen kann. Neben der sich im nichtlinearen Medium ausbreitenden zweiten Harmonischen tritt auch eine reflektierte Harmonische auf. Es werden Verallgemeinerungen des Reflexions- und Brechungsgesetzes und der Fresnelschen Formeln für den Fall abgeleitet, daß eine ebene Grundwelle aus einem unmagnetischen, nichtabsorbierenden, isotropen, linearen Medium auf die ebene Grenzfläche eines unmagnetischen, optisch einachsigen, optisch nichtlinearen (quadratischen) Kristalls fällt. Das Problem wird in parametrischer Näherung gelöst. Es wird vorausgesetzt, daß sich die Polarisationswellen nicht in einer Synchronisationsrichtung ausbreiten.

An intense, monochromatic, electromagnetic wave can generate a second harmonic by interaction with an optically non-linear crystal. In addition to the second harmonic which occurs in the non-linear medium, a reflected harmonic is also created. Generalized laws of reflection and refraction and generalized Fresnel formulae are derived to cover this situation. The generalization considers a plane wave travelling in a non-magnetic, non-absorbing, isotropic, linear medium meeting a plane boundary with a non-magnetic, optically non-linear (quadratic), uniaxial crystal. The problem is solved in parametric approximation. It is assumed, that the polarization waves are not propagated in a phase-matching direction.

1. Einleitung

Durch die Entwicklung der Laser wurde die Beobachtung einer Vielzahl neuer nichtlinearer optischer Effekte [1, 2] möglich. Die physikalischen Grundprozesse der nichtlinearen Optik sind Mehrphotonenprozesse [3, 4]. In der phänomenologischen Theorie ergeben sich diese Effekte bei der Lösung der Maxwellschen Gleichungen unter Verwendung nichtlinearer Materialgleichungen [1, 2, 5].

Die Wechselwirkung ebener, monochromatischer elektromagnetischer Wellen an der Grenzfläche eines quadratischen, isotropen Mediums wurde von Bloembergen und Pershan bereits 1962 theoretisch untersucht [6]. Diese Rechnungen sind nur für Kristalle der Klassen T_d ($\bar{4}3$ m), T (23), O (432)[2]) und wenige Spezialfälle optisch anisotroper Kristalle anwendbar.

[1]) Auszugsweise vorgetragen auf der Tagung „Laser und ihre Anwendungen", Berlin, 8. bis 11. 11. 1966.

[2]) Im Falle der Erzeugung der zweiten Harmonischen ist der Tensor der quadratischen Suszeptibilität für diese Klasse identisch Null [7].

Die einzige uns bekannt gewordene Arbeit zur Theorie der Erzeugung der zweiten Harmonischen in Reflexion an einem quadratischen, anisotropen Medium stammt von Kleinman [8]. In [8] wurde ein allgemeiner Ausdruck für die reflektierte Harmonische angeführt, in dem nur eine der sich im nichtlinearen Medium ausbreitenden Polarisationswellen berücksichtigt wurde. Außerdem wurde in [8] nur eine geometrische Methode zur Bestimmung der Lage der Refraktionsvektoren der Grundwellen und der Harmonischen angegeben.

Die Theorie von Bloembergen und Pershan [6] wurde inzwischen experimentell bestätigt [9]. Es wurden die in [6] gefundenen Verallgemeinerungen des Reflexions- und Brechungsgesetzes und der Fresnelschen Formeln für nichtlineare Dielektrika verifiziert und auch die Existenz eines Brewster-Winkels für die reflektierte zweite Harmonische nachgewiesen.

Maruschko und Maschkevich [10] diskutierten die Erzeugung der zweiten Harmonischen in Reflexion und Transmission an einer planparallelen Platte unter Verwendung kinetischer Gleichungen. Kürzlich wurde auch über die Erzeugung der zweiten Harmonischen an der Grenzfläche eines nichtlinearen Mediums mit Inversionssymmetrie berichtet [11].

In der vorliegenden Arbeit wird die Wechselwirkung ebener, monochromatischer Wellen der Form

$$\boldsymbol{E}(\boldsymbol{r}, t) = \boldsymbol{E}(\omega)\, \mathrm{e}^{\frac{i\omega}{c}(\boldsymbol{n}(\omega)\cdot\boldsymbol{r} - c t)} \tag{1}$$

betrachtet. Es wird angenommen, daß eine ebene Welle (1) aus einem optisch linearen, unmagnetischen, nichtabsorbierenden, isotropen Medium auf die ebene Grenzfläche eines sogenannten quadratischen, unmagnetischen, optisch einachsigen und zunächst als nichtabsorbierend vorausgesetzten Kristalls fällt. Der Normalen-Einheitsvektor $\boldsymbol{q}$ zur Grenzfläche zeige in das nichtlineare Medium hinein. Diese Annahmen haben zur Folge, daß der Tensor $\boldsymbol{\varepsilon}$ der linearen dielektrischen Permeabilität im linearen Medium ein reeller Skalar ist und daß dieser Tensor im quadratischen Kristall reell und symmetrisch ist; zwei seiner Eigenwerte sind identisch [12, 13]. Die Bezeichnung quadratischer Kristall bedeutet, daß der erste nichtverschwindende, nichtlineare Term in der Entwicklung der dielektrischen Polarisation $\boldsymbol{P}(\boldsymbol{r}, t)$ nach Potenzen der elektrischen Feldstärke $\boldsymbol{E}(\boldsymbol{r}, t)$ das Aussehen

$$\boldsymbol{P}^{(2)}(\boldsymbol{r}, t) = \boldsymbol{\chi} : \boldsymbol{E}(\boldsymbol{r}, t)\, \boldsymbol{E}(\boldsymbol{r}, t) \tag{2}$$

hat. Die Rechnungen sind deshalb nur auf Kristalle ohne Inversionssymmetrie anwendbar.

Auf Grund der Nichtlinearität (2) entstehen im nichtlinearen Kristall elektromagnetische Wellen der Frequenz $2\,\omega$. Es erweist sich jedoch, um die Grenzbedingungen für die elektrische und die magnetische Feldstärke erfüllen zu können, als notwendig, auch die Existenz einer in das lineare Medium reflektierten Harmonischen anzunehmen. In Abschnitt 2 der Arbeit werden die Verallgemeinerungen der aus der linearen Optik bekannten Fresnelschen Formeln und des Reflexions- und Brechungsgesetzes für diese zweite Harmonische abgeleitet. Die Rechnungen werden in der sogenannten parametrischen Näherung [1, 2] gewonnen, d. h., die Amplituden der Grundwellen werden als räumlich konstant vorausgesetzt. Die für die zweiten Harmonischen abgeleiteten Beziehungen erlauben auch die Bestimmung der Amplituden der Grundwellen. In Abschnitt 3 wird die Verallgemeinerung der in Abschnitt 2 angegebenen Beziehungen für absorbierende Medien diskutiert.

2. Die Verallgemeinerung der Fresnelschen Formeln und des Refraktionsgesetzes

Man steht zunächst vor der Aufgabe, Lösungen der Wellengleichung unter Berücksichtigung der Nichtlinearität (2) zu finden. Diese Lösungen sind der Grenzfläche anzupassen. Aus der Translationssymmetrie des Problems in der Grenzfläche folgt das Reflexions- und Brechungsgesetz für die zweite Harmonische. Die Amplituden der Harmonischen werden aus den Grenzbedingungen berechnet.

2.1 Lösung der Wellengleichung

Im isotropen, linearen Medium gilt für die reflektierte zweite Harmonische

$$\boldsymbol{E}_r(\boldsymbol{r}, t) = \boldsymbol{E}_r(2\,\omega)\, e^{\frac{i\,2\,\omega}{c}(\boldsymbol{n}_r(2\,\omega)\cdot\boldsymbol{r} - c\,t)}\,; \qquad \boldsymbol{n}_r(2\,\omega)\cdot\boldsymbol{E}_r(2\,\omega) = 0\,. \tag{3}$$

Im quadratischen Kristall lautet die Wellengleichung für die elektrische Feldstärke

$$[\nabla\times[\nabla\times\boldsymbol{E}(\boldsymbol{r}, t)]] + \frac{1}{c^2}\frac{\partial^2}{\partial t^2}\,\boldsymbol{\varepsilon}\cdot\boldsymbol{E}(\boldsymbol{r}, t) = -\frac{4\,\pi}{c^2}\frac{\partial^2}{\partial t^2}\boldsymbol{P}^{(2)}(\boldsymbol{r}, t)\,. \tag{4}$$

Die nichtlineare Polarisation (2) setzt sich aus den drei Anteilen (5) zusammen, die den drei möglichen Wechselwirkungen der Grundwellen im optisch einachsigen Kristall entsprechen, d. h. der Wechselwirkung innerhalb der ordentlichen ($\alpha = 1$), innerhalb der außerordentlichen ($\alpha = 2$) und zwischen ordentlicher und außerordentlicher ($\alpha = 3$) Grundwelle:

$$\boldsymbol{P}^{(2)}(\boldsymbol{r}, t) = \sum_{\alpha=1}^{3} \boldsymbol{P}_\alpha^{\mathrm{NL}}(2\,\omega)\, e^{\frac{i\,2\,\omega}{c}(\boldsymbol{n}_\alpha(2\,\omega)\cdot\boldsymbol{r} - c\,t)}\,. \tag{5}$$

$$\alpha = 1: \quad \boldsymbol{n}_1(2\,\omega) = \boldsymbol{n}_o(\omega)\,, \qquad \boldsymbol{P}_1^{\mathrm{NL}}(2\,\omega) = \boldsymbol{\chi}(2\,\omega):\, \boldsymbol{e}_o(\omega)\,\boldsymbol{e}_o(\omega)\,E_o^2(\omega)\,, \tag{6}$$

$$\alpha = 2: \quad \boldsymbol{n}_2(2\,\omega) = \boldsymbol{n}_{ao}(\omega)\,, \qquad \boldsymbol{P}_2^{\mathrm{NL}}(2\,\omega) = \boldsymbol{\chi}(2\,\omega):\, \boldsymbol{e}_{ao}(\omega)\,\boldsymbol{e}_{ao}(\omega)\,E_{ao}^2(\omega)\,, \tag{7}$$

$$\alpha = 3: \quad 2\,\boldsymbol{n}_3(2\,\omega) = \boldsymbol{n}_o(\omega) + \boldsymbol{n}_{ao}(\omega)\,, \qquad \boldsymbol{P}_3^{\mathrm{NL}}(2\,\omega) = \boldsymbol{\chi}(2\,\omega):\, \boldsymbol{e}_o(\omega)\,\boldsymbol{e}_{ao}(\omega)\,E_o(\omega)\,E_{ao}(\omega)\,. \tag{8}$$

In (6) bis (8) sind die $\boldsymbol{n}(\omega)$ die Refraktionsvektoren, während die $\boldsymbol{e}(\omega)$ die Polarisation der Wellen bestimmen. Es ist durchaus nicht günstig, unter $\boldsymbol{e}(\omega)$ stets einen Einheitsvektor zu verstehen. Die Frequenzabhängigkeit der einzelnen Größen geben wir üblicherweise als Argument an.

Da (5) für nichtmodulierte monochromatische Wellen, die hier betrachtet werden, eine gewöhnliche inhomogene Vektor-Differentialgleichung ist, besteht ihre Lösung aus der Summe der allgemeinen Lösung $\boldsymbol{E}_{\mathrm{hom}}(\boldsymbol{r}, t)$ der homogenen Gleichung $(\boldsymbol{P}^{(2)}(\boldsymbol{r}, t) \equiv 0)$ und einer speziellen Lösung der inhomogenen Gleichung $\boldsymbol{E}_{\mathrm{inh}}(\boldsymbol{r}, t)$. Die Lösung der homogenen Gleichung ist

$$\boldsymbol{E}_{\mathrm{hom}}(\boldsymbol{r}, t) = \left(\boldsymbol{e}_o(2\omega)\,E_o(2\,\omega)\, e^{\frac{i\,2\,\omega}{c}\boldsymbol{n}_o(2\,\omega)\cdot\boldsymbol{r}} + \boldsymbol{e}_{ao}(2\,\omega)\,E_{ao}(2\,\omega)\, e^{\frac{i\,2\,\omega}{c}\boldsymbol{n}_{ao}(2\,\omega)\cdot\boldsymbol{r}}\right) e^{-i\,2\,\omega\,t}. \tag{9}$$

Bei der Bestimmung der speziellen Lösung der inhomogenen Gleichung ist zu unterscheiden, ob sich die Polarisationswellen (5) in Synchronisationsrichtungen ausbreiten oder nicht. Bei Ausbreitung in Synchronisationsrichtungen gilt $\boldsymbol{n}_\alpha(2\,\omega) = \boldsymbol{n}_o(2\,\omega)$ bzw. $\boldsymbol{n}_{ao}(2\,\omega)$, d. h., in diesen Richtungen stimmen die

Phasengeschwindigkeiten von Polarisationswelle (und damit Grundwelle) und zweiter Harmonischer überein. Wir werden diesen Fall in einer folgenden Arbeit behandeln. Hier setzen wir voraus, daß die Matrizen $\mathbf{A}_\alpha$ in (11) nichtsingulär sind. Als spezielle Lösung von (4) mit (5) erhält man

$$\boldsymbol{E}_{\text{inh}}(\boldsymbol{r}, t) = \sum_{\alpha=1}^{3} (-1)\, \boldsymbol{p}_\alpha^{\text{NL}}(2\,\omega)\, \mathrm{e}^{\frac{i\,2\,\omega}{c}(\boldsymbol{n}_\alpha(2\,\omega)\cdot\boldsymbol{r} - c\,t)} \tag{10}$$

mit

$$\boldsymbol{p}_\alpha^{\text{NL}}(2\,\omega) = 4\,\pi\, \mathbf{A}_\alpha^{-1} \cdot \boldsymbol{P}_\alpha^{\text{NL}}(2\,\omega) \tag{11}$$

und

$$\mathbf{A}_\alpha = \boldsymbol{n}_\alpha(2\,\omega)\, \boldsymbol{n}_\alpha(2\,\omega) + \boldsymbol{\varepsilon}(2\,\omega) - \boldsymbol{n}_\alpha^2(2\,\omega)\, \boldsymbol{1}\,. \tag{12}$$

$\boldsymbol{n}_\alpha(2\,\omega)\, \boldsymbol{n}_\alpha(2\,\omega)$ in (12) ist das dyadische Produkt von $\boldsymbol{n}_\alpha(2\,\omega)$ mit sich selbst. Der Tensor $\boldsymbol{\varepsilon}$ der dielektrischen Permeabilität kann in optisch einachsigen Kristallen unter Verwendung seines zweifachen Eigenwertes ε_0 und des einfachen Eigenwertes ε_{ao} nach [12] geschrieben werden als[3])

$$\boldsymbol{\varepsilon}(2\,\omega) = \varepsilon_0(2\,\omega)\, \boldsymbol{1} + \big(\varepsilon_{\text{ao}}(2\,\omega) - \varepsilon_0(2\,\omega)\big)\, \boldsymbol{c}\, \boldsymbol{c}\,. \tag{13}$$

In (13) ist $\boldsymbol{c}$ ein reeller Einheitsvektor in Richtung der optischen Achse. $\mathbf{A}_\alpha^{-1}$ in (11) kann mit Hilfe des zu $\mathbf{A}_\alpha$ assoziierten Tensors $\overline{\mathbf{A}}_\alpha$ aus

$$\mathbf{A}_\alpha^{-1}\, |\mathbf{A}_\alpha| = \overline{\mathbf{A}}_\alpha \tag{14}$$

mit

$$|\mathbf{A}_\alpha| = \big(\boldsymbol{n}_\alpha^2(2\,\omega) - \varepsilon_0(2\,\omega)\big)\big(\boldsymbol{\varepsilon}(2\,\omega) : \boldsymbol{n}_\alpha(2\,\omega)\, \boldsymbol{n}_\alpha(2\,\omega) - \varepsilon_0(2\,\omega)\, \varepsilon_{\text{ao}}(2\,\omega)\big) \tag{15}$$

und

$$\overline{\mathbf{A}}_\alpha = (\boldsymbol{n}_\alpha^2 - \varepsilon_c)\, \boldsymbol{n}_\alpha\, \boldsymbol{n}_\alpha - \boldsymbol{1}\, \boldsymbol{\varepsilon} : \boldsymbol{n}_\alpha\, \boldsymbol{n}_\alpha + \boldsymbol{n}_\alpha\, \boldsymbol{n}_\alpha \cdot \boldsymbol{\varepsilon} + \boldsymbol{\varepsilon} \cdot \boldsymbol{n}_\alpha\, \boldsymbol{n}_\alpha + \bar{\boldsymbol{\varepsilon}} \tag{16}$$

berechnet werden. Aus (13) erhält man für die Spur von $\boldsymbol{\varepsilon}$

$$\varepsilon_c(2\,\omega) = 2\,\varepsilon_0(2\,\omega) + \varepsilon_{\text{ao}}(2\,\omega) \tag{17}$$

und für den zu $\boldsymbol{\varepsilon}$ assoziierten Tensor $\bar{\boldsymbol{\varepsilon}}$

$$\bar{\boldsymbol{\varepsilon}}(2\,\omega) = \varepsilon_0(2\,\omega)\, \big(\varepsilon_{\text{ao}}(2\,\omega)\, \boldsymbol{1} - (\varepsilon_{\text{ao}}(2\,\omega) - \varepsilon_0(2\,\omega))\, \boldsymbol{c}\, \boldsymbol{c}\big)\,. \tag{18}$$

2.2 Reflexions- und Brechungsgesetz

Die Lösungen (3), (9) und (10) der Wellengleichung sind der Grenzfläche anzupassen. An der Grenzfläche zweier unmagnetischer Medien lauten die Grenzbedingungen für die elektrische und die magnetische Feldstärke

$$[\boldsymbol{q} \times \boldsymbol{E}_1] = [\boldsymbol{q} \times \boldsymbol{E}_2]\,, \qquad \boldsymbol{H}_1 = \boldsymbol{H}_2\,. \tag{19}$$

Die Erfüllung von (19) mit (3), (9) und (10) führt dazu, daß die transversalen Komponenten der Refraktionsvektoren gleich sein müssen, d. h., es muß gelten

$$[\boldsymbol{q} \times \boldsymbol{n}_r(2\,\omega)] = [\boldsymbol{q} \times \boldsymbol{n}_0(2\,\omega)] = [\boldsymbol{q} \times \boldsymbol{n}_{\text{ao}}(2\,\omega)] = [\boldsymbol{q} \times \boldsymbol{n}_\alpha(2\,\omega)]\,. \tag{20}$$

Durch $\boldsymbol{n}_\alpha(2\,\omega)$ in (20) wird wegen (6) bis (8) die Verbindung zu den Refraktionsvektoren der Grundwellen hergestellt; für diese gilt (siehe z. B. [12])

$$[\boldsymbol{q} \times \boldsymbol{n}_e(\omega)] = [\boldsymbol{q} \times \boldsymbol{n}_r(\omega)] = [\boldsymbol{q} \times \boldsymbol{n}_0(\omega)] = [\boldsymbol{q} \times \boldsymbol{n}_{\text{ao}}(\omega)]\,. \tag{21}$$

---

[3]) Im Gegensatz zu [12] halten wir uns in der Schreibweise von Produkten zwischen Vektoren und Tensoren durchweg an die Richtlinien der IUPAP.

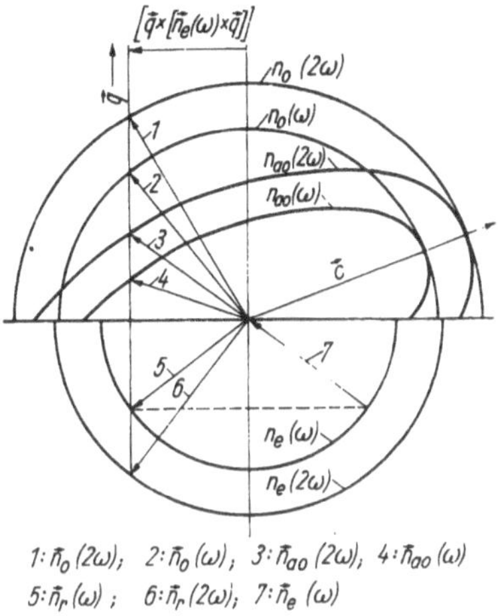

Fig. 1. Lage der Refraktionsvektoren der Grundwellen und der zweiten Harmonischen an der ebenen Grenzfläche isotropes Medium – optisch negativer einachsiger Kristall für den Fall, daß die optische Achse in der Einfallsebene liegt. Nach (20) und (21) liegen alle Refraktionsvektoren stets in der Einfallsebene und haben alle dieselbe zur Oberfläche transversale Komponente $[\boldsymbol{q}\times[\boldsymbol{n}_e(\omega)\times\boldsymbol{q}]]$. Sofern das isotrope Medium keine zeitliche Dispersion aufweist, sind $\boldsymbol{n}_r(2\,\omega)$ und $\boldsymbol{n}_r(\omega)$ parallel

$\boldsymbol{n}_e(\omega)$ in (21) ist der Refraktionsvektor für die aus dem linearen Medium einfallende Grundwelle.

(21) ist die vektorielle Formulierung des Reflexions- und Brechnungsgesetzes für die Grundwellen. Formal stimmt (21) mit dem Reflexions- und Brechungsgesetz an der Grenzfläche zweier isotroper Medien überein. Weil jedoch der Brechungsindex der außerordentlichen Welle $(n_{ao}(\omega))$ von der Ausbreitungsrichtung abhängt, besteht doch ein wesentlicher Unterschied zwischen diesen beiden Fällen. (20) ist die gesuchte Verallgemeinerung für die zweite Harmonische. Aus (20) und (21) folgt wegen (6) bis (8), daß alle Refraktionsvektoren in der Einfallsebene der Grundwelle liegen und daß sich die reflektierte zweite Harmonische nur dann in derselben Richtung wie die reflektierte Grundwelle ausbreitet, wenn das lineare, isotrope Medium keine zeitliche Dispersion zeigt, da

$$n_r(2\,\omega)\sin\big(\boldsymbol{n}_r(2\,\omega),\boldsymbol{q}\big) = n_r(\omega)\sin\big(\boldsymbol{n}_r(\omega),\boldsymbol{q}\big) \tag{22}$$

ist. In Fig. 1 sind die Refraktionsvektoren für die Grundwellen und die zweiten Harmonischen für den Spezialfall gezeichnet, daß der quadratische Kristall optisch negativ einachsig ist und daß die optische Achse in der Einfallsebene liegt.

2.3 Bestimmung der Amplituden

Berechnet man aus (3), (9) und (10) mit Hilfe der Maxwellschen Gleichungen die magnetische Feldstärke, so erhält man aus (19) folgendes Gleichungssystem zur Bestimmung der Amplituden der zweiten Harmonischen:

$$\begin{aligned}[\boldsymbol{q}\times\boldsymbol{E}_r(2\,\omega)] = [\boldsymbol{q}\times\boldsymbol{e}_o(2\,\omega)]\,E_o(2\,\omega) + [\boldsymbol{q}\times\boldsymbol{e}_{ao}(2\,\omega)]\,E_{ao}(2\,\omega) - \\ - \Big[\boldsymbol{q}\times\sum_\alpha \boldsymbol{p}_\alpha^{\mathrm{NL}}(2\,\omega)\Big]\,,\end{aligned} \tag{23}$$

$$\begin{aligned}[\boldsymbol{n}_r(2\,\omega)\times\boldsymbol{E}_r(2\,\omega)] = [\boldsymbol{n}_o(2\,\omega)\times\boldsymbol{e}_o(2\,\omega)]\,E_o(2\,\omega) + [\boldsymbol{n}_{ao}(2\,\omega)\times\boldsymbol{e}_{ao}(2\,\omega)]\,\boldsymbol{E}_{ao}(2\,\omega) \\ - \sum_\alpha [\boldsymbol{n}_\alpha(2\,\omega)\times\boldsymbol{p}_\alpha^{\mathrm{NL}}(2\,\omega)]\,.\end{aligned} \tag{24}$$

Die Polarisationsvektoren $\boldsymbol{e}_o(2\,\omega)$ und $\boldsymbol{e}_{ao}(2\,\omega)$ in (23), (24) können als bekannt angesehen werden, denn sie sind aus

$$\begin{aligned}\boldsymbol{e}_o(2\,\omega) &= [\boldsymbol{n}_o(2\,\omega)\times\boldsymbol{c}] \\ \text{und}\qquad \boldsymbol{e}_{ao}(2\,\omega) &= \big(\varepsilon_o(2\,\omega)\,\boldsymbol{1} - \boldsymbol{n}_{ao}(2\,\omega)\,\boldsymbol{n}_{ao}(2\,\omega)\big)\cdot\boldsymbol{c}\end{aligned} \tag{25}$$

zu berechnen [12], wenn man die Refraktionsvektoren kennt. Die Refraktionsvektoren $\boldsymbol{n}_o(2\,\omega)$ und $\boldsymbol{n}_{ao}(2\,\omega)$ sind jedoch bei bekannter kristallographischer Orientierung wegen (20) und (21) ebenso eindeutig festgelegt wie $\boldsymbol{n}_r(2\,\omega)$. Zur Bestimmung der Refraktionsvektoren wird der Ansatz

$$\boldsymbol{n}_i(2\,\omega) = \boldsymbol{b}(\omega) + \eta_i\,\boldsymbol{q}\,; \qquad i = r, o, ao$$

mit

$$\boldsymbol{b}(\omega) = [\boldsymbol{q}\times[\boldsymbol{n}_e(\omega)\times\boldsymbol{q}]] \tag{26}$$

gemacht. Die zu $\boldsymbol{q}$ parallelen Komponenten errechnet man aus den Gleichungen für die Normalflächen [12]

$$\boldsymbol{n}_o^2(2\,\omega) = \varepsilon_o(2\,\omega)\,, \quad \boldsymbol{\varepsilon}(2\,\omega):\boldsymbol{n}_{ao}(2\,\omega)\,\boldsymbol{n}_{ao}(2\,\omega) = \varepsilon_o(2\,\omega)\,\varepsilon_{ao}(2\,\omega) \tag{27}$$

zu

$$\eta_o(2\,\omega) = \left(\varepsilon_o(2\,\omega) - \boldsymbol{b}^2(\omega)\right)^{1/2}, \tag{28}$$

$$\eta_r(2\,\omega) = -\left(\boldsymbol{n}_e^2(2\,\omega) - \boldsymbol{b}^2(\omega)\right)^{1/2}, \tag{29}$$

$$\eta_{ao}(2\,\omega) = \frac{1}{\varepsilon(2\,\omega):\boldsymbol{q}\,\boldsymbol{q}}\{-\boldsymbol{\varepsilon}(2\,\omega) : \boldsymbol{q}\,\boldsymbol{b}(\omega) + [(\boldsymbol{\varepsilon}(2\,\omega) : \boldsymbol{q}\,\boldsymbol{b}(\omega))^2 - \\ - \boldsymbol{\varepsilon}(2\,\omega) : \boldsymbol{q}\,\boldsymbol{q}\left(\boldsymbol{\varepsilon}(2\,\omega) : \boldsymbol{b}(\omega)\,\boldsymbol{b}(\omega) - \varepsilon_o(2\,\omega)\,\varepsilon_{ao}(2\,\omega)\right)]^{1/2}\}\,. \tag{30}$$

Bei Totalreflexion sind η_o oder η_{ao} oder beide rein imaginär.

In (23), (24) sind deshalb nur $E_o(2\,\omega)$, $E_{ao}(2\,\omega)$ und $\boldsymbol{E}_r(2\,\omega)$ unbekannt, sie können durch geeignete Multiplikation von (23), (24) mit linear unabhängigen Vektoren eliminiert werden. Wählt man als linear unabhängige Vektoren $\boldsymbol{n}_o(2\,\omega)$, $\boldsymbol{n}_{ao}(2\,\omega)$, $\boldsymbol{e}_o(2\,\omega)$ und $\boldsymbol{e}_{ao}(2\,\omega)$, ergibt sich

$$\boldsymbol{E}_r(2\,\omega) = \frac{v_2\,[\boldsymbol{n}_r(2\,\omega)\times\boldsymbol{u}_1] - v_1\,[\boldsymbol{n}_r(2\,\omega)\times\boldsymbol{u}_2]}{\boldsymbol{n}_r(2\,\omega)\cdot[\boldsymbol{u}_1\times\boldsymbol{u}_2]}\,, \tag{31}$$

$$E_o(2\,\omega) = \frac{\boldsymbol{E}_r(2\,\omega)\cdot[\boldsymbol{n}_r(2\,\omega)\times\boldsymbol{n}_{ao}(2\,\omega)] - \sum_\alpha \boldsymbol{n}_{ao}(2\,\omega)\cdot[\boldsymbol{n}_\alpha(2\,\omega)\times\boldsymbol{p}_\alpha^{NL}(2\,\omega)]}{\boldsymbol{e}_o(2\,\omega)\cdot[\boldsymbol{n}_o(2\,\omega)\times\boldsymbol{n}_{ao}(2\,\omega)]}\,, \tag{32}$$

$$E_{ao}(2\,\omega) = \frac{\boldsymbol{E}_r(2\,\omega)\cdot[\boldsymbol{n}_o(2\,\omega)\times\boldsymbol{n}_r(2\,\omega)] + \sum_\alpha \boldsymbol{n}_o(2\,\omega)\cdot[\boldsymbol{n}_\alpha(2\,\omega)\times\boldsymbol{p}_\alpha^{NL}(2\,\omega)]}{\boldsymbol{e}_{ao}(2\,\omega)\cdot[\boldsymbol{n}_o(2\,\omega)\times\boldsymbol{n}_{ao}(2\,\omega)]}\,, \tag{33}$$

mit

$$\left.\begin{aligned} \boldsymbol{u}_1 &= \boldsymbol{e}_o\cdot[\boldsymbol{n}_o\times\boldsymbol{n}_{ao}]\,[\boldsymbol{e}_{ao}\times\boldsymbol{q}] - \boldsymbol{q}\cdot[\boldsymbol{e}_{ao}\times\boldsymbol{e}_o]\,[\boldsymbol{n}_{ao}\times\boldsymbol{n}_r]\,,\\ \boldsymbol{u}_2 &= \boldsymbol{e}_{ao}\cdot[\boldsymbol{n}_o\times\boldsymbol{n}_{ao}]\,[\boldsymbol{e}_o\times\boldsymbol{q}] - \boldsymbol{q}\cdot[\boldsymbol{e}_{ao}\times\boldsymbol{e}_o]\,[\boldsymbol{n}_o\times\boldsymbol{n}_r]\,, \end{aligned}\right\} \tag{34}$$

$$v_1 = [\boldsymbol{e}_{ao}\times\boldsymbol{e}_o]\left(\sum_\alpha [\boldsymbol{n}_\alpha\times\boldsymbol{p}_\alpha^{NL}]\right) : \boldsymbol{n}_{ao}\,\boldsymbol{q} - [\boldsymbol{n}_o\times\boldsymbol{n}_{ao}]\left[\boldsymbol{q}\times\sum_\alpha \boldsymbol{p}_\alpha^{NL}\right] : \boldsymbol{e}_{ao}\,\boldsymbol{e}_o\,, \tag{35}$$

$$v_2 = [\boldsymbol{e}_{ao}\times\boldsymbol{e}_o]\left(\sum_\alpha [\boldsymbol{n}_\alpha\times\boldsymbol{p}_\alpha^{NL}]\right) : \boldsymbol{n}_o\,\boldsymbol{q} - [\boldsymbol{n}_o\times\boldsymbol{n}_{ao}]\left[\boldsymbol{q}\times\sum_\alpha \boldsymbol{p}_\alpha^{NL}\right] : \boldsymbol{e}_o\,\boldsymbol{e}_{ao}\,. \tag{36}$$

Durch (31) bis (36) ist das Problem der Bestimmung der Amplituden der Harmonischen gelöst. Diese Gleichungen sind nicht brauchbar bei senkrechtem Einfall ($\boldsymbol{n}_o(2\,\omega)$ und $\boldsymbol{n}_{ao}(2\,\omega)$ sind parallel) und falls die optische Achse in der Einfallsebene liegt ($\boldsymbol{n}_o(2\,\omega)$, $\boldsymbol{n}_{ao}(2\,\omega)$ und $\boldsymbol{e}_{ao}(2\,\omega)$ sind linear abhängig). Unter Verwendung von (25) errechnet man für senkrechten Einfall, wenn man den bekannten Fall $\boldsymbol{q} = \boldsymbol{c}$ [6] ausschließt,

$$\boldsymbol{E}_r(2\,\omega) = \frac{\sum\limits_\alpha (n_\alpha - n_o)\, \boldsymbol{p}_\alpha^{\mathrm{NL}} \cdot [\boldsymbol{q}\times\boldsymbol{c}]\,[\boldsymbol{q}\times\boldsymbol{c}]}{(n_r + n_o)\,[\boldsymbol{q}\times\boldsymbol{c}]^2} - \frac{\sum\limits_\alpha (n_\alpha - n_{a\,o})\, \boldsymbol{p}_\alpha^{\mathrm{NL}} \cdot [\boldsymbol{q}\times[\boldsymbol{q}\times\boldsymbol{c}]]\,[\boldsymbol{q}\times[\boldsymbol{q}\times\boldsymbol{c}]]}{(n_r + n_{a\,o})\,[\boldsymbol{q}\times\boldsymbol{c}]^2}, \tag{37}$$

$$E_o(2\,\omega) = \frac{\left(\boldsymbol{E}_r(2\,\omega) + \sum\limits_\alpha \boldsymbol{p}_\alpha^{\mathrm{NL}}(2\,\omega)\right) \cdot [\boldsymbol{q}\times\boldsymbol{c}]}{n_o(2\,\omega)\,[\boldsymbol{q}\times\boldsymbol{c}]^2}, \tag{38}$$

$$E_{a\,o}(2\,\omega) = \frac{\left(\boldsymbol{E}_r(2\,\omega) + \sum\limits_\alpha \boldsymbol{p}_\alpha^{\mathrm{NL}}(2\,\omega)\right) \cdot [\boldsymbol{q}\times[\boldsymbol{q}\times\boldsymbol{c}]]}{n_o^2(2\,\omega)\,[\boldsymbol{q}\times\boldsymbol{c}]^2}. \tag{39}$$

In dem Fall, daß $\boldsymbol{c}$ in der Einfallsebene liegt, folgt aus (23), (24)

$$\boldsymbol{E}_r(2\,\omega) = \frac{v_3[\boldsymbol{n}_r(2\,\omega) \times [\boldsymbol{u}_3\times\boldsymbol{n}_r(2\,\omega)]]}{[\boldsymbol{n}_r(2\,\omega)\times\boldsymbol{u}_3]^2} - \frac{\sum\limits_\alpha \boldsymbol{n}_o(2\,\omega)\cdot[\boldsymbol{n}_\alpha(2\,\omega)\times\boldsymbol{p}_\alpha^{\mathrm{NL}}(2\,\omega)]\,[\boldsymbol{n}_o(2\,\omega)\times\boldsymbol{n}_r(2\,\omega)]}{[\boldsymbol{n}_o(2\,\omega)\times\boldsymbol{n}_r(2\,\omega)]^2}, \tag{40}$$

$$E_o(2\,\omega) = \frac{\boldsymbol{n}_o(2\,\omega)\cdot\left[\boldsymbol{q}\times\left(\boldsymbol{E}_r(2\,\omega) + \sum\limits_\alpha \boldsymbol{p}_\alpha^{\mathrm{NL}}(2\,\omega)\right)\right]}{\boldsymbol{n}_o(2\,\omega)\cdot[\boldsymbol{q}\times\boldsymbol{e}_o(2\,\omega)]}, \tag{41}$$

$$E_{a\,o}(2\,\omega) = \frac{\boldsymbol{e}_o(2\,\omega)\cdot\left[\boldsymbol{q}\times\left(\boldsymbol{E}_r(2\,\omega) + \sum\limits_\alpha \boldsymbol{p}_\alpha^{\mathrm{NL}}(2\,\omega)\right)\right]}{\boldsymbol{e}_o(2\,\omega)\cdot[\boldsymbol{q}\times\boldsymbol{e}_{a\,o}(2\,\omega)]}, \tag{42}$$

mit

$$v_3 = \left\{\left(\sum_\alpha [\boldsymbol{n}_\alpha\times\boldsymbol{p}_\alpha^{\mathrm{NL}}]\right)[\boldsymbol{q}\times\boldsymbol{e}_{a\,o}] - \left[\boldsymbol{q}\times\sum_\alpha \boldsymbol{p}_\alpha^{\mathrm{NL}}\right][\boldsymbol{n}_{a\,o}\times\boldsymbol{e}_{a\,o}]\right\} : \boldsymbol{e}_o\,\boldsymbol{e}_o\,, \tag{43}$$

$$\boldsymbol{u}_3 = \{[\boldsymbol{e}_o\times\boldsymbol{q}]\,[\boldsymbol{n}_{a\,o}\times\boldsymbol{e}_{a\,o}] - [\boldsymbol{e}_o\times\boldsymbol{n}_r]\,[\boldsymbol{q}\times\boldsymbol{e}_{a\,o}]\}\cdot\boldsymbol{e}_o\,. \tag{44}$$

Die Beziehungen (31) bis (44) erlauben auch die Bestimmung der Amplituden der Grundwellen, wenn in ihnen $2\,\omega$ durch ω und die spezielle Lösung der inhomogenen Gleichung durch die einfallende Grundwelle ersetzt werden. Sie sind sehr allgemein, denn sie gelten für alle optisch einachsigen Kristalle ohne Inversionssymmetrie, und werden wesentlich durch die konkrete Form des Tensors dritter Stufe χ in (2) bestimmt. Wir werden nun begründen, weshalb sie auch für absorbierende Kristalle gültig sind.

3. Berücksichtigung der Absorption

Es wird nun der Fall näher untersucht, in dem das quadratische Medium absorbierend ist, das lineare Medium sei weiterhin nichtabsorbierend. Dieser Fall ist vor allem deshalb von Interesse, weil man durch Messung der von einem absorbierenden Kristall reflektierten Harmonischen Aussagen gewinnen kann, die man in Transmission nicht erhält.

Die Ausbreitung des Lichtes in linearen, unmagnetischen, absorbierenden, anisotropen Kristallen wurde ausführlich in [12] diskutiert. In absorbierenden Kristallen breiten sich inhomogene, elliptisch polarisierte Wellen aus. Die Phasen- und Amplitudennormalen dieser Wellen stimmen (angesehen vom Fall des senkrechten Einfalls) nicht überein. Bei optisch einachsigen Kristallen ist es durch dieselbe orthogonale Transformation möglich, sowohl den Real- als auch den Imaginärteil der tensoriellen, komplexen, dielektrischen Permeabilität auf Hauptachsen zu transformieren [12], man kann in Analogie zu (13)

$$\boldsymbol{\varepsilon}'(2\,\omega) = \varepsilon_o'(2\,\omega)\,\boldsymbol{1} + \left(\varepsilon_{ao}'(2\,\omega) - \varepsilon_o'(2\,\omega)\right)\boldsymbol{c}\,\boldsymbol{c} \tag{45}$$

schreiben. In (45) ist $\boldsymbol{c}$ wieder ein reeller Einheitsvektor in Richtung der optischen Achse. $\varepsilon_o'(2\,\omega)$ und $\varepsilon_{ao}'(2\,\omega)$ lassen sich mit Hilfe der Eigenwerte der reellen dielektrischen Permeabilität $\boldsymbol{\varepsilon}(2\,\omega)$ und des Leitfähigkeitstensors $\boldsymbol{\sigma}$ ausdrücken zu

$$\left.\begin{aligned} \varepsilon_o'(2\,\omega) &= \varepsilon_o(2\,\omega) - \frac{4\,\pi\,i}{2\,\omega}\,\sigma_o(2\,\omega)\,, \\ \varepsilon_{ao}'(2\,\omega) &= \varepsilon_{ao}(2\,\omega) - \frac{4\,\pi\,i}{2\,\omega}\,\sigma_{ao}(2\,\omega)\,. \end{aligned}\right\} \tag{46}$$

Nach (45) ist $\boldsymbol{\varepsilon}'(2\,\omega)$ symmetrisch. Dies hat zur Folge, daß die Beziehungen (25) und (27) auch für absorbierende Kristalle richtig sind, wenn in ihnen nur die reellen durch die entsprechenden komplexen Größen ersetzt werden. Auch die Refraktionsvektoren (26) sind komplex. Aus (20), (21) folgt, daß die Extinktionsvektoren parallel zu $\boldsymbol{q}$ verlaufen. Auch bei Berücksichtigung der Absorption können η_o und η_{ao} aus (28), (30) berechnet werden. Dabei sind die Vorzeichen der Wurzeln so zu wählen, daß $\operatorname{Im}\eta_o < 0$ und $\operatorname{Im}\eta_{ao} > 0$ sind.

Formal ändert sich deshalb an den in Abschnitt 2 für die Amplituden abgeleiteten Beziehungen nichts, sie sind auch auf absorbierende Kristalle anwendbar. Allerdings ist bei Berücksichtigung der Absorption auch der Tensor χ in (2) komplex.

4. Zusammenfassung

Die angegebenen Beziehungen gelten für die Wechselwirkung ebener Wellen. Sie erlauben die Berechnung der Intensität und der Lage der Refraktionsvektoren der von der ebenen Grenzfläche eines unmagnetischen, quadratischen, optisch einachsigen Kristalls reflektierten zweiten Harmonischen und sind auch auf absorbierende Kristalle anwendbar. Die geometrische Lage der Grenzfläche bezüglich des kristallographischen Systems des nichtlinearen Kristalls ist beliebig. Es wurde vorausgesetzt, daß die ebene Grundwelle aus einem unmagnetischen, nichtabsorbierenden, linearen, isotropen Medium auf die Grenzfläche fällt und daß sich die Polarisationswellen im nichtlinearen Kristall in keiner Synchronisationsrichtung ausbreiten. Der Fall der Ausbreitung in einer Richtung räumlicher Synchronisation wird später diskutiert werden.

Herrn A. Wünsche danke ich für Diskussionen.

Literatur

[1] S. A. Akhmanov und R. V. Khokhlov, Probleme der nichtlinearen Optik, Moskau 1964.
[2] N. Bloembergen, Nonlinear Optics, W. A. Benjamin, New York 1965.
[3] A. M. Bonch-Bruevich und W. A. Khodovoi, Uspekhi fiz. Nauk **85**, 3 (1965).
[4] L. N. Ovander, Uspekhi fiz. Nauk **86**, 3 (1965).
[5] J. A. Armstrong, N. Bloembergen, J. Ducuing und P. S. Pershan, Phys. Rev. **127**, 1918 (1962).

[6] N. Bloembergen und P. S. Pershan, Phys. Rev. **128**, 606 (1962).
[7] J. F. Nye, Physical Properties of Crystals; their Representation by Tensors and Matrices, Oxford 1957, (Russische Übersetzung Moskau 1960, S. 354).
[8] D. A. Kleinman, Phys. Rev. **128**, 1761 (1962).
[9] N. Bloembergen und J. Ducuing, Phys. Letters (Netherlands) **6**, 5 (1963).
J. Ducuing und N. Bloembergen, Phys. Rev. Letters **10**, 474 (1963).
N. Bloembergen, R. K. Chang, J. Ducuing und P. Lallemand, in Proceedings of the 7th International Conference on Physics of Semiconductors, Dunod Cie., Paris 1964 (S. 121).
R. K. Chang und J. Ducuing, Phys. Rev. Letters **15**, 415 (1965).
R. K. Chang und N. Bloembergen, Phys. Rev. **144**, 775 (1966).
[10] I. A. Maruschko und W. S. Maschkevich, Optika i Spektroskopiya **20**, 117 (1966); in Quantenelektronik, Kiev 1966 (S. 34).
[11] S. S. Jha, Phys. Rev. **140**, A 2020 (1965); Phys. Rev. **145**, 145 (1966).
N. Bloembergen, R. K. Chang und C. H. Lee, Phys. Rev. Letters **16**, 986 (1966).
N. Bloembergen und R. K. Chang, in Physics of Quantum Electronics, New York 1966 (S. 80).
[12] F. I. Fjodorov, Optik anisotroper Medien, Minsk 1958.
[13] W. M. Fain und Ya. I. Khanin, Quanten-Radiophysik Moskau 1965.

(Received November 23, 1966)

phys. stat. sol. **19**, 767 (1967)

Subject classification: 18; 13

Service de Physique du Solide et de Résonance Magnétique, Centre d'Etudes Nucléaires de Saclay

Oscillations collectives de spins et ordre à courte distance

Par

J. VILLAIN

Nous étudions les oscillations collectives de spins à partir des deux premières dérivées non nulles de la fonction de corrélation $(\boldsymbol{S_k}(t), \boldsymbol{S_{-k}})$ pour $t = 0$, pour le calcul desquelles nous développons une approximation appropriée. Aux températures inférieures au point de transition on trouve pour la fréquence une formule identique à celle obtenue dans l'approximation des ondes de spin, mais l'approximation est en défaut pour $\boldsymbol{k}$ petit. Il semble que, dans les réseaux tridimensionnels isotropes, les oscillations de spins aient complètement disparu au point de transition. Par contre elles peuvent exister en l'absence d'ordre à grande distance si l'ordre à courte distance est intense, notamment dans un réseau plan ou linéaire.

Collective spin oscillations, of the spin wave type, are investigated using the first two non-vanishing derivatives of the spin correlation function $(\boldsymbol{S}_k(t), \boldsymbol{S}_{-k})$ for $t = 0$. These derivatives are calculated by means of an appropriate approximation. Below the transition point, the formula for the frequency is identical to that obtained from spin wave theory, but our approximation is found to fail for small wave vectors. In three dimensional isotropic lattices, spin oscillations are expected to have disappeared at the transition point, but they can exist (especially in a plane or linear lattice) when there is strong short range order but no long range order.

1. Introduction

Nous considérons un système de spins $\boldsymbol{S_R}$ de module s localisés aux noeuds $\boldsymbol{R}$ d'un réseau de Bravais et liés par un hamiltonien de Heisenberg:

$$\mathcal{H} = -\sum_{\boldsymbol{R}\boldsymbol{R}'} J_{\boldsymbol{R}\boldsymbol{R}'}\, \boldsymbol{S_R}\cdot\boldsymbol{S_{R'}}\,. \tag{1}$$

Des restrictions supplémentaires (spins classiques; interactions entre premiers voisins) seront introduites par la suite.

Nous définissons:

a) l'opérateur de Heisenberg

$$S_{\boldsymbol{R}}(t) = e^{\frac{it}{\hbar}\mathcal{H}}\, S_{\boldsymbol{R}}\, e^{-\frac{it}{\hbar}\mathcal{H}}\,, \tag{2}$$

qui est hermitique pour t réel:

$$[S_{\boldsymbol{R}}(t)]^* = S_{\boldsymbol{R}}(t^*)\,, \tag{3}$$

b) la transformée de Fourier

$$\boldsymbol{S_K}(t) = \frac{1}{\sqrt{N}}\sum_{\boldsymbol{R}} \boldsymbol{S_R}(t)\, e^{-i\boldsymbol{k}\cdot\boldsymbol{R}}\,, \tag{4}$$

c) la fonction de corrélation de 2 opérateurs A et B à température $T = 1/K_B\beta$

$$\langle A\, B\rangle_\beta = \frac{\mathrm{Tr}\, A\, B\, \mathrm{e}^{-\beta \mathcal{H}}}{\mathrm{Tr}\, \mathrm{e}^{-\beta \mathcal{H}}} \tag{5}$$

(nous omettrons le plus souvent l'indice β),

d) la fonction de corrélation intégrée [1, 2, 3]

$$(A, B)_\beta = \int_0^\beta \langle A(-i\,\hbar\,\lambda)\, B\rangle_\beta\, \mathrm{d}\lambda\,, \tag{6}$$

qui vérifie les propriétés suivantes:
Pour des spins classiques

$$(A, B)_\beta = \beta\langle A\, B\rangle_\beta\,, \tag{7}$$

$$(A, B) = (B, A)\,, \tag{8}$$

$$(A, B)^* = (A^*, B^*)\,, \tag{9}$$

$$\big(A(t), B\big) = \big(A, B(-t)\big)\,. \tag{10}$$

En particulier, la fonction de corrélation intégrée

$$\big(S^\alpha_{\boldsymbol{R}}(t), S^\gamma_{\boldsymbol{R}}\big) \qquad (\alpha, \gamma = x, y, z)$$

a la propriété

$$\big(S^\alpha_{\boldsymbol{R}}(t), S^\gamma_{\boldsymbol{R'}}\big) = \big(S^\alpha_{\boldsymbol{R'}}(t), S^\gamma_{\boldsymbol{R}}\big)\,, \tag{11}$$

valable si le réseau est centrosymétrique (restriction qui dans le cas d'un réseau de Bravais avec hamiltonien de Heisenberg, n'écarte que le cas des hélimagnétiques au-dessous du point de Néel). De (11) on déduit

$$\varphi^{\alpha\gamma}_{\boldsymbol{K}}(t) \equiv \big(S^\alpha_{\boldsymbol{K}}(t), S^\gamma_{-\boldsymbol{K}}\big) = \varphi^{\alpha\gamma^*}_{\boldsymbol{K}}(t)\,. \tag{12}$$

La méthode la plus simple pour étudier les oscillations de spins est l'approximation de Bogolyubov-Tyablikov, qui revient [2] à postuler

$$\big(S^+_{\boldsymbol{K}}(t), S^-_{\boldsymbol{K}}\big) \equiv \big[\varphi^{xx}_{\boldsymbol{K}}(t) + \varphi^{yy}_{\boldsymbol{K}}(t)\big] + i\big[\varphi^{yx}_{\boldsymbol{K}}(t) - \varphi^{xy}_{\boldsymbol{K}}(t)\big] = \big(S^+_{\boldsymbol{K}}, S^-_{\boldsymbol{K}}\big)\, \mathrm{e}^{i\omega_{\boldsymbol{K}} t}\,,$$

d'où l'on déduit

$$\omega_{\boldsymbol{K}} = \frac{S}{\big(S^+_{\boldsymbol{K}}, S^-_{\boldsymbol{K}}\big)}\,, \tag{13}$$

où S est l'aimantation par spin, supposée dirigée suivant $0z$. Ainsi, on ne prévoit pas d'oscillations de spins en l'absence d'ordre à grande distance alors qu'on peut penser qu'elles peuvent apparaître en présence d'un ordre à courte distance intense, par exemple dans un réseau plan. En outre certains auteurs pensent en avoir observé au point critique et au-dessus [4]. Ajoutons que la valeur (13) est trop petite et que la méthode s'applique mal aux antiferromagnétiques.

Il est donc intéressant d'étudier, au lieu de $\big(S^+_{\boldsymbol{K}}(t), S^-_{\boldsymbol{K}}\big)$, la fonction

$$\varphi_{\boldsymbol{K}}(t) = \sum_\alpha \varphi^{\alpha\alpha}_{\boldsymbol{K}}(t) = \big(\boldsymbol{S}_{\boldsymbol{K}}(t), \boldsymbol{S}_{-\boldsymbol{K}}\big)\,. \tag{14}$$

2. Relaxation oscillante de $\varphi_K(t)$

En effet, même en l'absence d'ordre à longue distance — donc quand $(S_K^+(t), S_K^-)$ est réelle — il se peut que l'ordre à courte distance soit suffisant pour que les fluctuations de longueur d'onde assez courte voient une aimantation locale dont les fluctuations spatiales sont faibles sur une distance de quelques longueurs d'onde, et dont les fluctuations temporelles sont faibles sur un temps de quelques périodes. La fonction $\varphi_K(t)$ peut alors, *pour t petit*, présenter une relaxation de type oscillatoire, plus ou moins représentable par une équation de la forme

$$\varphi_K(t) = \varphi_K(0) \cos \omega_K t . \tag{15}$$

D'où

$$\omega_K = \sqrt{\frac{-\ddot{\varphi}_K(0)}{\varphi_K(0)}} . \tag{16}$$

$\ddot{\varphi}_K$ désigne la dérivée seconde de φ_K par rapport à t, qui est relativement facile à calculer pour $t = 0$.

Remarquons que, même en présence d'ordre à grande distance, avec une aimantation dirigée suivant $0z$, c'est bien $\varphi_K(t)$ qu'il faut considérer, et en aucun cas $\varphi_K^{xx}(t)$, car l'aimantation locale n'est pas forcément parallèle à $0z$ [5].

La fréquence définie par (16), contrairement à celle définie par (13), ne s'annule jamais; il est donc nécessaire de se fixer un critère pour que la relaxation soit oscillatoire, critère qui devra être vérifié pour que la relation (16) ait un sens. Le critère le plus correct, et le meilleur pour l'expérimentateur serait que la transformée de Fourier

$$\Phi_K(\omega) = \int_{-\infty}^{\infty} \varphi_K(t)\, e^{-i\omega t}\, dt$$

doit avoir 2 maxima en $\omega = \pm\omega_K$ et un minimum en $\omega = 0$. Mais il est plus commode pour nous de dire que $\varphi_K(t)$ a un comportement oscillatoire si elle a au moins un zéro. Ce critère est évidemment moins restrictif que le précédent, mais on peut espérer compenser ce défaut en limitant $\varphi_K(t)$ aux 3 premiers termes de son développement de Taylor (le terme suivant est négatif). Le polynôme

$$\varphi_K(0) + \frac{t^2}{2}\ddot{\varphi}_K(0) + \frac{t^4}{24}\ddddot{\varphi}_K(0) \sim \varphi_K(t)$$

a des zéros réels si

$$\frac{\varphi_K(0)\,\ddddot{\varphi}_K(0)}{\ddot{\varphi}_K^2(0)} < \frac{3}{2} \tag{17}$$

relation qui sera notre critère de relaxation oscillatoire (bien entendu $\ddot{\varphi}_K(0)$ doit aussi être négatif, mais nous verrons que c'est toujours le cas).

3. Calcul de $\ddot{\varphi}_K(0)$ et $\ddddot{\varphi}_K(0)$ en fonction des fonctions de corrélation à 2 et 4 spins

On a d'après (1)

$$\dot{S}_R = \frac{2}{\hbar} \sum J_{RR'}\, S_R \wedge S_{R'} . \tag{18}$$

Cette relation, jointe à la relation générale [1]

$$(\dot{A}, B) = \frac{i}{\hbar} \langle [B, A] \rangle , \tag{19}$$

permet de calculer $\ddot{\varphi}_{\boldsymbol{K}}(0)$ et $\ddddot{\varphi}_{\boldsymbol{K}}(0)$.

D'abord, utilisant (4), (18), et les relations de commutation entre les composantes d'un spin, on a

$$[S^\alpha_{\boldsymbol{R}}, \dot{S}^\gamma_{\boldsymbol{K}}] = \frac{2\,i}{\hbar \sqrt{N}} \sum J_{\boldsymbol{R}\boldsymbol{R}'} (\mathrm{e}^{i\boldsymbol{k}\cdot\boldsymbol{R}} - \mathrm{e}^{i\boldsymbol{k}\cdot\boldsymbol{R}'}) [\delta_{\alpha\gamma} (\boldsymbol{S}_{\boldsymbol{R}} \cdot \boldsymbol{S}_{\boldsymbol{R}'}) - S^\gamma_{\boldsymbol{R}} S^\alpha_{\boldsymbol{R}'}] . \tag{20}$$

D'où d'après [19]

$$\ddot{\varphi}^{\alpha\alpha}_{\boldsymbol{K}}(0) = -\frac{2}{\hbar^2} \sum_{\boldsymbol{R}'} J_{\boldsymbol{R}\boldsymbol{R}'} \langle \boldsymbol{S}_{\boldsymbol{R}} \cdot \boldsymbol{S}_{\boldsymbol{R}'} - S^\alpha_{\boldsymbol{R}} S^\alpha_{\boldsymbol{R}'} \rangle [1 - \mathrm{e}^{i\boldsymbol{k}\cdot(\boldsymbol{R}-\boldsymbol{R}')}]$$

et

$$\ddot{\varphi}_{\boldsymbol{K}}(0) = -\frac{4}{\hbar^2} \sum_{\boldsymbol{R}'} J_{\boldsymbol{R}\boldsymbol{R}'} \langle \boldsymbol{S}_{\boldsymbol{R}} \cdot \boldsymbol{S}_{\boldsymbol{R}'} \rangle [1 - \mathrm{e}^{i\boldsymbol{k}\cdot(\boldsymbol{R}-\boldsymbol{R}')}] . \tag{21}$$

De (4) et (18) on déduit encore

$$\ddot{S}_{\boldsymbol{K}} = \frac{2}{\hbar^2 \sqrt{N}} \sum_{\boldsymbol{R}\boldsymbol{R}'\boldsymbol{R}''} J_{\boldsymbol{R}\boldsymbol{R}'} J_{\boldsymbol{R}'\boldsymbol{R}''} (\mathrm{e}^{i\boldsymbol{k}\cdot\boldsymbol{R}} - \mathrm{e}^{i\boldsymbol{k}\cdot\boldsymbol{R}'}) [\boldsymbol{S}_{\boldsymbol{R}'} (\boldsymbol{S}_{\boldsymbol{R}} \cdot \boldsymbol{S}_{\boldsymbol{R}''}) - \boldsymbol{S}_{\boldsymbol{R}''} (\boldsymbol{S}_{\boldsymbol{R}} \cdot \boldsymbol{S}_{\boldsymbol{R}'})] , \tag{22}$$

relation valable pour des spins *classiques* ($s = \infty$). Nous nous limiterons par la suite, sauf dans la Section 4, à des spins classiques.[1])

Utilisant (19), (22) et (20) on trouve alors

$$\begin{aligned}
\frac{\hbar^4 N}{16} \ddddot{\varphi}_{\boldsymbol{K}}(0) = \\
= \sum_{\boldsymbol{R}\boldsymbol{R}'\boldsymbol{R}''\boldsymbol{R}'''} & J_{\boldsymbol{R}\boldsymbol{R}'} J_{\boldsymbol{R}'\boldsymbol{R}'''} J_{\boldsymbol{R}\boldsymbol{R}''} [(\mathrm{e}^{-i\boldsymbol{k}\cdot\boldsymbol{R}'''} - \mathrm{e}^{-i\boldsymbol{k}\cdot\boldsymbol{R}'}) (\mathrm{e}^{i\boldsymbol{k}\cdot\boldsymbol{R}} - \mathrm{e}^{i\boldsymbol{k}\cdot\boldsymbol{R}''}) \times \\
& \times \langle (\boldsymbol{S}_{\boldsymbol{R}} \cdot \boldsymbol{S}_{\boldsymbol{R}''}) (\boldsymbol{S}_{\boldsymbol{R}'} \cdot \boldsymbol{S}_{\boldsymbol{R}'''}) \rangle + \\
& + J_{\boldsymbol{R}\boldsymbol{R}'} J_{\boldsymbol{R}\boldsymbol{R}''} J_{\boldsymbol{R}\boldsymbol{R}'''} (1 - \mathrm{e}^{i\boldsymbol{k}\cdot(\boldsymbol{R}-\boldsymbol{R}''')}) (1 - \mathrm{e}^{i\boldsymbol{k}\cdot(\boldsymbol{R}'-\boldsymbol{R})}) \times \\
& \times \langle (\boldsymbol{S}_{\boldsymbol{R}} \cdot \boldsymbol{S}_{\boldsymbol{R}'''}) (\boldsymbol{S}_{\boldsymbol{R}'} \cdot \boldsymbol{S}_{\boldsymbol{R}''}) \rangle + \frac{1}{2} J_{\boldsymbol{R}\boldsymbol{R}'} J_{\boldsymbol{R}'\boldsymbol{R}'''} J_{\boldsymbol{R}\boldsymbol{R}''} \times \\
& \times (\mathrm{e}^{-i\boldsymbol{k}\cdot\boldsymbol{R}'''} - \mathrm{e}^{-i\boldsymbol{k}\cdot\boldsymbol{R}'}) (\mathrm{e}^{i\boldsymbol{k}\cdot\boldsymbol{R}} - \mathrm{e}^{i\boldsymbol{k}\cdot\boldsymbol{R}''}) \langle (\boldsymbol{S}_{\boldsymbol{R}} \wedge \boldsymbol{S}_{\boldsymbol{R}'''}) (\boldsymbol{S}_{\boldsymbol{R}'} \wedge \boldsymbol{S}_{\boldsymbol{R}''}) \rangle + \\
& + \frac{1}{2} J_{\boldsymbol{R}\boldsymbol{R}'} J_{\boldsymbol{R}\boldsymbol{R}''} J_{\boldsymbol{R}\boldsymbol{R}'''} \mathrm{e}^{i\boldsymbol{k}\cdot(\boldsymbol{R}'-\boldsymbol{R}''')} \langle (\boldsymbol{S}_{\boldsymbol{R}} \wedge \boldsymbol{S}_{\boldsymbol{R}'}) (\boldsymbol{S}_{\boldsymbol{R}''} \wedge \boldsymbol{S}_{\boldsymbol{R}'''}) \rangle] - \\
& - \frac{1}{2} \sum_{\boldsymbol{R}\boldsymbol{R}'\boldsymbol{R}''} J^2_{\boldsymbol{R}\boldsymbol{R}'} J_{\boldsymbol{R}\boldsymbol{R}''} |1 - \mathrm{e}^{i\boldsymbol{k}\cdot(\boldsymbol{R}-\boldsymbol{R}')}|^2 \langle (\boldsymbol{S}_{\boldsymbol{R}} \wedge \boldsymbol{S}_{\boldsymbol{R}'}) \cdot (\boldsymbol{S}_{\boldsymbol{R}} \wedge \boldsymbol{S}_{\boldsymbol{R}''}) \rangle .
\end{aligned} \tag{23}$$

---

[1]) Toutefois nous donnerons à titre documentaire la formule pour les spins quantiques (s fini):

$$\begin{aligned}
\ddddot{\varphi}^{\alpha\alpha}_{\boldsymbol{K}} = \frac{8}{\hbar^3 N} \sum_{\substack{\boldsymbol{R}\boldsymbol{R}'\boldsymbol{R}''\boldsymbol{R}''' \\ \eta\lambda\mu\nu\zeta}} & J_{\boldsymbol{R}\boldsymbol{R}'} J_{\boldsymbol{R}\boldsymbol{R}'} [\delta_{\alpha\eta} \delta_{\lambda\mu} \mathrm{e}^{i\boldsymbol{k}\cdot\boldsymbol{R}'} + \delta_{\alpha\lambda} \delta_{\eta\mu} (\mathrm{e}^{i\boldsymbol{k}\cdot\boldsymbol{R}} - \mathrm{e}^{i\boldsymbol{k}\cdot\boldsymbol{R}'}) - \\
& - \delta_{\alpha\mu} \delta_{\eta\lambda} \mathrm{e}^{i\boldsymbol{k}\cdot\boldsymbol{R}}] [J_{\boldsymbol{R}''\boldsymbol{R}'''} (\delta_{\alpha\eta} \delta_{\nu\zeta} - \delta_{\alpha\zeta} \delta_{\nu\eta}) (\mathrm{e}^{-i\boldsymbol{k}\cdot\boldsymbol{R}''} - \mathrm{e}^{i\boldsymbol{k}\cdot\boldsymbol{R}'''}) S^\lambda_{\boldsymbol{R}} S^\mu_{\boldsymbol{R}'} S^\zeta_{\boldsymbol{R}''} S^\nu_{\boldsymbol{R}'''} + \\
& + J_{\boldsymbol{R}'\boldsymbol{R}'''} (\delta_{\alpha\mu} \delta_{\nu\zeta} - \delta_{\alpha\zeta} \delta_{\mu\nu}) (\mathrm{e}^{-i\boldsymbol{k}\cdot\boldsymbol{R}'} - \mathrm{e}^{-i\boldsymbol{k}\cdot\boldsymbol{R}'''}) S^\lambda_{\boldsymbol{R}} S^\zeta_{\boldsymbol{R}'} S^\nu_{\boldsymbol{R}'''} S^\eta_{\boldsymbol{R}''} + \\
& + J_{\boldsymbol{R}\boldsymbol{R}'''} (\delta_{\alpha\lambda} \delta_{\nu\zeta} - \delta_{\alpha\zeta} \delta_{\mu\nu}) (\mathrm{e}^{-i\boldsymbol{k}\cdot\boldsymbol{R}} - \mathrm{e}^{-i\boldsymbol{k}\cdot\boldsymbol{R}'''}) S^\zeta_{\boldsymbol{R}} S^\nu_{\boldsymbol{R}'''} S^\mu_{\boldsymbol{R}'} S^\eta_{\boldsymbol{R}''}] .
\end{aligned}$$

4. Fréquence caractéristique (cas d'un ferromagnétique)

La fréquence caractéristique $\omega_{\boldsymbol{K}}$ résulte de (16) et (21) lorsque l'on connaît en outre $\varphi_{\boldsymbol{K}}(0)$, qui peut être calculé dans l'approximation du champ moléculaire [6].

Par exemple, au-dessous du point de Curie T_C et en l'absence de champ, on a, si $0z$ est la direction d'aimantation,

$$\varphi_{\boldsymbol{K}}^{xx}(0) = \varphi_{\boldsymbol{K}}^{yy}(0) = \frac{1}{2\,[\mathcal{J}(0) - \mathcal{J}(\boldsymbol{k})]}\,, \tag{24}$$

$$\varphi_{\boldsymbol{K}}^{zz}(0) = \frac{1}{\dfrac{1}{\varphi_0^{zz}(0)} + 2\,[\mathcal{J}(0) - \mathcal{J}(\boldsymbol{k})]}\,. \tag{25}$$

Il existe 2 cas limites simples:

1. Pour $T \ll T_C$, $\varphi_0^{zz}(0)$ est un infiniment petit d'ordre T et

$$\varphi_{\boldsymbol{K}}(0) = \frac{1}{\mathcal{J}(0) - \mathcal{J}(\boldsymbol{k})}\,, \tag{26}$$

2. pour $T = T_C$, $1/\varphi_0^{zz}(0) = 0$ et

$$\varphi_{\boldsymbol{K}}(0) = \frac{3/2}{\mathcal{J}(0) - \mathcal{J}(\boldsymbol{k})}\,. \tag{27}$$

Pour simplifier les formules qui suivent, nous nous limiterons dans la suite au cas *d'interactions entre premiers voisins*. Il y a alors une seule intégrale d'échange J, et on peut définir un paramètre d'ordre à courte distance

$$\tau = \langle \boldsymbol{S_R} \cdot \boldsymbol{S_{R'}} \rangle \qquad (\boldsymbol{R} \text{ et } \boldsymbol{R}' \text{ voisins})\,.$$

Les équations (16), (21), (26) et (27) donnent alors

$$\omega_{\boldsymbol{K}} = \frac{2}{\hbar}\,[\mathcal{J}(0) - \mathcal{J}(\boldsymbol{k})]\,\sqrt{\tau} \qquad (T \ll T_C)\,, \tag{28}$$

$$\omega_{\boldsymbol{K}} = \frac{2}{\hbar}\,[\mathcal{J}(0) - \mathcal{J}(\boldsymbol{k})]\,\sqrt{\frac{2}{3}\,\tau} \qquad (T = T_C)\,. \tag{29}$$

Notons que ces formules sont valables indifféremment pour des spins classiques ou quantiques.

En tenant compte de $\varphi_{\boldsymbol{K}}^{zz}(0)$, on ne modifierait dans (28) que les termes en k^4. Les principaux enseignements à tirer de la formule (29) sont: a) que la fréquence

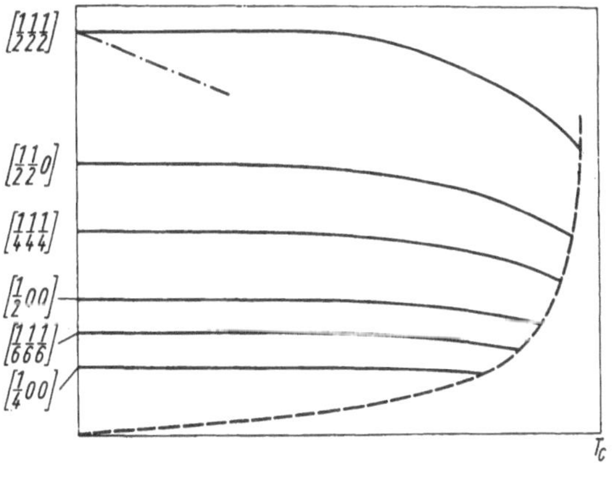

Fig. 1. Fréquences des „ondes de spin" en fonction de la température pour diverses valeurs de $\boldsymbol{k}$ (réseau cubique simple ferromagnétique). On a utilisé les formules (16), (21), (24) et (25) et pris les valeurs de τ données pour $s = 1/2$ par l'approximation du couplage constant [8]. L'approximation du Section 5 prévoit une relaxation non oscillatoire à droite de la courbe pointillée. La courbe en traits discontinus s'applique à une chaine linéaire pour $k = \pi/a$, a étant la distance interatomique, et $T_C = 4\,J\,s^2/3\,K_B$ désignant dans ce cas la température de Curie donnée par l'approximation du champ moléculaire. On voit que les oscillations de spins ont complètement disparu à $T = 0{,}3\,T_C$

ne s'annule pas à T_C, ce qui a déjà été dit, mais nous verrons plus loin que la relaxation cesse généralement à T_C d'être oscillatoire, non par annulation de la fréquence, mais par élargissement des raies. b) La formule (29) ne diffère de la formule (28) que par un facteur de l'ordre de 1, 2, dont il n'y a pas lieu de tenir compte en raison de l'élargissement de la raie au voisinage de T_C.

La relation (28) pourra donc être considérée comme une approximation très satisfaisante (quand elle a un sens, ce que nous examinerons plus loin) dans tout l'intervalle de 0 à T_C.

Notons que la relation (28) est identique à celle que l'on obtient par une approximation d'ondes de spin poussée au second ordre [5].

Notons aussi qu'il ne faut pas se faire trop d'illusions sur la possibilité pour l'expérimentateur de vérifier la formule (28) quand la largeur de raie $\Delta\omega_{\boldsymbol{K}}$ est de l'ordre de $\omega_{\boldsymbol{K}}$. L'expérimentateur mesure, ***habituellement***[2]) le maximum $\omega'_{\boldsymbol{K}}$ de $\Phi_{\boldsymbol{K}}(\omega)$, qui, lui, peut très bien s'annuler. La dépendance de $\omega'_{\boldsymbol{K}}$ vis-à-vis de $\boldsymbol{k}$ et T est notablement plus compliquée que celle de $\omega_{\boldsymbol{K}} \cdot \omega_{\boldsymbol{K}}$ est en tout cas une limite inférieure de $\omega'_{\boldsymbol{K}}$.

5. Une approximation pour le calcul de la dérivée quatrième

(Cas d'interactions entre premiers voisins)

Des fonctions de corrélation à 4 spins de la formule (23), nous allons donner une approximation par des polynomes homogènes du second degré en s^2 et τ, définis de facon que

a) les valeurs à $T = 0$ soient correctes;

b) les valeurs à haute température soient correctes jusqu'aux termes d'ordre 1 en τ (les termes d'ordre 0 se compensent dans (23)).

On pourrait obtenir une approximation plus raffinée (notamment à basse température) en utilisant des polynômes du 3ème degré ou plus.

Les 5 fonctions de corrélation qui apparaissent successivement dans (23) s'écrivent alors

$$\langle(\boldsymbol{S_R}\cdot\boldsymbol{S_{R''}})\,(\boldsymbol{S_{R'}}\cdot\boldsymbol{S_{R'''}})\rangle = \tau^2 + \frac{1}{3}(s^4-\tau^2)\,\delta_{\boldsymbol{R}\boldsymbol{R}'''}\,\delta_{\boldsymbol{R}'\boldsymbol{R}''} + \frac{1}{3}\tau\,(s^2-\tau)\times$$
$$\times\left[\delta_{\boldsymbol{R}''\boldsymbol{R}'''} + \frac{1}{J}\left(J_{\boldsymbol{R}'\boldsymbol{R}''}\,\delta_{\boldsymbol{R}\boldsymbol{R}'''} + J_{\boldsymbol{R}\boldsymbol{R}'''}\,\delta_{\boldsymbol{R}'\boldsymbol{R}''}\right)\right] \qquad \text{[F]}$$

---

[2]) Il faut insister sur la méthode classique [4] qui consiste à orienter un monocristal de façon qu'un noeud $\boldsymbol{\tau}$ du réseau réciproque ait un léger déplacement („missetting") par rapport à la sphère de réflexion, et, à étudier le „volume de diffusion", c'est-à-dire le volume de l'espace réciproque où la diffusion est appréciable. Dans le domaine des ondes de spin ce volume se réduit [4] à une surface sphérique de centre $\boldsymbol{\tau}$ (cas des ferromagnétique à point de Curie élevé).

Dans le domaine paramagnétique [7], le volume de diffusion est défini par une inéquation de la forme

$$\frac{1}{\varkappa^2+q^2}\,\frac{\Lambda\,q^2}{\omega^2+\Lambda^2 q^4} > A\,,$$

où $\varkappa$, Λ, A sont des constantes, $\boldsymbol{q} = \boldsymbol{k} - \boldsymbol{\tau}$, où $\boldsymbol{k}$ est le transfert d'impulsion et le transfert d'énergie ω est défini par

$$\hbar\,\omega = \hbar^2\,|(\boldsymbol{k}_0+\boldsymbol{\tau}+\boldsymbol{q})^2 - k_0^2|\,,$$

où $\boldsymbol{k}_0$ est le vecteur d'onde des neutrons incidents. Le volume de diffusion peut avoir grossièrement la forme d'une boule creuse qui peut faire croire à la présence d'ondes de spin.

ou $\tau^2 + \frac{1}{3}(s^4 - \tau^2)\,\delta_{\boldsymbol{R}\,\boldsymbol{R}'''}\,\delta_{\boldsymbol{R}'\,\boldsymbol{R}''}$; [AF]

$$\langle(\boldsymbol{S_R}\cdot\boldsymbol{S_{R'''}})\,(\boldsymbol{S_{R'}}\cdot\boldsymbol{S_{R''}})\rangle =$$

$$= \tau^2 + \tau\,(s^2 - \tau)\left(\delta_{\boldsymbol{R}'\,\boldsymbol{R}''} + \frac{1}{3}\delta_{\boldsymbol{R}'\,\boldsymbol{R}'''} + \frac{1}{3}\delta_{\boldsymbol{R}''\,\boldsymbol{R}'''} - \frac{2}{3}\delta_{\boldsymbol{R}'\,\boldsymbol{R}'''}\,\delta_{\boldsymbol{R}'\,\boldsymbol{R}''}\right) \quad \text{[F]}$$

ou $$-\tau^2 + \tau\,(s^2 + \tau)\left(\delta_{\boldsymbol{R}'\,\boldsymbol{R}''} + \frac{1}{3}\delta_{\boldsymbol{R}'\,\boldsymbol{R}'''} + \frac{1}{3}\delta_{\boldsymbol{R}''\,\boldsymbol{R}'''} - \frac{2}{3}\delta_{\boldsymbol{R}'\,\boldsymbol{R}'''}\,\delta_{\boldsymbol{R}'\,\boldsymbol{R}''}\right); \text{[AF]}$$

$$\langle(\boldsymbol{S_R}\wedge\boldsymbol{S_{R'''}})\cdot(\boldsymbol{S_{R'}}\wedge\boldsymbol{S_{R''}})\rangle = \frac{2}{3}\tau\,(s^2 - \tau)\,\delta_{\boldsymbol{R}''\,\boldsymbol{R}'''} \quad \text{[F]}$$

ou 0; [AF]

$$\langle(\boldsymbol{S_R}\wedge\boldsymbol{S_{R'}})\cdot(\boldsymbol{S_{R''}}\wedge\boldsymbol{S_{R'''}})\rangle = \frac{2}{3}\tau\,(s^2 - \tau)\,(\delta_{\boldsymbol{R}'\,\boldsymbol{R}'''} - \delta_{\boldsymbol{R}'\,\boldsymbol{R}''}) \quad \text{[F]}$$

ou $$\frac{2}{3}\tau\,(s^2 + \tau)\,(\delta_{\boldsymbol{R}'\,\boldsymbol{R}'''} - \delta_{\boldsymbol{R}'\,\boldsymbol{R}''}); \quad \text{[AF]}$$

$$\langle(\boldsymbol{S_R}\wedge\boldsymbol{S_{R'}})\cdot(\boldsymbol{S_R}\wedge\boldsymbol{S_{R''}})\rangle = \frac{2}{3}(s^4 - \tau^2)\,\delta_{\boldsymbol{R}'\,\boldsymbol{R}''} + \frac{2}{3J}\tau\,(s^2 - \tau)\,J_{\boldsymbol{R}'\,\boldsymbol{R}''} \quad \text{[F]}$$

ou $$\frac{2}{3}(s^4 - \tau^2)\,\delta_{\boldsymbol{R}'\,\boldsymbol{R}''}\,. \quad \text{[AF]}$$

Les valeurs marquées [F] sont relatives à un ferromagnétique avec interactions entre premiers voisins, les valeurs marquées [AF] sont relatives à un antiferromagnétique avec réseau *non compact* et interactions entre premiers voisins. Ces formules, portées dans (23), donnent

$$\frac{\hbar^4}{16}\ddot{\ddot{\varphi}}_{\boldsymbol{K}}(0) = [\mathcal{J}(0) - \mathcal{J}(\boldsymbol{k})]^3\,\tau^2 + J\,\tau\,(s^2 - \tau)\,[\mathcal{J}(0) - \mathcal{J}(\boldsymbol{k})]\left[\frac{7}{3}\mathcal{J}(0) - \mathcal{J}(\boldsymbol{k}) - \frac{4}{3}J\right] -$$
$$- \frac{8}{3}\eta\,J^2\,\tau\,(s^2 - \tau)\,[\mathcal{J}(0) - \mathcal{J}(\boldsymbol{k})] \quad \text{[F]} \quad (30)$$

ou

$$- [\mathcal{J}(0) - \mathcal{J}(\boldsymbol{k})]^2\,[\mathcal{J}(0) + \mathcal{J}(\boldsymbol{k})]\,\tau^2 + J\,\tau\,(s^2 + \tau)\,[\mathcal{J}(0) - \mathcal{J}(\boldsymbol{k})]\times$$
$$\times\left[\frac{7}{3}\mathcal{J}(0) - 3\,\mathcal{J}(\boldsymbol{k}) - \frac{4}{3}J\right]. \quad \text{[AF]}$$

On a posé

$$\mathcal{J}(\boldsymbol{k}) = \sum_{\boldsymbol{R}'} J_{\boldsymbol{R}\,\boldsymbol{R}'}\cos\big(\boldsymbol{k}\cdot(\boldsymbol{R}' - \boldsymbol{R})\big);$$

$$\eta = \begin{cases} 1 \text{ pour un réseau compact tridimensionnel,} \\ 1/2 \text{ pour le réseau hexagonal plan,} \\ 0 \text{ autrement.} \end{cases}$$

Nous avons testé la formule (30) pour une chaîne linéaire de spins classiques. Le second terme est nettement trop petit à basse température, mais son signe (au moins pour k petit) et son ordre de grandeur sont donnés correctement, ce qui est l'essentiel comme on le verra plus loin.

6. Condition de relaxation oscillatoire (cas d'un ferromagnétique)

6.1 $T < T_C$

Utilisant (17), (26), (21), (30) et (28), on trouve que la condition de relaxation oscillatoire est

$$\hbar^2 \omega_K^2 > \frac{32}{3} J^2 (s - \tau)(z - 1 - 2\eta) . \quad (31)$$

On trouve donc que la relaxation n'est jamais oscillatoire au voisinage de $\boldsymbol{k} = 0$. Ce résultrat contraire à celui qu'on obtient à basse température par la théorie de ondes de spin est manifestement erroné. Cela tient à ce que les 2 derniers termes de (30), dont le signe joue un rôle déterminant, deviennent très petits pour T petit et pour $\boldsymbol{k}$ petit. Or ils sont la somme de termes finis sur lesquels on a pu commettre des erreurs finies qui faussent complètement le résultat final. L'approximation de la section 5 n'est donc utilisable que pour T et $\boldsymbol{k}$ assez grands.

6.2 $T = T_C$

L'inégalité (17) est incompatible avec les équations (27), (21) et (17) et nous en concluons qu'avec nos approximations, les oscillations de spins ont complètement disparu à T_C, et même un peu au-dessous.

En fait, si l'approximation du Section 5 nous paraît qualitativement raisonnable, il convient de discuter l'approximation du champ moléculaire.

Dans le cas d'un réseau tridimensionnel où notre théorie s'applique (soit les 3 réseaux cubiques, les réseaux quadratique et orthorhombique centré, le réseau rhomboédrique), l'équation (27) semble être une très bonne approximation — de toute façon bien difficile à améliorer. Il nous paraît douteux qu'un raffinement ultérieur puisse faire apparaître une relaxation oscillatoire nettement marquée au point de Curie et au-dessus. En tout cas c'est surtout dans les réseaux à faible coordination (réseau cubique simple ou du type du diamant) qu'il faudra chercher ce phénomène, car c'est dans ce cas que l'écart avec la théorie du champ moléculaire est le plus important, et que l'ordre à courte distance au point de Curie est le plus intense.

Venons-en aux réseaux linéaires ou plans. Lorsque la température est très inférieure au point de Curie prévu par l'approximation du champ moléculaire, soit $2\, z\, J\, s^2/3\, K_B$, on peut, pour $\boldsymbol{k}$ assez grand, appliquer la formule (26) bien qu'il n'y ait pas d'ordre à grande distance. Le lecteur le vérifiera aisément dans le cas de la chaine linéaire de spins classiques, pour laquelle:

$$\varphi_K(0) = \beta\, s^2 \frac{1 - \mathcal{L}^2(2\,\beta\, J\, s^2)}{1 - 2\,\mathcal{L}(2\,\beta\, J\, s^2) \cos k + \mathcal{L}^2(2\,\beta\, J\, s^2)} ,$$

où $\mathcal{L}$ est la fonction de Langevin. Un argument général est le suivant: Pour $\boldsymbol{k}$ assez grand, $\varphi_K(0)$ dépend surtout des valeurs de $(\boldsymbol{S_R}, \boldsymbol{S_{R'}})$ pour $|\boldsymbol{R} - \boldsymbol{R}'|$ *petit*, et ces valeurs ne dépendent que de l'ordre à courte distance et sont à peu près indépendantes de l'ordre à grande distance.

Ainsi, dans le cas d'un réseau plan ou linéaire, les oscillations de spins peuvent exister en l'absence d'ordre à grande distance. Il en sera de même chaque fois qu'il y aura un ordre à courte distance intense.

7. Cas des antiferromagnétiques

Nous nous bornerons, pour simplifier, au cas des réseaux non compacts. L'extension des résultats précédents à un ferromagnétique à 2 sous-réseaux avec interactions entre premiers voisins ne présente aucune difficulté. J et τ sont définis comme précédemment et sont maintenant négatifs (pour un réseau compact, τ n'a pas une valeur unique et cela complique la question).

Les susceptibilités au-dessous du point de Néel T_N dans l'approximation du champ moléculaire sont

$$\varphi_K^{xx}(0) = \varphi_K^{yy}(0) = -\frac{1}{2\,[\mathcal{J}(0) + \mathcal{J}(\boldsymbol{k})]},$$

$$\varphi_K^{zz}(0) = \frac{1}{A(T) - 2\,[\mathcal{J}(0) + \mathcal{J}(\boldsymbol{k})]},$$

où $A(0) = \infty$, $A(T_N) = 0$. Ces formules remplacent (24) et (25). D'après (16) et (21), on voit que (28) est remplacée par

$$\omega_{\boldsymbol{K}} = \frac{2}{\hbar}\sqrt{[\mathcal{J}(0) - \mathcal{J}(\boldsymbol{k})]\,[\mathcal{J}(0) + \mathcal{J}(\boldsymbol{k})]}\,\sqrt{-\tau}\,.$$

D'autre part, utilisant les relations de la Section 5, le critère de relaxation oscillatoire (31) devient, pour $T \ll T_N$,

$$\hbar^2\,\omega_{\boldsymbol{K}}^2 > \frac{32}{3}\,J^2\,(s^2 + \tau)\,(z - 1)\,.$$

La discussion du Section 6 sur la possibilité d'oscillations de spin au point critique et au-dessus reste valable.

8. Cas d'un réseau ferromagnétique cubique simple à $z/2$ dimensions dans la limite des grands z

Cet exemple académique donne un cas limite intéressant. On fait tendre J vers 0 et z vers ∞ de façon que

$$z\,J = \frac{3\,K_B\,T_C}{2\,s^2} \tag{32}$$

reste constant.

La théorie du champ moléculaire donne alors des résultats exacts. Pour le voir, une méthode possible consiste à le vérifier d'abord pour $T \gg T_C$ en écrivant le développement de $(\boldsymbol{S_K}, \boldsymbol{S_{-K}})$ en puissances de β. Il est également utile d'avoir la valeur du paramètre d'ordre à courte distance

$$\tau = \frac{2\,J\,s^4}{3\,K_B\,T} \qquad (T \geqq T_C)\,, \tag{33}$$

qui est un infiniment petit d'ordre z^{-1} d'après (32).

Pour $T < T_C$, nous allons vérifier que, si l'on admet l'approximation du champ moléculaire, l'énergie $E_{\text{fluct}}(T)$ due aux fluctuations est négligeable par rapport à

$$E_{\text{C M}}(T) = \sum J_{\boldsymbol{R}\boldsymbol{R}'} \langle \boldsymbol{S_R} \rangle \langle \boldsymbol{S_{R'}} \rangle = N\,z\,J\,S^2\,.$$

On aura ainsi démontré l'auto-consistance de la théorie du champ moléculaire. Nous poserons $\boldsymbol{S_R} = \boldsymbol{S} + \boldsymbol{\delta S_R}$ et nous appellerons $\boldsymbol{\delta S_R^{\perp}}$ la composante de $\boldsymbol{\delta S_R}$ orthogonale à $\boldsymbol{\sigma}$. On a

$$E_{\text{fluct}}(T) = K_B\, T \sum_{\boldsymbol{R}\boldsymbol{R}'} J_{\boldsymbol{R}\boldsymbol{R}'}(\delta \boldsymbol{S}_{\boldsymbol{R}}, \delta \boldsymbol{S}_{\boldsymbol{R}'}) =$$

$$= \frac{1}{N} K_B\, T \sum_{\boldsymbol{k} \neq 0} \sum_{\boldsymbol{R}\boldsymbol{R}'} J_{\boldsymbol{R}\boldsymbol{R}'}(\boldsymbol{S}_{\boldsymbol{k}}, \boldsymbol{S}_{-\boldsymbol{k}})\, e^{i\boldsymbol{k}\cdot(\boldsymbol{R}-\boldsymbol{R}')} =$$

$$= K_B\, T \sum_{\boldsymbol{k} \neq 0} \mathcal{J}(\boldsymbol{k})\, (\boldsymbol{S}_{\boldsymbol{K}}, \boldsymbol{S}_{-\boldsymbol{K}}) < \frac{3}{2} K_B\, T \sum_{\boldsymbol{K}} \mathcal{J}(\boldsymbol{k})\, (\boldsymbol{S}^{\perp}_{\boldsymbol{K}}, \boldsymbol{S}^{\perp}_{-\boldsymbol{K}}) ,$$

$$E_{\text{fluct}}(T) < \frac{3}{2} K_B\, T \sum_{\boldsymbol{K}} \frac{\mathcal{J}(\boldsymbol{k})}{\mathcal{J}(0) - \mathcal{J}(\boldsymbol{k})} ,$$

$$E_{\text{fluct}}(T) < \frac{T}{T_C} E_{\text{fluct}}(T_C) = \frac{T}{T_C} N\, z\, J\, \tau(T_C) ,$$

qui est un infiniment petit d'ordre $1/z$ d'après (32) et (33).

8.1 Région ferromagnétique

De la même façon, on voit que la partie de τ due aux fluctuations est un infiniment petit d'ordre z, et par suite

$$\tau = S^2 \qquad (T < T_C) .$$

L'inégalité (17) est toujours vérifiée et l'équation (16) devient identique à l'équation (13): La diffusion des ondes de spin est négligeable à toute température $T < T_C$ et l'approximation de Bogolyubov-Tyablikov devient rigoureuse.

8.2 Région paramagnétique

On peut calculer rigoureusement la dérivée quatrième:

$$\frac{\hbar^4}{16} (\ddot{\boldsymbol{S}}_{\boldsymbol{K}}, \ddot{\boldsymbol{S}}_{-\boldsymbol{K}}) = [\mathcal{J}(0) - \mathcal{J}(\boldsymbol{k})]^3\, \tau^2 - \frac{4}{3} \mathcal{J}(0)\, [\mathcal{J}(0) - \mathcal{J}(\boldsymbol{k})]^2\, \tau^2 +$$

$$+ s^2\, \tau\, [\mathcal{J}(0) - \mathcal{J}(\boldsymbol{k})] \left[\frac{7}{3} \mathcal{J}(0) - \mathcal{J}(\boldsymbol{k})\right] . \tag{34}$$

D'autre part

$$(\boldsymbol{S}_{\boldsymbol{K}}, \boldsymbol{S}_{-\boldsymbol{K}}) = \frac{3}{2} \frac{1}{\frac{T}{T_C} \mathcal{J}(0) - \mathcal{J}(\boldsymbol{k})} . \tag{35}$$

Les équations (35), (21), (34) et (33) sont incompatibles avec (17): La relaxation n'est jamais oscillatoire.

Bibliographie

[1] R. KUBO, dans Lectures in Theoretical Physics, Vol. 1, Ed. W. BRITTIN et L. DONHAM, Interscience Publishers, 1959 (p. 120).

[2] H. MORI et K. KAWASAKI, Progr. theor. Phys. **27**, 529 (1962).

[3] J. VILLAIN, J. Phys. Radium **25**, 618 (1964).

[4] B. JACROT et T. RISTE, dans Thermal Neutron Scattering, Ed. P. EGELSTAFF, Academic Press 1965 (p. 251).

[5] F. KEFFER et R. LOUDON, J. appl. Phys. **32**, 25 (1961).

[6] P. G. DE GENNES, Thèse, Rapport C.E.A. n° 925 (1959).

[7] L. VAN HOVE, Phys. Rev. **95**, 1374 (1954).

[8] P. W. KASTELEIJN et J. VAN KRANENDONK, Physica **22**, 317 (1956).

(Received October 22, 1966)

phys. stat. sol. 19, 777 (1967)
Subject classification: 20.2; 13.4

Department of Applied Mathematics and Mathematical Physics, University College, Cardiff

The Condition for Negative Absorption in Semiconductors

By
P. T. LANDSBERG

For a given photon energy $h\nu$ the condition that the stimulated emission rate exceeds the absorption rate is $F_c - F_v > h\nu$ when the F's are quasi-Fermi levels for conduction and valence bands. This holds also if phonons are involved. The condition is generalised here to the case where several electrons and phonons are involved and localised or band states play a part. The condition becomes: The drop in the quasi-Gibbs free energy of the electron system as a result of one radiative transition must exceed $h\nu$.

Für eine gegebene Photonenenergie $h\nu$ ist die Bedingung, daß die stimulierte Emissionsrate die Absorptionsrate übertrifft $F_c - F_v > h\nu$, wobei F_c und F_v die Quasiferminiveaus für das Leitungs- bzw. Valenzband sind. Dies bleibt auch bei Einbeziehung von Phononen gültig. Die Bedingung wird in der vorliegenden Arbeit für den Fall verallgemeinert, daß verschiedene Elektronen und Phononen beteiligt sind und daß lokalisierte Zustände oder Zustände im Band eine Rolle spielen. Die Bedingung lautet dann: das Absinken der Quasi-Gibbs'schen Freien Energie des Elektronensystems als Ergebnis eines strahlenden Übergangs muß $h\nu$ übersteigen.

In this note a general form of the condition for negative absorption in a semiconductor is given. For this purpose suppose that the electron states can be divided into groups, such that a quasi-Fermi level can be associated with each group. These levels will be denoted by F_l $(l = 1, 3, 5, \ldots)$ or F_g $(g = 2, 4, 6, \ldots)$ depending on whether in the radiative transition studied these groups lose or gain an electron. If $h\nu$ is the energy of the photon emitted in this transition, then in the simplest case the condition for negative absorption takes the form [1]

$$F_1 - F_2 > h\nu \,. \tag{1}$$

This result also holds for indirect transitions in which a phonon is involved. This relation is most useful if $h\nu$ is the dominant photon energy.

Recently the condition for negative absorption has been given for the interesting case when an electron collision takes up a part a of the energy which would otherwise have gone into the emitted photon (see Fig. 1). The condition has, in this case, been stated in the form [2]

$$n\,p > N_c\,P_v \exp\left(-\frac{a}{kT}\right). \tag{2}$$

Here N_c, P_v are effective densities of states in the conduction and valence bands. It was pointed out [3] that this condition is for non-degenerate bands again equivalent to equation (1). A general argument is, therefore, called for which will give an indication of the range of validity of equation (1). This is attempted in the following lines.

The electron states which lose an electron as a result of the radiative transition considered will be labelled by odd integers $l = 1, 3, 5, \ldots$. States which gain an electron will be labelled by even integers $g = 2, 4, 6, \ldots$. The expression for the radiative transition rate between a pair of levels then involves the

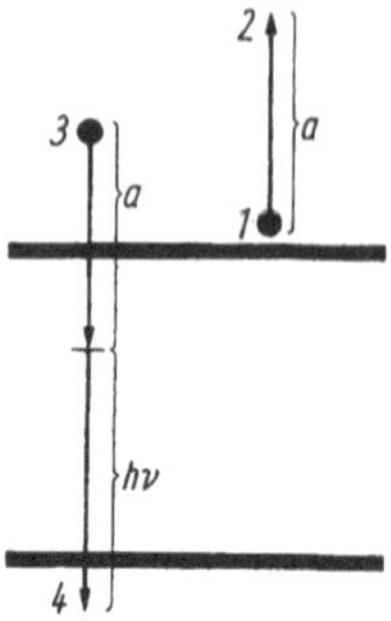

Fig. 1. Illustration of energies introduced in the text. The energies marked a are equal energies

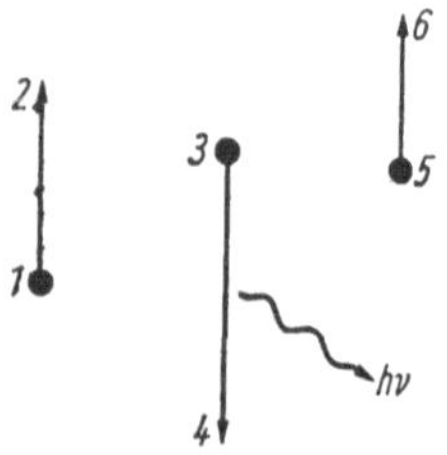

Fig. 2. Electronic transitions associated with a generalised radiative transition in which phonon assistance and electron collisions can play a part

probabilities of finding a state l in a condition to lose an electron, and another state in a condition to be converted into a state g under consideration by capturing an electron. These probabilities will be denoted by p_l and q_g respectively. For states in a band the separate labels are not needed, since the rates of emission and absorption are proportional to

$$p_l\, q_g = p_l\,(1 - p_g)\,, \qquad p_g\, q_l = p_g\,(1 - p_l)\,.$$

For localised states the situation is more complicated, and $p_i + q_i = 1$ $(i = l$ or $g)$ is not necessarily true. If one of the states is localised one needs p_l and q_g for the emission rate, while the competing absorption rate is proportional to $p_g\, q_l$. The expressions are given in Table 1, assuming a state β of an $(r-1)$-electron centre and a state α of an r-electron centre to be involved. The partition function of an r-electron centre is denoted by Z_r and its quasi-Gibbs free energy is denoted by

$$F^{(1)} + F^{(2)} + \cdots + F^{(r)} \text{ instead of } r\, F^{(r)}\,.$$

This is merely a matter of convention, which turns out to be convenient and is more fully discussed in [4]. Normally one needs only one state l and one state g. In Fig. 1, $l = 1$ or 3 and $g = 2$ or 4. In Fig. 2 more values of l and g are needed.

In addition phonons of modes a $(= 1, 3, 5, \ldots)$ of energies ε_a may be absorbed during the radiative process, while phonons of modes e $(= 2, 4, \ldots)$ of energies ε_e may be emitted. The numbers of phonons in mode a or e will be denoted by n_a and n_e. Cancelling matrix elements for the radiative and the inverse absorptive transition, consider now

$$R \equiv \frac{\text{rate of stimulated emission of photons of energy } h\,\nu}{\text{rate of absorption of photons } h\,\nu \text{ by the inverse process}}\,.$$

For negative absorption we need $R > 1$.

With this notation, from phonon statistics, and from Table 1,

$$R = \prod_a \frac{n_a}{n_a + 1} \prod_e \frac{n_e + 1}{n_a} \prod_l \frac{p_l}{q_l} \prod_g \frac{q_g}{p_g}\,, \tag{3}$$

$$R = \exp\left\{\frac{1}{kT}\left[\sum_e \varepsilon_e - \sum_a \varepsilon_a + \sum_l (F_l - E_l) - \sum_g (F_g - E_g)\right]\right\}, \tag{4}$$

where all electron states have been assumed to be in bands. Energy balance yields

$$\sum_a \varepsilon_a + \sum_l E_l = \sum_e \varepsilon_e + \sum_g E_g + h\,\nu\,. \tag{5}$$

Table 1

The form of the probabilities

$p_l\ (l = 1, 2, \ldots)$	$q_l\ (l = 1, 3, \ldots)$	$p_g\ (g = 2, 4, \ldots)$	$q_g (g = 2, 4, \ldots)$
If l is a state in a band:		If g is a state in a band:	
$\dfrac{1}{1 + \exp\left\{\dfrac{1}{kT}[E_l - F_l]\right\}}$	$1 - p_l$	$\dfrac{1}{1 + \exp\left\{\dfrac{1}{kT}[E_g - F_g]\right\}}$	$1 - p_g$
If l is a localised state:		If g is a localised state:	
$\dfrac{\exp\left\{\dfrac{1}{kT}[F^{(1)} + \cdots + F^{(r)} - E(r, \alpha)]\right\}}{D}$	$\dfrac{\exp\left\{\dfrac{1}{kT}[F^{(1)} + \cdots + F^{(r-1)} - E(r-1, \beta)]\right\}}{D}$	as p_l on the left	as q_l on the left

$$D \equiv Z_0 + e^{\frac{F^{(1)}}{kT}} Z_1 + \cdots + e^{\frac{1}{kT}[F^{(1)} + F^{(2)} + \ldots + F^{(M)}]} Z_M$$

M is the number of electrons a centre is capable of capturing

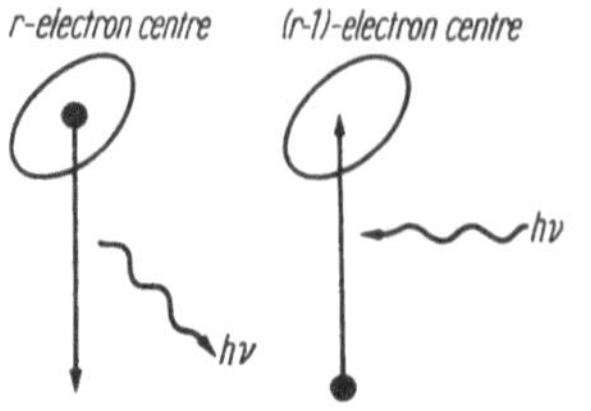

Fig. 3. The part of the emission process of Fig. 2 in which an electron transition has an *initial* state which is localised. The corresponding part of the absorption process is also shown

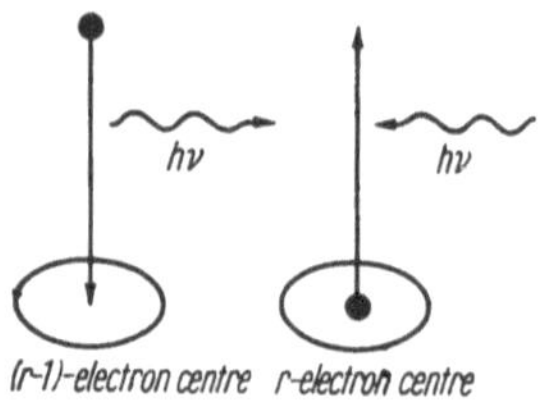

Fig. 4. The part of the emission process of Fig. 2 in which an electron transition has a *final* state which is localised. The corresponding part of the absorption process is also shown

Hence

$$R = \exp\left\{\frac{1}{kT}\left[\sum_l F_l - \sum_g F_g - h\nu\right]\right\}.$$

For negative absorption therefore

$$\sum_l F_l - \sum_g F_g > h\nu\,. \tag{6}$$

As a result of the radiative transition the electron system must therefore decrease its quasi-Gibbs free energy by an amount in excess of $h\nu$.

If one of the states is localised the situation arising is that of Fig. 3 of Fig. 4: An electron transition either begins or ends in a localised state. If the former holds, the replacement in equation (4) is

$$\frac{p_l}{q_l} = \exp\left\{\frac{1}{kT}[F_l - E_l]\right\} \to \exp\left\{\frac{1}{kT}[F^{(r)} - E(r,\alpha) + E(r-1,\beta)]\right\}.$$

In the case of Fig. 4 the replacement needed is

$$\frac{q_g}{p_g} = \exp\left\{\frac{1}{kT}[E_g - F_g]\right\} \to \exp\left\{\frac{1}{kT}[E(r,\alpha) - E(r-1,\beta) - F^{(r)}]\right\}.$$

Thus in either case the replacement has the form

$$E_i \to E(r,\alpha) - E(r-1,\beta); \qquad i = l \text{ or } g\,. \tag{7}$$

The replacement (7) is also needed in the energy balance (5). Thus the condition (6) remains unaffected. This is, therefore, a general condition for negative absorption to be associated with a radiative transition, possibly accompanied by single or multiple electron collisions, and absorption or emission of phonon(s). Momentum conservation has no explicit effect on condition (6); also electron and phonon energies do not occur by virtue of (5).

References

[1] M. G. A. BERNARD and B. DURAFFOURG, phys. stat. sol. 1, 699 (1961).
[2] A. A. GRINBERG, A. A. ROGACHEV, and S. M. RYVKIN, Fiz. tverd. Tela 7, 2206 (1965); Soviet Phys. – Solid State Phys. 7, 1774 (1966). Also discussed in P. T. LANDSBERG, phys. stat. sol. 15, 623 (1966).
[3] P. T. LANDSBERG, Internat. Conf. Physics of Semiconductors, Tokyo 1966, discussion on paper I–12a-3, to be published.
[4] P. T. LANDSBERG, in Festkörperprobleme, Ed. O. MADELUNG Vol. 6, Vieweg, Braunschweig, 1967 (to be published).

(Received November 18, 1966)

phys. stat. sol. 19, 781 (1967)
Subject classification: 6; 22.9

Institut für Neutronenphysik der Kernforschungsanlage Jülich

Lattice Dynamics of Solid Adamantane

By

R. STOCKMEYER and H. H. STILLER

The frequency distributions of the translational and rotational motions of H-atoms in $C_{10}H_{16}$ and their corresponding amplitudes are determined by means of neutron time-of-flight spectroscopy for several temperatures. An estimate of the frequency distribution for the low temperature phase with the statement of James and Keenan (octopole-octopole interactions) leads to an agreement with the experimental results.

Die Frequenzverteilungen von Schwerpunkts- und Rotationsbewegungen und die Amplituden der H-Atome in $C_{10}H_{16}$ werden neutronenspektroskopisch bei mehreren Temperaturen gemessen. Eine Abschätzung der Frequenzverteilung für die Tieftemperaturphase mit dem Wechselwirkungsansatz von James und Keenan (Oktopol–Oktopol-Wechselwirkung) führt zur Übereinstimmung mit den experimentellen Ergebnissen.

Owing to its unusual physical properties adamantane ($C_{10}H_{16}$) is particularly interesting for a study of intermolecular forces in the solid state. It exhibits a specific heat anomaly at 208 °K [1] which has been interpreted as resulting from reorientation of the molecules within the lattice (see also [2], [3], and [4]). NMR data [5] have revealed a sharp decrease of the line width between 143 and 163 °K; this has been interpreted as indicating the onset of "free" rotations.[1])

Experimentally the intermolecular forces can be studied most directly by inelastic scattering of slow neutrons [6]. The differential incoherent scattering cross section[2]) for a cubic molecular crystal has been derived by Hahn [7] in the harmonic approximation:

$$\frac{d^2\sigma}{d\Omega\, d\omega} = \frac{\sigma_{\text{inc}}}{4\pi}\frac{k'}{k_0}\, e^{-2W}\, \frac{\hbar K^2}{2 m_{\text{H}}\, \omega}\left[\exp\frac{\hbar\omega}{k_{\text{B}} T} - 1\right]^{-1} \cdot V(\omega)\,. \tag{1}$$

σ_{inc}	incoherent scattering cross section,
$\hbar\omega$	energy transfer,
$\hbar \boldsymbol{K} = \hbar(\boldsymbol{k}_0 - \boldsymbol{k}')$	momentum transfer,
$\exp(-2W)$	Debye-Waller factor,
$k_{\text{B}} T$	thermal energy,
m_{H}	proton mass.

The frequency distribution $V(\omega)$ can be split into the frequency spectrum of the vibrations of molecular centres of masses, $g_{\text{vibr}}(\omega)$, and the spectrum of librations, $g_{\text{libr}}(\omega)$, if the molecule has the symmetry of the lattice formed by the centres of masses [8]:

$$V(\omega) = 3.92\,[g_{\text{vibr}}(\omega) + 1.04\, g_{\text{libr}}(\omega)]\,. \tag{2}$$

[1]) By "free" we mean rotations of 360°.

[2]) In the case of adamantane the neutrons are scattered predominantly by the protons; this scattering is almost entirely incoherent.

The numerical factors in (2) have been calculated with the molecular data given in [2].

The measurements were done at the FRJ-2 reactor in Jülich with a time-of-flight spectrometer for several sample temperatures between 133 and 300 °K. Neutrons of 4 Å wavelength were selected from the reactor beam by a Pb single crystal. The monochromatic beam was pulsed at a repetition rate of 145 rps with a pulse length of 20 μs. The spectra of scattered neutrons were measured at scattering angles of 15° and 60° with flight pathes of 3.12 m length. Fig. 1 shows typical time-of-flight spectra obtained at

a) 133 °K (below the transition point observed in NMR measurements),
b) 198 °K (below the transition point observed in specific heat measurements),
c) 293 °K (above the latter transition point).

The peak positions in all spectra did not exhibit any K-dependence, they are hence not due to interference effects, but represent frequency distributions in the samples. The large peak at channel number 400 represents elastically scattered neutrons. The half-width of this peak (4 channels of 8 μs width for 15° scattering) cannot be seen on the figures.

The small peak between channels number 145 and 114 (35 to 60 meV energy gain) probably represents the lowest E, F_1, and F_2 internal vibrations of the molecule, unobserved in optical measurements but predicted theoretically [9]. The spectrum between channels number 366 and 235 (1 to 10 meV energy gain) is then due to the deexcitation of rotational motions and of vibrations of the molecular centres of masses (phonons). This spectrum shows more structure below the transition point at 208 °K. Moreover, below the transition point the centre of the spectrum is shifted to higher energies. In contrast, no qualitative difference is observed between the data below and above 143 °K. Fig. 2 shows $V(\omega)$ as calculated with the help of equation (1) for the 198 °K data (15° scattering).

Fig. 3 shows the mean square displacement of the H-atoms as a function of temperature. The values are calculated from the Debye-Waller factor as determined from the ratio of intensities of the elastic peak for two values of K, i.e. assuming that the K-dependence of the intensity is determined by the Debye-Waller factor only. This assumption appears justified in view of the very slight broadening of the elastic peak. The rotation amplitude α is given under the assumption that the displacement is entirely due to rotational motions. The values indicated in Fig. 3 show that up to room temperature the rotations do not become free. This is confirmed by the small width of the elastic peak. Using a simple relation between the half-width ΔE of the elastic peak and the diffusion constant D_r, $\Delta E = 2\,\hbar\,D_r\,K^2$, we get at 293 °K $D_r \approx 7\times10^{-6}\,\mathrm{cm^2\,s^{-1}}$. This rotational diffusion constant is connected with the frequency of librations ω_{libr} and the squared average displacement x^2 by [10]

$$D_r \approx \frac{x^2}{8\,\pi}\,\omega_{\mathrm{libr}}\,\exp\left[-\frac{7}{8}\,\frac{T_b}{T}\right], \tag{3}$$

where T_b is the temperature above which the crystal has no uniform orientational order. Using the measured data we can estimate from (3) a transition point at $T_b \approx 525$ °K with $\hbar\,\omega_{\mathrm{libr}} = 5$ meV.

The phase transition at 208 °K hence seems to be very similar to the lower transition in solid methane [11]. Then the band of frequencies of the librational motions may be estimated in the same way as in this case [12]. Using the

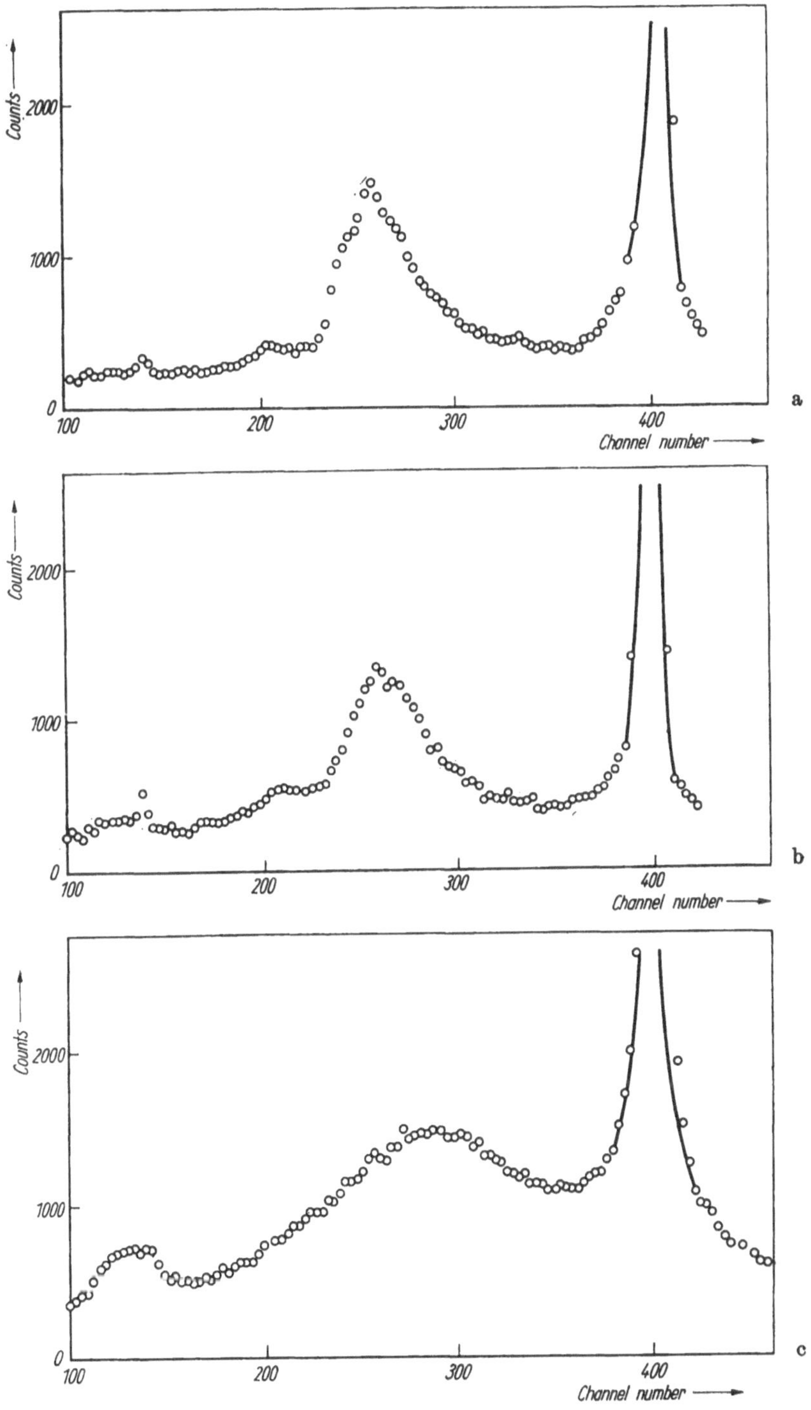

Fig. 1. Time-of-flight spectrum at 133 °K (a), at 198 °K (b), at 293 °K (c)

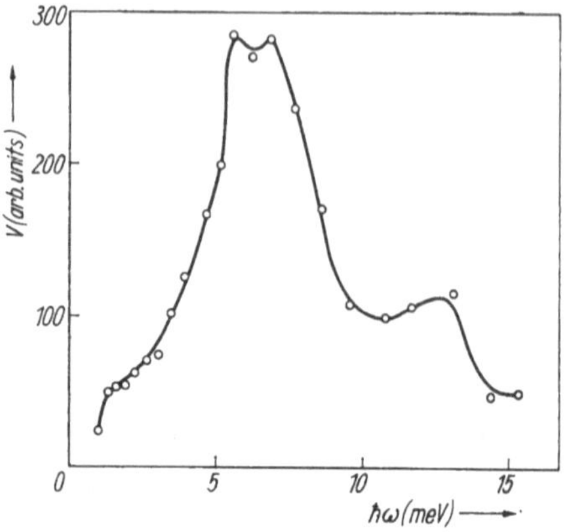

Fig. 2. $V(\omega)$ at 198 °K

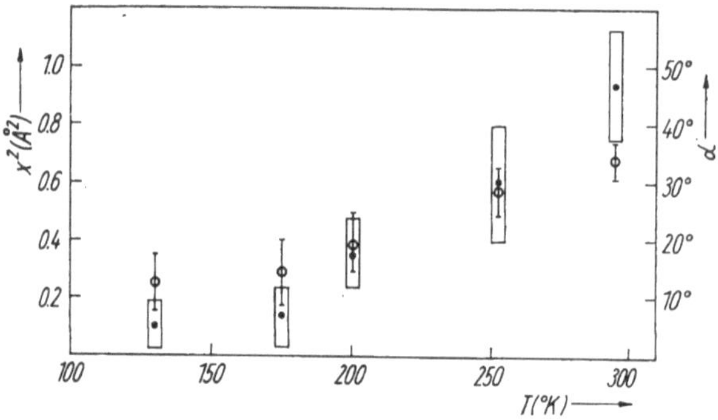

Fig. 3. Average square displacement x^2 and upper limit for the libration amplitude α: $x^2 =$, $\alpha =$

estimated temperature T_b and assuming octopole–octopole interaction we get for the librational motions

$$4 \text{ meV} = 14\, \hbar\, \omega_0 \lessapprox \hbar\, \omega_{\text{libr}} \lessapprox 28\, \hbar\, \omega_0 = 8 \text{ meV}\,,$$

with $\omega_0^2 = \frac{28}{351} \frac{k_B T_b}{\theta}$ (θ moment of inertia).

For the phonon frequencies one obtains from the elastic constants [13] or the Debye constant [14] the main contribution within 5 meV $\lessapprox \hbar\, \omega_{\text{vibr}} \lessapprox$ 8 meV. The observed spectra, indeed, fall into these regions of frequencies. Details shall be discussed later by comparison with dispersion relations, calculated with the help of the dynamic coefficient matrix [14].

Acknowledgements

The authors wish to thank Professor H. Stetter for provision of the samples and Dr. T. Springer for his continuous encouragement and support.

References

[1] SHU-SING CHANG and E. F. WESTRUM, J. phys. Chem. **64**, 1547 (1960).
[2] W. NOWACKI, Helv. chim. Acta **28**, 1233 (1945).
[3] G. GIACOMOLLO and G. ILLUMINATI, Gazz. chim. ital. **75**, 246 (1945).

[4] C. E. NORDMAN and D. L. SCHMITKONS, Acta crystall. **18**, 764 (1965).
[5] D. W. MCCALL and D. C. DOUGLASS, J. chem. Phys. **33**, 777 (1960).
[6] Survey articles in: Inelastic Scattering of Neutrons, IAEA, Wien 1963.
[7] H. HAHN, Inelastic Scattering of Neutrons, IAEA, Wien 1965 (p. 279).
[8] W. BIEM, phys. stat. sol. **3**, 1927 (1963).
[9] R. G. SNYDER and I. H. SCHACHTSCHNEIDER, Spectrochim. Acta (Berlin) **21**, 169 (1965).
[10] H. M. JAMES and T. A. KEENAN, J. chem. Phys. **31**, 12 (1959).
[11] B. DORNER and H. STILLER, phys. stat. sol. **5**, 511 (1964).
[12] B. DORNER, Thesis, TH Aachen 1966.
R. STOCKMEYER, Report JÜL-292-RW (1965).
[13] W. LUDWIG, private communication.
[14] A. I. KITAIGORODSKY, B. D. KORESHKOV, and A. G. KULKIN, Soviet Phys. – Solid State Phys. **7**, 511 (1965).

(Received November 30, 1966)

phys. stat. sol. 19, 787 (1967)

Subject classification: 13; 6; 22.6

Theoretical Physics Division, Atomic Energy Research Establishment, Harwell

Interference Effects in Spin Lattice Relaxation by Two-Phonon Processes

By

A. M. STONEHAM

The usual theory of relaxation by two-phonon processes assumes that the relaxation rate is the sum of a non-resonant Raman contribution and resonant Orbach contributions. This neglects terms which may be regarded as interference between these contributions and which appear when several excited states contribute to the relaxation. These terms affect the temperature dependence of the relaxation rate and will affect the values of energy level splittings obtained from the Orbach terms. Further the contributions of the several excited states to the relaxation rate may themselves tend to cancel. The usual approximation of omitting phonon energies in energy denominators is then invalid, as the cancellation is sensitive to their presence. A comparison of the exact theory and the usual theory is made for $MgO:Co^{2+}$. Agreement with experiment is rather better in the exact treatment, but poor in both cases. Possible reasons for the discrepancy are suggested.

Die gebräuchliche Theorie der Relaxation durch Zwei-Phononenprozesse geht von der Annahme aus, daß die Relaxationsgeschwindigkeit die Summe eines nichtresonanten Ramananteils und eines resonanten Orbachanteiles ist. Hierbei werden Terme vernachlässigt, die als Interferenz zwischen diesen Anteilen angesehen werden können und die auftreten, wenn verschiedene angeregte Zustände zur Relaxation beitragen. Diese Terme beeinflussen die Temperaturabhängigkeit der Relaxationsgeschwindigkeit und die Werte der Aufspaltung der Energieniveaus, die aus den Orbachtermen erhalten werden. Ferner können sich die Anteile zur Relaxationsgeschwindigkeit von den verschiedenen angeregten Zuständen gegenseitig kompensieren. Die gewöhnliche Näherung durch Weglassen der Phononenenergien im Nenner der Energie ist dann nicht zulässig, da die Kompensation empfindlich von deren Anwesenheit abhängt. Ein Vergleich der exakten Theorie und der gewöhnlichen Theorie wird für $MgO:Co^{2+}$ durchgeführt. Die Übereinstimmung mit dem Experiment ist bei der exakten Behandlung etwas besser, aber in beiden Fällen schlecht. Mögliche Gründe für diese Diskrepanz werden angegeben.

1. Introduction

The spin lattice relaxation time for two spin states $|a\rangle$ and $|b\rangle$ measures the rate at which the populations of these states return to their equilibrium values after a disturbance. One may distinguish between the transitions which contribute to the direct process (in which a phonon of energy $\delta = E_b - E_a$ is absorbed) and those which contribute to the two-phonon processes in which a phonon of energy E' is absorbed and one of energy $E' - \delta$ emitted. If E' is equal to the energy of some state $|c\rangle$ above $|a\rangle$ there is a resonant contribution to the transition probability, leading to the Orbach process in addition to the non-resonant Raman process. Both these two-phonon contributions may be obtained from a common approach given by Orbach [1]. In practice, however, the only case discussed is that in which only one excited state (or Kramer's conjugate pair of states) is important. More complicated systems have features which are not obvious from the results for such simple systems; in this article we consider some of these effects.

It is usual to assume that the relaxation rate may be split into a Raman term and a sum of Orbach terms, one for each possible resonance. This neglects terms which may be considered as arising from interference between these contributions. The interference terms are also given by Orbach's approach, but introduce effects analogous to the resonance and anti-resonance phenomena familiar in such fields as the scatter of electrons by atoms [2]. The present case differs from these in two respects. First, the incident phonons have a thermal distribution rather than a single energy. Secondly, the transition probabilities measured experimentally are those between $|a\rangle$ and $|b\rangle$ and not between $|a\rangle$ and $|c\rangle$ or between $|b\rangle$ and $|c\rangle$ for which the resonance occurs.

A second approximation is that in the energy denominators which occur in the transition probability either the phonon energies are negligible [1] or the electronic energy splittings are negligible [3]. The resulting errors may often be tolerated because of the limited accuracy of relaxation measurements. In certain cases this is not so. For example, the relaxation rate when there are two excited levels $|c\rangle$ and $|d\rangle$ may be much smaller than if one of these levels were completely omitted from the calculation. Thus in this case the contributions of $|c\rangle$ and $|d\rangle$ tend to cancel. The degree of cancellation, and hence the relaxation rate, will depend critically on the presence of the phonon energies.

In addition to the approximations mentioned there are, of course, more basic approximations, notably in the use of a Debye density of states and the assumption that the phonon modes are those of the perfect lattice. These approximations will be retained in the present calculations.

The relaxation rate is written explicitly as an integral over the phonon spectrum of the host crystal in Section 2. The complexity of this integral prevents its analytic evaluation in terms of simple functions, although qualitative conclusions can be deduced. For this reason we present numerical calculations for one system, $MgO:Co^{2+}$, in Section 3. These are compared with experiment and with the usual approach, which neglects the interference terms and the phonon energies in the energy denominators. Although the "exact" calculations show slightly improved agreement with experiment the agreement is still poor. Possible reasons for this are outlined; it is almost certain that the use of a Debye density of states for the phonon system is an inadequate approximation.

2. The Transition Probabilities

In this section the results of Orbach's calculations of the two-phonon rate are outlined and our notation established. Throughout we assume that the Debye approximation for the phonon density of states is valid, with a Debye temperature θ_D. To be specific the magnetic ion will be taken to be a Kramer's ion with effective spin 1/2, whose lowest states are $|a\rangle$ and $|b\rangle$. Other systems can be treated in exactly analogous ways.

It is convenient to introduce dimensionless variables in the expressions which follow. Thus if E_j is the energy of an excited state $|j\rangle$ with respect to the ground state we write $\varepsilon_j = E_j/kT$. Similarly we write $x = E'/kT$, where E' is the energy of a phonon, and $x_D = \theta_D/T$. The matrix elements of the spin–lattice interaction, H_{SL}, are written

$$M_j = \frac{\langle b\,|H_{SL}|\,j\rangle\,\langle j|\,H_{SL}\,|a\rangle}{G^2},$$

where G is a coupling coefficient chosen so that the M_j are dimensionless.

The two-phonon relaxation time, τ, is shown in Section 5 of [1] to be

$$\frac{1}{\tau} = \frac{9}{4\pi^3\hbar^7}\left(\frac{G^2}{\varrho\,v^5}\right)\frac{(kT)^9}{E_c^4}\,I_c\,, \tag{1}$$

where ϱ and v are the density and velocity of sound of the host lattice respectively. I_c is the integral

$$I_c = \varepsilon_c^4\int_0^{x_D} dx\,\frac{x^8\,e^{-x}}{(1-e^{-x})^2}\left|\sum_{j\neq a,b}\frac{M_j}{\varepsilon_j^2\,(1-i\,\sigma_j)^2 - x^2}\right|^2, \tag{2}$$

where the excited state, $|c\rangle$, has been singled out quite arbitrarily; in practice a particular choice of $|c\rangle$ is made for algebraic convenience. The σ_j in (2) are introduced to take account of the finite lifetimes $\hbar/2\,\sigma_j\,E_j$, of the states $|j\rangle$ due to the spin–phonon interaction [4].[1]) σ_j is proportional to the factor $(G^2/\varrho\,v^5)$ which enters in (1). Strictly the σ_j should be temperature dependent. This dependence is negligible as the finite lifetimes are almost entirely due to spontaneous phonon emission, at least when E_j exceeds kT.

3. Explicit Calculation of the Relaxation Rate

Here we compare the usual approximate treatment of equations (1) and (2) with their exact evaluation. As mentioned earlier some of the various integrals must be evaluated numerically. To be explicit, therefore, we assume that only two excited levels, $|c\rangle$ and $|d\rangle$, are important and that σ_d is zero because E_d exceeds $k\,\theta_D$. The system $MgO:Co^{2+}$ has these properties; no essential changes are introduced by the complication that for this system $|c\rangle$ is two degenerate Kramers' doublets and $|d\rangle$ is three degenerate Kramers' doublets.

To describe the usual approximate treatment we introduce variables

$$M = M_c\,,\quad m = M_d\,(\varepsilon_c/\varepsilon_d)^2\,,\quad x_0^2 = (1-\sigma_c^2)\,\varepsilon_c^2\,,\quad \text{and}\quad \alpha = 2\,\sigma_c\,\varepsilon_c^2\,.$$

Neglecting the terms in x^2 in the non-resonant denominators in (2) we have

$$\begin{aligned} I_c &= \left(m + \frac{M}{1-\sigma_c^2}\right)^2\int_0^{x_D} dx\cdot x^8\,e^{-x} + M^2\,\varepsilon_c^4\int_0^{x_D} dx\cdot x^8\,e^{-x}\left\{\frac{1}{(x_0^2-x^2)^2+\alpha^2} - \frac{1}{x_0^4}\right\} + \\ &\quad + 2\,m\,M\,\varepsilon_c^2\int_0^{x_D} dx\cdot x^8\,e^{-x}\left\{\frac{x_0^2 - x^2}{(x_0^2-x^2)^2+\alpha^2} - \frac{1}{x_0^2}\right\} \\ &= \left(m + \frac{M}{1-\sigma_c^2}\right)^2 I_R + M^2\,\varepsilon_c^4\,I_0 + 2\,m\,M\,\varepsilon_c^2\,I_{\text{int}}\,. \end{aligned}$$

---

[1]) Spin–spin interactions do not contribute to the σ_j. There is sometimes confusion on this point, so it may be worth emphasizing that in principle the transitions of interest are between many-spin states which diagonalize the sum of the total Zeeman energy, H_Z, and the spin–spin interaction, H_{SS}. Clearly H_{SS} causes no transitions between the stationary states of $H_Z + H_{SS}$, and so does not affect σ_j. In practice an isolated spin is considered, so the matrix elements of the spin–lattice interaction are evaluated between the eigenstates of H_Z alone. This is acceptable if error of order (H_{SS}/H_Z) can be tolerated.

This apparently clumsy form of I_c is exactly analogous to the usual treatment, where I_R gives the Raman contribution, I_0 gives the Orbach contribution, and I_{int} is neglected. The term in I_{int} is, of course, identically zero when $M_d \sim m$ is zero, that is when only one excited state is important. The first terms in the integrands of I_0 and I_{int} are dominated by the integrands near $x \approx x_0$ near $x \approx 0$. The $x \approx 0$ contribution is more naturally considered as a part of I_R, and the second terms in the integrands of I_0 and I_{int} subtract off this contribution, which is then included in I_R.

I_R and I_0 can be evaluated analytically when $x_D \to \infty$, $\alpha \to 0$, and $\sigma \to 0$. I_R becomes 8! and

$$I_0 \approx x_0^{12} e^{-x_0} \int_0^\infty \frac{dx}{(x_0^2 - x^2)^2 + \alpha^2} \approx \frac{\pi}{2\alpha} x_0^{11} e^{-x_0},$$

noting that for small α the resonance at $x \approx x_0$ is very sharp. Finally, neglecting I_{int}, we obtain the usual result

$$\frac{1}{\tau} = \frac{9}{4\pi^3 \hbar^7} \left(\frac{G^2}{\varrho v^5}\right)^2 \frac{(kT)^9}{E_c^4} \left\{(m + M)^2 \cdot 8! + \frac{\pi}{4\sigma_c} M^2 \varepsilon_c^9 e^{-\varepsilon_c}\right\}. \tag{3}$$

This has also been used to calculate the relaxation rate.

I_{int} may be calculated explicitly in the limit $\alpha = 0$, when

$$I_{int} = x_0^8 \left[\frac{1}{2} x_0 \{\overline{Ei(x_0)}\, e^{-x_0} + Ei(-x_0)\, e^{x_0}\} - \left\{1 + \frac{2!}{x_0^2} + \frac{4!}{x_0^4} + \frac{6!}{x_0^6} + \frac{8!}{x_0^8}\right\}\right],$$

in which $Ei(-x)$ and $\overline{Ei(x)}$ are the exponential integrals (Jahnke and Emde, 1945). These integrals cannot be expressed in terms of simpler functions, demonstrating that even in this especially simple cas numerical evaluation is essential. We can shown, however, that the asymptotic behaviour at low temperatures is $I_{int} \approx 10!\,(kT/E_c)^2$, whereas at high temperatures $I_{int} \approx - 1440\,(E_c/kT)^2$. This term shows the typical dispersion-like behaviour expected from other resonance–antiresonance phenomena. In Fig. 1 I_{int} is calculated for $MgO:Co^{2+}$, for which $\sigma = 0.023$, $\varepsilon_c = 438°/T$, and $x_D \approx 946°/T$. The value of σ_c is found from the theory of Foglio [5] using the experimental coupling coefficients of Tucker [6]. A reduction in σ_c would sharpen the peaks in I_{int}. One general conclusion is that E_c, as estimated from (3), will be incorrect as a result of the interference terms. These apparent shifts of energies measured in this fashion are quite distinct from those discussed in [7].

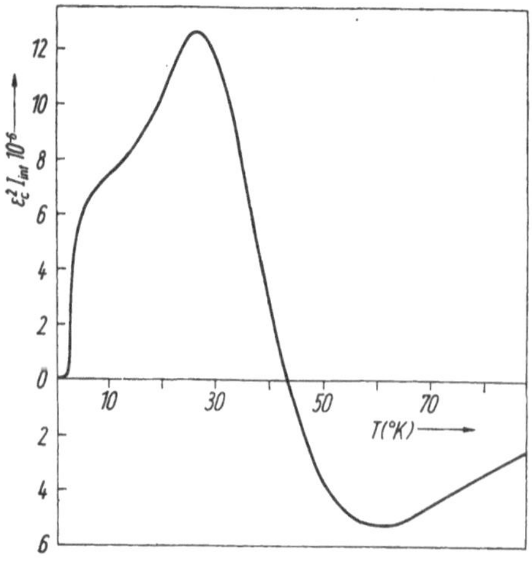

Fig. 1. The interference term $2\,\varepsilon_c^2\, I_{int}\, 10^{-6}$ as a function of temperature. The results are for $MgO:Co^{2+}$. The contribution of the interference terms to the relaxation rate varies with temperatures as T^9 multiplied by the term shown

We will not calculate the corrections from the neglect of the phonon energies in the energy denominators in the same detail. These corrections do not have a simple characteristic temperature dependence like I_{int}, and it is difficult to divide the various contributions to the relaxation rate between the different interference terms unambiguously. We may indicate when the phonon energy corrections are likely to be important by considering a simple example where σ_c and σ_d are both zero — i.e. where there are no resonant contributions. The integrand of (2) is then proportional to the square modulus of

$$\frac{M_c}{\varepsilon_c^2 - x^2} + \frac{M_d}{\varepsilon_d^2 - x^2} = \frac{(M_c\,\varepsilon_d^2 + M_d\,\varepsilon_c^2) - (M_c + M_d)\,x^2}{(\varepsilon_c^2 - x^2)\,(\varepsilon_d^2 - x^2)}\,.$$

The x^2 terms in the denominators on the right-hand side of this equation can usually be neglected, exactly as in the case of one level alone. However, it is not always correct to drop the $(M_c + M_d)\,x^2$ term in the numerator. The corrections are necessary if

$$\frac{(M_c + M_d)\,\overline{x^2}}{M_c\,\varepsilon_d^2 + M_d\,\varepsilon_c^2} = \frac{\overline{x^2}}{\varepsilon_d^2}\,, \qquad \frac{1 + \dfrac{m}{M}\dfrac{\varepsilon_c^2}{\varepsilon_d^2}}{1 + \dfrac{m}{M}} = F\,\frac{\overline{x^2}}{\varepsilon_d^2}$$

is appreciable; as an example we note that $F = -32$ for MgO:Co^{2+}.

In Fig. 2 the relaxation rate for MgO:Co^{2+} calculated from (1) and (2) and from (3) is shown. The experimental results [8, 9] are also given. The "exact" calculations agree with experiment marginally better than the results of the usual approach. In both cases, however, the agreement is disappointing. There are two main reasons for this. First, (1), (2), and (3) all assume a Debye spectrum and use the long-wavelength approximation for the phonon system. Both assumptions should be poor in this case as the phonons of energy E_c, for which $x \approx x_0$, have energies of order $1/2\,k\theta_D$. Secondly, at lower temperatures the two-phonon rate may have been influenced by other magnetic impurities. A symptom of this is the slight sample dependence of the two-phonon rate observed at liquid hydrogen temperatures.

4. Conclusion

In spin lattice relaxation by two-phonon processes two types of interference terms appear when more than one excited spin state is important. These

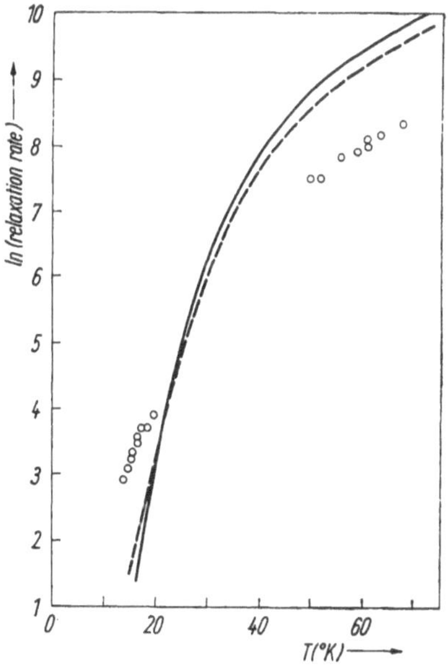

Fig. 2. The two-phonon relaxation rate obtained exactly from equation (1) und (2) (dotted line) and with the usual approximation (equation (3); full line) compared with experiment

are usually neglected, the neglect being justified by the rather low accuracy with which two-phonon relaxation rates are measured. One of the interference terms gives interference between the resonant Orbach and non-resonant Raman contributions; the other comes from the cancellation of the contributions of the several excited states to the relaxation rate. In both cases the relative signs of the matrix elements of the spin–phonon interaction (that is the M_f defined previously) are important. These terms are discussed in general. The MgO:Co^{2+} system is treated numerically, giving a slight improvement in the agreement of theory and experiment.

References

[1] R. Orbach, Proc. Roy. Soc. **A264**, 458 (1961).
[2] P. G. Burke, Adv. Phys. **14**, 521 (1965).
[3] R. Orbach and M. Blume, Phys. Rev. Letters **8**, 478 (1962).
[4] W. Heitler, The Quantum Theory of Radiation, Oxford University Press 1954.
[5] M. E. Foglio, Ph. D. Thesis, University of Bristol 1962.
[6] E. B. Tucker, Phys. Rev. **143**, 264 (1966).
[7] B. A. Young and H. J. Stapleton, Phys. Letters (Netherlands) **21**, 498 (1966).
[8] P. R. Whittlestone, Ph. D. Thesis, University of Bristol 1964.
[9] M. H. L. Pryce, Proc. Roy. Soc. **A283**, 433 (1965).

(Received October 11, 1966)

phys. stat. sol. **19**, 793 (1967)

Subject classification: 10.1; 21.1

Institut für Werkstoffkunde I der Technischen Hochschule Karlsruhe

The Yield Point of α-Copper–Tin Alloys

By

O. VÖHRINGER and E. MACHERAUCH

Dynamics tensile tests ($\dot{\varepsilon} = 1.6 \times 10^{-4}\ \mathrm{s}^{-1}$) are performed on polycrystalline α-Cu-Sn alloys. The effects of temperature (T) and composition (c) on the yield point σ_{St} are investigated. In the temperature range $T \lessapprox 300$ °K the relation $\sigma_{St} = (a_1 - b_1\, T^{1/3}) \sqrt{c}$ holds where a_1 and b_1 are constants. This dependence can be estimated theoretically from the elastic interaction of Sn-atoms with dislocations. Theory and experiment agree quite well.

An vielkristallinen α-Cu-Sn-Legierungen wurden dynamische Zugversuche ($\dot{\varepsilon} = 1{,}6 \times$ $\times 10^{-4}\ \mathrm{s}^{-1}$) zur Ermittlung des Einflusses von Temperatur (T) und Konzentration (c) auf die Streckgrenze σ_{St} durchgeführt. Für $T \lessapprox 300$ °K gilt $\sigma_{St} = (a_1 - b_1\, T^{1/3}) \sqrt{c}$. a_1 und b_1 sind Konstanten. Auf Grund der Wechselwirkung von Zinnatomen mit Versetzungen wird dieser Zusammenhang theoretisch begründet. Zwischen Theorie und Experiment besteht befriedigende Übereinstimmung.

1. Introduction

Our knowledge about the mechanisms of plasticity of f.c.c. polycrystalline pure metals has been extended by a large number of principle investigations during the last years [1 to 5]. Many considerations and results of the plasticity of single crystals have been useful in this respect. However, a lot of typical "phenomena of polycrystals" occurred. Therefore, directed experiments with polycrystals had been necessary. In the meantime, plasticity of single crystals of alloys has made remarkable progress due to experimental work [6, 7] as well as to theoretical considerations [8 to 14]. Thus, systematic experimental studies on polycrystals of alloys seem to be desirable, too. Simple conditions for basic research exist in the primary solid solution of pure f.c.c. metals.

This paper deals with the influence of temperature, composition, and grain size on the yield point of polycrystalline α-copper–tin alloys. With one exception [15, 16] there have not beeen published any newer experimental results on this substitutional type of solid solution.

2. Experimental Procedures

Tensile tests have been done with the alloys specified in Table 1. The α-copper–tin alloys with 0.17 and 0.40 at% were produced by ourselves. The other alloys have been placed to our disposal by the Vereinigte Deutsche Metallwerke, Frankfurt-Heddernheim, Germany.[1]) From the rods cylindrical specimens were fabricated with a diameter of 3.5 mm and a gauge length of 60 mm. After suitable recristallisation heat treatment all alloys showed mean grain sizes of 50 μm. Only in the case of the 1.27 at% tin alloy different grain sizes have been produced.

[1]) The help of Dr. Kirmser and Dr. Boden is thankfully acknowledged.

Table 1
Composition and grain size of the alloys investigated

wt% Cu	at% Sn	Investigated grain sizes (μm)
99.98	0	50
99.68	0.17	50
99.26	0.40	50
98.36	0.88	50
97.65	1.27	20, 50, 250
95.23	2.62	50
94.21	3.12	50
91.38	4.90	50

The tensile tests were done in a Zwick-machine Z 1.3 D. The low temperature tests (78 °K $\leqq T \leqq$ 270 °K, $\Delta T <$ 2 °K) were carried out with a selfmade cryostat. The experiments in the high-temperature region were made either with a Wobser-thermostat ($T <$ 500 °K, $\Delta T <$ 0.5 °K) or a Amsler-furnace (500 °K $< T <$ 800 °K). The force–time scripts of the machine were converted in stress–strain diagrams with the aid of a computer Z 22. The usual corrections have been made.

3. Definitions

The transition region from elastic to plastic deformation is strongly dependent on test temperature and alloy composition. Therefore, the yield point definitions, shown in Fig. 1, were used. The yield point of pure copper was defined as the intersection point of the extended Hookian straight line and the extended stage II straight line (Fig. 1a). This value corresponds with the 0.03% offset yield strength. Specimens with 0.17 at% tin show an intimation of a yield point elongation (Fig. 1b). Pronounced yield point elongations occur with alloys of 0.40 at% $\leqq c \leqq$ 1.27 at% tin (Fig. 1c). Specimens with $c \geqq$ 2.62 at% tin show something like an upper yield point in the yield point elongation area (Fig. 1d). At temperatures of about $T \approx$ 600 °K the specimens behave elastically up to the yield point; then the yield elongation passes from a straight or curved to a zig-zag course (Fig. 1d). At higher temperatures similar stress–strain curves occur as in Fig. 1a.

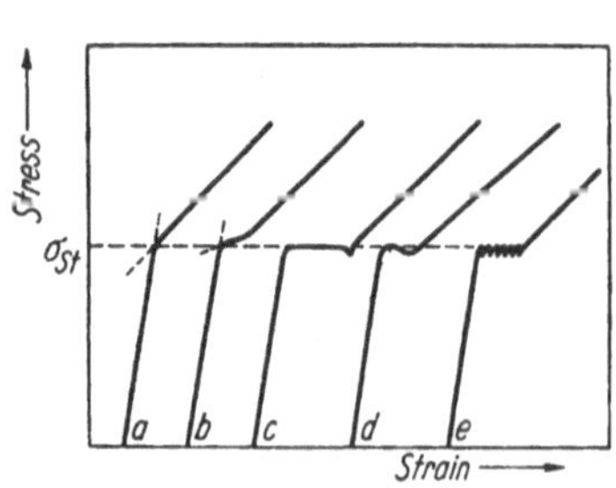

Fig. 1. Definitions of the yield point

4. Experimental Results

4.1 The temperature dependence of the yield points

Fig. 2 shows the observed temperature dependence of the yield points of alloys with different compositions. At low temperatures an addition of 0.17 at% tin already raises the yield point of pure copper markedly (two times at 78 °K). The important feature, to notice from Fig. 2, is that in the low-temperature region all yield points increase with decreasing temperature relatively in the same extent. At temperatures higher than 300 °K only a small temperature dependence appears. Quantitatively the yield point is described by

$$\sigma_{St} = A - B\,T^{1/3}\,. \tag{1}$$

Table 2

Influence of composition and grain size on A and B

c (at%)	d (μm)	A (kp/mm²)	B (kp/mm² °K$^{1/3}$)
0.17	50	9.2	0.90
0.40	50	15.6	1.58
0.88	50	22.2	2.20
1.27	50	25.5	2.55
1.27	20	28.1	2.73
1.27	250	21.8	2.28
2.62	50	38.9	4.05
3.12	50	43.8	4.60
4.90	50	53.1	5.65
3.2 [15]	50	41	4.5

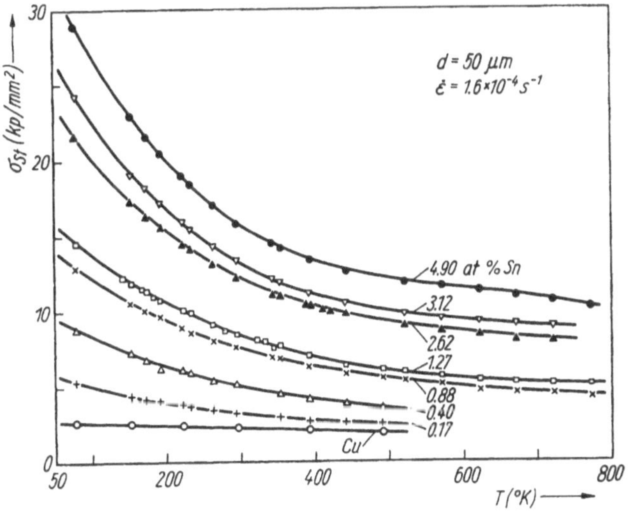

Fig. 2. Temperature dependence of the yield point (d = 50 μm)

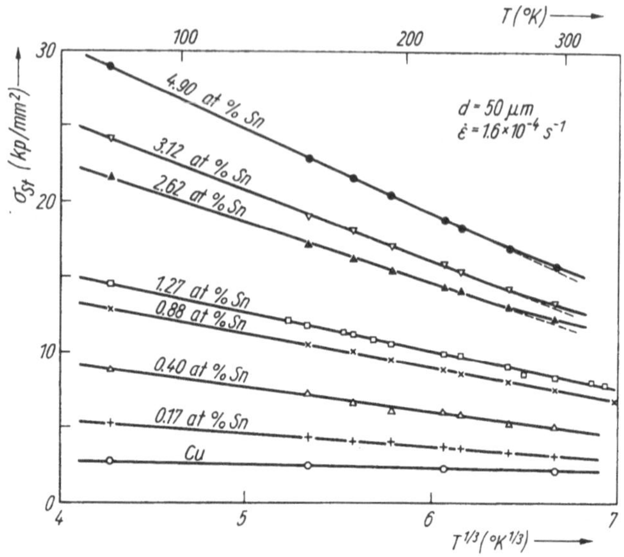

Fig. 3. σ_{St}–$T^{1/3}$ diagrams of α-copper–tin alloys [20]

This equation holds for 0.17 at% $\leqq c \leqq$ 1.27 at% Sn up to $T <$ 300 °C and for 2.62 at% $\leqq c \leqq$ 4.90 at% Sn up to $T <$ 260 °C (see Fig. 3). As can be seen from Table 2 the A- and B- values increase with increasing tin content. A and B decrease with increasing grain size as could be shown with specimens of the 1.27 at% tin alloy. The quantities A and B contain the influence of concentration and temperature on the shear modulus (see also equations (8), (9), and (11)). All the experimental data available point to an influence of smaller than about 10% in the low-temperature region. This is unimportant with respect to the accuracy of the considerations which we will do later on.

A power law of the form

$$\sigma_{St} \sim T^{-n_i}$$

describes the temperature dependence of the yield points in the high-temperature region investigated. It is

$$n_1 = 0.6 \quad \text{for} \quad 300\ °\text{K} < T < 460\ °\text{K}$$

and

$$n_2 = 0.4 \quad \text{for} \quad T > 460\ °\text{K}\,.$$

The relatively small influence of temperature is not only due to the temperature dependence of the shear modulus. Likewise holds

$$\frac{\sigma_{St}}{G(T)} \sim T^{-n_i}$$

with $\quad n_1 = 0.43 \quad \text{for} \quad 300\ °\text{K} < T < 460\ °\text{K}$

and $\quad n_2 = 0.15 \quad \text{for} \quad T > 460\ °\text{K}\,.$

$G(T)$ has been approximated as

$$G(T) = \left[\frac{G(295\ °\text{K})}{E(295\ °\text{K})}\right] E(T)$$

using literature values of $G(295\ °\text{K})$ and Young's modulus $E(T)$ [17 to 19].

4.2 The yield points of polycrystals with a mean grain size of 50 μm

The yield points increase with increasing tin content (see Fig. 2). At tin contents greater than 0.17 at% the yield points are proportional to $\sqrt{c}$, as shown in Fig. 4. A and B of equation (1) are linearly dependent on $\sqrt{c}$ as can be seen from Fig. 5. The relations $A = a_1 \sqrt{c} = 240 \sqrt{c}$ (kp/mm²) and $B = b_1 \sqrt{c} = 25 \sqrt{c}$ (kp/mm² °K^{1/3}) are fulfilled. A- and B-values calculated from experimental investigations on a 3.2 at% copper–tin alloy [15] with a grain size of 50 μm agree quite well with our findings (see Table 2). Summarizing all the results we get in the low-temperature region

$$\sigma_{St}(T, c) = (a_1 - b_1 T^{1/3}) \sqrt{c}\,. \tag{2}$$

At higher temperatures

$$\sigma_{St}(T, c) = \alpha_i \sqrt{c}\, T^{-n_i} \tag{3}$$

yields, with

$$\alpha_1 \approx 2.3 \text{ kp/mm}^2 \text{ °K}^{0.6} \quad \text{and} \quad n_1 \approx 0.6 \quad \text{for} \quad 300 \text{ °K} < T < 460 \text{ °K}$$

and

$$\alpha_2 \approx 6.8 \text{ kp/mm}^2 \text{ °K}^{0.4} \quad \text{and} \quad n_2 \approx 0.4 \quad \text{for} \quad T > 460 \text{ °K}\,,$$

respectively.

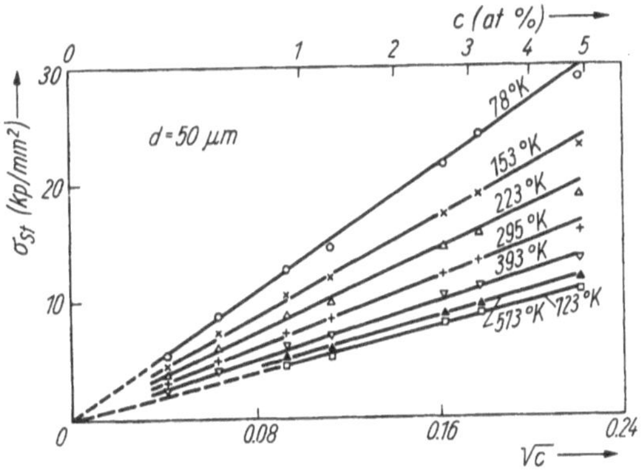

Fig. 4. Influence of tin concentration on the yield point

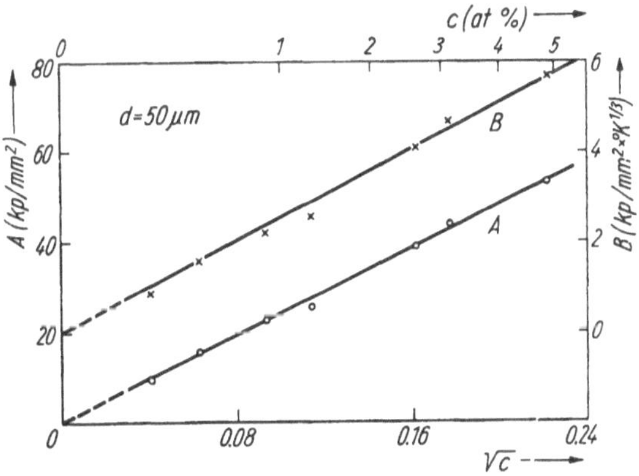

Fig. 5. Influence of tin concentration on A and B

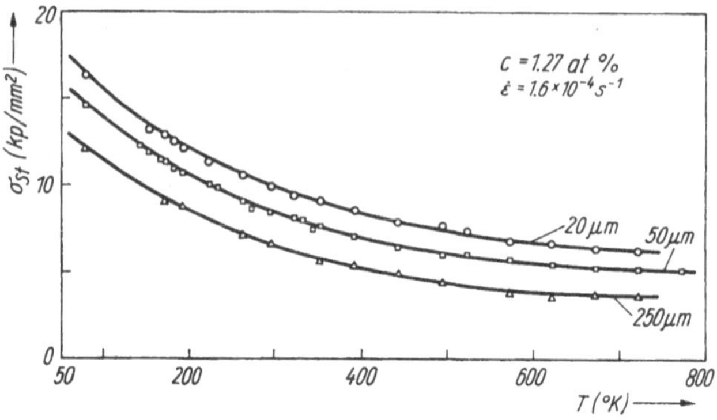

Fig. 6. Influence of grain size on the yield point

4.3 The influence of grain size of the yield point of an α-copper–tin alloy with 1.27 at % tin

According to Fig. 6, there is a strong dependence of the yield point on the grain size up to 773 °K. σ_{St} decreases with increasing grain diameter. Below 300 °K, a $T^{1/3}$-relation is fulfilled independently of grain size. With the grain diameter the quantity A changes its value in a somewhat stronger manner than B (see Table 2).

5. Discussion

Equation (2) leads to the conclusion that $\sigma_{St}(T)/\sigma_{St}(78\ °K)$ of α-copper–tin alloys is practically independent of the tin content. Fig. 7 shows that this conclusion is true: A single curve describes the temperature dependence of the yield point ratio of all alloys with a tin content greater than 0.17 at%. Therefore, in the intervall $0 \lessapprox c \lessapprox 0.17$ at% Sn a very pronounced temperature influence on the yield point ratio has to be expected. That means that different mechanisms are responsible for the temperature dependence of the yield points of pure copper and α-copper–tin alloys, respectively. In the case of copper dislocations intersections are the predominant processes. The α-copper–tin alloys under investigation are homogeneous solid solutions without precipitations [23].

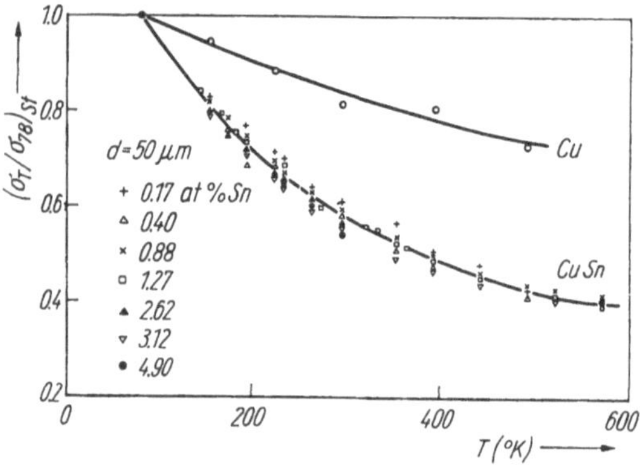

Fig. 7. Influence of temperature on the yield point ratio of pure copper and α-copper–tin alloys

Fig. 8. $\frac{\sigma_{St}}{G\sqrt{c}}$ in dependence of temperature

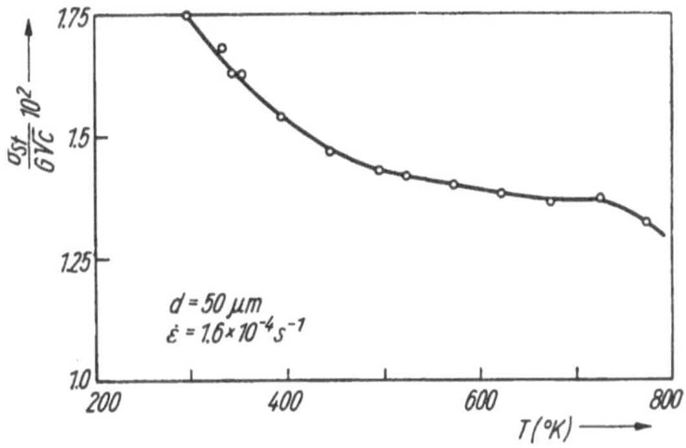

In the α-copper–tin alloys on the other hand the short-range elastic interaction of the alloying atoms with dislocations seems to be the most important process. Glide dislocations may be pinned locally by statistically distributed tin atoms and break off again by thermal activation. The part of the yield point which is only temperature dependent as the shear modulus, may be due to the long-range par- and dielastic interaction [12] of tin atoms with screw dislocations.

5.1 Yield point at $T > 460$ °K

Considering the influence of the temperature on the shear modulus we get for $T > 300$ °K the $\sigma_{St}/G(T)$-curve, shown in Fig. 8. Beyond 460 °K, the yield points are practically temperature independent. We try to calculate this plateau stress [14] σ_G, using the idea of par- and dielastic interactions between screw dislocations and tin atoms.

In a homogeneous solid solution of solute concentration c the mean distance L' between two atoms is $L' = b/\sqrt[3]{c}$. However, the interaction between a dislocation and a solute atom is determined by a quantity L. L is given by the probability $w(L) = b\,F/(L')^3$ that a bowed out dislocation meets an atom of the solute. With $w(L) = 1$ in the case of Fig. 9 is

$$L = \frac{b}{\sqrt{\theta\, c}}\,. \tag{4}$$

According to Fleischer [12] θ is equal to K_w/E_L where E_L is the line tension of the dislocation and K_w is the interaction power $G\,b^2\,\varepsilon_s/120$. ε_s is given by $\varepsilon_s = \left|\frac{\varepsilon_G}{1+\frac{\varepsilon_G}{2}} - 3\,\varepsilon_b\right|$ with $\varepsilon_G = \frac{1}{G}\frac{dG}{dc}$ and $\varepsilon_b = \frac{1}{b}\frac{db}{dc}$, where b is the Burgers vector.

The dislocation line can be moved if the applied shear stress $\tau \geqq \tau_G = \frac{K_w}{b\,L}$. If

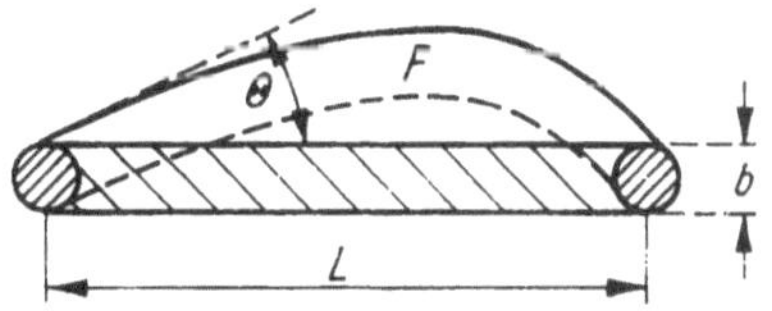

Fig. 9. Bowing out of a pinned dislocation ($F = L^2\,\theta$)

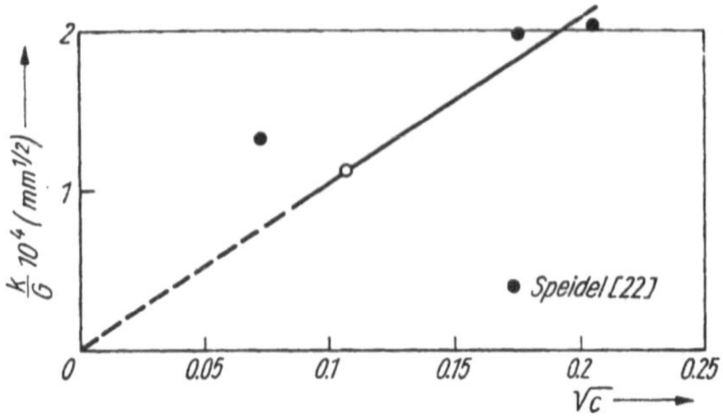

Fig. 10. Concentration dependence of the k-factor at 673 °K

$E_L \approx \frac{G\,b^2}{3}$ is the line tension the plateau stress is

$$\tau_G = \frac{G\,\varepsilon_s^{3/2}}{760}\sqrt{c}\,. \tag{5}$$

Considering the mean orientation factor m of the grains of the polycrystal and the influence of the grain size which can be described by a Petch relation, one obtains

$$\sigma_G = \frac{m\,G\,\varepsilon_s^{3/2}}{760}\sqrt{c} + k\,d^{-1/2}\,.$$

In the case of atomic concentrations of 1.27 at% and greater [22] we take from Fig. 10 for a temperature of 673 °K $k = k^*\,G\sqrt{c}$ with $k^* \approx 10^{-3}\,\mathrm{mm}^{1/2}$. Hence

$$\sigma_G = G\left(\frac{m\,\varepsilon_s^{3/2}}{760} + k^*\,d^{-1/2}\right)\sqrt{c} = a_G\sqrt{c}\,. \tag{6}$$

Using the value for α-copper–tin alloys [12] $\varepsilon_s = 2.03$ then gives

$$\frac{\sigma_G}{G\sqrt{c}} = (1.16 + 0.46) \times 10^{-2} = 1.62 \times 10^{-2}\,.$$

This holds for polycrystals of a mean grain size of 50 μm and is in good agreement with the experimental value

$$\frac{\sigma_{St}(700\ °K)}{G(700\ °K)\sqrt{c}} = 1.36 \times 10^{-2}\,.$$

5.2 Yield point at $T < 300$ °K

In this temperature area it is believed that the short range interaction between solutes and dislocations determines the yield point. The binding energy of a solute atom with the dislocation core [8 to 10] is given by

$$U = \frac{1}{3\pi}\,\frac{1+\nu}{1-\nu}\,G\,b\,|\Delta v|\,\frac{\sin\alpha}{r}\,. \tag{7}$$

ν is Poisson's ratio, Δv is the difference in atomic volums of the solvent and the substitutional solute atoms, r is the distance between the solute atom and the core of the dislocation, and α is the angle between r and the glide plane.

The largest binding energy occurs with $(b \sin \alpha)/r = 3\, b/4\, r_0$ and $r_0 = b/2$. One has

$$U_0 = \frac{1}{2\pi} \frac{1+\nu}{1-\nu} G\, |\Delta v| \,. \tag{8}$$

The binding energy for a tin atom deduced from this equation is $U_0 = 0.96$ eV ($\nu_{Cu} = 0.35$, $G(T=0) = 4700$ kp/mm², and [25] $\Delta v = 10.1 \times 10^{-24}$ cm³).

With U_0 and L the termal part of the flow stress at $T = 0$°K can now be written according to Cottrell [9, 10] as

$$\sigma_s(0) = \frac{m\, U_0}{6\, r_0\, b\, L} \,. \tag{9}$$

If one assumes that equation (4) still holds for the bowing of the dislocations one gets with $\theta = 1/20$ $L = 4.5\, b/\sqrt{c}$. Using this value and putting in $m = 3.06$ and $r_0 = b/2$ one obtains

$$\sigma_s(0) = a_s^0 \sqrt{c} = 215 \sqrt{c}\ (\text{kp/mm}^2)\,.$$

This short-range stress is independent of the grain size. The total stress at $T = 0$ is then, when making use of equation (6),

$$\sigma_{St}^0 = a_s^0 \sqrt{c} + a_G^0 \sqrt{c} = 290 \sqrt{c}\ (\text{kp/mm}^2)\,.$$

This theoretical calculation agrees with the experimental value within 20%:

$$\sigma_{St}^{0\,exp} = A = 240 \sqrt{c}\ (\text{kp/mm}^2)$$

which we have got by extrapolation of our measured yield points.

Friedel's suggestion [24] that the dislocation lines always take the form of zig-zags in solid solutions, leads to $\sigma_s(0) = (m\, U_0/b^3)\, c$ with the same value for U_0 as given by (8). Using these $\sigma_s(0)$-values we calculate yield points which are in the case of $c = 0.17$ at% about a factor two smaller, in the case of $c = 4.90$ at% about a factor three larger than the extrapolated experimental values.

To calculate the flow stress which is necessary to move the dislocation a distance x from the pinned solute, up to now the dependences shown in Fig. 11

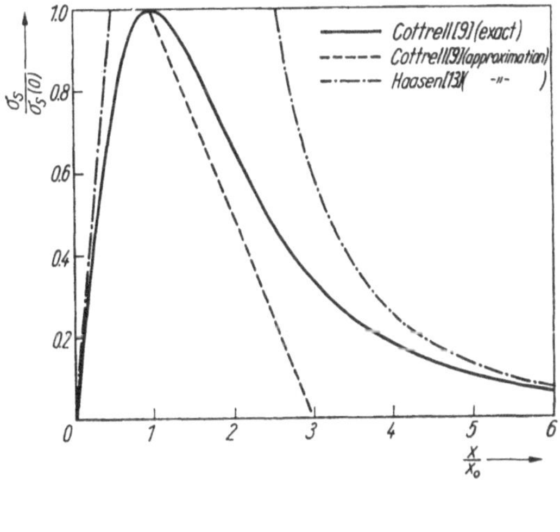

Fig. 11. Exact stress-distance dependence of the interaction between an edge dislocation and a solute atom $\dfrac{\sigma}{\sigma_s(0)} = \dfrac{16\,\frac{x}{x_0}}{\left[\left(\frac{x}{x_0}\right)^2 + 3\right]^2}$.

The approximations after Cottrell [9] and Haasen [13] are

for $\dfrac{x}{x_0} < 1$: $\dfrac{x}{x_0} = \dfrac{1}{2} \dfrac{\sigma_s}{\sigma_s(0)}$ [9, 13]

and for $\dfrac{x}{x_0} > 1$: $\dfrac{x}{x_0} = 3 - 2 \dfrac{\sigma_s}{\sigma_s(0)}$ [9]

or $\dfrac{x}{x_0} = 2 \sqrt[3]{\dfrac{2\, \sigma_s(0)}{\sigma_s}}$ [13],

respectively

have been discussed [9, 13]. Cottrell's approximation [9] of the problem seems to be most useful for our purposes. The stress dependence of the activation energy then is given by

$$U = U_0\left(1 - \frac{\sigma_s(T)}{\sigma_s(0)}\right)^3 . \tag{10}$$

With $\sigma_{St} = \sigma_s(T) + \sigma_G = (a_s + a_G)\sqrt{c}$ equation (10) can be written as

$$U = U_0\left(1 - \frac{\sigma_{St}(T) - a_G\sqrt{c}}{a_s^0\sqrt{c}}\right)^3 .$$

If we assume that the strain rate is only a function of this U-value we get

$$\dot{\varepsilon} = \dot{\varepsilon}_0\, e^{-\frac{U}{kT}} .$$

From this we then obtain for the yield point

$$\sigma_{St} = \left[a_s^0 + a_G^0 - a_s^0\left(\frac{k\ln\frac{\dot{\varepsilon}_0}{\dot{\varepsilon}}}{U_0}\right)^{1/3} T^{1/3}\right] = (a_1 - b_1\, T^{1/3})\sqrt{c} . \tag{11}$$

The yield point should be proportional to $\sqrt{c}$ and $T^{1/3}$, again in good qualitative agreement with our experimental findings in equation (2). Quantitatively we calculate from equation (11) $b_1 = 28$ to 30 kp/mm² °K$^{1/3}$ if we use $U_0 = 0.96$ eV and $\ln \dot{\varepsilon}_0/\dot{\varepsilon} = 25$ to 30. From our experiments we have determined $b_1 =$ $= 25$ kp/mm² °K$^{1/3}$.

5.3 The grain size dependence of the yield point at $T < 300$ °K

Without regard to very large grains the Petch relation

$$\sigma_{St} = \sigma_{St}^{\infty} + k\, d^{-1/2} \tag{12}$$

describes nearly correct the influence of the grain size on the yield point. Taking into account the extensive experimental results of Speidel [22] the quantities a_1 and b_1 in equation (2) can be written

$$a_1(c, d) = (a_0 + k_a\, d^{-1/2})\sqrt{c} = (250 + 7.2\, d^{-1/2})\sqrt{c}\,(\text{kp/mm}^2)$$

and

$$b_1(c, d) = (b_0 + k_b\, d^{-1/2})\sqrt{c} = (30 + 0.37\, d^{-1/2})\sqrt{c}\,(\text{kp/mm}^2\ {}^\circ\text{K}) .$$

Then equation (2) can be rewritten as

$$\sigma_{St} = [a_0 + k_a\, d^{-1/2} - (b_0 + k_b\, d^{-1/2})\, T^{1/3}]\sqrt{c} =$$
$$= (a_0 - b_0\, T^{1/3})\sqrt{c} + (k_a - k_b\, T^{1/3})\sqrt{c}\, d^{-1/2} . \tag{13}$$

Comparing (13) with (12) one can see that $k = (k_a - k_b\, T^{1/3})\sqrt{c}$ is an explicit function of temperature. Between 78 and 300 °K the dependence on temperature is comparable with that of the shear modulus. Therefore, in this T-interval k/G is independent of temperature. This considerations confirm the conclusions drawn above that the grain boundaries do not contribute to the temperature dependence of the yield point.

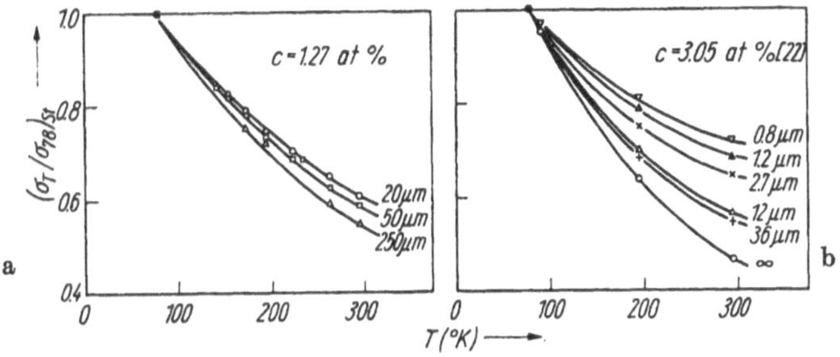

Fig. 12. Grain size influence on the yield point ratio. a) Measured quantities at $c = 1.27$ at%, b) calculated from measurements of Speidel [22]

Furthermore, the yield point ratio calculated from equation (13) will be independent of concentration. This corresponds with our earlier findings. However, it should decrease with decreasing grain size. Indeed, as can be seen from Fig. 12, there exists a strongly marked influence of grain size on the ratio.

References

[1] E. Macherauch, Z. Metallk. **55**, 60 (1964).
[2] E. Macherauch and O. Vöhringer, phys. stat. sol. **6**, 491 (1964).
[3] H. Knöll and E. Macherauch, Z. Metallk. **55**, 638 (1964).
[4] G. Zankl, Z. Naturf. **18a**, 795 (1963).
[5] Ch. Schwink, phys. stat. sol. **8**, 457 (1965).
[6] E. Peissker, Acta metall. **13**, 419 (1965).
[7] E. Peissker, Z. Metallk. **56**, 155 (1965).
[8] A. H. Cottrell and B. A. Bilby, Proc. Phys. Soc. **62A** 49 (1949).
[9] A. H. Cottrell, Conf. High Rates of Strain, IME, 1957 (p. 448).
[10] A. H. Cottrell, Trans. AIME **10**, 192 (1958).
[11] R. L. Fleischer, Acta metall. **9**, 996 (1961); **11**, 203 (1963).
[12] R. L. Fleischer and W. R. Hibbard, The Relation between Structure and Mechanical Properties of Metals, Her Majesty's Stationary Office, Teddington 1963 (p. 262).
[13] P. Haasen, Internal Stresses and Fatigue in Metals, Elsevier, New York 1959 (p. 205).
[14] P. Haasen, Z. Metallk. **55**, 55 (1964).
[15] B. Russell and D. Jaffrey, Acta metall. **13**, 1 (1965).
[16] B. Russell, Acta metall. **13**, 11 (1965).
[17] W. Köster, Z. Metallk. **39**, 1 (1948).
[18] R. Berner, Z. Naturf. **15a**, 689 (1960).
[19] W. C. Overton and J. Gaffney, Phys. Rev. **91**, 816 (1953).
[20] O. Vöhringer and E. Macherauch, Z. Naturf. **20a**, 1079 (1965).
[21] D. Munz, Dissertation TH Stuttgart 1966.
[22] W. Köster and M. O. Speidel, Z. Metallk. **56**, 585 (1965).
[23] W. Bell, P. Okamoto, and G. Thomas, Acta metall. **13**, 559 (1965).
[24] J. Friedel, Dislocations, Pergamon Press 1964.
[25] T. B. Massalski and H. W. King, Progr. Metal. Sci. **10**, 1 (1961).

(Received December 3, 1966)

phys. stat. sol. 19, 805 (1967)
Subject classification: 14.4.2

Solid State Laboratory, Physics Research Center "Patronato Juan de la Cierva", Madrid

Orientation Dependence of the Ferroelectricity of T.G.S.

By
B. JIMÉNEZ, J. MENDIOLA, and E. MAURER

The ferroelectric behaviour of T.G.S. single crystals is studied for slabs cut along directions which form gradual orientations with the polar axis [010]. Distinct ferroelectric phenomena are observed for several directions, even for orientations well away from the polar axis, where perfectly defined ferroelectric properties are obtained (hysteresis loop, λ-curve of the dielectric constant, and maximum switching current). The results are interpreted on the basis of a simple consideration of H-bond contributions to the ferroelectricity. The theoretical conclusions are in good agreement with the experimental results.

Es wird das ferroelektrische Verhalten von T.G.S. Einkristallen an tafelförmigen Proben untersucht, die in verschiedenen Orientierungen zur polaren Achse [010] geschnitten wurden. Deutliche ferroelektrische Phänomene werden auch für Orientierungen beobachtet, die beträchtlich von der polaren Achse entfernt sind, wobei vollständig definierte, ferroelektrische Eigenschaften erhalten werden (Hysteresisschleife, λ-Kurve der Dielektrizitätskonstante und maximaler Schaltstrom). Die Ergebnisse werden mit einer einfachen Berücksichtigung des Beitrags der H-Bindungen zur Ferroelektrizität gedeutet. Die theoretischen Schlußfolgerungen befinden sich in guter Übereinstimmung mit den experimentellen Ergebnissen.

1. Introduction

It is well known that T.G.S. belongs to the ferroelectrics of the order–disorder type. The ferroelectric activity in this material was discovered by Mathias and coworkers [1] in 1956. Its Curie temperature lies at 49 °C. At room temperature it has monoclinic symmetry, and its space group is $P2_1$, the twofold axis being the polar one. The phase above the transition point belongs to the centrosymmetrical class 2/m because of the appearance of a mirror plane.

According to Hoshino and coworkers [2], the dielectric constant along the [010] axis exhibits a lambda peak, and no anomaly has been obtained for his *a*- and *c*-directions. These other two directions lie in the (001) plane.

T.G.S. is a typical H-bond ferroelectric compound in which different groups are bound through O–H–O. These bonds play an important role in the transition mechanism as well as in the ferroelectric properties.

All investigations have been made for the direction parallel to the unit cell axis, but no investigations have been made for different directions.

It must be emphasized that it is very important to cut the samples accurately normal to the polar *b*-axis in order to get typical ferroelectric effects, such as a good hysteresis loop, Curie-Weiss law, and all ferroelectric properties. On different occasions crystals have been cut in several other directions for different purposes, i.e. for studying piezoelectricity, domain properties, etc., but never in order to study ferroelectric phenomena. We started the present work to show that these phenomena critically depend upon direction, and thus to contribute to the explanation of ferroelectricity.

2. Experimental Procedures

T.G.S. crystals were grown in our laboratory by the classical method (from an aqueous solution), and good, large crystals were obtained.

We cut samples from one large crystal in directions which form different angles with the ferroelectric axis. Two different sets of samples were prepared in order to get information on the directional behaviour of the ferroelectric phenomena. The set of slabs referred to as No. 1 was cut with a common edge in the (001) plane as shown in Fig. 1, which forms angles of 20, 40, 50, 70, and 90 degrees with the polar axis. The set referred to as No. 2 was cut with a common edge in the (100) plane as also shown in Fig. 1, forming angles of 20, 45, 55, and 85 degrees with the polar axis.

Samples were ground to get a thickness of about 0.7 mm. Silver electrodes were attached to the plates.

The apparatus employed for the experiments consisted of a General Radio 716-C capacitance bridge to measure the dielectric constant, a typical Sawyer Tower with compensated circuit to measure hysteresis loops, and a double-pulse electric field for switching studies. In order to measure temperature dependences, a sample of each set was put into similar containers and introduced into an oil bath whose temperature could be varied and regulated by a contact-thermometer thermostat.

3. Experimental Results

Our first operation was to cut and polish the crystals using the information already available about morphology. From Fig. 1 the proper orientations to cover the full range from the [010] direction to [001] and [100] were chosen.

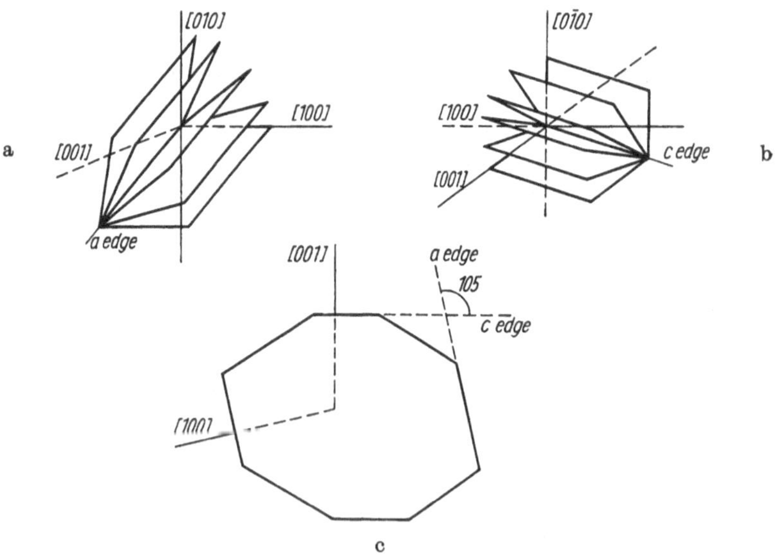

Fig. 1. a) Set of slabs with common edge "a". b) Set of slabs with common edge "c". c) Cross section of the crystal normal to the polar axis [010]. The edges "a" and "c" are shown

Fig. 2. Spontaneous polarization of T.G.S. as a function of temperature for several slab orientations

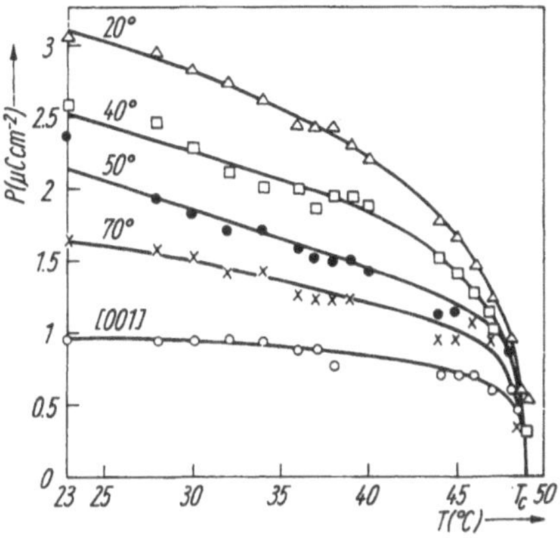

In both sets the edges of the faces (100) and (001) were taken as common. This is clearly shown in Fig. 1 where the slabs are represented as well as the cross section of the crystal normal to the polar axis (Konstantinova [3]).

Properly cut slabs were painted with silver, as described previously, in order to obtain the hysteresis loop and switching current. In both experiments neither hysteresis loop nor switching current were found for the slabs normal to [001] direction of the set No. 1. Then, temperature was varied with the purpose of measuring the behaviour of the dielectric constant. Subsequently the hysteresis loop and switching curve were again measured, and this time the curves shown in Fig. 2 and 3a were obtained. The broken curve in Fig. 3a indicates directional variations of the maximum switching current. The switching curve is still absent for the [001] direction. Notice the pronounced change of i_{max} above 20 degrees; the corresponding polarization P is quite different for all temperatures.

Fig. 4a shows the thermal variation of the dielectric constant for the slabs of set No. 1. It is particularly interesting to note that a sharp lambda peak appears for all slabs. It should be emphasized that this behaviour of the dielectric constant exists for any orientation, and not for one orientation alone.

These results stimulated us to make a similar experiment with the other set referred to as No. 2. The same behaviour was observed as in the first set, as shown in Fig. 3b and 4b. However, the quite different directional variation of i_{max} must be noted. In this case no sharp changes with orientation occur. Notice the different shapes of the curves shown in Fig. 4a and 4b and that, in the latter case, the curves are closer to each other because the thermal variation of the dielectric constant in the [001] direction is greater than in the [100] direction.

4. Discussion of the Results

As mentioned previously, from the crystallochemical point of view, the T.G.S. molecule is formed by three glycine groups and one sulphate group. From the bond configuration given by Hoshino and coworkers [4] it is shown that the H-bond between the glycine groups II and III are strongly connected with each other. In a first approximation, it may be assumed that this H-bond is responsible for the dipolar configuration of the polar phase because of the par-

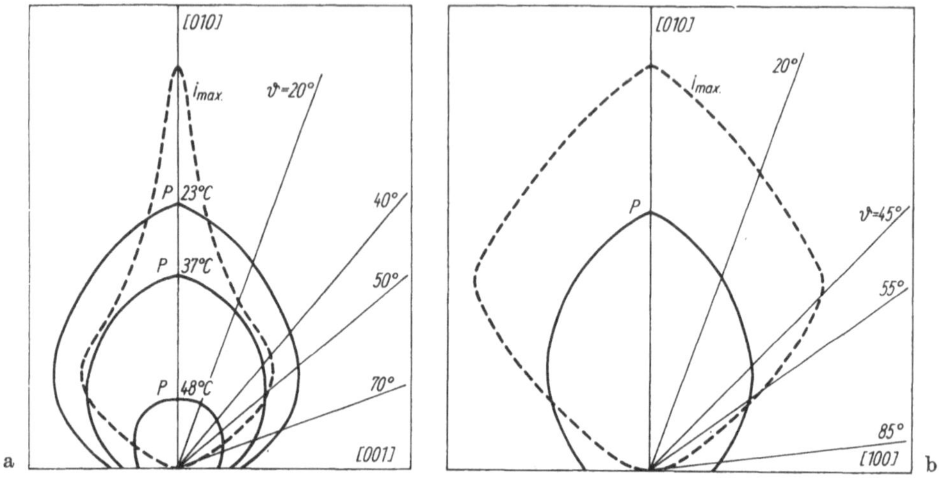

Fig. 3. a) Polar curves of spontaneous polarization for different temperatures. The broken line represents the maximum switching current at room temperature. b) Polar curves of spontaneous polarization (solid line) and maximum switching current (broken line) at room temperature (set No. 2)

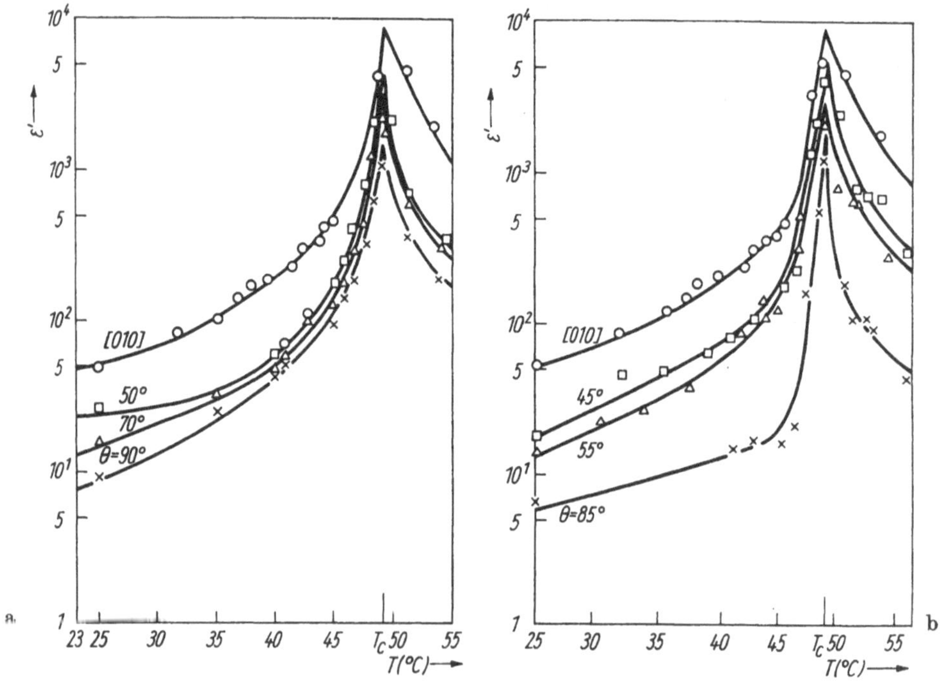

Fig. 4. a) Dielectric constant as a function of temperature for different orientations corresponding to the set of slabs with common edge "c". b) Dielectric constant as a function of temperature for different orientations corresponding to the set of slabs with common edge "a"

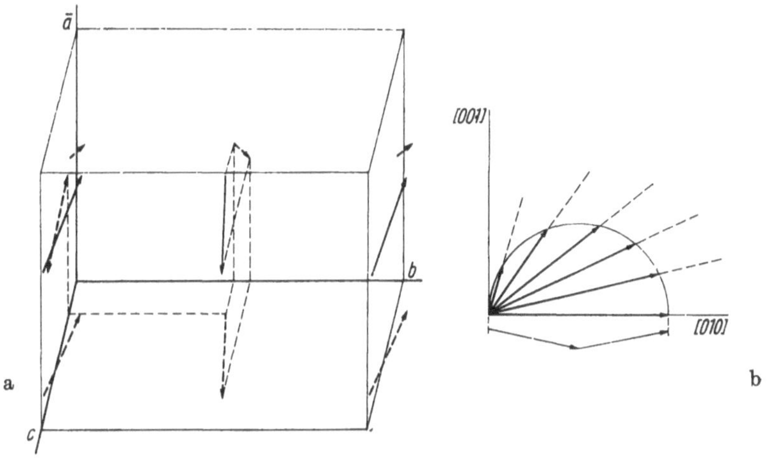

Fig. 5. a) Unit cell of T.G.S. showing the dipoles (solid arrows) according to the model described. The broken arrows show the projections on cell axis. b) Projection of the dipole contribution of the H-bond (dipole) on different orientations. The circle is the amount of these values

ticularity that in the transition its direction strongly changes. Hence, we assumed an H-bond chain as shown in Fig. 5a, without bringing any other elements into the structure. Referring to the last figure, it is clear that the dipoles in the slabs are arranged in such a way that the only non-zero projection on the axis is the polar one [010].

On the basis of this simple model, assuming the dipole has been projected on different orientations. This results in a circle as shown in Fig. 5b. At the same time the effect of an electric field applied in the direction of the dipoles was taken into account and a similar circle pattern was obtained. By adding both effects, an other circle results. These features seem to indicate that the polar representation of the polarization distribution must be a circle. Now, we can refer to Fig. 3a where a pattern similar to the last one is obtained for several temperatures. This is actually true for both sets of plates studied. For both of them we have emphasized the disagreement for [100] and [001] directions because it must be noted that, at the beginning, before varying temperature or applying an electric field, no hysteresis loop and no switching pulses were obtained. But after temperature was varied and an electric field was applied, both phenomena are present though being of small magnitude. Therefore it is confirmed that "history" has a remarkable effect on the bulk behaviour.

The most striking experimental result is certainly the appearance of the peak in the temperature dependence of the dielectric constant (Fig. 4a and 4b) and the fairly marked hysteresis loop for orientations far from the polar direction [010] as shown in Fig. 2. It may be suggested that this behaviour is the result of the presence of an H-bonds chain in the structure as mentioned. This is to say, ferroelectric properties are strongly present not only along the polar axis, but in a wide range of directions as well.

The thermal behaviour may be explained on the basis of a mixture of structures of opposite polarities as assumed by Shibuya and Mitsui [5]. This hypothesis is not incompatible with the H-bond dipolar model introduced in this work.

5. Conclusions

We have shown that the ferroelectric phenomena are not limited to a narrow range around the polar axis, but are present even at 90 degrees from the polar axis. This behaviour has never been studied in ferroelectrics and the results seem very interesting.

A simple dipolar model has been assumed to interpret our experimental results and is in fair agreement with them. This model can explain that ferroelectricity is found not only in the polar direction but in a very wide range around it.

This behaviour can be observed for other similar ferroelectrics such as T.G.F.B. (Triglycine fluorberyllate) which is also H-bonded, and has a crystal structure similar to T.G.S.

Acknowledgements

The authors wish to express their sincere gratitude to Dr. R. de Lucas for his kind encouragement throughout this study.

References

[1] B. T. MATHIAS, C. E. MILLER, and J. P. REMEIKA, Phys. Rev. **104**, 849 (1956).
[2] S. HOSHINO, T. MITSUI, F. JONA, and R. PEPINSKY, Phys. Rev. **107**, 1255 (1957).
[3] V. P. KONSTANTINOVA, I. M. SILVESTROVA, and K. S. ALEKSANDROV, Kristallografiya **4**, 125 (1959).
[4] S. HOSHINO, Y. OKAYA, and R. PEPINSKY, Phys. Rev. **115**, 323 (1959).
[5] I. SHIBUYA and T. MITSUI, J. Phys. Soc. Japan **16**, 479 (1961).

(Received December 5, 1966)

phys. stat. sol. **19**, 811 (1967)
Subject classification: 1.4; 18.2; 21.1.1

Service Lames Minces,
Laboratoire d'Électrostatique et de Physique du Métal, Grenoble

Preferred Orientation and Domain Structure of Iron Films Deposited on (100) Copper Substrates

By
G. A. Jones

Thin films of iron are deposited both by electroplating and by evaporation, on (100) foils of copper. They display a well-defined preferred texture with (110) planes parallel to the substrate. The shape, orientation, and size of the grains are discussed. Magnetic observations of the domain structure using electron microscopy show that the layers are effectively biaxial. Near thicknesses of 500 Å cross-tie walls appear, which is in accordance with energy considerations. Finally, torque measurements lead to a value for K_1.

Des couches minces de fer préparées, soit par électrolyse, soit par évaporation, ont été déposées sur des feuilles de cuivre (100). Nous obtenons une texture préférentielle à plan (110) parallèle au plan du substrat. On discute la forme, l'orientation et la taille des cristallites. L'observation des domaines magnétiques par microscopie électronique montre que les lames sont effectivement biaxiales. Au voisinage d'une épaisseur de 500 Å, des parois barbelées apparaissent, ce qui concorde bien avec les calculs d'énergie. Finalement, les courbes de couple donnent la valeur de K_1.

1. Introduction

In a previous communication, Jones et al. [1] discussed the domain structures of iron, nickel, and their alloys when deposited electrolytically on copper substrates. The results were especially interesting in the case of pure iron which showed a very pronounced texture having (110) planes parallel to the substrate, together with a remarkably simple overall domain structure. The main aim of this paper is to describe the extension of this work to evaporated films of iron, although some further results on electroplated specimens will also be presented. It can be remarked now that even with identically prepared substrates the two types of films, i.e. electro-deposited and evaporated, possess different features especially with regard to grain size and orientation.

The domain structure of the films has been investigated with Lorentz microscopy using the Fresnel out-of-focus method [2] and the orientation of the films with the aid of electron diffraction. Further, subsidiary measurements have been made with a hysteresiscope and torque magnetometer to define more clearly the magnetic behaviour.

2. Experimental Procedure

Electro-deposited films were made at room temperature from a ferrous ammonium sulphate bath (250 g/l) together with an addition of boric acid (6 g/l) acting as buffer agent. Current densities were in the region of 20 mA/cm^2 and pH values in the range 2 to 4.

Evaporated specimens were deposited at a pressure of 10^{-6} Torr, the substrate temperature being 350 °C. The rate of evaporation was maintained between 1 Å/s and 4 Å/s.

The substrate was commercially hot rolled copper leaf of thickness 40 µm possessing a well defined cube texture. Prior to deposition, the copper was electropolished in a methanol–nitric acid solution in order to present an atomically smooth surface. For examination in the microscope the films were stripped with a solution of chromic acid and then washed before mounting.

3. Preferred Textures of the Layers

3.1 Layers of type I: electro-deposited on copper foil

Although the epitaxy of these films has already been partly discussed [1] the intimate orientation relationship between substrate and deposit was not made at all clear. This is due to the fact that no principal axes of either the copper or iron are found parallel to each other. The most informative way of describing the epitaxy is as follows:

$$(110)_{\mathrm{Fe}} \parallel (100)_{\mathrm{Cu}}\,, \quad [1\bar{1}1]_{\mathrm{Fe}} \parallel \begin{matrix}[011]\\ [0\bar{1}1]\end{matrix}_{\mathrm{Cu}} \pm \frac{19°30' - \psi}{2}\,.$$

As demonstrated in Fig. 1 this gives rise to two sets of orthogonal orientations, thus making four in all, which are displaced through an angle ψ with respect to each other. When $\psi = 19°30'$ the $[1\bar{1}1]_{\mathrm{Fe}}$ and $\langle 011\rangle_{\mathrm{Cu}}$ axes are parallel and there is no rotational displacement. The measured values of ψ lie between 15 and 18° thus leading to departures from the ideal epitaxy of up to 2.5°. The cause of the disorientation is probably constraints imposed by the lattice mismatch which produce rotational slip [3].

Direct measurements of the crystallite size indicate a dimension of 600 Å. They were too small to reveal whether slip traces might exist.

3.2 Layers of type II: evaporated on copper foil

3.2.1 Diffraction pattern

The most striking difference of these films by contrast with those discussed above is the presence of a fifth orientation, namely

$$(100)_{\mathrm{Fe}} \parallel (100)_{\mathrm{Cu}}\,, \quad [\bar{1}10]_{\mathrm{Fe}} \parallel [001]_{\mathrm{Cu}}$$

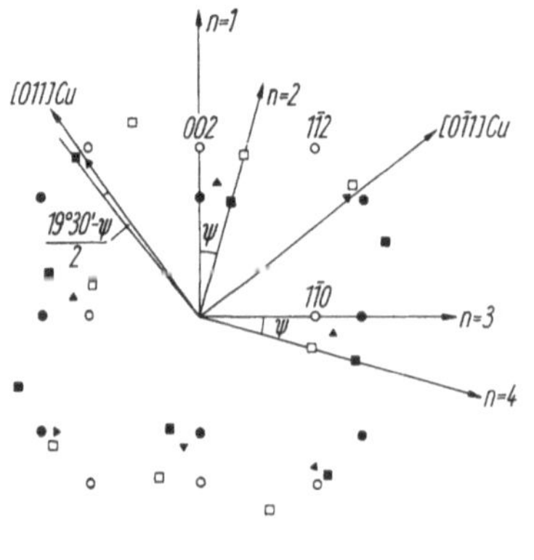

Fig. 1. A diagram illustrating the fourfold symmetry of iron electro-deposited on (100) copper. The orientation relationship may be expressed as

$$(110)_{\mathrm{Fe}}\,(100)_{\mathrm{Cu}}; \quad [1\bar{1}1]_{\mathrm{Fe}} \left\| \begin{matrix}[011]\\ [0\bar{1}1]\end{matrix}\mathrm{Cu} \pm \frac{19°30' - \psi}{2}\right. .$$

$n = 1, 2, \ldots$ indicate the [001] azimuths of each crystallite orientation. The quantity $(19°30' - \psi)/2$ determines the deviation from the ideal epitaxy and is about 2.5°, i.e. $\psi = 15°$. Triangles are (100) copper spots

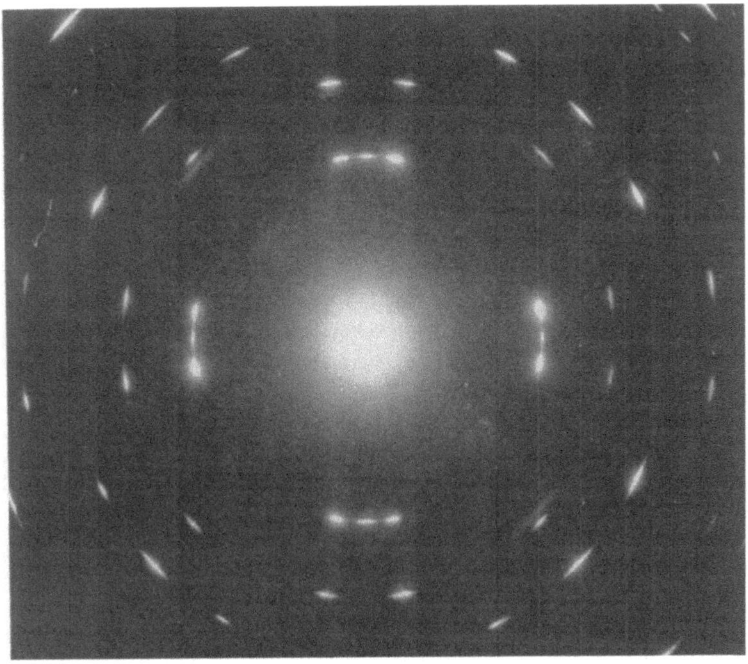

Fig. 2. A diffraction pattern from iron evaporated at 350 °C on (100) copper foil. In addition to the four (100) orientations there is also $(100)_{Fe} \parallel (100)_{Cu}$, $[\bar{1}10]_{Fe} \parallel [001]_{Cu}$

Fig. 3. A micrograph of an 800 Å type II film showing the distribution of grains. The edges of the grains are parallel to the copper $[1\bar{1}0]$ and [110] axes

which is the disposition normally observed for iron deposited directly on rock-salt (Fig. 2). This parallel orientation always appeared regardless of the evaporation rate or substrate temperature (always $< 400\ ^\circ C$). However, it is doubtful whether the (100) crystallites constitute a very great fraction of the total number. Precise measurements on the arrangement of the crystallite axes were made difficult on account of the imperfect alignment of the grains. However, except in one or two isolated cases where values of 2.5° were recorded, the rotational displacement was comparatively small, being of the order of 1° or less.

3.2.2 Crystallite structure

By contrast with electro-deposition, evaporated films have very much larger crystallites (up to 1 μm) which exhibit a fairly regular form. Fig. 3 shows clearly their rectangular shape. In the film plane the edges of the crystallites lie along two perpendicular directions which correspond with the $[1\bar{1}0]$ and [110] axes of the copper substrate. The grains are thus bounded by iron directions of the types $\langle 112 \rangle$ and $\langle 111 \rangle$. It is interesting to note the role played by the $\langle 110 \rangle$ copper axes: not only do they influence the initial mode of epitaxy but also the actual form of the grains.

Although the characteristics outlined above persist down to the lowest thicknesses (250 Å) above 1000 Å a curious feature appears, namely the presence of many fine lines within the individual iron crystals (Fig. 4). Despite a wavy aspect, their spacing is approximately constant with a value of 250 Å. Within a grain the direction of the fringes is the same but will change abruptly at a crystal boundary. In all, four distinct directions are observed (marked A, B, C, D in Fig. 4) one each being contributed by a discrete iron orientation. Analysis of the diffraction pattern indicates that the fringes follow either $\pm$ [001] or $\pm$ $[1\bar{1}0]$

Fig. 4. A 1500 Å evaporated film revealing fine traces within the grains. The traces follow four distinct directions which are found at A, B, C, D

axes. It will be recalled that within the film plane the [001] axis of one crystallite always coincides with the [1$\bar{1}$0] of another. Thus to determine unequivocally the direction of the fringes it is necessary to isolate a single grain but because of their small size this has proved impossible with selected area diffraction.

The source of the fringes remains obscure. Certainly moiré effects can be discounted, only one film being present, and it is more likely that their origin lies in some type of slipping or dislocation network. Although the slip direction in b.c.c. metals is $\langle 111 \rangle$ the slip systems in iron are so numerous that it would not be impossible for traces to appear either along $\langle 001 \rangle$ or $\langle 110 \rangle$. More work is continuing on this topic to elucidate their origin.

4. Magnetic Properties

4.1 Domain structure

In both types of films the domains show a very high degree of order approaching the configuration expected for a single crystal of iron with a (100) orientation, i.e., two apparently perpendicular easy directions of magnetization. Fig. 5 is the micrograph of a type II film of thickness 800 Å. In appearance it resembles the examples given by Sato et al. [4] of iron deposited on (100) rocksalt. However, the domain structure observed here is consistent with the fourfold disposition of the crystallites as can be shown by the following anisotropy considerations.

For the purposes of calculation x and y coordinates are constructed along the [001] and [010] axes of the copper substrate as illustrated in Fig. 6. The [001]

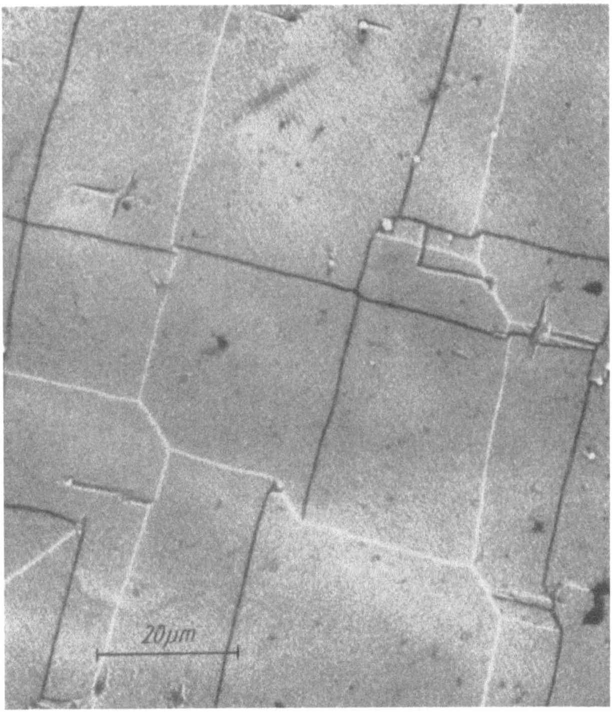

Fig. 5. The domain structure of a demagnetized 800 Å type II film clearly revealing two easy axes

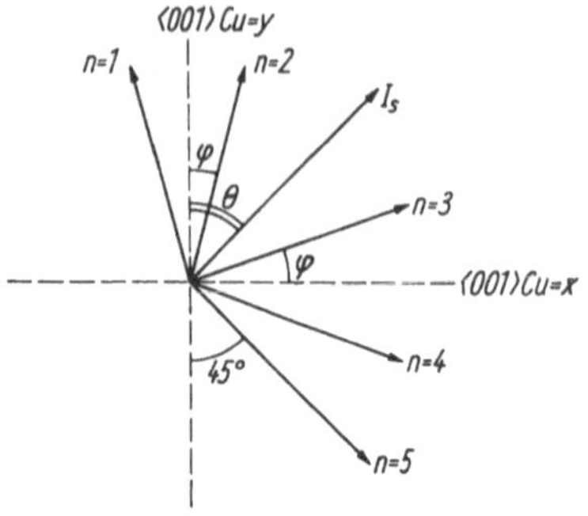

Fig. 6. A diagram for the calculation of the combined anisotropy energy of the iron crystallite configuration. The [001] azimuths are numbered $n = 1 \cdots 5$, the latter being that of the (100) orientation. (Read Φ instead of φ)

axes of the four (110) crystallite orientations then make angles of $\pm \Phi$, $90° \pm \Phi$ $\left(\text{where } \Phi = \frac{\psi}{2}\right)$ with the x- and y-axes. Account may also be taken of the (100) orientation $n = 5$ for which the [001] direction lies at 45° to the y-axis. Further more we define a parameter L, as

$$L = \frac{\text{total number of crystallites of (100) orientation}}{\text{total number of crystallites of (110) orientation}} .$$

For any general direction of magnetization, I_s, making an angle θ with the y-axis, the anisotropy energy density, E_K, is given by (assuming equal distribution of (110) grains)

$$E_K = \frac{K_1}{1+L}\left\{L\left(\alpha_{51}^2 \alpha_{52}^2 + \alpha_{51}^2 \alpha_{53}^2 + \alpha_{52}^2 \alpha_{53}^2\right) + \frac{1}{4}\sum_{n=1}^{4} \alpha_{n1}^2 \alpha_{n2}^2 + \right.$$
$$\left. + \alpha_{n1}^2 \alpha_{n3}^2 + \alpha_{n2}^2 \alpha_{n3}^2\right\} + \frac{K_2}{1+L}\left\{L\left(\alpha_{51}^2 \alpha_{52}^2 \alpha_{53}^2\right) + \frac{1}{4}\sum_{n=1}^{4} \alpha_{n1}^2 \alpha_{n2}^2 \alpha_{n3}^2\right\}. \quad (1)$$

The values of the direction cosines α_{n1}, etc., are

$$\alpha_{11} = \alpha_{12} = \frac{1}{\sqrt{2}} \sin(\theta + \Phi), \quad \alpha_{13} = \cos(\theta + \Phi),$$

$$\alpha_{21} = \alpha_{22} = \frac{1}{\sqrt{2}} \sin(\theta - \Phi), \quad \alpha_{23} = \cos(\theta - \Phi),$$

$$\alpha_{31} = \alpha_{32} = \frac{1}{\sqrt{2}} \cos(\theta + \Phi), \quad \alpha_{33} = \sin(\theta + \Phi),$$

$$\alpha_{41} = \alpha_{42} = \frac{1}{\sqrt{2}} \cos(\theta - \Phi), \quad \alpha_{43} = \sin(\theta - \Phi),$$

$$\alpha_{51} = \cos(45° - \theta), \quad \alpha_{52} = \sin(45° - \theta), \quad \alpha_{53} = 0.$$

Substituting for α_{11}, etc., equation (1) becomes

$$E_K = \frac{K_1}{8(1+L)} + \left\{\frac{K_2}{64(1+L)} + \frac{3K_1}{32(1+L)}\right\}\left(\sin^2 2(\theta + \Phi) + \sin^2 2(\theta - \Phi)\right) +$$
$$+ \frac{K_1 L}{4(1+L)} \sin^2 2(45° - \theta). \quad (2)$$

To investigate the energy dependence on θ equation (2) is differentiated, whence

$$\frac{\partial E_K}{\partial \theta} = \left\{\frac{K_2}{16\,(1+L)} + \frac{3\,K_1}{8\,(1+L)}\right\} \sin 4\,\theta \cos 4\,\Phi - \frac{K_1\,L}{2\,(1+L)} \sin 4\,\theta\,. \tag{3}$$

Equation (3) thus shows that E_K is always a maximum, minimum, or turning point for $\theta = \frac{m\,\pi}{4}$, depending on the sign of $\frac{\partial^2 E_K}{\partial \theta^2}$. With $\theta = 0$ we find that

$$\frac{\partial^2 E_K}{\partial \theta^2} = \left\{\frac{K_2}{4\,(1+L)} + \frac{3\,K_1}{2\,(1+L)}\right\} \cos 4\,\Phi - \frac{2\,K_1\,L}{1+L}\,. \tag{4}$$

For electro-deposited films where $L = 0$ and $\Phi = 7.5°$ equation (4) predicts that for $0 < \Phi \leqq \frac{\pi}{8}$, the x–y coordinates become the magnetic easy axes.

In the case of type II films, the effect of a non-zero L is to reduce the range of Φ for which the x–y coordinates are favoured. Although it is impossible to determine L, an examination of the diffracted intensities would indicate an upper limit of $L = \frac{1}{4}$. Taking this value equation (4) predicts the x–y coordinates to be easy in the range $0 < \Phi < 18°$. Experimentally $\Phi \approx 10°$ and the microscope observations confirm that once again the x–y coordinates are the favoured magnetic axes.

Alternatively assuming a value of $\Phi = 10°$ the value of L at which the $\langle 001 \rangle$ copper directions become hard is given by

$$L = \frac{(K_2 + 6\,K_1)}{8\,K_1} \cos 40°\,, \tag{5}$$

whence $L = 0.6$. In none of the films did the magnetic hard axes coincide with the $\langle 001 \rangle$ copper directions. Values of K_1 and K_2 for this calculation and the one preceding were quoted from Bozorth [5].

Several torque curves of type II films were obtained each revealing, as expected, a strong biaxial term, as well as a smaller uniaxial component. The latter was measured as 10^4 erg/cm³. In principle the biaxial torque can be found using equation (3) but owing to the disposition of the crystallites no explicit expressions can be obtained for K_1 and K_2. Furthermore there still remains the unknown value of L. At best the term in K_2 can be neglected and L assumed zero. With these approximations the value of K_1, turns out to be $\approx 5.4 \times 10^5$ erg/cm³ somewhat higher than the bulk figure. A finite value of L increases this further. In other recent work published on epitaxial iron films, Tsukahara and Kawakatsu [6] claim enhanced values of K_1 in one case recording 6.4×10^5 erg/cm³. This tendency may be induced by stresses acting in the films as is the case with nickel [7].

Hysteresis loops of type I films are isotropic indicating that the domain structure is not ordered over macroscopic dimensions. On the other hand, films of type II possess loops with distinctive biaxial properties. The ratio of the 90 and 180° switching fields $\frac{H_{90}}{H_{180}}$ yields values in the range 0.6 to 0.8 with H_{180} itself equal to ≈ 20 Oe. For a biaxial element which is switching only by rotation, the theoretical value of $\frac{H_{90}}{H_{180}}$ is 0.27 and the discrepancy must be assigned to the nucleation of domain walls.

Apparently the most important factor influencing the size of the domains is the presence of holes which tend to subdivide the film into small units. Moreover, Néel spikes always form in the neighbourhood of isolated holes in an attempt to reduce demagnetizing energy. Frequently inclusions are actually found embedded in domain walls. Fig. 7a illustrates these points.

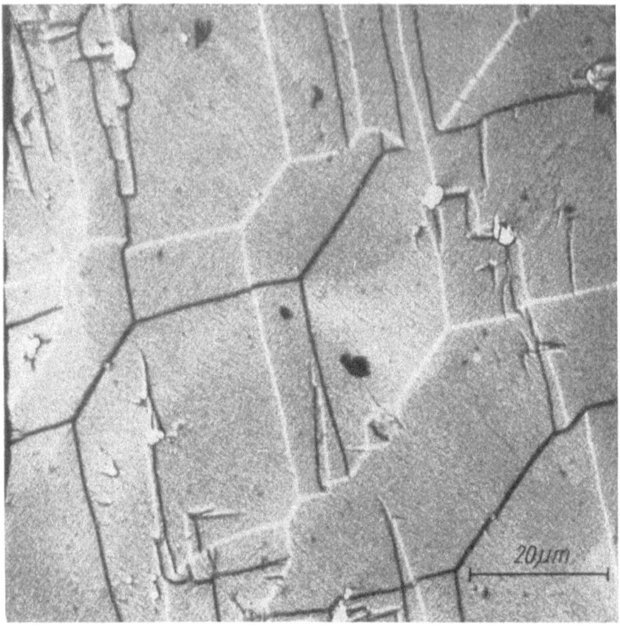

Fig. 7a. The domain structure of a type II 1500 Å film. The domain size is largely determined by the distribution of inclusions

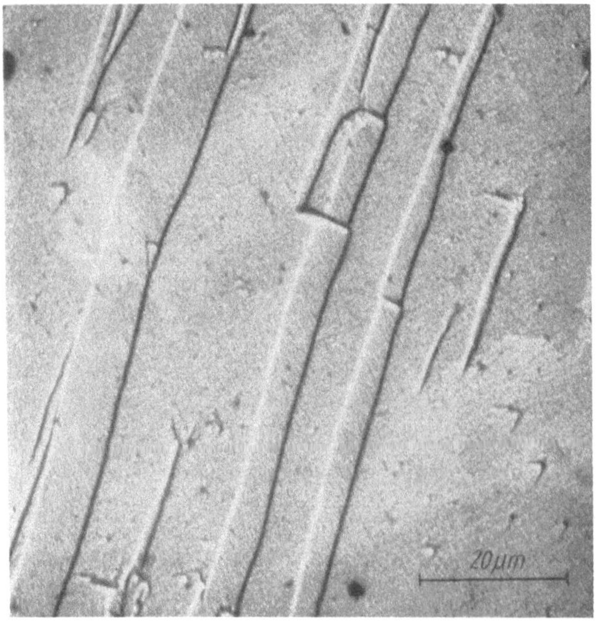

Fig. 7b. Parallel 90° walls in a 1300 Å film identified by the presence of ripple within the domains

Finally, there is a ripple pattern observed within each region of the evaporated specimens. This is clearly seen in Fig. 7b where the walls are distinguished as 90° despite their being parallel. Since the magnetocrystalline anisotropy axes are not randomly oriented it seems likely that the ripple must be due to the effects of an isotropic stress component induced during fabrication of the films. Upon heating the ripple becomes more intense as observed by Cohen [8] in polycrystalline permalloy films. After re-crystallization the grains become so large that ripple is no longer observed.

4.2 Domain walls

On account of their lower energy 90° walls are much more prevalent than those of 180°. The distinction between the two is made easy on two grounds — firstly by the presence of ripple and secondly by their different image widths. The latter is ultimately dependent upon the component of magnetization parallel to the wall and is smaller for a 90° wall by a factor of $\sqrt{2}$.

A readily obtainable estimate of 180° wall energies can be found by using Middelhoek's equations [9], viz.

$$\gamma_N = \frac{A\,\pi^2}{a} + E_{KN}\,a + \frac{\pi\,I_s^2\,a\,D}{a+D}\,; \quad \frac{\partial\gamma_N}{\partial a} = 0\,, \tag{6}$$

$$\gamma_B = \frac{A\,\pi^2}{a} + E_{KB}\,a + \frac{\pi\,I_s^2\,a^2}{a+D}\,; \quad \frac{\partial\gamma_B}{\partial a} = 0\,, \tag{7}$$

$$\gamma_{CT} = 0.6\,\gamma_N\,, \tag{8}$$

where γ_N, γ_B, γ_{CT} are the energies/cm² of Néel, Bloch, and cross-tie walls, A the exchange constant, a the wall width, D the film thickness, E_K the anisotropy energy density, and I_s the magnetization. The crystallite texture of the layers makes the calculation of E_{KB} difficult since the plane of rotation of the spins is not rational. In order to simplify this situation we can assume the presence of only two orthogonal (110) orientations (i.e. $\psi = 0$), as shown in Fig. 8. Thus a wall follows either [001] or [$1\bar{1}0$] directions, the planes of rotation being (110) for the Néel wall and alternatively (001) or ($1\bar{1}0$) for the Bloch wall. Substituting for the direction cosines and assuming a linear rotation of spins across the wall,

$$E_{KN} = \frac{K_1}{8\,\pi}\int_{-\pi/2}^{\pi/2} (\cos^4\theta + \sin^4\theta + 8\sin^2\theta\cos^2\theta)\,d\theta\,,$$

$$E_{KB} = \frac{K_1}{8\,\pi}\int_{-\pi/2}^{\pi/2} \left[\cos^4\theta + \sin^2 2\,\theta + \sin^2 2\left(\frac{3\pi}{4} - \theta\right)\right] d\theta\,,$$

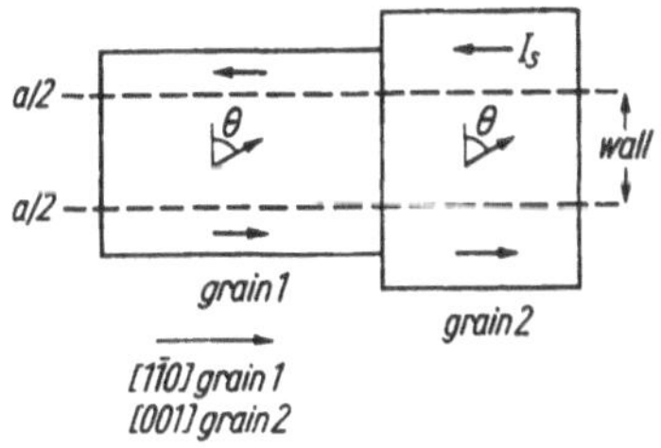

Fig. 8. A simplified model of the crystallite texture (with $\psi = 0$) for the determination of 180° wall energies. The walls follow [001] and [$1\bar{1}0$] directions, the planes of rotation being (110) for the Néel wall and (001) and ($1\bar{1}0$) for the Bloch wall

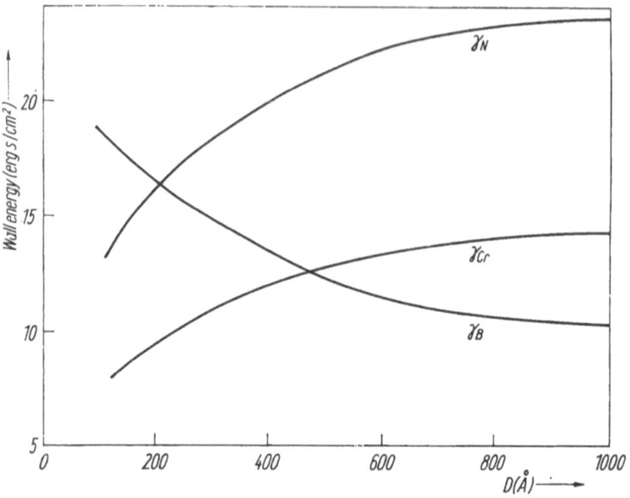

Fig. 9. A plot of film thickness D against energy of Bloch (γ_B), Néel (γ_N), and cross-tie (γ_{CT}) walls. (Read γ_{CT} instead of γ_{Cr})

whence

$$E_{K\mathrm{N}} = \frac{7\,K_1}{32}, \qquad E_{K\mathrm{B}} = \frac{11\,K_1}{64},$$

terms in K_2 having been neglected. With these values for E_K equations (6) and (7) may be solved numerically for γ and a: having computed γ_N, relation (8) gives the cross-tie wall energy. The results are plotted in Fig. 9 for the various types of boundary. Although the anisotropy energy contribution to the wall

Fig. 10. The domain structure of a 500 Å type I film showing the formation of cross-ties on 180° walls

energy is a hundredfold greater than that, say for permalloy, the determining factor is still the stray field energy which varies as I_s^2. Thus the simplification made for the calculation of E_K is probably not too serious.

The most interesting result of Fig. 9 is the occurrence of cross-tie walls in the thickness region near 500 Å, a prediction which is actually confirmed experimentally. Fig. 10 is a micrograph from a 500 Å type I film in which 180° cross-tie walls are visible. Unfortunately at this thickness the orientation of the domains tends to be less marked. Such walls are rarely seen in electropolished foils of iron [10] probably due to the difficulty of obtaining sufficiently thin specimens rather than a prohibitive anisotropy constant. 90° walls never possess cross-ties.

5. Conclusions

It has been demonstrated that iron grows epitaxially on (100) copper with (110) planes parallel to the substrate. Electro-deposited layers have a small crystallite size and display a rotational displacement about an axis perpendicular to the film. In the case of evaporated specimens the grains are very much larger ($\approx 1\ \mu$m) with a definite form. The origin of many fine lines within the individual grains has tentatively been assigned to a slip process.

Lorentz microscopy of the magnetic domains has shown that the films are biaxial in nature, an observation which follows from the anisotropy distribution of the crystallites. Torque curves have verified this and enabled a value of K_1 to be arrived at. Further calculations of wall energies predict the occurrence of cross-tie walls at a thickness of 500 Å. This has again been confirmed.

Acknowledgements

This work has been carried out under a contract provided by D.G.R.S.T. (65 FR 241) and under the direction of Dr. R. Montmory. The author wishes to thank his colleagues for stimulating discussions, as well as Mr. J. Devenyi who assisted with the electron microscopy and Mrs. C. Schlenker who performed the torque measurements.

References

[1] G. A. Jones, D. P. Oxley, and R. S. Tebble, Phil. Mag. **11**, 993 (1965).
[2] H. W. Fuller and M. E. Hale, J. appl. Phys. **31**, 238 (1960).
[3] H. Wilman, Proc. Phys. Soc. **A64**, 329 (1951).
[4] H. Sato, S. Shinozaki, and R. S. Toth, Proc. Intern. Magnet. Conf., Nottingham 1964 (p. 798).
[5] R. M. Bozorth, Ferromagnetism, van Nostrand Comp. 1951.
[6] S. Tsukahara and H. Kawakatsu, J. Phys. Soc. Japan **21**, 313 (1966).
[7] J. F. Freedman, IBM J. Res. Developm. **6**, 449 (1962).
[8] M. Cohen, J. appl. Phys. **34**, 1841 (1963).
[9] S. Middelhoek, Thesis, University of Amsterdam 1961.
[10] J. M. Rodgers, Thesis, University of Sheffield 1964.

(Received December 6, 1966)

phys. stat. sol. 19, 823 (1967)

Subject classification: 13.1; 22.4

Physikalisch-Technisches Institut der Deutschen Akademie der Wissenschaften zu Berlin, Bereich elektrischer Durchschlag, and IV. Physikalisches Institut der Humboldt-Universität zu Berlin

Spin-Orbit Splitting of the Valence Band of Wurtzite Type Crystals

By

E. GUTSCHE and E. JAHNE

A 12×12 representation is given of the valence band Hamiltonian at $\boldsymbol{k} = 0$ which includes the spin-orbit interaction. Expressions are derived for the energies and the linear combinations of the wave functions for the Γ_7, Γ_8, and Γ_9 subbands. The whole valence band is completely determined by eight parameters: three energies, two parameters coupling states of equal spin, and three parameters coupling states of opposite spin. In principle, all parameters can be obtained experimentally from the energy separations and oscillator strengths for interband transitions for polarization $\perp c$ and $\parallel c$. Anisotropic spin-orbit interaction and mixing of states originating from the two anions per unit cell are discussed.

Für das Valenzband bei $\boldsymbol{k} = 0$ wird eine 12×12 Matrix-Darstellung, die die Spin-Bahn-Wechselwirkung einschließt, gegeben. Die Energien und die Linearkombinationen der Wellenfunktionen der Γ_7, Γ_8 und Γ_9 Subbänder werden angegeben. Das gesamte Valenzband bei $\boldsymbol{k} = 0$ ist durch acht Parameter vollständig bestimmt: Drei Energien, zwei Parameter, welche Zustände mit gleichem, und drei Parameter, welche Zustände mit entgegengesetztem Spin koppeln. Im Prinzip können alle Parameter aus den gemessenen Energieabständen und den Oszillatorenstärken für Interbandübergänge für die Polarisationen $\perp c$ und $\parallel c$ bestimmt werden. Effekte wie anisotrope Spin–Bahn-Wechselwirkung und Mischung der Zustände, die von den beiden Anionen in der Elementarzelle herrühren, werden diskutiert.

1. Introduction

It is well known that the valence band states in diamond and zincblende type materials originate from atomic p orbitals. According to our present knowledge, the maximum of the valence band lies at $\boldsymbol{k} = 0$ in all these materials. In wurtzite type crystals the valence band states also originate from p orbitals, but the situation is more complicated due to the lower symmetry and the fact that the number of atoms per unit cell is doubled with respect to the diamond and zincblende type materials. Nevertheless, according to our present knowledge, the valence band maximum is also located at or very close to the point Γ in wurtzite type materials. Therefore, the band structure in the vicinity of this point is of special interest.

From the theoretical point of view, the energy band structure at and near $\boldsymbol{k} = 0$ has been treated by group theoretical and perturbational methods by a number of investigators [1 to 6]. The most fruitful treatment for the point Γ was given by Hopfield [4] which was based, however, on a "quasi-cubic model" and on the assumption that the two molecules in the unit cell are crystallographically equivalent. He assumed that the energy states of a wurtzite crystal can be treated as the states of a cubic crystal strained in the [111] direction. The magnitude of the resultant splitting of the p_z state from the p_x, p_y states was denoted as "crystal field splitting energy", and the accuracy of this approximation was assumed to be of the order (splitting energy/energy gap). Expressions have been derived in [4] and [5] for the energy separations of the subbands observed experimentally, which take into account the combined action of crystal field splitting and spin-orbit interaction.

However, all these papers suffer from the fact that the presence of two molecules per unit cell has not been correctly taken into consideration. The relevance of the second molecule which leads to another set of states at $\boldsymbol{k} = 0$ was pointed out in [7] and [8].

In the present paper we treat the band structure at $\boldsymbol{k} = 0$ from a more general point of view. The symmetry of wurtzite is rigorously taken into account as well as the fact that the existence of a second anion per unit cell doubles the number of valence bands. The Bloch-Hamiltonian of the correct symmetry C_{6v}^4 (including spin-orbit interaction) is represented by a set of wave functions which are the solutions of the Schrödinger equation without spin for $\boldsymbol{k} = 0$. A representation of the Hamiltonian with respect to the valence band states at $\boldsymbol{k} = 0$ by a 12×12 matrix is obtained which involves complete information about level separation and mixing between the various states. Special attention is given to interesting details of the valence band structure which have not been discussed so far, such as anisotropic spin-orbit interaction and mixing between the states resulting from the non-equivalent anions in the unit cell. The results of earlier works appear as limiting cases.

2. Eigenvalues and Wave Functions at $\boldsymbol{k} = 0$

Let

$$H = \frac{\boldsymbol{p}^2}{2m} + V(\boldsymbol{r}) \tag{1}$$

be the Bloch-Hamiltonian without spin which possesses the full symmetry of wurtzite C_{6v}^4. Every solution of the corresponding Schrödinger equation for the point Γ must then belong to one of the six irreducible representations Γ_1 to Γ_6 of the single group. The corresponding wave functions can be written in most general form in cylindrical coordinates r, φ, z as Fourier series:

$$\Gamma_1: \quad \sum_{\mu=-\infty}^{+\infty} f_\mu^{(1)}(r, z)\, e^{i 6 \mu \varphi}\,, \tag{2a}$$

$$\Gamma_2: \quad \sum_{\mu=-\infty}^{+\infty}{}' f_\mu^{(2)}(r, z)\, e^{i 6 \mu \varphi}\,, \tag{2b}$$

$$\left.\begin{aligned} \Gamma_3: &\quad \sum_{\mu=-\infty}^{+\infty}{}' f_\mu^{(3)}(r, z)\, e^{i(6\mu \mp 3)\varphi} && \text{(2c)} \\ \Gamma_4: &\quad \sum_{\mu=-\infty}^{+\infty}{}' f_\mu^{(4)}(r, z)\, e^{i(6\mu \mp 3)\varphi} && \text{(2d)} \end{aligned}\right\} \mp \text{ for } \mu \gtrless 0$$

with $f_\mu^{(i)} = f_{-\mu}^{(i)}$ for $i = 1, 3$ and $f_\mu^{(i)} = - f_{-\mu}^{(i)}$ for $i = 2, 4$,

$$\Gamma_5: \quad \sum_{\mu=-\infty}^{+\infty} f_\mu^{(5)}(r, z)\, \mathrm{e}^{\pm i(6\mu+1)\varphi}, \tag{2e}$$

$$\Gamma_6: \quad \sum_{\mu=-\infty}^{+\infty} f_\mu^{(6)}(r, z)\, \mathrm{e}^{\pm i(6\mu+2)\varphi} \tag{2f}$$

(see for instance [4]). The coefficients are arbitrary functions of r and z and can be taken real without loss of generality. They have to be chosen so that the wave functions constructed according to (2) are normalized over the unit cell. Note that the functions belonging to the twofold degenerate states Γ_5 and Γ_6 are the complexes conjugate of each other.

As was done in the earlier treatments, we assume that the valence band states at $\boldsymbol{k} = 0$ originate from the anions and are p-like in the tight binding limit. The states resulting from both anions in the unit cell must then belong to the representations Γ_5, Γ_1 and Γ_6, Γ_3, which transform like $(x \pm i\,y)$, z and $(x \pm i\,y)^2$, $x\,(x^2 - 3\,y^2)$, respectively (Fig. 1). The corresponding wave functions will be denoted as u_5, u_5^*, u_1, and u_6, u_6^*, u_3.

If we include spin the number of wave functions doubles. The representations are changed in $2\,\Gamma_7$, $2\,\Gamma_8$, and $2\,\Gamma_9$, since $\Gamma_5, \Gamma_1, \Gamma_6, \Gamma_3 \times D_{1/2} = \Gamma_7 + \Gamma_9, \Gamma_7, \Gamma_8 + \Gamma_9, \Gamma_8$, respectively. Simultaneously we have to add to (1) the spin-orbit interaction operator

$$H_{\mathrm{so}} = \frac{\hbar}{4\,m^2\,c^2}\,(\boldsymbol{\sigma} \times \nabla V) \cdot \boldsymbol{p} \equiv \boldsymbol{\sigma} \cdot \boldsymbol{N}\,. \tag{3}$$

Neglecting spin-orbit interaction of the valence band states with conduction band states as well as with lower lying states, the Hamiltonian $H + H_{\mathrm{so}}$ is diagonal except for the part referring to the valence band states.

In order to obtain for this part the Hamiltonian matrix in a form as simple as possible, we take the basis functions in the sequence

$$\left.\begin{array}{l} u_5^*\,\alpha,\ u_1\,\beta,\ u_6\,\alpha,\ u_3\,\beta\,,\ u_5\,\alpha,\ u_6\,\beta\,; \\ u_5\,\beta,\ -u_1\,\alpha,\ u_6^*\,\beta,\ -u_3\,\alpha,\ u_5^*\,\beta,\ -u_6^*\,\alpha\,. \end{array}\right\} \tag{4}$$

α, β are the spin functions $\begin{pmatrix}1\\0\end{pmatrix}$ and $\begin{pmatrix}0\\1\end{pmatrix}$, respectively.

The last six functions in (4) can be derived from the first six by applying the time reversal operator $K = -\,i\,\sigma_y\,K_0$. The basis is a Kramers basis which has the advantage that the Hamiltonian matrix takes the simple form

$$\begin{pmatrix} S & T \\ -T^* & S^* \end{pmatrix}.$$

In the present case S and T are 6×6 matrices in the representation for $H + H_{\mathrm{so}}$.

Using the explicit angular dependence of the wave functions according to (2), it is found that T vanishes, and S has the

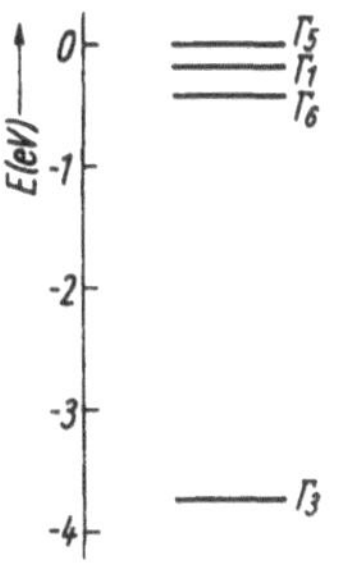

Fig. 1. Valence band states at $\boldsymbol{k} = 0$ without spin. The energy scale on the left hand side applies to CdS after OPW calculations by Collins et al. ([9] (Fig. 3))

following form:

$$S = \left(\begin{array}{cc|cc|cc} |u_5^* \alpha\rangle & |u_1 \beta\rangle & |u_6 \alpha\rangle & |u_3 \beta\rangle & |u_5 \alpha\rangle & |u_6 \beta\rangle \\ E_5 - \frac{\Delta_5}{3} & \sqrt{2}\,\frac{\Delta_7^*}{3} & \multicolumn{2}{c|}{0} & \multicolumn{2}{c}{0} \\ \sqrt{2}\,\frac{\Delta_7}{3} & E_1 & & & & \\ \hline \multicolumn{2}{c|}{0} & E_6 + \frac{\Delta_6}{3} & \sqrt{2}\,\frac{\Delta_8^*}{3} & \multicolumn{2}{c}{0} \\ & & \sqrt{2}\,\frac{\Delta_8}{3} & E_3 & & \\ \hline \multicolumn{2}{c|}{0} & \multicolumn{2}{c|}{0} & E_5 + \frac{\Delta_5}{3} & \sqrt{2}\,\frac{\Delta_9^*}{3} \\ & & & & \sqrt{2}\,\frac{\Delta_9}{3} & E_6 - \frac{\Delta_6}{3} \\ \multicolumn{2}{c}{\Gamma_7} & \multicolumn{2}{c}{\Gamma_8} & \multicolumn{2}{c}{\Gamma_9} \end{array}\right). \tag{5}$$

E_5, E_1, E_6, and E_3 are the eigenvalues of the Hamiltonian H corresponding to the single group representations Γ_5, Γ_1, Γ_6, and Γ_3, respectively. The five non-equivalent spin-orbit coupling parameters are defined as follows:

$$\frac{\Delta_5}{3} = \langle u_5 | N_z | u_5 \rangle, \tag{6a}$$

$$\frac{\Delta_6}{3} = \langle u_6 | N_z | u_6 \rangle, \tag{6b}$$

$$\frac{\Delta_7}{3} = \langle u_1 | N_+ | u_5^* \rangle, \tag{7a}$$

$$\frac{\Delta_8}{3} = \langle u_3 | N_+ | u_6 \rangle, \tag{7b}$$

$$\frac{\Delta_9}{3} = \langle u_6 | N_+ | u_5 \rangle. \tag{7c}$$

Here $N_+ = \frac{1}{\sqrt{2}} (N_x + i\, N_y)$, N_x, N_y, N_z being the cartesian components of the operator $\boldsymbol{N}$ defined by (3) which have to be taken here in cylindrical coordinates. $\Delta_5/3$ and $\Delta_6/3$ are expectation values for N_z whereas $\Delta_7/3$, $\Delta_8/3$, and $\Delta_9/3$ are coupling parameters. The factor 1/3 is used in the definition of the Δ's in order to obtain the well-known expressions in the limiting case.

The secular matrix is step-like and breaks into six 2×2 matrices. The eigenvalues are

$$E_7 = \frac{1}{2}\left\{(E_5 + E_1) - \frac{\Delta_5}{3} \pm \sqrt{\left[(E_5 - E_1) - \frac{\Delta_5}{3}\right]^2 + \frac{8}{9}\,|\Delta_7|^2}\right\}, \tag{8a}$$

$$E_8 = \frac{1}{2}\left\{(E_6 + E_3) + \frac{\Delta_6}{3} \pm \sqrt{\left[(E_6 - E_3) + \frac{\Delta_6}{3}\right]^2 + \frac{8}{9}\,|\Delta_8|^2}\right\}, \tag{8b}$$

$$E_9 = \frac{1}{2}\left\{(E_6 + E_5) + \frac{\Delta_5 - \Delta_6}{3} \pm \sqrt{\left[(E_5 - E_6) + \frac{\Delta_5 + \Delta_6}{3}\right]^2 + \frac{8}{9}\,|\Delta_9|^2}\right\}. \tag{8c}$$

Each of these six energy states is twofold degenerate because of time reversal symmetry. The indices correspond to the representations Γ_7, Γ_8, and Γ_9 of the double group.

The linear combinations of wave functions which diagonalize the Hamiltonian $H + H_{so}$ are of the form

$$\sqrt{1-q^2}\,|u'\,\alpha\rangle + q\,|u''\,\beta\rangle\,,$$
$$-\,q\,|u'\,\alpha\rangle + \sqrt{1-q^2}\,|u''\,\beta\rangle\,.$$

The q's are given by

$$\left.\begin{matrix} q_7^2 \\ 1-q_7^2 \end{matrix}\right\} = \frac{1}{2}\left[1 \mp \frac{E_5 - E_1 - \dfrac{\Delta_5}{3}}{\sqrt{\left[(E_5 - E_1) - \dfrac{\Delta_5}{3}\right]^2 + \dfrac{8}{9}\,|\Delta_7|^2}}\right], \tag{9a}$$

$$\left.\begin{matrix} q_8^2 \\ 1-q_8^2 \end{matrix}\right\} = \frac{1}{2}\left[1 \mp \frac{E_6 - E_3 + \dfrac{\Delta_6}{3}}{\sqrt{\left[(E_6 - E_3) + \dfrac{\Delta_6}{3}\right]^2 + \dfrac{8}{9}\,|\Delta_8|^2}}\right], \tag{9b}$$

$$\left.\begin{matrix} q_9^2 \\ 1-q_9^2 \end{matrix}\right\} = \frac{1}{2}\left[1 \mp \frac{E_5 - E_6 + \dfrac{\Delta_5 + \Delta_6}{3}}{\sqrt{\left[(E_5 - E_6) + \dfrac{\Delta_5 + \Delta_6}{3}\right]^2 + \dfrac{8}{9}\,|\Delta_9|^2}}\right]. \tag{9c}$$

q_7^2, q_8^2, and q_9^2 measure the fraction of u_1, u_3, and u_6 states in the upper Γ_7, Γ_8, and Γ_9 band, respectively, admixed by the spin-orbit interaction. The amount of this mixing is controlled by the parameters $|\Delta_7|$, $|\Delta_8|$, $|\Delta_9|$ which couple states of opposite spin.

For the complete description of the valence band at $\boldsymbol{k} = 0$ eight energy band parameters are required: three energies E_5, E_1, E_6, E_3 (one of them can arbitrarily be set equal to zero); two parameters Δ_5, Δ_6 coupling states of the same spin and three parameters coupling states of opposite spin: The situation is therefore much more complex than, for instance, in zincblende crystals, for which the valence band at $\boldsymbol{k} = 0$ is completely determined by only one parameter, viz. the spin-orbit splitting energy Δ. In wurtzite type crystals the spin-orbit coupling is anisotropic. The number of valence bands doubles in view of the fact that states belonging to a point L in zincblende appear as additional states at the point Γ in wurtzite. This accounts for the parameters Δ_5 to Δ_8. The additional parameter Δ_9 couples states arising from the two anions per unit cell and has no counterpart in zincblende.

Therefore the valence band parameters for wurtzite type materials cannot be determined by measuring the five energetic positions of the states only. Complete information can be obtained, however, by measuring the three q^2 values in addition to the energies. The q^2's determine the relative oscillator strengths and can be obtained, in principle, from the relative intensities for optical transitions for polarization parallel and perpendicular to the c-axis.

The allowed electrical dipole transitions between the valence band states and the lowest conduction band states Γ_7 and Γ_8 (s states) can be seen from

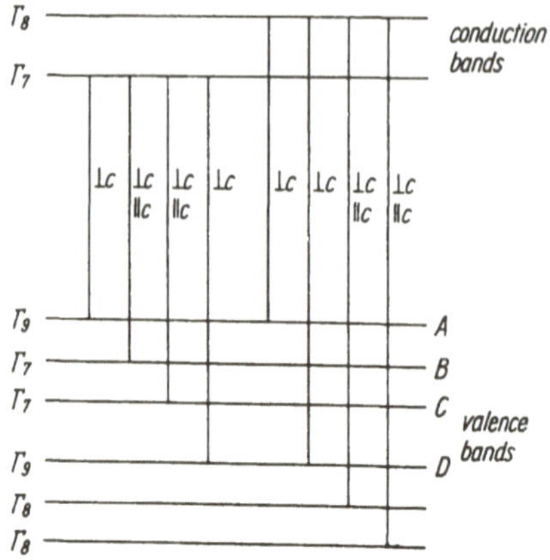

Fig. 2. Allowed electrical dipole transitions between the valence bands and the conduction bands at $\boldsymbol{k} = 0$.

Fig. 2. At present, only transitions to the lowest (Γ_7) conduction band state have been identified. Since transitions from the Γ_8 valence bands to the Γ_7 conduction band are forbidden, no informations about the Γ_8 valence bands states are available. For this reason we shall not discuss these Γ_8 states here.

Transitions from the Γ_9 states to the lowest Γ_7 conduction band are allowed for polarization $\perp c$. However, they have been detected so far only for the upper Γ_9 state. Since these transitions are controlled by the share of u_5 states in the Γ_9 bands, it must be concluded that the lower Γ_9 band contains only a small amount of u_5 states which means that q_9^2 is small. We can then expand the square root in (8c) and (9c). Furthermore we denote the Γ_9 bands by A and D with $E_A > E_D$ and the Γ_7 bands by B and C with $E_B > E_C$. This corresponds to the sequence $E_A > E_B > E_C$ observed in most wurtzite type materials. Assuming $E_5 > E_6$, we get from (8) and (9)

$$q_9^2 = \frac{2}{9} \frac{|\Delta_9|^2}{\left[(E_5 - E_6) + \dfrac{\Delta_5 + \Delta_6}{3}\right]^2}, \tag{10a}$$

$$E_A - E_B = \frac{1}{2}\left\{(E_5 - E_1) + \Delta_5 - \sqrt{\left[(E_5 - E_1) - \frac{\Delta_5}{3}\right]^2 + \frac{8}{9}|\Delta_7|^2 + \frac{2}{9} \frac{|\Delta_9|^2}{E_5 - E_6 + \dfrac{\Delta_5 + \Delta_6}{3}}}\right\}, \tag{10b}$$

$$E_A - E_D = E_5 - E_6 + \frac{\Delta_5 + \Delta_6}{3} + \frac{4}{9} \frac{|\Delta_9|^2}{E_5 - E_6 + \dfrac{\Delta_5 + \Delta_6}{3}}, \tag{10c}$$

$$E_B - E_C = \sqrt{\left[(E_5 - E_1) - \frac{\Delta_5}{3}\right]^2 + \frac{8}{9}|\Delta_7|^2}\,. \tag{10d}$$

In the equations (9) and (10) the measurable quantities q_7^2, q_9^2, $E_A - E_B$ etc. are expressed in therms of the valence band parameters. The q's can be determined from the relative (exciton-) line intensities as follows: Denoting the oscillator strengths for polarized light by $f_A^\perp$, $f_B^\perp$ etc., one can easily see that

$$q_7^2 = \frac{1}{1 + \dfrac{f_B^\perp}{f_C^\perp}} = \frac{1}{1 + \dfrac{f_C^\parallel}{f_B^\parallel}}$$

and

$$q_9^2 = 1 - \frac{f_B^\perp + f_C^\perp}{f_A^\perp}\,.$$

In any case, the relations $f_B^\perp f_B^\parallel = f_C^\perp f_C^\parallel$ and $f_A^\perp \leqq f_B^\perp + f_C^\perp$ must hold. The sign of equality in the last relation applies only if the admixture of u_6 states to the band A is negligible. Furthermore, the ratios $f_B^\perp/f_B^\parallel$ and $f_C^\parallel/f_C^\perp$ differ only by the ratio of the square of the interband matrix elements, $(|\langle u_1^c|p_x|u_5^v\rangle/\langle u_1^c|p_z|u_1^v\rangle|)^2$ which should approximately be equal to unity.

Solving (10) for the valence band parameters, we find:

$$E_5 - E_1 = \frac{1}{2}(E_A - E_B) + \frac{1}{2}(E_B - E_C)(2 - 3q_7^2) - \frac{1}{4}(E_A - E_D)q_9^2, \tag{11a}$$

$$\Delta_5 = \frac{3}{2}(E_A - E_B) + \frac{3}{2}(E_B - E_C)q_7^2 - \frac{3}{4}(E_A - E_D)q_9^2, \tag{11b}$$

$$|\Delta_7| = \frac{3}{\sqrt{2}}(E_B - E_C)\sqrt{(1 - q_7^2)q_7^2}, \tag{11c}$$

$$|\Delta_9| = \frac{3}{\sqrt{2}}(E_A - E_D)\sqrt{q_9^2}, \tag{11d}$$

$$E_5 - E_6 + \frac{\Delta_5 + \Delta_6}{3} = \frac{E_A - E_D}{1 + 2q_9^2}. \tag{11e}$$

3. Discussion

The results to be discussed now differ from those of the earlier work in two respects: a) a mixture between the Γ_9 bands and b) an anisotropic spin-orbit interaction are taken into consideration. The expressions derived in earlier papers can easily be reproduced in the limiting cases. For instance, the well known formulae for the energy separation of the A, B, and C bands ((6) in [4], (9) in [5]) are obtained from (10) by setting $q_9^2 = 0$ and $\Delta_5 = \Delta_7$, and the "mixing rate" $(2 - 3E/\delta)^2$ introduced by Hopfield [4] corresponds to $2q_7^2/(1 - q_7^2)$ in the same limit. The "crystal field splitting energy" appears as the energy separation between the Γ_5 and Γ_1 states without restriction as to the magnitude of this separation and the c/a-ratio. Furthermore, one obtains $E_A = E_B$ and $E_B - E_C = \Delta$ as in zincblende materials by putting $E_5 = E_1$ and $\Delta_5 = \Delta_7 = \Delta$.

Let us now discuss the relevance of our results with respect to the experimental situation. The energy separations $E_A - E_B$ and $E_B - E_C$ are known for most wurtzite materials. They are summarized in [10] and [11] and the values used in the present paper are indicated in Fig. 3a to d. It should be stressed, however, that these results are derived from the positions of exciton lines, which can introduce some uncertainty at least for the $(E_A - E_C)$ separation, discussed for CdS in [12]. This may turn out to be quite remarkable from the present point of view.

Oscillator strengths, on the other hand, have neither been measured nor discussed so far for wurtzite crystals, apart from [12] and [13]. The energy separation $E_A - E_D$ has not yet been identified at all. We shall focus our interest, therefore, on a brief discussion of the band parameters $E_5 - E_1$, Δ_5, and Δ_7 for tho various wurtzite type materials. For this purpose it is useful to plot these parameters vs. q_7^2, and possibly with $q_0^0(E_A - E_D)$ as a parameter according to equations (11a to c).

a) CdS: Of the wurtzite type materials this is the best understood one at present. The earlier works are based on the assumption $|\Delta_7| = \Delta_5 = \Delta$ which corresponds to the two intersections of the strong lines in Fig. 3a. In view of the

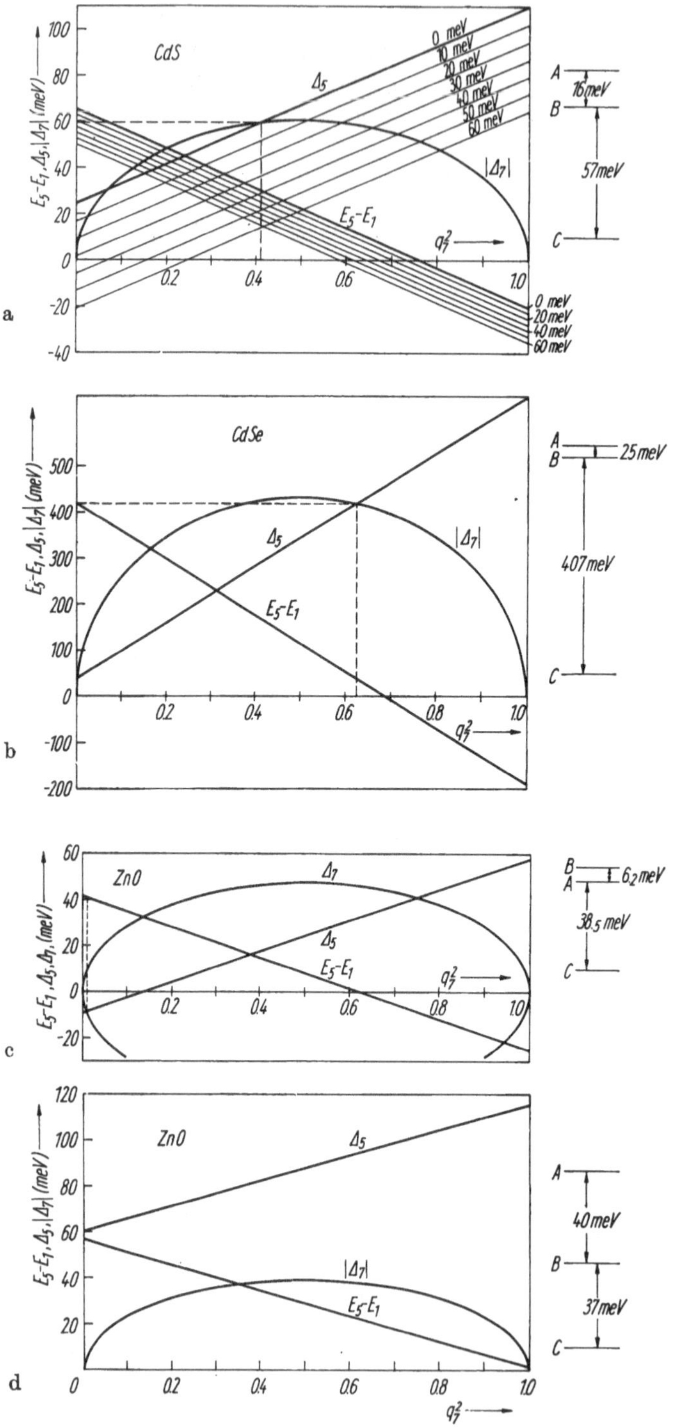

Fig. 3. Energy band parameters as a function of the band mixing parameter q_7^2 evaluated after equation (11a to c) using the experimentally observed energy separations.
a) CdS; after [12]. Parameter in this graph is q_9^2 $(E_A - E_D)$,
b) CdSe; after [14],
c) ZnO; after [13] (note the reversed notation for the A and B bands compared with [13]),
d) after [18]

increasing spin-orbit splitting in the sequence S, Se, Te, the values $E_5 - E_1 = 29$ meV, $\Delta = 60$ meV were adopted, corresponding to $q_7^2 = 0.43$. However, recent line intensity measurements performed in this laboratory seem to indicate, a somewhat smaller value for q_7^2, resulting in a spin-orbit splitting anisotropy of about four percent.

This result refers to the limit $q_9^2 = 0$. The influence of q_9^2 can be estimated from the weak lines in Fig. 3a. Since the coupling parameter Δ_7 is independent of q_9^2, any appreciable amount of q_9^2 means an increased anisotropy of spin-orbit interaction. Because $E_5 - E_6$ is certainly at least several tenth of one eV, the effect of q_9^2 can be quiter emarkable even if q_9^2 is only about few percent. Assuming, for instance, $E_A - E_D = 0.5$ eV, as is suggested by the result of recent band structure calculations [9] which give $E_5 - E_6 \approx 0.48$ eV, and furthermore $q_9^2 = 0.04$, this would result in $E_5 - E_1 = 24$ meV, $\Delta_5 = 45$ meV, giving a spin-orbit interaction anisotropy of 25%. However, the accuracy of line intensity measurements available at present is not sufficient to determine precise values for the parameters.

b) CdSe: This material exhibits the largest spin-orbit splitting of the wurtzite family and is therefore of special interest. The previous assignment is $E_5 - E_1 = 39$ meV, $\Delta_5 = |\Delta_7| = \Delta = 418$ meV. According to Fig. 3b this corresponds to $q_7^2 = 0.625$ which is close to the upper limit 2/3 attained for $\Delta_5 = |\Delta_7| = \Delta$ and $E_5 - E_1 \ll \Delta$. Unfortunately, only a rough estimate can be made from line intensities at present, indicating that q_7^2 is somewhat less than 0.6. This would result in a spin-orbit interaction anisotropy larger than in CdS which seems to be reasonable.

c) ZnS and ZnSe: The energy separations $E_A - E_B$ and $E_B - E_C$ are about 26 meV and 82 meV for ZnS [15] and 16 meV and 50 meV for ZnSe [16], respectively. The situation is therefore similar to that illustrated in Fig. 3a for CdS. No line intensity measurements are reported so far and little can be said about these materials at present.

d) ZnO: This material is especially interesting in view of some ambiguity with respect to line assignments and even the sequence of the valence band states. From the fact that transitions from the uppermost valence band are partially polarized $\parallel c$ in absorption measurements, Thomas [13] concluded that the sequence of the valence band states should be Γ_7, Γ_9, Γ_7 (Fig. 3c). This assignment is supported by theoretical calculations due to Shindo et al. [17] who obtained a negative value for the spin-orbit splitting energy for ZnO which is due to a large influence of d electrons in this case.

On the other hand, Park et al. [18] in a recent paper reported another line assignment which adopts the "normal" sequence, but yields a value for $E_A - E_B$ that is almost one order of magnitude larger than that of Thomas. The corresponding situation is outlined by Fig. 3d. The striking fact is that no intersections exist at all between Δ_5 and Δ_7. This fact corresponds to the statement of [18] that the spin-orbit energy "is a complex number". However, the conclusion that more than one crystal field splitting parameter should be introduced is certainly wrong. The treatment presented in this paper does not refer to any "quasi-cubic model" and shows quite clearly that only one crystal field parameter enters the problem for one group of states. In principle, it is possible to explain the respective data of [18] by means of an anisotropic spin-orbit interaction. In this case, however, one would be forced to adopt a very large anisotropy and rather large values for Δ_5 and $|\Delta_7|$. This seems to be highly improbable in view of the light anion. On the contrary, considering this fact and the comparatively large deviation of $c/a = 1.60$ from the ideal value $\sqrt{8/3}$, one would expect Δ_5, $|\Delta_7| \ll |E_5 - E_1|$. According to equation (9a) this results in a small q_7^2. Therefore the assignment of Fig. 3c which corresponds to $q_7^2 = 0.01$ (for the isotropic case) seems to be quite reasonable. Furthermore, this assignment is strongly supported by the observed line intensities which have been properly discussed on the basis of a very small "mixing ratio" in [13].

4. Concluding Remarks

From the above discussion one should draw the conclusion that the mixing of the states of the two anions per cell and the anisotropy of the spin-orbit interaction is probably not a large effect in all wurtzite type materials, but can lead, in some cases, to appreciable modifications of the valence band parameters adopted so far. In order to clarify this situation and to come to decisive results, careful measurements of oscillator strengths are necessary. From such results one can expect to gain deeper insight into the details of the valence band structures of wurtzite type crystals. Moreover, such results would yield reliable data for fitting or checking non-relativistic band structure calculations. Measurements of this kind with CdS and CdSe are in progress in this laboratory.

Acknowledgements

The authors are indebted to Dr. H. W. Streitwolf for critical remarks on the manuscript, to Dr. J. Voigt for valuable discussions, and to Dr. T. C. Collins for making available details of the work reported in [9].

References

[1] G. DRESSELHAUS, Phys. Rev. **105**, 135 (1957).
[2] R. C. CASELLA, Phys. Rev. **114**, 1514 (1959).
[3] J. L. BIRMAN, Phys. Rev. Letters **2**, 157 (1959); Phys. Rev. **115**, 1493 (1959); J. Electrochem. Soc. **107**, 409 (1960).
[4] J. J. HOPFIELD, J. Phys. Chem. Solids **15**, 97 (1960).
[5] M. BALKANSKI and J. DES CLOIZEAUX, J. Phys. Radium **21**, 825 (1960).
[6] E. B. RASHBA, Fiz. tverd. Tela **1**, 407 (1959).
E. B. RASHBA and V. I. SHEKA, Fiz. tverd. Tela, Supplement II, 1959 (p. 162).
[7] S. A. MOSKALENKO and M. J. SHMIGLYUK, Fiz. tverd. Tela **6**, 3535 (1964).
[8] M. CARDONA and G. HARBEKE, Phys. Rev. **137**, A1467 (1965).
[9] T. C. COLLINS, R. N. EUWEMA, and J. S. DE WITT, Proc. Intern. Conf. Semiconductor Physics, Kyoto 1966.
[10] V. V. SOBOLEV, Fiz. tverd. Tela **6**, 906 (1964).
[11] D. C. REYNOLDS, C. W. LITTON, and T. C. COLLINS, phys. stat. sol. **9**, 645 (1965); **12**, 3 (1965).
[12] D. G. THOMAS and J. J. HOPFIELD, Phys. Rev. **116**, 573 (1959).
J. J. HOPFIELD and D. G. THOMAS, Phys. Rev. **122**, 35 (1961).
[13] D. G. THOMAS, J. Phys. Chem. Solids **15**, 86 (1960).
[14] R. G. WHEELER and J. O. DIMMOCK, Phys. Rev. **125**, 1805 (1962).
[15] A. LEMPICKI, J. BIRMAN, H. SAMELSON, and G. NEUMARK, Proc. Intern. Conf. Semiconductor Physics, Prague 1960 (p. 768).
[16] E. F. GROSS, L. G. SUSLINA, and P. A. KONKOV, Fiz. tverd. Tela **4**, 396 (1962).
[17] K. SHINDO, A. MORITA, and H. KAMIMURA, J. Phys. Soc. Japan **20**, 2054 (1965).
[18] Y. S. PARK, C. W. LITTON, T. C. COLLINS, and D. C. REYNOLDS, Phys. Rev. **133**, 512 (1966).

(Received December 7, 1966)

phys. stat. sol. 19, 833 (1967)

Subject classification: 13.3; 16; 22.1.2

2. Physikalisches Institut der Technischen Hochschule Aachen

Leitfähigkeit und Feldeffekt reiner Siliziumspaltflächen[1])

Von

M. HENZLER

An Siliziumkristallen (etwa 10^4 Ωcm, p-leitend), die im Ultrahochvakuum gespalten worden waren, wurden Oberflächenleitfähigkeit und Feldeffekt gemessen. Durch eine Abschirmung der Proben und Messung des Feldeffektes für verschiedene Abstände der Feldelektrode konnte der verfälschende Beitrag der oxidbedeckten Seitenflächen ausgeschaltet werden. Die reine Oberfläche ergibt sich als schwach p-leitend ($\Delta\sigma = (4{,}2 \pm 1{,}3) \times 10^{-7}$ A/V) mit einer hohen Dichte von Oberflächenzuständen ($S_E = 10^{14}$ eV^{-1} cm^{-2}). Durch Wasserdampf wird die Spaltfläche stark n-leitend, während durch Sauerstoff die Oberflächenleitfähigkeit nahezu den Minimalwert zwischen p- und n-Leitung annimmt.

Surface conductivity and field effect mobility are observed in p-type silicon crystals (about 10^4 Ωcm) cleaved in ultra high vacuum. The effect of the uncleaved surfaces is eliminated by shielding the crystal and measuring the field effect for various distances of the field electrode. The clean surface is found to be weakly p-type ($\Delta\sigma = (4.2 \pm 1.3) \times 10^{-7}$ Ω^{-1}) with a high density of surface states ($S_E = 10^{14}$ eV^{-1} cm^{-2}). Admission of water vapour causes the cleaved surface to become strongly n-type whereas with oxygen treatment the surface conductivity tends towards the minimum between p-type and n-type.

1. Einleitung

Bei Experimenten mit reinen Oberflächen findet man elektronische Oberflächenzustände, die nicht von unkontrollierbaren Verunreinigungen stammen, sondern vom Halbleiter selbst. Die Messung von Oberflächenleitfähigkeit und Feldeffektbeweglichkeit wird vielfach zur Untersuchung von Oberflächenzuständen benützt (zusammenfassende Berichte [1 bis 3]). Wenn man neue Oberflächen durch Spalten im Ultrahochvakuum herstellt, so hat man sicher eine reine Oberfläche, aber die nicht gespaltenen Seitenflächen können erheblich stören. In den bisher bekannt gewordenen Arbeiten über reine Spaltflächen wurde diese Möglichkeit nicht berücksichtigt.

Bei den hier beschriebenen Experimenten wurden an reinen Siliziumspaltflächen Bandverbiegung und Dichte der Oberflächenzustände gemessen [21], wobei der verfälschende Beitrag der oxidbedeckten Seitenflächen durch Präparation und Meßmethode ausgeschaltet wurde.

2. Experimentelles

2.1 Meßverfahren

Um während der Messung wirklich reine Oberflächen zu haben, wurden die Kristalle im Ultrahochvakuum gespalten und sofort untersucht. Fig. 1 zeigt schematisch die verwendete Vakuumapparatur. Nach einer Ausheizzeit von 2 bis 3 Stunden (bei ca. 300 °C) wurden Enddrucke von einigen 10^{-10} bis 10^{-11} Torr erreicht.

---

[1]) Gekürzte Dissertation, T. H. Aachen (D 82). Teilweise vorgetragen auf der Arbeitstagung des Fachausschusses „Halbleiter“ in Bad Pyrmont, März 1966.

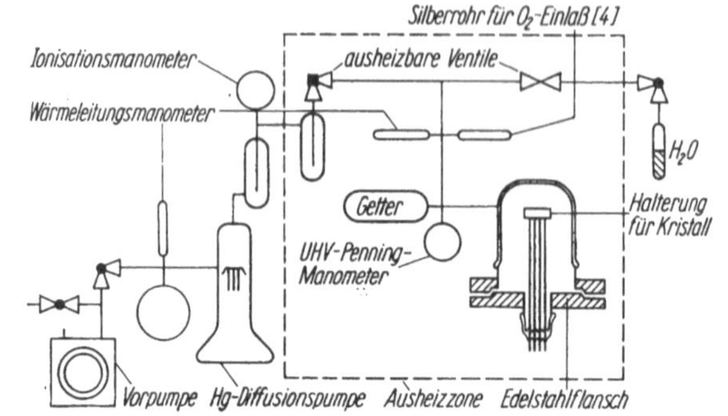

Fig. 1. Schematische Darstellung de Vakuumapparatur

Der Leitwert der Probe wurde mit Wechselstrom in einer Kompensationsschaltung gemessen (Genauigkeit 10^{-4} bis 10^{-6}). Die beiden Siliziumspaltstücke dienten gegenseitig als Feldelektroden. Die Influenzladung für die Feldeffektmessungen wurde mit einer Gleichspannung erzeugt.

2.2 Präparation der Kristalle

Für die Herstellung der Proben wurde p-leitendes Silizium (3 bis 14×10^3 Ωcm) verwendet. In Fig. 2 ist ein fertig vorbereiteter Kristall dargestellt. Die Kristalle von etwa $14 \times 5 \times 5$ mm³ Größe wurden an den Schmalseiten mit Kerben versehen. Die Orientierung der Proben wurde so gewählt, daß die Ebene der beiden Kerben eine (111)-Fläche war (Spaltfläche des Diamantgitters). Nach Schleifen und Ätzen der Probe wurden durch Aufdampfen von Aluminium und Legieren auf zwei Seiten je vier Kontakte angebracht, die durch aufgedrückte Golddrähte mit den Flanschdurchführungen verbunden waren. Die aufgedampfte Siliziumoxidschicht machte einerseits den Leitwert des Kristalls weitgehend unempfindlich gegen Gasadsorption, isolierte andererseits die darüber liegende elektrostatische Abschirmung aus Aluminium gegen den Kristall. Abschließend aufgedampftes Aluminium oder eingebranntes Glanzgold stellten einen leitenden Kontakt zur Abschirmung her (in Fig. 2 nicht eingezeichnet). Durch diese Schutzschichten konnte der Einfluß der Seitenflächen auf die Meßergebnisse stark reduziert werden.

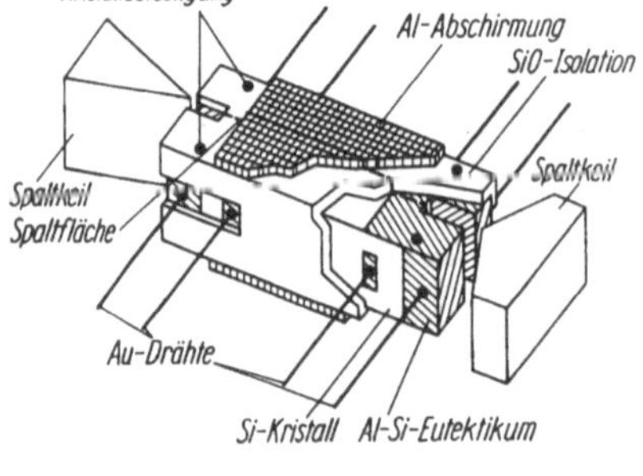

Fig. 2. Fertig vorbereiteter Siliziumkristall. Die Aufdampfschichten sind teilweise entfernt und in ihrer Dicke stark übertrieben eingezeichnet

2.3 Kristallhalterung und Spaltvorrichtung

In Fig. 3 ist zu sehen, wie der fertig vorbereitete Kristall in die Klammern der Kristallhalterung eingespannt wurde. Die Glasperlen in den Haltestäben isolierten den Kristall von der Halterung. Als verspannungsfreie Befestigung wurde das äußere Ende der Kristallhalterung (in Fig. 3 als ,,Lötstelle" bezeichnet) mit dem Spaltgestell durch Induktionserwärmung hart verlötet.

Für das Spalten wurden die zwei Spaltkeile (Fig. 2) mit einer magnetisch gedrehten Schraubspindel in die vorgesägten Kerben gepreßt. Der Spaltvorgang begann immer auf der Seite des spitzen Keiles.

Nach dem Spalten war die eine Kristallhälfte (die vordere in Fig. 3) durch die Drehachse beweglich. Der Drehwinkel und damit der Abstand der Kristallhälften wurde über den Auflagearm und eine Spiralscheibe magnetisch eingestellt, und mit Hilfe eines Lichtzeigers über den Spiegel auf $1-2 \cdot 10^{-2}$ mm genau abgelesen.

In Fig. 4 sind zwei zusammengehörige Spaltstücke nebeneinander gelegt. Die meisten sichtbaren Streifen gehen von der Stelle aus, an der der Spaltvorgang begann. Weitergehende Untersuchungen der Gestalt der Spaltflächen wurden nicht vorgenommen. Jedoch liegen einige Arbeiten anderer Autoren zu dieser Frage vor:

Auch in Bereichen, in denen mit dem Lichtmikroskop keine Unebenheiten zu erkennen sind, kann man mit elektronenmikroskopischen Aufnahmen treppenartig angeordnete Stufen (Höhe 50–500 Å) in (110)-Richtung, plastische Deformationen und Gleitlinien finden [5 bis 8].

Aus den Beugungsexperimenten mit langsamen Elektronen [9] folgt, daß die Anordnung der Atome in den obersten Atomlagen stark von der im Innern

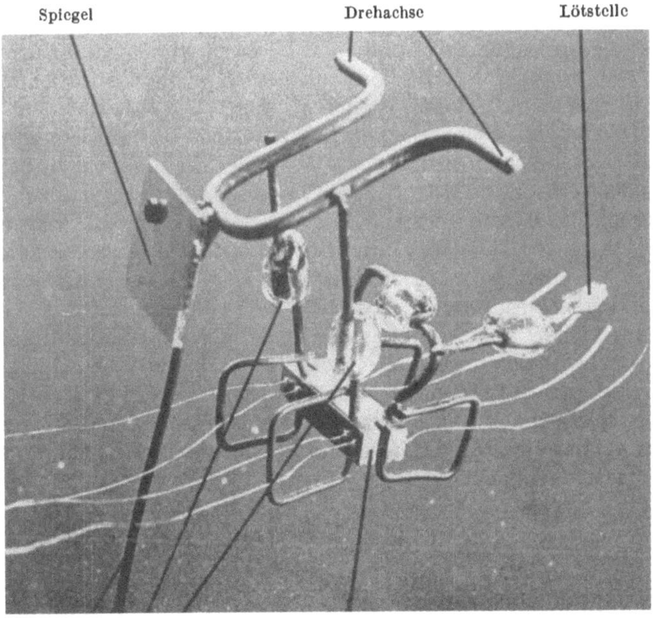

Fig. 3. Siliziumkristall, der in die Halterung eingespannt wurde

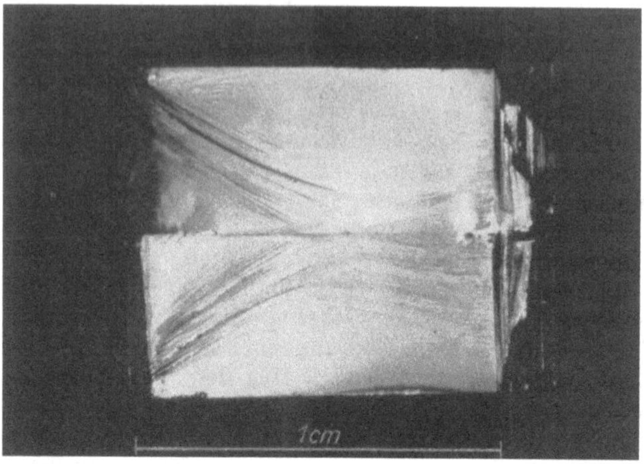

Fig. 4. Siliziumspaltflächen. Zwei zusammengehörige Spaltstücke sind nebeneinandergelegt

abweicht. Hiernach hat die Oberflächenstruktur eine 2-zählige Achse senkrecht zur Unterlage (das Volumen hat eine 3-zählige Achse). Durch die zahlreichen Einstellmöglichkeiten zum Volumen (Translation, Rotation) ist die Überstruktur dieser obersten Atomlagen im allgemeinen nicht einheitlich, besonders wenn Stufen und andere Störungen an der Oberfläche auftreten. Bei guten Spaltstücken waren die Bereiche einheitlicher Oberflächenstruktur maximal 0,5 mm groß [9].

2.4 Abstandsabhängigkeit des Feldeffektes

Da zur Bestimmung der Feldeffektbeweglichkeit die influenzierte Ladungsdichte wie beim idealen Plattenkondensator berechnet wird [1], wurden in einem Modellexperiment (elektrolytischer Trog) die Kapazitäten von Spalt- und Seitenflächen gemessen. Daraus läßt sich für die verwendeten Kristallabmessungen berechnen, wie die Feldeffektbeweglichkeit scheinbar vom Abstand abhängt, wenn man den idealen Plattenkondensator zugrunde legt. In Fig. 8 unten ist diese Abhängigkeit für die zwei Grenzfälle dargestellt, daß entweder nur die Spaltfläche oder nur die Seitenflächen zur Leitwertsänderung beitragen. Im allgemeinen setzt sich die gemessene Beweglichkeit aus Anteilen von beiden Flächen zusammen. Erst die Extrapolation auf den Abstand null liefert also eine sichere Aussage über die Spaltfläche, weil hier der Einfluß der Seitenflächen verschwindet. Die in Abschnitt 2.2. beschriebene elektrostatische Abschirmung wurde aufgedampft, um den Einfluß der Seitenfläche auszuschalten. Als Kriterium für die Brauchbarkeit einer Feldeffektmessung blieb aber stets die Abhängigkeit vom Abstand bestehen, um die Fälle mangelhafter Abschirmung auszuschließen.

3. Meßergebnisse

3.1 Ablauf einer Messung

An den Kristallen wurden folgende Messungen durchgeführt:

1. Leitwert vor dem Spalten,

2. Leitwert nach dem Spalten, wobei die Stromsonden der beiden Spaltstücke parallel geschaltet waren. Aus der Differenz der beiden Leitwerte folgt die Auswertung nach Handler [12].
3. Leitwert der einzelnen reinen Spaltstücke.
4. Feldeffekt der reinen Spaltfläche als Funktion des Abstandes an beiden Spaltstücken.
5. Kontinuierliche Messung der Leitwertänderung eines Spaltstückes durch Gaseinwirkung (Wasserdampf oder Sauerstoff).
6. Feldeffekt der definiert verunreinigten Spaltfläche.

3.2 *Oberflächenleitfähigkeit*

Die Oberflächenleitfähigkeit wurde aus den Leitwertsänderungen beim Gaseinlaß bestimmt. Denn der Leitwert der Probe zeigte beim Einlaß von Wasserdampf stets ein eindeutiges Minimum (19 Proben). Ein Beispiel ist in Fig. 5 zu sehen. Ab etwa 10^{-6} Torr sank der Leitwert und nahm bei etwa 10^{-4} bis 10^{-3}Torr einen Minimalwert an. Bei weiter steigendem Druck (bis etwa 0,2 Torr) erreichte die Leitfähigkeit Werte zwischen 10^{-6} und 10^{-5} A/V. Der Nullpunkt der Ordinate in Fig. 5 ist zunächst willkürlich. Er wurde hier unter der mit Hilfe des Feldeffekts bewiesenen Annahme festgelegt, daß das beobachtete Minimum mit dem theoretischen Minimum zwischen p- und n-Leitung zusammenfällt (Fig. 6). In der Zeit zwischen Spalten und Gaseinlaß (30 bis 60 Minuten) trat häufig bereits eine Leitwertänderung auf, die meistens eine Erniedrigung war. Ein Zusammenhang mit dem herrschenden Druck war nicht erkennbar, so daß die Leitwertänderung wohl nicht auf die Restgase zurückzuführen ist (siehe auch Abschnitt 4.1).

Beim Einlaß von Sauerstoff nahm der Leitwert stets ab. In Fig. 7 ist eine von den sieben gemessenen Kurven zu sehen. Ein nachfolgender Wasserdampfeinlaß brachte stets einen Leitwertsanstieg (bei einem Druck $> 10^{-3}$ Torr), manchmal vorher noch eine geringfügige Leitwerterniedrigung. Durch den Sauerstoff wurde also (ebenso wie durch Wasserdampf von 10^{-4} Torr) das Minimum der Oberflächenleitfähigkeit ganz oder zumindest bis auf einige 10^{-8} A/V erreicht.

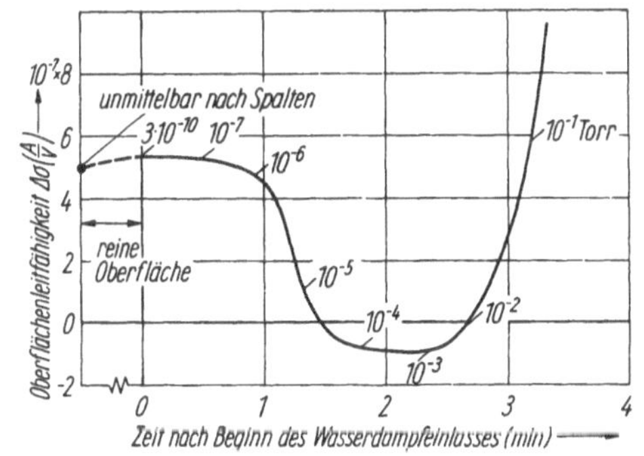

Fig. 5. Zeitlicher Verlauf der Oberflächenleitfähigkeit beim Einlaß von Wasserdampf auf die reine Spaltfläche

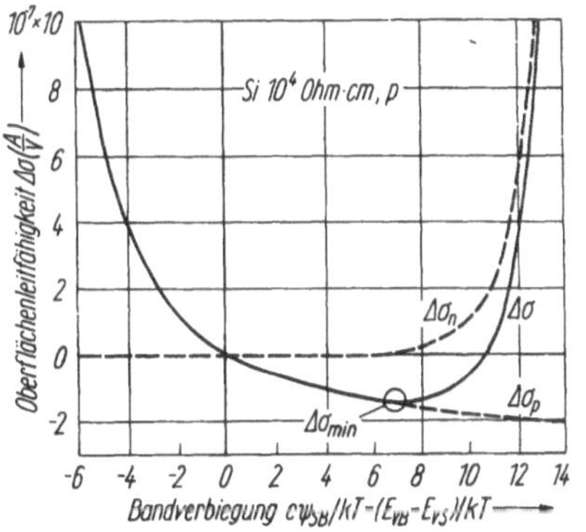

Fig. 6. Oberflächenleitfähigkeit berechnet als Funktion der Bandverbiegung $e\,\psi_{SB}/kT$. Eine Korrektur für Oberflächenstreuung ist wegen der geringen Bandverbiegungen nicht erforderlich. E_{VB} und E_{VS} sind die Energien der Valenzbandkante im Volumen bzw. an der Oberfläche

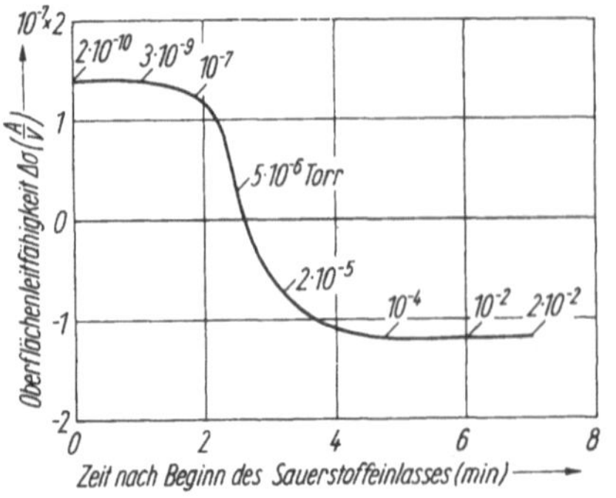

Fig. 7. Zeitlicher Verlauf der Oberflächenleitfähigkeit beim Einlaß von Sauerstoff auf die reine Spaltfläche

3.3 *Kontrollexperimente zur Oberflächenleitfähigkeit*

Durch Kontrollexperimente wurde überprüft, inwieweit die beobachteten Leitwertänderungen allein auf die Spaltflächen oder teilweise auch auf die Seitenflächen zurückzuführen seien. Dazu wurden Kristalle schon vor dem Spalten den Gasen ausgesetzt. Beim Wasserdampfeinlaß (5 Proben) zeigte sich erst bei einem Druck über 3×10^{-2} Torr eine Leitwertänderung. Beim Sauerstoffeinlaß lagen die Leitfähigkeitänderungen (umgerechnet auf eine fiktive Änderung der Spaltfläche) stets unter $d\Delta\sigma = 10^{-7}$ A/V (4 Proben). Hier lag der Druckbereich, in dem sich der Leitwert änderte, etwa eine Größenordnung höher als nach dem Spalten. Die Kontrollexperimente zeigen, daß die oben berichteten Ergebnisse innerhalb der Meßgenauigkeit allein auf die Spaltfläche zurückzuführen sind.

3.4 *Oberflächenleitfähigkeit aus dem Leitwert vor und nach dem Spalten*

In Tabelle 1 sind die Ergebnisse zusammengestellt, die aus der Leitwertänderung durch das Spalten [12] gewonnen wurden. Da die Meßwerte stark streuen, wurden sie in Bereiche eingeordnet, wodurch ein Vergleich mit den anderen Messungen erst möglich wird (siehe Abschn. 4.5).

Tabelle 1
Die Ergebnisse nach der Methode von Handler [12].
Die Proben stammen aus 3 Einkristallen

Zahl der Proben bei einem spezifischen Widerstand von 3×10^3 Ω cm	1×10^4 Ω cm	$1{,}4\times10^4$ Ω cm	Bereich der gemessenen Oberflächenleitfähigkeit
1	4	1	$10^{-7} < \Delta\sigma < 7\cdot10^{-7}$ A/V
0	3	1	$0 < \Delta\sigma < 10^{-7}$
1	2	0	$\Delta\sigma_{min} < \Delta\sigma < 0$
4	1	0	$\Delta\sigma < \Delta\sigma_{min}$

3.5 *Feldeffekt an der reinen Spaltfläche*

Bei drei Proben war die Feldeffektbeweglichkeit unabhängig vom Abstand und das Vorzeichen übereinstimmend negativ (p-Typ). Die Meßwerte von $\mu_{fe} = 1$ bzw. 1,5 bzw. 0,3 cm²/Vs lagen an der Nachweisgrenze. Sie sind deshalb

Fig. 8. Oben: Feldeffektbeweglichkeit, gemessen als Funktion des Abstandes d der beiden Spaltstücke.
Unten: Feldeffektbeweglichkeit, berechnet aus den gemessenen Teilkapazitäten (siehe Abschn. 2.4) und den Kristalldimensionen für die beiden Grenzfälle, bei denen nur die Spaltfläche bzw. die Seitenflächen zur Leitwertänderung beitragen

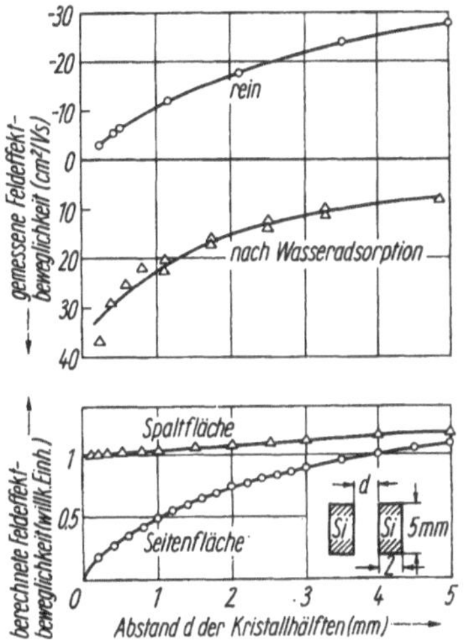

nur bis auf etwa $\pm 50\%$ sicher. Bei weiteren drei Kristallen konnte kein Feldeffekt gefunden werden, d. h. $|\mu_{fe}| \lesssim 1$ cm²/Vs für alle Abstände. Bei 19 Proben nahm die Feldeffektbeweglichkeit nichtlinear mit dem Abstand zu. In diesen Fällen war die Abschirmung der Seitenflächen offensichtlich nicht ausreichend. Die Extrapolation auf den Abstand Null (siehe Abschn. 2.4) ergab nur eine obere Grenze des Betrages ($|\mu_{fe}| \lesssim 2$ cm²/Vs), aber keine Entscheidung über das Vorzeichen. Ein Beispiel zeigt die oberste Kurve in Fig. 8. In diesem Fall war die Beweglichkeit für endliche Abstände negativ, bei anderen Proben jedoch meistens positiv. Alle Messungen sind also nur mit dem Ergebnis verträglich, daß die reine Spaltfläche p-leitend ist und die Feldeffektbeweglichkeit etwa 1 cm²/Vs beträgt.

3.6 Feldeffekt nach Gaseinlaß

Nach Einlaß von Wasserdampf (0,2 Torr) und Evakuieren war die Feldeffektbeweglichkeit stets positiv (n-Typ), unabhängig vom Abstand und etwa 2 Größenordnungen höher als vorher. Bei 12 Kristallen lag ihr Wert zwischen 200 und 1000 cm²/Vs, also relativ nahe bei der Volumenbeweglichkeit der Elektronen (1500 cm²/Vs), nur bei 5 Kristallen darunter. Wie wesentlich beim Feldeffekt die Abhängigkeit vom Abstand ist, sieht man aus Fig. 8. Hier ist die Feldeffektbeweglichkeit vor und nach dem Einlaß von Wasserdampf (10^{-2} Torr) als Funktion des Abstandes aufgetragen. Durch Vergleich mit den unteren Kurven der Figur erkennt man, daß durch den Wasserdampf die Feldeffektbeweglichkeit der Spaltfläche (abstandsunabhängiger Anteil der Kurve) auf etwa 35 cm²/Vs gestiegen ist, während der abstandsabhängige Anteil der Seitenfläche annähernd konstant geblieben ist.

Bei den hohen Feldeffektbeweglichkeiten nach Wasserdampfeinwirkung war die Leitwertänderung nicht mehr proportional zur Feldstärke. Wie in Fig. 9 zu sehen ist, konnten hier das Minimum der Oberflächenleitfähigkeit und ein sehr flacher Anstieg bei negativer Influenzladung beobachtet werden (8 Proben). Durch das Minimum ist eine neue Absoluteichung der Oberflächenleitfähigkeit möglich.

Nach Sauerstoffeinlaß wurde kein Feldeffekt beobachtet, was nicht überraschend ist, da die Oberflächenleitfähigkeit dann nahe dem Minimum lag (Abschn. 3.2). Die nachfolgende Wasseradsorption ergab jedoch stets eine deutlich positive Feldeffektbeweglichkeit (n-Leitung) zwischen 20 und 100 cm² pro Vs (5 Proben).

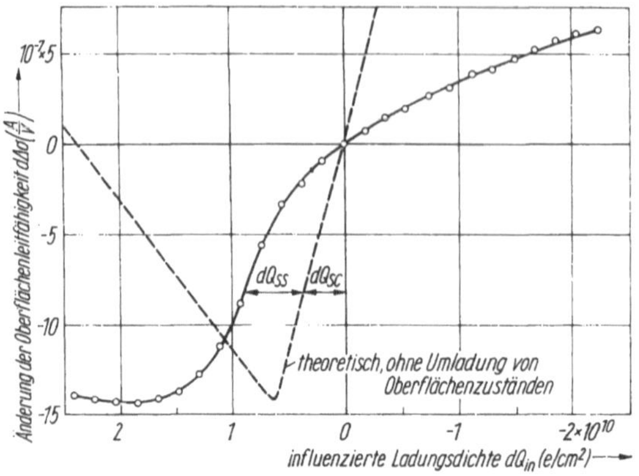

Fig. 9. Änderung der Oberflächenleitfähigkeit durch Influenz von Ladung (nach Wasseradsorption). dQ_{SS} Ladungsänderung in den Oberflächenzuständen, dQ_{SC} Ladungsänderung in der Raumladungsschicht

Bei zwei Kristallen wurde durch Feldeffekt bestätigt, daß das Minimum beim Gaseinlaß das theoretische Minimum zwischen p- und n-Leitung darstellt. Dazu wurde nur so viel Wasserdampf (5×10^{-3} Torr) eingelassen, daß zwar die Feldeffektbeweglichkeit schon groß geworden war, die Seitenflächen sich aber noch kaum verändert hatten. Nach Evakuieren wurde dann mit dem Feldeffekt ein Minimum erreicht, das mit dem durch Gaseinlaß gefundenen Minimum fast zusammenfiel (Differenz 1×10^{-7} bzw. $0{,}2\times10^{-7}$ A/V).

4. Diskussion

4.1 Reinheit der Oberfläche

Bei den Messungen an der reinen Oberfläche betrug der Druck etwa 10^{-10} Torr. Dann könnte mit dem für Sauerstoff gemessenen Haftkoeffizienten von 10^{-3} bis 10^{-2} [13] in einer halben Stunde höchstens 0,01 bis 0,1% einer Monoschicht adsorbiert werden. Wegen der Getterpumpe sollte außerdem der Anteil der leicht chemisorbierbaren Gase gering sein. Die Spaltflächen waren also bis zum Gaseinlaß etwa nach 30 min rein. Beim Gaseinlaß (siehe Fig. 5 und Fig. 7) wurde erst beim Druck von etwa 10^{-6} Torr eine Leitwertänderung beobachtet. Deshalb können die vor dem Gaseinlaß auftretenden Leitwertänderungen nicht durch die Restgase hervorgerufen sein. Aber die Oberflächenatome der Spaltflächen haben eine vom Innern stark abweichende Anordnung [9 bis 11] und befinden sich nicht im thermischen Gleichgewicht. Da diese Oberflächenstruktur in verschieden orientierte Bereiche aufgeteilt ist [9], wäre durch eine Veränderung der Bereichsgrenzen eine Beeinflussung der Oberflächenzustände und damit der Oberflächenleitfähigkeit denkbar. Systematische Untersuchungen über den Zusammenhang zwischen der Oberflächenstruktur und den Oberflächenzuständen fehlen noch.

4.2 Fehlerquellen bei Leitfähigkeits- und Feldeffektmessungen

Der Einfluß der Seitenflächen auf den Feldeffekt ist ausgeschaltet durch die Abschirmung des Kristalls und durch die Messung der Beweglichkeit für verschiedene Abstände der Feldelektrode, der Einfluß der Seitenflächen auf die Leitwertänderung beim Gaseinlaß wurde vor dem Spalten kontrolliert.

Liegt an der Oberfläche eine Inversionsschicht vor (in Fig. 6 z. B. $e\,\psi_{SB} = 11\,kT$), dann ist es wichtig festzustellen, ob diese Inversionsschicht nicht elektrisch isoliert ist, d. h. ob die Ladungsträger in der Schicht auch zum Gesamtleitwert beitragen. Dies kann man mit Hilfe des Feldeffektes entscheiden. Denn bei einer isolierten n-Inversionsschicht ergibt der Feldeffekt p-Leitung, während man bei nicht isolierter Schicht n-Leitung findet. Da nach Wasseradsorption die Feldeffektbeweglichkeit n-Leitung anzeigte und nahe der Volumenbeweglichkeit lag, konnte die Randschicht keine erheblichen Übergangswiderstände zu den Stromkontakten haben. Durch die experimentell gefundenen Leitwertserniedrigungen (Fig. 5 und Fig. 7) läßt sich auch ausschließen, daß an der reinen Spaltfläche eine isolierte Inversionsschicht vorlag, deren Übergangswiderstände erst durch die Wasseradsorption aufgehoben wurden. Dann hätte der Leitwert bei Wasserdampfeinlaß nur ansteigen dürfen, da sich die Löcherkonzentration nur noch geringfügig verringern kann (Kurve $\Delta\sigma_p$ in Fig. 6).

4.3 Auswertung für die reine Spaltfläche

Aus der Differenz des Leitwertes unmittelbar nach dem Spalten und des Minimalleitwertes beim Gaseinlaß $dG = G_{rein} - G_{min}$ wurde die Änderung der Oberflächenleitfähigkeit $d\Delta\sigma = \Delta\sigma_{rein} - \Delta\sigma_{min}$ und daraus die Oberflächenleitfähigkeit der reinen Spaltfläche $\Delta\sigma_{rein}$ bestimmt, da die Größe $\Delta\sigma_{min}$ sich aus den Volumeneigenschaften berechnen läßt (Fig. 6). Die Kontrollversuche mit dem Feldeffekt bestätigen, daß das Minimum beim Gaseinlaß das theoretische Minimum zwischen p- und n-Leitung darstellt. Die Zahlenwerte sind in der Tabelle 2 zusammengestellt. Bei den gewählten Volumendotierungen liegt also eine schwache Anreicherungsschicht vor. Da der Abstand Ferminiveau-Valenzbandkante an der Oberfläche nur sehr wenig von der Dotierung abhängt (siehe weiter unten), kann man sagen, daß für einen spezifischen Widerstand von 200 Ωcm (p-Leitung) die Flachbandsituation erreicht wäre.

Aus der Feldeffektbeweglichkeit ($\mu_{fe} \approx 1$ cm²/Vs) kann man die Dichte der Oberflächenzustände ableiten. Es gibt viele Verteilungen der Zustände über die Energieskala, die mit den Meßergebnissen verträglich sind. Nimmt man ein Kontinuum an, so ergibt sich eine Dichte von etwa 10^{14} Zustände/eV cm². Von allen möglichen Verteilungen erfordert diejenige die geringste Gesamtzahl von Oberflächenzuständen, bei der alle Zustände auf der Höhe des Ferminiveaus liegen. Der daraus folgende Wert (10^{13} cm^{-2}) ist wesentlich größer als die Zahl der Löcher in der Randschicht (etwa 5×10^9 e/cm²). Die Oberflächenzustände müssen also aus „Donatoren" und „Akzeptoren" bestehen, deren Gesamtladung jedoch sehr gering ist. Wegen der hohen Zustandsdichte hängt der Abstand Ferminiveau–Valenzbandkante an der Oberfläche $(E_F - E_V)_S$ nur in sehr geringem Maße von der Dotierung ab (siehe z. B. [2]). Deshalb ist es sinnvoll, diesen Energieabstand zur Charakterisierung der reinen Oberflächen anzugeben.

4.4 Auswertung für die Spaltfläche nach Gaseinwirkung

Nach Adsorption von Wasserdampf ist die Oberfläche n-leitend. Wegen der großen Feldeffektbeweglichkeit kann die Bandverbiegung durch Influenz von Ladung um etwa $\pm$ 0,1 eV verändert werden. Wenn man dabei das Minimum der Leitfähigkeit erreicht, kann man mit dieser Messung allein Oberflächenleitfähigkeit und Oberflächenzustände bestimmen (Fig. 9). Durch Vergleich mit der theoretischen Kurve, die man mit Hilfe der Raumladungstheorie ableitet und mit dem Minimalwert an die Meßkurve anpaßt, erhält man direkt die Ladungsänderung dQ_{SS} in den Oberflächenzuständen. Man kann also die Dichte und Energie der Oberflächenzustände bis auf die durch die Fermische Besetzungsfunktion bedingte Unschärfe angeben. Die Auswertung für Fig. 9 ergibt für den Bereich von der Mitte der verbotenen Zone bis etwa 8 kT darüber eine konstante Zustandsdichte ($S_E \approx 10^{11}$ eV^{-1} cm^{-2}), die für höhere Energien steil ansteigt.

4.5 *Auswertung der Leitwertänderung durch das Spalten*

Bei der von Handler vorgeschlagenen Methode [12] können die für die Oberflächenleitfähigkeit gefundenen Werte wegen Unebenheiten der Spaltfläche nur zu klein sein. Unter Berücksichtigung dieser Fehlerquelle kann man aus Tab. 1 sehen, daß die Ergebnisse gut zu den in Tab. 2 berichteten passen. Denn die höchsten Werte (Zeile 1) stimmen mit den durch Gaseinlaß gewonnenen Ergebnissen überein. Die Meßwerte in den Zeilen 2 bis 4 liegen offensichtlich wegen der Unebenheiten zu tief, in Zeile 4 sogar unter dem theoretisch zulässigen Minimalwert. Da die Störungen durch die Unebenheiten der Spaltfläche proportional zur Volumenleitfähigkeit sind, ist es verständlich, daß bei den hochohmigen Kristallen mehr Proben den richtigen Wert liefern. Mit diesen Messungen allein läßt sich allerdings keine Entscheidung zwischen Anreicherung und Inversion treffen.

5. Vergleich mit Ergebnisssen anderer Autoren

Palmer u. a. [14] maßen den Feldeffekt und die Channelleitfähigkeit an im Ultrahochvakuum hergestellten Spaltflächen. Ihre Ergebnisse wurden bereits mehrfach diskutiert [15, 2]. Sie fanden bei den gewählten Dotierungen keine Channelleitfähigkeit über 10^{-8} A/V.

Bei den Feldeffektmessungen fanden Palmer u. a. sowohl für n- als auch für p-leitendes Material eine Feldeffektbeweglichkeit von 50 cm^2/Vs, n-leitend. Sie bestimmten jedoch diese Größe nur für einen einzigen Abstand und hatten die Seitenflächen nicht geschützt. Da dieses Ergebnis sich mit keinem Modell für die Oberflächenzustände verstehen läßt, vermutete Handler [15] hier den Einfluß der Seitenflächen. Durch unsere Messungen wird diese Vermutung bestätigt. Denn bei ungeschützten Seitenflächen und geeigneten Abständen wurde die gleiche Beweglichkeit nach Vorzeichen und Betrag gefunden. Durch Variation des Abstandes wurde nachgewiesen, daß diese Feldeffektbeweglichkeit eindeutig durch die Seitenflächen bestimmt war.

In Tab. 3 sind die Ergebnisse zusammengestellt, die bis jetzt an der reinen Siliziumoberfläche gwonnen wurden. Allen und Gobeli [16] bestimmten an reinen Spaltflächen aus Messung der optischen und thermischen Austrittsarbeit den Abstand Valenzbandkante–Ferminiveau. Die hier aus der Oberflächenleitfähigkeit gewonnenen Werte stimmen damit überein. Aus der geringen Dotierungsabhängigkeit der thermischen Austrittsarbeit folgerten Allen und Gobeli eine hohe Dichte der Oberflächenzustände ($S_E = 10^{14}$ cm^{-2} eV^{-1}). Aus den hier berichteten Feldeffektmessungen folgt der gleiche Wert. Die Ergebnisse beider Arbeiten stimmen also trotz der verschiedenen Methoden überein.

Aspnes und Handler [17] maßen die Änderung des Leitwerts durch das Spalten. Weitere Informationen erhielten sie aus dem Temperatur- und Sauerstoffeinfluß auf diese Änderung. Die von ihnen gefundene Oberflächenleitfähigkeit und Bandverbiegung stimmen für die reine sowie für die mit Sauerstoff bedeckte Spaltfläche weitgehend mit den vorliegenden Messungen überein. Leider wird von den Autoren weder der Einfluß des Sauerstoffs auf die Seitenflächen dis-

Tabelle 2

Zusammenfassung der Ergebnisse und der Auswertung für die reine Spaltfläche

Ausgangsmaterial	Spezifischer Widerstand ϱ	3	10	14	$\cdot 10^{3}$ Ωcm
	Theoretischer Minimalwert der Oberflächenleitfähigkeit $\Delta\sigma_{min}$	−2,5	−1,4	−0,9	$\cdot 10^{-7}$ A/V
Meßwerte	Zahl der gemessenen Proben	8	10	3	
	Änderung von $\Delta\sigma$ (Mittelwert) $\Delta\sigma_{rein} - \Delta\sigma_{min}$	6,6 ± 1,2	5,9 ± 1,2	4,3 ± 1,7	$\cdot 10^{-7}$ A/V
	Feldeffektbeweglichkeit μ_{fe}	—	≈1; p *)	—	cm²/Vs
Auswertung	Oberflächenleitfähigkeit der reinen Spaltfläche $\Delta\sigma_{rein}$	4,1	4,5	3,4	$\cdot 10^{-7}$ A/V
	Zahl der Löcher in der Randschicht	5,1	5,6	4,3	$\cdot 10^{9}$ cm^{-2}
	Abstand Ferminiveau-Valenzband an der Oberfläche	0,35 ± 0,02	0,36 ± 0,02	0,36 ± 0,02	eV
		14,0 ± 1	14,4 ± 1	14,3 ± 1	kT
	Bandverbiegung	−3,1	−4,4	−4,1	kT
	Dichte der Oberflächenzustände S_E	—	10^{14} *)	—	eV^{-1} cm^{-2}
	Zahl der Oberflächenzustände N_S	—	$>10^{13}$ *)	—	cm^{-2}

*) Mittelwert aus drei Messungen.

Tabelle 3

Gegenüberstellung der bisherigen Ergebnisse an der reinen Siliziumoberfläche. Alle Experimente wurden an Spaltflächen ausgeführt, nur Heiland und Lamatsch reinigten ihre Proben durch Argonbeschuß und Glühen.

Die angegebenen Größen bedeuten folgendes:

μ_{fe} = Feldeffektbeweglichkeit

S_E = Zahl der Oberflächenzustände pro Energie und Fläche

N_S = Zahl der Oberflächenzustände pro Fläche

$(E_F - E_V)_S$ = Abstand Ferminiveau-Valenzbandkante an der Oberfläche.

Autoren	Meßgröße	μ_{fe} cm^2/Vs	S_E $eV^{-1}\,cm^{-2}$	N_S cm^{-2}	$(E_F - E_V)_S$ eV
F. G. Allen und G. W. Gobeli [16]	thermische und optische Austrittsarbeit	–	10^{14}	$>10^{13}$	$0{,}34 \pm 0{,}1$
G. Heiland und H. Lamatsch [18, 19]	Oberflächenleitfähigkeit und Feldeffekt	$\approx 0{,}1$; p	10^{15}	$>10^{14}$	$0{,}30 \pm 0{,}03$
D. E. Aspnes und P. Handler [17]	Oberflächenleitfähigkeit	–	$>10^{11}$	$>10^{10}$	$0{,}28 \pm 0{,}08$
M. Henzler (vorliegende Arbeit)	Oberflächenleitfähigkeit und Feldeffekt	≈ 1; p	10^{14}	$>10^{13}$	$0{,}36 \pm 0{,}02$

kutiert, noch wird angegeben, ob die Meßwerte nicht auch mit einer n-leitenden Oberfläche verträglich sind.

Heiland und Lamatsch [18, 19] untersuchten Feldeffekt und Oberflächenleitfähigkeit an durch Ionenbeschuß und Glühen gereinigten Siliziumoberflächen. Obwohl die untersuchten Kristalle eine andere Oberfläche (verschiedene kristallographische Flächen durch Zylinderform, andere Oberflächenstruktur durch Glühbehandlung) hatten, finden sie etwa die gleiche Bandverbiegung. Die Dichte der Oberflächenzustände liegt allerdings etwa eine Größenordnung höher. Der Einfluß des Sauerstoffs auf die Oberflächenleitfähigkeit ist ähnlich wie bei den hier in Abschnitt 3.2 berichteten Messungen.

6. Schluß

Aus allen Untersuchungen geht hervor, daß auch an der von Fremdatomen freien Oberfläche viele Oberflächenzustände vorhanden sind. Als Ursache kommen die vom Inneren stark abweichende Oberflächenstruktur [9] in Frage (z. B. wegen vervielfachter Periodizitätskonstante) oder auch die Grenzen einheitlicher Orientierung in der Oberflächenstruktur. Bei der Strukturänderung durch kurzzeitiges Erhitzen auf etwa 700 °C [9] wurde eine Änderung der elektrischen Eigenschaften festgestellt [20]. Auch die submikroskopischen und sichtbaren Störungen der Oberfläche (z. B. Stufen und Versetzungen) verursachen vielleicht Oberflächenzustände. So fanden Allen und Gobeli [16] einen Einfluß der sichtbaren Unebenheiten auf die thermische Austrittsarbeit. Nach den theoretischen Vorstellungen von Tamm und von Shockley (siehe z. B. [1], Seite 398) kann bereits das Abschneiden der Volumenperiodizität der Grund für die Oberflächenzustände sein. Um die gefundenen Oberflächenzustände einer dieser Ursachen zuordnen zu können, sind jedoch weitere Experimente erforderlich.

Herrn Prof. Dr. G. Heiland danke ich aufrichtig für die Anregung zu dieser Arbeit und die hilfreichen Diskussionen bei ihrer Durchführung.

Herrn Dr. E. Spenke von den Siemens-Schuckertwerken, Laboratorium Pretzfeld und Herrn Dr. H. Striebel von der Firma Wacker-Chemie Burghausen danke ich für die großzügige Überlassung des Kristallmaterials.

Die Deutsche Forschungsgemeinschaft unterstützte die Arbeit durch zahlreiche Leihgaben.

Literatur

[1] G. Heiland, Fortschr. Phys. **9**, 393 (1961).
[2] G. Heiland, Elektronische Zustände und freie Valenzen an Halbleiteroberflächen, in Festkörperprobleme Bd. III, Herausg. F. Sauter, Vieweg, Braunschweig 1964 (S. 125).
[3] A. Many, Y. Goldstein und N. B. Grover, Semiconductor Surfaces, North-Holl. Publ. Comp., Amsterdam 1965.
[4] N. R. Whetten und J. R. Young, Rev. sci. Instrum. **30**, 472 (1959).
[5] G. W. Gobeli und F. G. Allen, J. Phys. Chem. Solids **14**, 23 (1960).
[6] D. R. Frankl, J. appl. Phys. **34**, 3514 (1963).
[7] W. Mehl und M. C. Coutts, J. appl. Phys. **34**, 2120 (1963).
[8] M. Inoue, Japan. J. appl. Phys. **3**, 57 (1964).
[9] J. J. Lander, G. W. Gobeli und J. Morrison, J. appl. Phys. **34**, 2298 (1963).
[10] J. J. Lander und J. Morrison, J. appl. Phys. **34**, 1403 (1963).
[11] J. J. Lander, Surface Sci. **1**, 125 (1964).
[12] P. Handler, Appl. Phys. Letters **3**, 96 (1963).
[13] R. J. Archer und G. W. Gobeli, J. Phys. Chem. Solids **26**, 343 (1965).
[14] D. R. Palmer, S. R. Morrison und C. E. Dauenbaugh, Phys. Rev. Letters **6**, 170 (1961).
[15] P. Handler, Phys. Rev. **126**, 971 (1962).
[16] F. G. Allen und G. W. Gobeli, Phys. Rev. **127**, 150 (1962).
[17] D. E. Aspnes und P. Handler, Surface Sci. **4**, 353 (1966).
[18] G. Heiland und H. Lamatsch, Surface Sci. **2**, 18 (1964).
[19] H. Lamatsch, phys. stat. sol. **9**, 119 (1965).
[20] F. G. Allen und G. W. Gobeli, J. appl. Phys. **35**, 597 (1964).
[21] M. Henzler und G. Heiland, Solid State Comm. **4**, 499 (1966).

(Received December 3, 1966)

phys. stat. sol. 19, 847 (1967)

Subject classification: 18.2; 21.1; 21.1.1

C.E.N., Grenoble, Isère

Electron Diffraction from a Magnetic Phase Grating

By

R. H. WADE

It is shown that periodic magnetic structures in thin ferromagnetic specimens produce diffraction effects in a transmitted electron beam due to the action of such structures as phase gratings.

Nous montrons qu'une structure magnétique périodique dans des échantillons ferromagnétiques produit un effet de diffraction dans le faisceau d'électrons transmis, ceci par suite du comportement de la structure en réseau de phase.

1. Introduction

Aharanov and Bohm [1] show that in the presence of a magnetic field there is a phase difference (Δ) between two parts of an electron beam, due to the difference in the magnetic vector potential along the two paths, which is given by the total magnetic flux $\int B \, da$ enclosed between the paths $\Delta = e \int B \, da/c \, \hbar$; e is the electronic charge in e.s.u.; c is the velocity of light, $2 \pi \hbar$ is Planck's constant, B is given in e.m.u., and da is an element of area.

Experimental investigations using an electrostatic biprism to divide an electron beam so as to pass either side of a flux carrying element (ferromagnetic whisker [2, 3], solenoid [4]) have verified the existence of the predicted phase effects which do not depend on the electron beam actually passing through the field region. A complementary experiment by Boersch et al. [5] shows that interference fringes observed in the transmission electron microscope images of domain walls in thin ferromagnetic films are properly explained in terms of the magnetic vector potential.

A further consequence of the phase effects of magnetic fields on an electron beam appears to have escaped attention: a ferromagnetic film possessing a magnetic repeat structure will act as a phase grating for a transmitted electron beam. The purpose of the present paper is to present an experimental and theoretical account of electron diffraction by magnetic phase gratings.

2. Choice of Specimens

Two conditions must be fulfilled if clear diffraction effects are to be observed. The repeat structure must be extremely regular; the periodicity must not be greater than about 1 μm since this is about the limit of the resolution in the diffraction image. These conditions appear to be fulfilled by the following magnetic structures:

(i) Cobalt foils in which the anisotropy axis makes a large angle with the foil plane show a regular structure of antiparallel domains with spacings in the range 10^{-5} to 10^{-4} cm [6].

(ii) Cross-tie walls in vacuum condensed nickel-iron films show a repeat structure along the wall direction with a minimum spacing around 10^{-4} cm.

(iii) Stripe domain structures [7, 8] show spacings upwards of about 10^{-5} cm.

In the present paper results obtained from cobalt foils and from cross-tie walls in Ni–Fe are presented.

3. Specimen Preparation

Cobalt foils were prepared from 0.1 mm sheet (Johnson, Matthey, "specpure") annealed for 24 h at 800 °C and electropolished in 5% perchloric acid, 45% glacial acetic acid, 50% butyl cellosolve.

Permalloy films (80 Ni–20 Fe) were prepared by vacuum condensation onto cleaved mica held at room temperature. The anisotropy constant which controls the periodicity of the cross walls was varied using oblique incidence condensation.

4. Microscope Operation

A Philips EM 200 microscope which has three lenses below the objective lens is operated as a small angle diffraction camera in the following way. The objective lens is inoperative; the condenser lenses are strongly overfocussed to give a small effective illumination source well above the specimen plane: the projector lens is at maximum strength; the camera length is determined by the excitation of the intermediate lens; the diffraction image is focussed by varying the strength of the diffraction lens.

The range of camera lengths used in the present work was 25 to 75 m, whilst a maximum value of about 150 m could be achieved at 100 kV. The details of the image formation are as described by Wade and Silcox [9] with an additional magnification by the extra lens used in the present case. The resolution in the diffraction image, determined by the angle subtended by the illumination source at the specimen plane, was determined experimentally by using a carbon replica of a 1200 lines/mm optical diffraction grating. The diffraction image of the grating shown in Fig. 1 has an angular resolution of around 2×10^{-6} rad corresponding to a spacing of 2 μm. The same grating was used to calibrate the camera length as a function of the intermediate lens strength.

The specimen area contributing to the diffraction image is limited by the objective aperture. An image of the object with a maximum magnification of about 4000 is obtained by a slight increase in the strength of the diffraction lens.

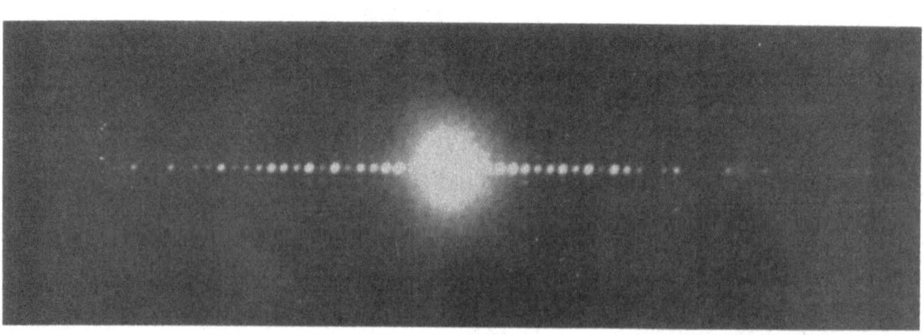

Fig. 1. Diffraction image from a carbon replica of an optical grating with 1200 lines/mm. The camera length on the original plate was 130 m, the beam potential was 80 kV. The resolution in the image is about 2 μm. The structure perpendicular to the main axis of the pattern is due to a regular faulting in the grating

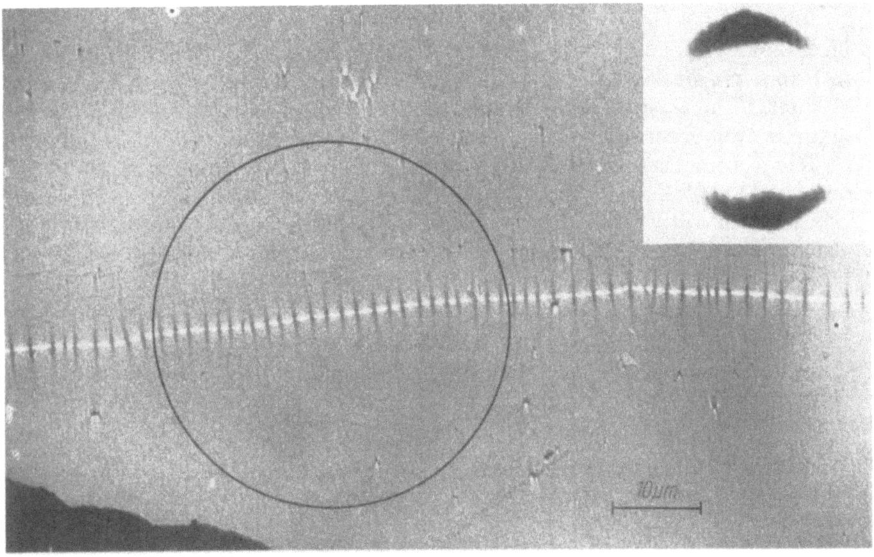

Fig. 2. Off-focus image of a cross-tie wall in a Ni–Fe film about 500 Å thick condensed at an oblique incidence angle onto a mica substrate. The inset shows the diffraction image of the encircled region obtained at an original camera length of 40 m at 80 kV, and with the specimen tilted 30° about an axis perpendicular to the wall length

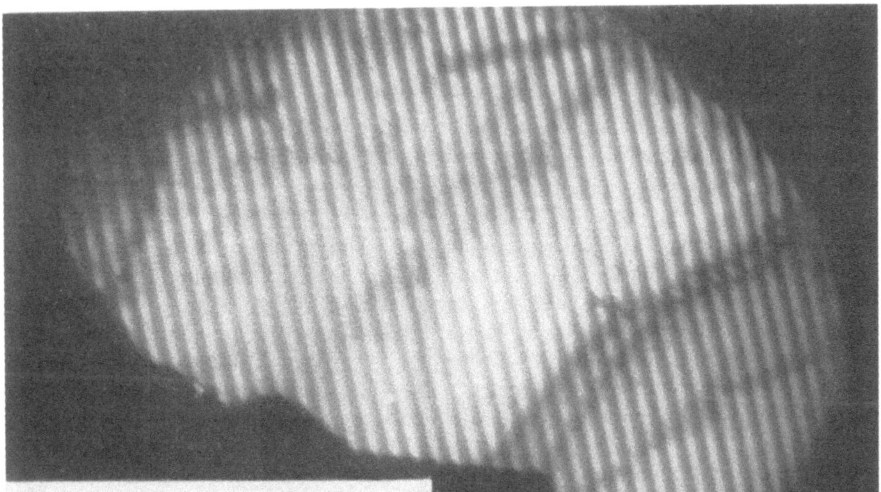

Fig. 3. Off-focus image of magnetic domains in a cobalt foil with a period 2.7×10^{-5} cm. The inset shows the diffraction image corresponding to this region obtained at 100 kV with an original camera length of 30 m. In both figures the diffraction images are shown correctly oriented with respect to the direct images

5. Experimental Results

Fig. 2 shows the image and electron diffraction pattern obtained from a cross-tie wall in a permalloy film prepared at an oblique incidence angle and about 500 Å thick. The separation of the diffraction fringes corresponds to a distance of 1.6 μm in the direct image, the average separation of the cross walls is 1.7 μm.

The image and the corresponding diffraction pattern from a single crystal region of a cobalt foil are shown in Fig. 3. The diffraction image yields a spacing of 2.9×10^{-5} cm whilst the direct image shows a periodicity of 2.7×10^{-5} cm. Results obtained from the direct and diffraction images are compared in Table 1.

Table 1

Direct image	2.7	2.3	3.5	3.8	6.0	2.1	2.7	2.7	2.7
Diffraction image	2.9	2.4	3.6	3.4	5.9	1.9	2.4	2.7	3.0

A comparison of the domain periodicities measured from direct images with those calculated from the diffraction patterns from the same regions of cobalt foils. The values are given in units of 10^{-5} cm.

Continuous defocussing of the wall images in the cobalt foils was noticed to produce successive reversals in the image contrast as shown in Fig. 4. The defocussing distance corresponding to one contrast reversal was found to be about 0.1 cm by measuring the separation of Fresnel fringes [10] produced around holes in a collodion film.

6. Discussion

6.1 Diffraction effects

Only the observations on cobalt will be discussed in detail. There are two possible interpretations of the diffraction patterns of the type shown in Fig. 3.

6.1.1 Lorentz deflection interpretation

Adjacent domains deflect the transmitted electron beam in opposite senses through the angle $\Phi = B\,e\,D/c\,m\,v$, where B is the flux density (e.m.u.) in the foil of thickness D,m and v are the relativistic mass and velocity of the electron, respectively. Each system of parallel domains can be considered to be separately illuminated along the appropriate direction $\pm\Phi$ so that the two sets of domains form independent amplitude gratings of period $2a$ (a is the width of an individual domain), which will produce distinct diffraction images separated by the angle 2Φ.

The periodic structures can be described by

$$f(x) = \delta\,(x - 2\,a\,n) * b(x)\,, \tag{1}$$

where n is a positive or negative integer, $*$ denotes the convolution operation, and $b(x) = 1$ for $|x| < a/2$, $b(x) = 0$ for $|x| > a/2$.

The amplitude in the diffraction image is given by the Fourier transform of $f(x)$

$$F(s) = \delta\left(s - \frac{n}{2\,a}\right) \cdot B(s)\,. \tag{2}$$

$B(s)$ is the Fourier transform of $b(x)$.

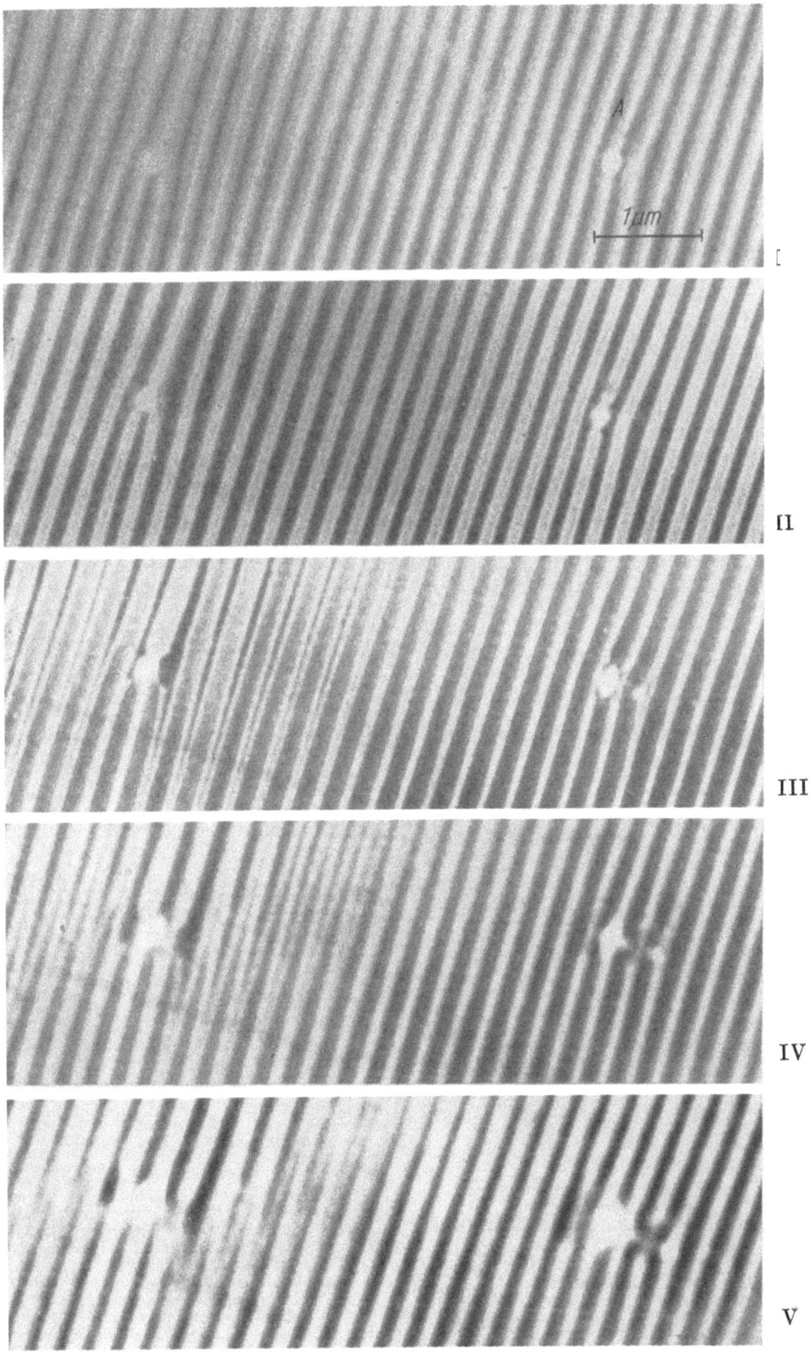

Fig. 4. The domain wall images in a cobalt foil showing the change in image contrast as the off-focus distance is increased from *I* to *V* in steps of about 0.1 cm. The hole A serves as a reference point

Since $B(s)$ has zeros at $s = 1/a, 2/a, 3/a, \ldots$ these orders will be missing in the diffraction image which will show maxima of orders 0, 1, 3, 5, 7, etc.

The entire diffraction pattern consists of the sum of the two systems $F\,(s - g)$ and $F\,(s + g)$ where $g = \Phi/\lambda$ (λ is the electron wave-length). Since g and a are independent the two patterns will show a mismatch in the overlap region unless $|g| = n/2\,a$.

6.1.2 Phase grating interpretation

Fig. 5a shows a cross section of the magnetic structure in a cobalt foil. The magnetic phase difference between a point x and the origin 0 is given by $\Delta(x) = \frac{e\,D}{c\,\hbar}\int_0^x B_y\,dx$ and has the form shown in Fig. 5b. This repeat structure can be expressed in the same form as in (1) with $b_1(x) = A(x)\exp[i\,\Delta(x)]$ where $A(x)$ is the amplitude and $\Delta(x)$ is the phase. In the present case the object does not impose any amplitude variations on the transmitted electron beam. The amplitude in the diffraction image is again given by equation (2) with

$$B_1(s) = A\int_{-a}^{a}\exp\left[i\left(\Delta(x) - 2\,\pi\,x\,s\right)\right]dx\,, \tag{3}$$

where

$\Delta(x) = \Delta_0\,(a - x)/a$ for $0 \leqq x \leqq a$, and $\Delta(x) = \Delta_0(a + x)/a$ for $-a \leqq x \leqq 0$.

$B_1(s)$ can be found explicitely and the intensity (I) in the diffraction image is found to be given by

$$I \sim \delta\left(-s\,\frac{n}{2\,a}\right)\left\{\left(\frac{2\sin(\Delta_0 - 2\,\pi\,s\,a)/2}{\Delta_0 - 2\,\pi\,s\,a}\right)^2 + \left(\frac{2\sin(\Delta_0 + 2\,\pi\,s\,a)/2}{\Delta_0 + 2\,\pi\,s\,a}\right)^2 + \right.$$
$$\left. + \frac{2\cos 2\,\pi\,s\,a\cdot\sin(\Delta_0 - 2\,\pi\,s\,a)/2\cdot\sin(\Delta_0 + 2\,\pi\,s\,a)/2}{(\Delta_0 - 2\,\pi\,s\,a)/2\cdot(\Delta_0 + 2\,\pi\,s\,a)/2}\right\}, \tag{4}$$

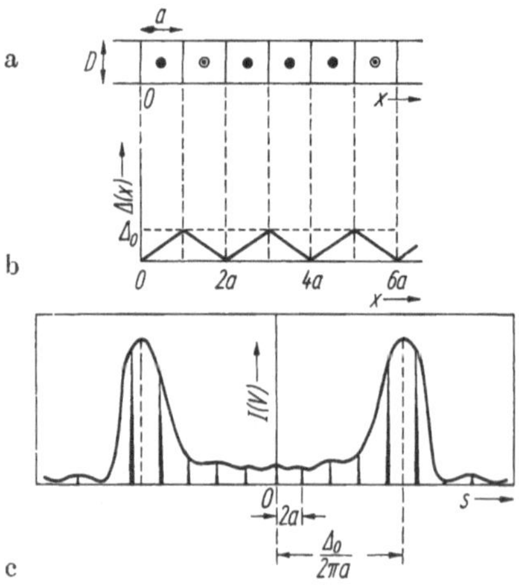

Fig. 5.
a) Cross section of the repetition structure in a cobalt foil of thickness D and domain width a, only the components of the magnetic induction B_y into the plane of the paper are indicated
b) The variation of the phase $\Delta(x)$ between the origin 0 and x for the structure (a)
c) The form of the intensity distribution given by (4) for the phase grating (b). The maxima in the intensity correspond to the Lorentz deflections within the foil, the substructure of diffraction peaks of separation $2a$ is due to the action of the domain structure as a phase grating. (In Fig. 5c rea $\frac{1}{2a}$ instead of $2a$)

with $\Delta_0 = D\, e\, B_y\, a/c\, \hbar$. The expression (4) shows intensity maxima at $s_i = \pm \Delta_0/2\,\pi\, a$ which correspond to the peaks obtained from the Lorentz deflection since $\Phi = \lambda\, s_i = \lambda\, D\, e\, B_y/c\, \hbar$. The intensity distribution has the form shown in Fig. 5c. All orders of reflection occur and since the position of the diffraction peaks is determined by $\delta\,(s - n/2\,a)$ there can be no mismatch as in the Lorentz interpretation. The last term on the R.H.S. of (4) mainly adds intensity in the region $|s| < \Delta_0/2\,\pi\, a$ and reduces the intensity for $|s| > \Delta_0/2\,\pi\, a$. This is in accord with the experimental observations.

In the present analysis the effect of the finite thickness of the domain walls has been neglected. It is to be noted that if the resolution in the diffraction image is inadequate to resolve the diffraction maxima the image degenerates into two peaks at the positions given by the classical Lorentz deflections joined by a straight streak of reduced intensity. It is often assumed that the nature of the domain wall determines the form of the streak. The present result shows that in diffraction from repeat structures the intensity distribution in the streak is partly due to the action of the structure as a grating. Taking account of the domain walls in the diffraction calculations will add intensity to the region between the Lorentz deflection positions.

6.2 Defocussing effect

The direct images of a phase grating show a reversal of contrast [1] for defocussing distances $df = (2\,a)^2/\lambda$. In the cobalt foil shown in Fig. 4, $2\,a = 2.7 \times 10^{-5}$ cm yielding $df \approx 2$ cm for 100 kV electrons.

It is readily shown that the beam deflections within the domains yield similar intensity reversal effects in the case of a repeat structure and that in this case $df = a/\Phi \approx 0.1$ cm for $\Phi = 2 \times 10^{-4}$ rad.

Since the two effects are independent the contrast reversal due to the beam deflections will mask those due to the phase effects in the present case where the repeat distances are large [12]. The observed contrast reversals are clearly due to the beam deflections.

7. Conclusions

The observed diffraction images from cobalt foils show a substructure within the intensity distribution normally considered as being due to the Lorentz deflections. This additional diffraction effect is due to the action of the periodic magnetic structure as a phase grating and constitutes another demonstration of the physically observable effects due to the magnetic vector potential.

The observed diffraction images are accounted for using a kinematic analysis in which the Fraunhofer diffraction image is regarded as the Fourier transform of the object [13]. The diffraction images from cross-tie walls are due to similar phase effects and have been accounted for by Bourret [12] by a direct use of the Kirchhoff diffraction integral [14].

Acknowledgements

The author is grateful to I. Melnick and E. Fontanet for the preparation of the permalloy films, to J. M. Penisson for his help in preparing the cobalt foils, to D. Dautreppe and A. Bourret for the use of the electron microscope, and is particularly indebted to A. Bourret for several enlightening discussions.

After the present manuscript was submitted for publication I have learned of some rather similar work being carried out by M. J. Goringe and J. P. Jakubovics an account of which is to appear in the Philosophical Magazine.

References

[1] Y. AHARANOV and D. BOHM, Phys. Rev. **115**, 485 (1959).
[2] R. G. CHAMBERS, Phys. Rev. Letters **5**, 3 (1960).
[3] H. A. FOWLER, L. MARTON, J. A. SIMPSON, and J. A. SUDDETH, J. appl. Phys. **32**, 1153 (1961).
[4] W. BAYH, Z. Phys. **169**, 492 (1962).
[5] H. BOERSCH, H. HAMISCH, K. GROHMANN, and D. WOHLLEBEN, Z. Phys. **167**, 72 (1962).
[6] J. SILCOX, Phil. Mag. **8**, 1395 (1963).
[7] R. J. SPAIN, Appl. Phys. Letters **3**, 208 (1963).
[8] A. BOURRET and D. DAUTREPPE, phys. stat. sol. **13**, 559 (1966).
[9] R. H. WADE and J. SILCOX, phys. stat. sol.
[10] R. D. HEIDENREICH, Fundamentals of Transmission Electron Microscopy, Interscience Publishers, New York 1964 (p. 113).
[11] C. FERT, Traité de Microscopie Electronique, ed. C. MAGNAN, Hermann, Paris 1961 (p. 374).
[12] A. BOURRET, private communication.
[13] P. P. EWALD, Proc. Phys. Soc. **52**, 167 (1940).
[14] H. BOERSCH, H. HAMISCH, D. WOHLLEBEN, and K. GROHMANN, Z. Phys. **164**, 55 (1961).

(Received November 23, 1966)

phys. stat. sol. 19, 855 (1967)

Subject classification: 13.4; 10.1; 14.1; 21

Materials Research Laboratory, Martin Company, Orlando, Florida

Electron Charge Distribution about Zinc Ions in Aluminum

By

L. I. VAN TORNE

The electron charge density distribution about a zinc solute ion in a dilute aluminum alloy is computed by three different methods. These are: a) an exact method where the solute ion potential is derived from the model potential difference between zinc and aluminum, b) a semiempirical method where the partial wave phase shifts are obtained from alloy specific electrical resistivity, and c) an approximate method where the solute ion potential is represented by a screened Coulomb potential. All three methods are in agreement with the experimentally observed solute ion clustering. Methods a and b also are in agreement with the observed decrease in lattice parameter with increase in zinc content.

Die Ladungsdichteverteilung der Elektronen um ein gelöstes Zinkion in einer Aluminiumlegierung wird mit drei unterschiedlichen Methoden berechnet. Diese sind: a) eine exakte Methode, bei der das Ionenpotential aus der Modellpotentialdifferenz zwischen Zink und Aluminium abgeleitet wird, b) eine halbempirische Methode, bei der die Phasenverschiebungen der Partialwellen aus dem spezifischen elektrischen Widerstand der Legierung erhalten wird und c) eine Näherungsmethode, bei der das Ionenpotential durch ein abgeschirmtes Coulombpotential dargestellt wird. Alle drei Methoden befinden sich in Übereinstimmung mit der experimentell beobachteten Komplexionenbildung. Die Methoden a) und b) sind auch in Übereinstimmung mit der beobachteten Verkleinerung der Gitterkonstante mit wachsendem Zinkgehalt.

1. Introduction

It is well known that the perturbing potential produced by the presence of a solute atom or other point imperfection in a metal is completely screened by the conduction electrons at distances which are large compared to unit cell dimensions [1]. Furthermore, the sharp cutoff in the occupation of electron states which occurs at the Fermi surface produces an oscillatory behavior in the charge density which in turn decreases as an inverse power of the distance from the point imperfection. The sign and magnitude of the charge density at neighboring ion sites determines whether solute atoms in an alloy will cluster and in the limit of clustering form an immiscibility gap in the alloy series or take up ordered positions with respect to the solvent atoms and in the limit will form an intermetallic compound. An appreciation for the consequences of the charge density distribution about a solute atom can be obtained by the following consideration.

The interaction energy of two solute ions spaced r distance apart is given [2] approximately by

$$W(\alpha_1, \alpha_2, r) \approx \frac{\pi\,\alpha_1\,\alpha_2}{k_F\,(2\,k_F + 1)^2}\left(\frac{\cos 2\,k_F\,r}{r^3}\right), \tag{1}$$

where α_1 and α_2 are the signed amplitudes of the charge density oscillation about

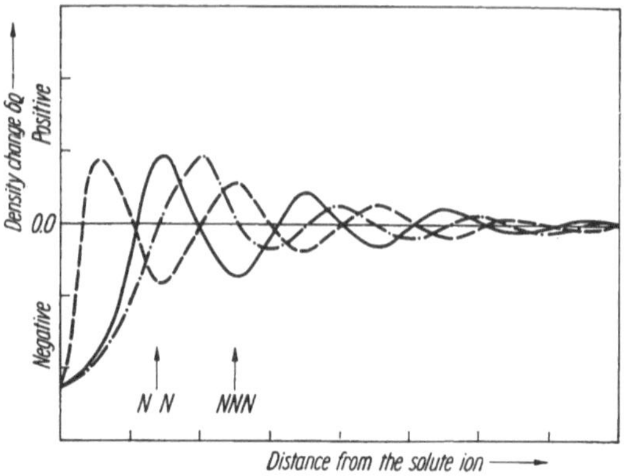

Fig. 1. A schematic representation of the charge density behavior near a solute ion which possesses a negative charge with respect to the solvent ions. Near neighbor and next nearest neighbor positions are indicated by NN and NNN, respectively. Curve ——— corresponds to solute clustering, – – – – to solute ordering, and —— – —— to a random solid solution

the solute ions, 1 and 2, and $\boldsymbol{k}_F$ is the wave vector of electrons at the Fermi surface of the solvent metal. A schematic representation of the charge density behavior near a solute ion which possesses a negative charge with respect to the solvent ions is shown in Fig. 1. Near neighbor and next nearest neighbor positions are indicated by NN and NNN, respectively. With regard to Fig. 1, one notes that for the solid curve the sign of the amplitude of the charge density at the NN position is opposite that of the solute ion and, by (1), NN solute ions are energetically favorable. Furthermore, when two solutes are NN's, their NN's would likely be solute ions. Therefore, the solid curve of Fig. 1 corresponds to a charge density distribution which is conducive to solute ion clustering. Similar arguments indicate that the charge density distribution represented by the dashed curve corresponds to a condition conducive to solute ion ordering and the third curve to a random solid solution. For the case of a solute ion which is positive with respect to the solvent ions, the curves are just mirrored across the abscissa of Fig. 1 and the arguments for clustering, ordering and random solid solutions are the same.

There are several methods of calculating the charge density distribution near a solute ion (for nontransition solute and solvent elements). The most significant are:

1. An exact method derived by Langer and Vosko [3] from an application of many-body perturbation theory and a representation of the solute ion by a self-consistent Hartree potential.
2. A semiempirical method developed by Kohn and Vosko [4] which requires a knowledge of the increase in specific dc electrical resistivity as a function of solute concentration and satisfaction of the Friedel sum rule.
3. A good approximate method due to Friedel [5] which applies formal scattering theory to a screened Coulomb potential with an adjustable parameter as a representation of the solute ion. This parameter is adjusted to satisfy the Friedel sum rule and the resulting phase shifts, η_l, are inserted into an asymptotic form for the charge density.

The charge density distributions about a zinc solute ion in an aluminum solvent are calculated by each of the above three methods with a modification to the exact method. In method 1, the Hartree self-consistent potential is replaced by the difference in the model potentials for zinc and aluminum. These model potentials have recently been derived — in a self-consistent manner — by Heine and Animalu [6]. The aluminum–zinc system was chosen as a suitable subject for computation because the presence of solute atom clustering is experimentally verified [7] and resistivity data are in the literature [8].

2. Discussion

The electron charge distribution to first order in the charge has been derived by Langer and Vosko [3] using many-body perturbation theory methods developed by Gell-Mann and Brueckner [9]. Only the resultant expression for the charge density about the solute ion or imperfection is given, since the derivation is quite lengthy and is clearly presented in [3]. The exact expression for the charge density is

$$\delta\varrho(r) = \frac{Z\,e}{2\,\pi^2}\int\limits_0^\infty \frac{k\sin k\,r}{r}\left[\frac{-4\,\pi\,e^2\,V(k,0)}{k^2 - 4\,\pi\,e^2\,V(k,0)}\right]\mathrm{d}k\,, \tag{2}$$

where Z is the ionic charge difference between solute and solvent ions, e is the electron charge, k is the solvent metal electron wave vector, and $4\,\pi\,e^2\,V(k, 0)$ is the representative potential of the solute ion in the solvent metal. When $\delta\varrho(r)$ is calculated from "first principles", the representative potential inserted in (2) is a Hartree self-consistent potential.

Recently, Animalu and Heine [6] have derived the model potentials for a host of nontransitional metals and semimetals in a self-consistent manner. The model potentials for aluminum and zinc are given in Fig. 2 and 3. The figures were

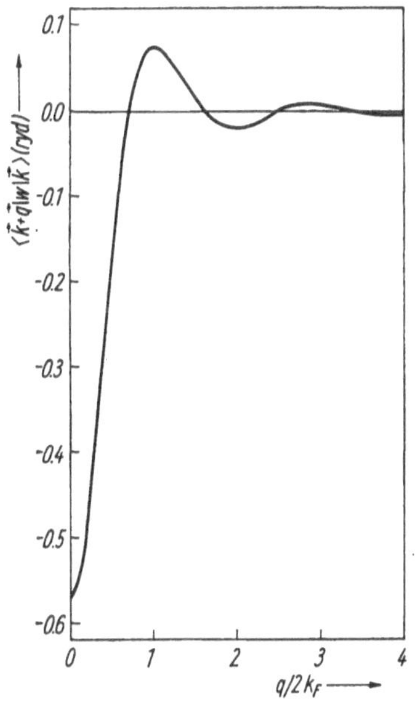

Fig. 2. The Heine and Animalu model potential for aluminum

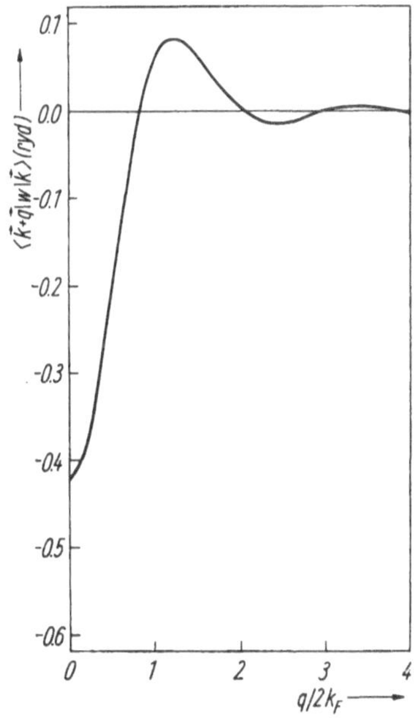

Fig. 3. The Heine and Animalu model potential for zinc

plotted from values appearing in Table 8-4, [14]. These model potentials are given here in terms of the matrix element for electron scattering from the state $\boldsymbol{k} + \boldsymbol{q}$ to state $\boldsymbol{k}$ as a function of the dimensionless quantity $\boldsymbol{q}/2\,\boldsymbol{k}_F$, where $\boldsymbol{k}_F$ is the wave vector for electrons at the Fermi surface. As a first approximation, one takes the difference of the model potentials for zinc and aluminum as being representative of the potential of a zinc solute ion in an aluminum solvent, i.e. the model potential for the zinc solute ion is obtained by subtracting the value of the model potential of aluminum from that of zinc at the same value of q. This is consistent with the concept that a zinc solute ion in aluminum is a repulsive potential to electrons, since a zinc ion is less positive than an aluminum ion. The difference potential representation obtained from the model potentials are also of a repulsive nature. This difference potential is inserted in (2) to determine $\delta\varrho(r)$ about a zinc solute ion by one of the three methods employed here for calculating $\delta\varrho(r)$.

The second method used to determine $\delta\varrho(r)$ about a zinc solute ion in aluminum is a semiempirical method presented by Kohn and Vosko [4]. The increase in the specific electrical resistivity resulting from the presence of zinc ions in dilute aluminum alloys is given by

$$\Delta\varrho = \left(\frac{\hbar\, k_F\, c}{e^2}\right) Q\,, \tag{3}$$

where c is the atom fraction of zinc solute and Q is the total scattering cross section of a zinc ion for electrons at the Fermi surface of aluminum. A result obtainable from an application of partial wave analyses in formal scattering theory — which finds use here for relating Q to the phase shifts of the l'th partial wave η_l — is

$$Q = \frac{4\pi}{k_F^2} \sum_{l=1}^{\infty} l \sin^2 (\eta_{l-1} - \eta_l)\,. \tag{4}$$

The η_l are required to satisfy the Friedel sum rule

$$Z = \frac{2}{\pi} \sum_{l=0}^{\infty} (2\,l + 1)\, \eta_l(k_F)\,, \tag{5}$$

where Z is the valence difference between zinc and aluminum. When the first two phase shifts η_0 and η_1 are considered as being the most significant, they can clearly be determined by an experimental evaluation of $\Delta\varrho$ as a function of zinc content and application of (3) through (5). The phase shifts so determined are then inserted into an expression for $\delta\varrho(r)$ in terms of η_l which is given below. However, the use of only the first two phase shifts, η_0 and η_1, gave results which were in disagreement with the experimental observation of zinc atom clustering in aluminum, e.g. the resultant $\delta\varrho(r)$ indicated a tendency for unlike near neighbors. However, for the electron energies and scattering potentials considered here, one cannot neglect η_2. Therefore, a third relation for the η_l's is required. This relation can be obtained from the ratios of the η_l's which can be determined from an estimate of the height and width of the zinc solute ion potential barrier and the energy of the incident electrons [10]. For the case considered here, one finds

$$\eta_0 + \eta_1 + \eta_2 \approx 1.3 \tag{6}$$

which, in conjunction with (3) through (5), allows a semiempirical determination of the first three phase shifts.

The third method used for computing $\delta\varrho(r)$ about a zinc solute ion is due to Friedel [5]. This method utilizes a partial wave analysis from formal scattering theory and the asymptotic form of $\delta\varrho(r)$ at large r. An outline of the method is presented. In the presence of a spherical potential, $U(r)$, there exist, at a distance r from the potential, spherical waves of the form

$$\Psi_{k,l} \sim \frac{1}{r} \sin\left(k r - \frac{l\pi}{2} + \eta_l\right) P_l(\cos\theta) , \tag{7}$$

where $P_l(\cos\theta)$ is a spherical harmonic of order l and η_l are the phase shifts. In this same range of r, there will be an extra charge

$$\begin{aligned}\delta\varrho(r) &= Z e \sum_l (2l+1) \int_0^{k_F} \{|\Psi_{k,l}(\eta_l)|^2 - |\Psi_{k,l}(0)|^2\} \frac{2R}{\pi} \mathrm{d}k \\ &= \frac{Ze}{\pi^2} \sum_l (2l+1) \int_0^{k_F} \left\{\sin^2\left(k r - \frac{l\pi}{2} + \eta_l\right) - \sin^2\left(k r - \frac{l\pi}{2}\right)\right\} \frac{1}{r^2} \mathrm{d}k \\ &= \frac{Ze}{\pi^2} \sum_l (2l+1) \int_0^{k_F} \sin(2kr - l\pi + \eta_l) \sin\eta_l \frac{1}{r^3} \mathrm{d}(k r) \\ &\sim \frac{Ze}{\pi^2} \sum_l (2l+1)(-1)^l \sin\eta_l \frac{\cos(2k_F r + \eta_l)}{r^3} ,\end{aligned} \tag{8}$$

where the normalization factor $1/\sqrt{2\pi R}$ is used for the wave function of (7) in the sphere of radius R, and terms going to zero more rapidly than $1/r^3$ are neglected. One can employ the following rather clever technique to determine the phase shifts. One represents the solute ion by a screened Coulomb potential

$$U(r) = \frac{Z e^2}{r} \exp(-\lambda r) , \tag{9}$$

where λ is an adjustable parameter. Using this potential, the phase shifts are computed from

$$\eta_l = -\frac{\pi m}{\hbar^2} \int_0^{\infty} U(r) [J_{l+1/2}(k_F r)]^2 r \, \mathrm{d}r , \tag{10}$$

where m is the electron mass and $J_{l+1/2}$ are spherical Bessel functions of fractional order. Equation (10) is valid [11] for all orders l and magnitudes of $|\eta_l| < \pi/2$. The values of η_l which are used in (8) are determined by varying the adjustable parameter λ in (9) until the η_l's of (10) satisfy the Friedel sum rule, i.e. (5). Then, $\delta\varrho(r)$ is computed using the asymptotic form, (8), with the η_l's which satisfy the Friedel sum rule.

The aluminum–zinc phase diagram [12] shows a solid solubility of zinc in aluminum in excess of 20 atomic precent zinc in the temperature range from the solidus down to about 300 °C. The solid solubility persists to lower temperatures for more dilute alloys. Below these temperatures, the alloy separates into two solid solutions, one being aluminum rich and the other zinc rich. A 10 at% solid solution zinc alloy has been extensively investigated by diffuse X-ray scattering techniques [7]. The results of this investigation give conclusive evidence for the presence of zinc and aluminum rich clusters. However, the excess of like ion neighbors does not appear to extend much beyond the first shell of near neighbor ions. Clustering in these dilute alloys is also accompanied by a decrease in the alloy lattice parameter with increase in zinc content [13].

3. Numerical Evaluation and Results

The evaluation of $\delta\varrho(r)$ by the three methods examined here is most readily obtained by numerical methods employing a high speed computer. Hence, three separate computer programs were written in Fortran IV language by the author and were run on a Control Data Corp. 3600 computer. The numerical integrations on r and k were carried out in increments of 1×10^{-10} cm and $0.2\times\times10^{7}$ cm^{-1}, respectively. The Fermi surface of aluminum was approximated as being spherical, which is a quite valid approximation in this case [14]. The values used for k_F of aluminum (1.77×10^{8} cm^{-1}) and zinc (1.07×10^{8} cm^{-1}) were obtained from geometric considerations given in [14] with the following lattice parameters [15]: aluminum $a_0 = 4.049$ Å, zinc $a = 2.664$ Å, $c = 4.945$ Å. The mass m is taken as the free electron mass and cgs, esu units are used. The experimentally determined incremental resistivity, ($\Delta\varrho$/at fraction Zn) $= 8.73\times\times10^{2}$ stat Ωcm, is given in [8].

The charge density distributions, $\delta\varrho(r)$, about a zinc solute ion in aluminum, computed by the three different methods, are shown in Fig. 4. The interrupted dashed curve is from the exact method using the difference in model potentials for the solute ion potential. The solid curve results from the semiempirical determination of the phase shifts and the asymptotic form of $\delta\varrho(r)$. The dashed curve corresponds to the Friedel method using a screened Coulomb potential and the asymptotic form of $\delta\varrho(r)$. The near neighbor and next nearest neighbor positions are noted by NN and NNN, respectively.

It is quite evident that the sign of $\delta\varrho(r)$ in the vicinity of solute ion near neighbors for all three methods is in agreement with the experimental evidence

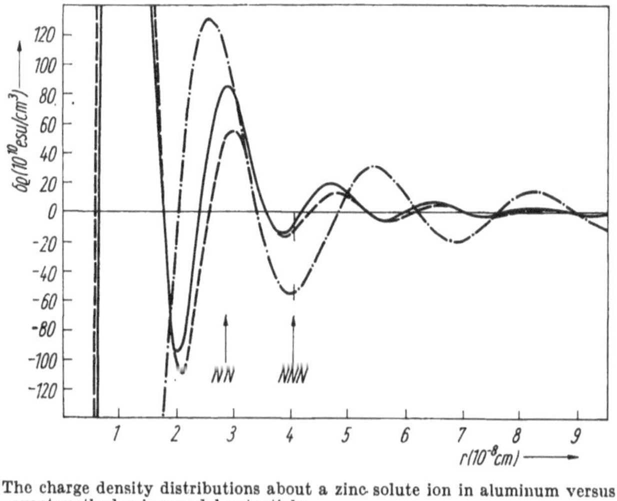

Fig. 4. The charge density distributions about a zinc solute ion in aluminum versus distance.
Curves: — · — exact method using model potentials,
——— semiempirical determination of phase shifts and asymptotic form of $\delta\varrho(r)$,
– – – – screened Coulomb potential with adjustable parameter and asymptotic form of $\delta\varrho(r)$

for clustering. One notes immediately that the oscillations of $\delta\varrho(r)$ corresponding to the exact method with model potentials are larger in magnitude and longer ranged than those corresponding to the other two methods. This behavior of $\delta\varrho(r)$ was previously noted by Langer and Vosko. If one considers that near neighbor interactions are the most significant, then the observed maximum of $\delta\varrho(r)$ occurring at a position nearer the solute ion than the near neighbor site is in qualitative agreement with the observed decrease in the lattice parameters of aluminum with increasing zinc concentration. The maximum of $\delta\varrho(r)$ corresponding to the semiempirical method occurs exactly at the near neighbor site. However, the $\delta\varrho(r)$ behavior at the next nearest neighbor site would produce a decrease in lattice parameter with increase in zinc content. Hence, the experimentally observed decrease in lattice parameter must be interpreted in terms of non-negligible next nearest neighbor interactions for $\delta\varrho(r)$ computed by the semiempirical method. Next nearest neighbor interactions must be significant in metals and alloys, since the energy of a stacking fault arises from next nearest neighbor interactions [16], and if such interactions were negligible, the presence of high stacking fault densities in metals and alloys would be the rule rather than the exception. Therefore, $\delta\varrho(r)$ computed by the semiempirical method is also in qualitative agreement with lattice parameter contraction. The behavior of $\delta\varrho(r)$ computed by the Friedel method at near neighbor and next nearest neighbor sites indicates a lattice parameter expansion for near neighbor interactions and a contraction for next nearest neighbor interactions. One would certainly expect near neighbor interactions to be stronger than next nearest neighbor. Hence, a net lattice parameter expansion is indicated by $\delta\varrho(r)$ computed by the latter method which is in disagreement with experimental observation.

4. Conclusions

1. All three methods employed to compute $\delta\varrho(r)$ about a zinc solute ion in aluminum are in qualitative agreement with the experimental evidence for clustering in this alloy system.
2. The exact method with a solute ion potential derived from the difference in model potentials of zinc and aluminum and the semiempirical method are in qualitative agreement with the experimentally observed decrease in lattice parameter with increase in zinc content.
3. All three methods for computing $\delta\varrho(r)$ appear to predict the existence of ordering, clustering and random solid solution in nontransition metal alloys.

Acknowledgement

The stimulating comments provided by Dr. T. R. Cass and helpful information given by Prof. J. Hren, University of Florida, are appreciated.

References

[1] J. Friedel, Phil. Mag. **43**, 153 (1952).
[2] A. Blandin and J. L. Déplanté, Metallic Solid Solutions, Ed. J. Friedel and A. Guinier, W. A. Benjamin, Inc., New York 1963.
[3] J. S. Langer and S. H. Vosko, J. Phys. Chem. Solids **12**, 196 (1959).
[4] W. Kohn and S. H. Vosko, Phys. Rev. **119**, 912 (1960).
[5] J. Friedel, Adv. Phys. **3**, 446 (1954).

[6] A. O. E. Animalu and V. Heine, Phil. Mag. **12**, 1249 (1965).
[7] P. S. Rudman and B. L. Averbach, Acta metall. **2**, 576 (1954).
[8] C. Panseri and T. Federighi, Acta metall. **8**, 217 (1960).
[9] M. Gell-Mann and K. Brueckner, Phys. Rev. **106**, 364 (1957).
[10] N. F. Mott and H. S. W. Massey, The Theory of Atomic Collisions, 3rd Edition, Clarendon Press, Oxford 1965 (p. 29).
[11] See [10] (pp. 28, 89).
[12] M. Hansen, Constitution of Binary Alloys, McGraw-Hill Publ. Comp. 1958 (p. 148).
[13] H. J. Axon and W. Hume-Rothery, Proc. Roy. Soc. **A193**, 1 (1948).
[14] W. H. Harrison, Pseudo-potentials in the Theory of Metals, Chap. 3, W. A. Benjamin, Inc., New York 1966.
[15] C. S. Barrett, Structure of Metals, App. IX, McGraw-Hill Publ. Comp. 1952.
[16] W. T. Read, Dislocations in Crystals, McGraw-Hill Publ. Comp. 1953 (p. 94).

(Received November 17, 1966)

phys. stat. sol. 19, 863 (1967)

Subject classification: 6; 21.2

Institute of Theoretical Physics, Göteborg

Polarization Vectors for Lattice Vibrations in Sodium

By

G. Björkman and G. Grimvall

The polarization vectors of the lattice vibrations in sodium are calculated for 46 points in the first Brillouin zone using Krebs's model.

Die Polarisationsvektoren für Gitterschwingungen in Na werden unter Benutzung des Krebsschen Modells für 46 Punkte in der ersten Brillouin-Zone berechnet.

1. Introduction

In calculations involving the electron–phonon interaction the phonons are for simplicity assumed to be purely longitudinal or transverse. The aim of this note is to investigate numerically the validity of this approximation for sodium.

In the theory of electron–phonon interaction expressions of the type

$$\sum_{\lambda} F(\boldsymbol{q}, \omega_{\lambda}(\boldsymbol{q})) \cdot (\hat{e}_{\lambda, \boldsymbol{q}} \cdot \hat{\boldsymbol{q}})^2 \tag{1}$$

occur, where F is some function of the wave vector $\boldsymbol{q}$ and the phonon frequency $\omega_{\lambda}(\boldsymbol{q})$. λ denotes the three phonon branches and $\hat{e}_{\lambda, \boldsymbol{q}}$ is a unit polarization vector. For these unit vectors we have the relation

$$\sum_{\lambda} (\hat{e}_{\lambda, \boldsymbol{q}} \cdot \hat{\boldsymbol{q}})^2 = 1 \,. \tag{2}$$

We see that if the frequencies of the three branches are almost equal, the approximation mentioned is good. This is however far from true in cases like sodium (we do not discuss the case of umklapp processes where transverse branches are always important).

2. Calculation

We have used Krebs's [1] model to calculate the polarization vectors. Krebs's model is a three-parameter model which fulfils the symmetry requirements of the lattice. As parameters Krebs uses the force constants to the neighbours of first and second order and a constant giving the strength of the electron–phonon coupling. These three parameters are in turn related to the elastic constants C_{11}, C_{12}, and C_{44} for which experimental values are used. We have taken the elastic constants used by Krebs.

The calculation was carried through for 46 points in an "equivalent part" of the first Brillouin zone making 1/48 of the total zone volume (Fig. 2). In Fig. 1 we give the polarization vectors of the longitudinal branch for points on three sides of the above mentioned "equivalent part". In Fig. 2 we map the deviation of $(\hat{e}_{\lambda, \boldsymbol{q}} \cdot \hat{\boldsymbol{q}})^2$ from unity for the longitudinal branch.

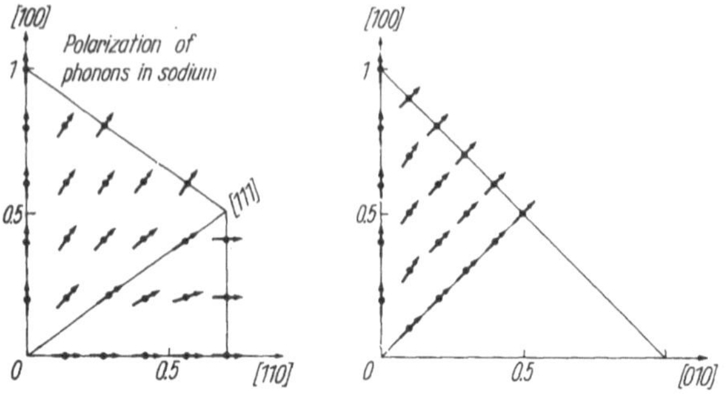

Fig. 1. Some polarization vectors for the longitudinal branch

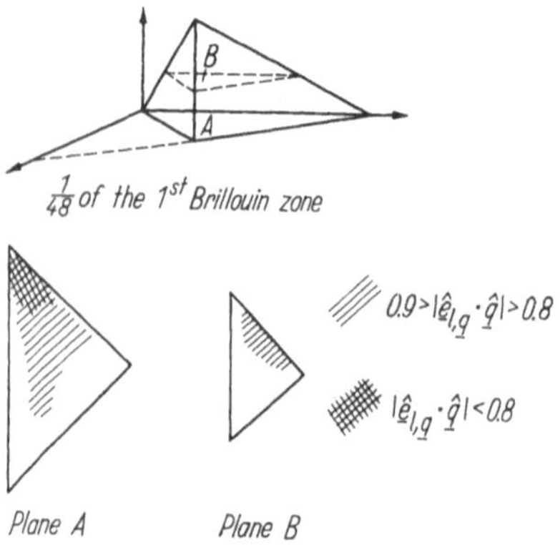

Fig. 2. Deviation of the scalar product $|\hat{e}_{\lambda,\,q} \cdot \hat{q}|$ from unity

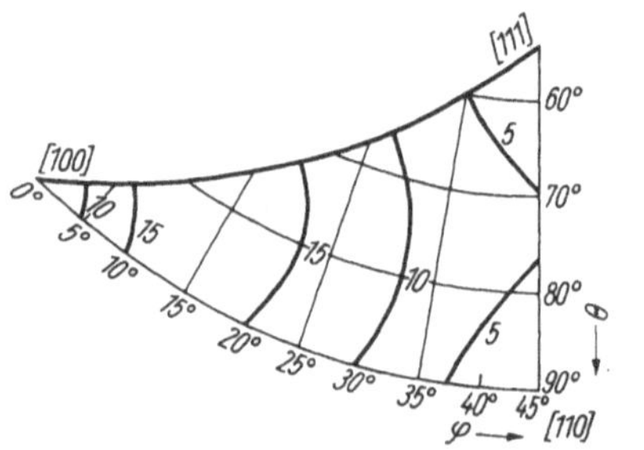

Fig. 3. The angle of deviation α for the longitudinal sound wave

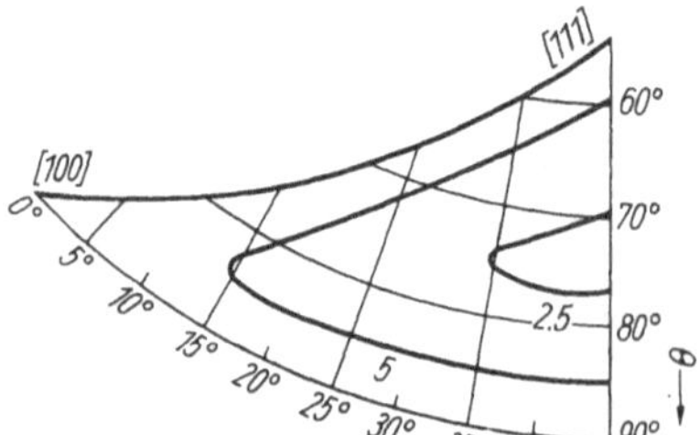

Fig. 4. The angle of deviation α for the higher transverse sound wave

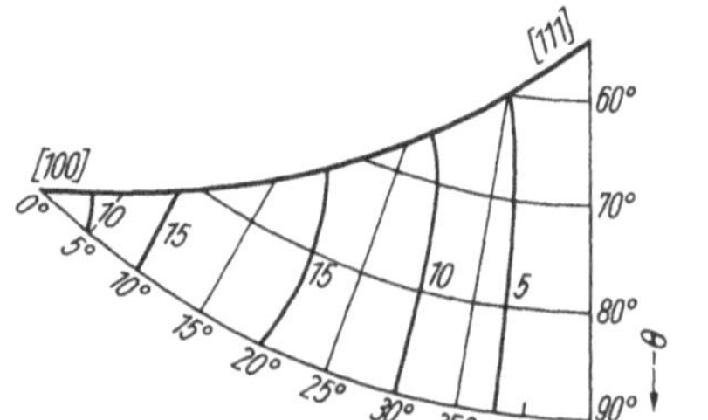

Fig. 5. The angle of deviation α for the lower transverse sound wave

Finally we have solved the problem in the simple sound wave limit. In Fig. 3 to 5 we plot the angle α of deviation from an isotropic case. The plot is made on a polar coordinate paper and the $\boldsymbol{q}$-direction is specified by the usual spherical coordinates φ and θ. The lines are for constant deviation α.

3. Discussion

Woods, Brockhouse et al. [2, 3], have measured the polarization vectors in a few non-symmetry directions and also given theoretical values from a force constant model based on their experiment. At some points the agreement is only qualitative but the experimental uncertainties of their measurements can easily explain the discrepancy. The usage of a screening constant according to Langer and Vosko or according to Pines leads to a slight difference in the calculated energy spectrum. This point has been dealt with in detail by Mahesh and Dayal [4]. We have used both of the screening constants. The angular difference of the polarization vectors is always very small, except in a narrow region enclosing the points $\boldsymbol{q} = \frac{2\pi}{a}(0.7, 0.1, 0)$ and $\boldsymbol{q} = \frac{2\pi}{a}(0.5, 0.1, 0)$, where the deviation is about 1.7°. This is also to be expected because the larger deviations in frequency occur near the zone boundaries, where the polarization vectors are determined by symmetry conditions.

The result in the long-wavelength limit is in qualitative agreement with a work by Fein and Smith [5] for Ni. We note that the maximum angle of deviation is considerable, about 18° for both the longitudinal branch and the lower transverse branch. We also note that the higher transverse branch always stays transverse to good approximation.

The answer to the question concerning the validity of approximating the phonons as purely longitudinal and transverse depends on the structure of F in (1). The deviations are large enough to require a detailed attention to this point in many cases, e.g. the electrical conductivity of sodium.

References

[1] K. Krebs, Phys. Rev. **138**, A143 (1965).
[2] A. D. B. Woods, B. N. Brockhouse, R. H. March, A. T. Stewart, and R. Bowers, Phys. Rev. **128**, 1112 (1962).
[3] B. N. Brockhouse et al., Inelastic Scattering of Neutrons in Solids and Liquids II, International Atomic Energy Agency, Vienna 1963.
[4] P. S. Mahesh and B. Dayal, phys. stat. sol. **9**, 351 (1965).
[5] A. E. Fein and C. S. Smith, J. appl. Phys. **23**, 1212 (1952).

(Received November 7, 1966)

phys. stat. sol. 19, 867 (1967)

Subject classification: 13

Institut de Physique, Université de Liège

The Pair Distribution Function in the R.P.A. and in the Hubbard Approximation

By

F. Brouers

The pair distribution function of the metallic electron gas is calculated exactly in the R.P.A. and in the Hubbard approximation. The results obtained are discussed and compared with those of previous approximate calculations and a slight modification to Hubbard's interpolation method is proposed.

On a calculé exactement la fonction de distribution de paire du gaz électronique des métaux réels dans la R.P.A. et dans l'approximation de Hubbard. On compare les résultats à ceux obtenus à l'aide de calculs approchés. On propose ensuite une légère modification à la méthode d'interpolation de Hubbard.

If in the instantaneous density-density correlation function $S(\boldsymbol{r})$ defined by

$$S(\boldsymbol{r}) = \frac{1}{N} \langle \psi_0 | \varrho\, (\boldsymbol{r} + \boldsymbol{r}')\, \varrho(\boldsymbol{r}') |\, \psi_0 \rangle\,, \tag{1}$$

where ψ_0 is the exact ground state wave function, we separate the part associated with the correlation of a particle with itself, we have

$$S(\boldsymbol{r}) = \delta(\boldsymbol{r}) + (N-1)\, g(\boldsymbol{r})\,. \tag{2}$$

In a translationally invariant system the pair distribution function $g(\boldsymbol{r})$ is given by

$$g(\boldsymbol{r}) = \frac{1}{N\,(N-1)} \langle \psi_0 \left| \sum_{i \neq j} \delta\,(\boldsymbol{r} + \boldsymbol{r}_i - \boldsymbol{r}_j) \right| \psi_0 \rangle\,. \tag{3}$$

It is the probability that, when one particle is observed at a point (say r_0), another particle will be found at a distance r from r_0. $S(\boldsymbol{r})$ is the Fourier transform of the static form factor $S(\boldsymbol{k}) = \frac{1}{N} \langle \psi_0 | \varrho_{\boldsymbol{k}}^{+}\, \varrho_{\boldsymbol{k}} |\, \psi_0 \rangle$ and a straightforward calculation shows that

$$g(\boldsymbol{r}) = \frac{1}{N-1} \sum_{k \neq 0} \left(S(\boldsymbol{k}) - 1\right) \mathrm{e}^{i \boldsymbol{k} \boldsymbol{r}} + 1\,. \tag{4}$$

In the Hartree approximation there is no correlation between the electrons and $g_{\mathrm{H}}(r) = 1$. In the Hartree-Fock approximation $g(\boldsymbol{r})$ is given by

$$g_{\mathrm{H.F.}}(r) = 1 - \frac{9}{2} \left[\frac{\sin k_{\mathrm{F}}\, r - k_{\mathrm{F}}\, r \cos k_{\mathrm{F}}\, r}{k_{\mathrm{F}}^3\, r^3} \right]^2, \tag{5}$$

where $\boldsymbol{k}_{\mathrm{F}}$ is the Fermi wave vector.

The function $g_{\mathrm{H.F.}}(r)$ is splotted in Fig. 1. The Pauli principle is here taken into account: two electrons of the same spin cannot be found at the same point and $g_{\mathrm{H.F.}}(0) = 0.5$. As r increases from zero, $g_{\mathrm{H.F.}}(r)$ increases corresponding to the fact that the Pauli principle restriction gradually weakens.

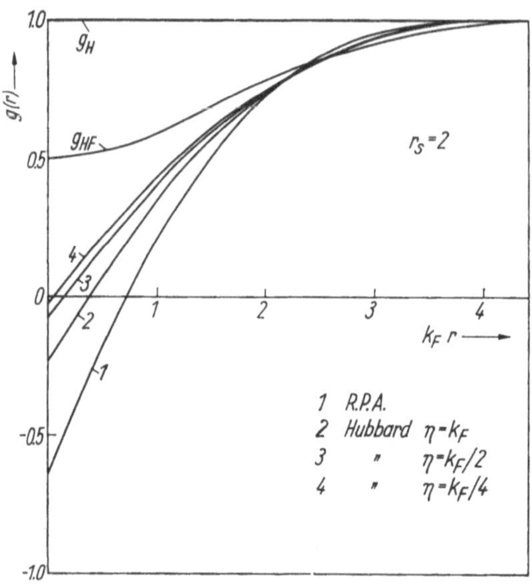

Fig. 1. Pair distribution function in the Hartree and Hartree-Fock approximations and for $r_s = 2$, in the R.P.A. (1), in the H.A. (2), and in the modified Hubbard approximation (3, 4)

If the Coulomb repulsion among electrons is taken into account, $g(r)$ is expected to show a considerably different behaviour. Approximate calculations of $g(r)$ in the R.P.A. have been performed by Glick and Ferrell [1] and Ueda [2]. Glick and Ferrell have found that for the electron density of aluminium ($r_s \approx 2$) $g_{\text{R.P.A.}}(r) < 0$ for small r. Ueda has confined himself to the case of high electron densities ($r_s =$ 0.1, 0.5, and 1) and he has computed the contribution from parallel spin with an approximate expression.

More recently, Gaskell [3] carried out a calculation of $g(r)$ by means of a rather difficult variational method using a wave function of the form $\psi = D \prod_{i<j} f(r_{ij})$, where D is a determinant of plane waves and $f(r_{ij})$ is a correlation function. The product $\Pi\, f(r_{ij})$ is then expressed in terms of the Fourier component of the density ϱ_k and a variational function $c(k)$ determined by minimizing the energy. Gaskell's method may be used to derive the pair distribution function $g(r)$. He finds a well behaved $g(r)$ for $r_s \leqq 2.66$, but for $r_s > 2.66$ $g(r)$ becomes negative for small r.

Till now, no exact calculation of $g(r)$ has been performed neither in the R.P.A. nor using the interpolation procedure that Hubbard has proposed to take into account the exchange terms in the calculation of the ground state energy.

In this report, we give the results of an exact calculation of $g(r)$ in the R.P.A. and in the Hubbard approximation for the electron density range ($r_s = 2$, 3, and 4) of real metals.

We have used the simple relationship between $S(\boldsymbol{k}, \omega)$ and the dielectric constant $\varepsilon(\boldsymbol{k}, \omega) = \varepsilon_1(\boldsymbol{k}, \omega) + i\, \varepsilon_2(\boldsymbol{k}, \omega)$, namely

$$S(\boldsymbol{k}, \omega) = -\frac{\hbar\, k^2}{4\,\pi\, e^2}\, \mathrm{Im}\, \frac{1}{\varepsilon(\boldsymbol{k}, \omega)} = \frac{\hbar\, k^2}{4\,\pi\, e^2}\, \frac{\varepsilon_2(\boldsymbol{k}, \omega)}{\varepsilon_1^2(\boldsymbol{k}, \omega) + \varepsilon_2^2(\boldsymbol{k}, \omega)} . \tag{6}$$

$S(\boldsymbol{k})$ is then obtained making use of the relation

$$S(\boldsymbol{k}) = \frac{1}{N} \int\limits_0^\infty \mathrm{d}\omega\; S(\boldsymbol{k}, \omega) . \tag{7}$$

In the R.P.A. $\varepsilon(\boldsymbol{k}, \omega)$ is given by

$$\varepsilon_{\text{R.P.A.}}(\boldsymbol{k}, \omega) = 1 + \frac{4\,\pi\, e^2}{k^2}\, Q_0(\boldsymbol{k}, \omega) , \tag{8}$$

where $Q_0(\boldsymbol{k}, \omega)$ is the pair electron–hole propagator. $S(\boldsymbol{k}, \omega)$ can be separated into two parts [4]. One is the pair contribution and comes from the frequency region in which $\varepsilon_2(\boldsymbol{k}, \omega)$ is different from zero. The second, the plasmon contribution $S_P(\boldsymbol{k}, \omega)$, comes from the frequency region defined by $\varepsilon_1(\boldsymbol{k}, \omega) = \varepsilon_2(\boldsymbol{k}, \omega) = 0$. We may write for $k \neq 0$

$$S(\boldsymbol{k}) = \frac{\hbar\, k^2}{4\,\pi^2\, N\, e^2} \int\limits_0^\infty \mathrm{d}\omega\, \frac{\varepsilon_2(\boldsymbol{k}, \omega)}{\varepsilon_1^2(\boldsymbol{k}, \omega) + \varepsilon_2^2(\boldsymbol{k}, \omega)} + \frac{\hbar\, k^2}{2\, m\, \omega_P(\boldsymbol{k})}\,, \tag{9}$$

where $\omega_P(\boldsymbol{k})$ is the wavelength-dependent plasmon frequency.

By means of equation (4), (8), and (9), we have computed numerically $g_{\text{R.P.A.}}(r)$ for $r_s = 2$, 3, and 4 (cf. Fig. 1, 2, and 3). For the particular choice $r_s = 2.66$, the pair function is exactly that obtained by Gaskell making use of the R.P.A. in his variational procedure.

Here the breakdown of the R.P.A. for small values of r is much more pronounced than in the approximate calculation of Glick and Ferrell who have found $g(0) \approx -0.2$ for $r_s = 2$. The failure of $g(r)$ to remain positive definite for small value of r is due to the fact that in the R.P.A. the exchange terms are neglected.

The Hubbard interpolation procedure [5] (H.A.) is equivalent to the introduction of a more elaborate dielectric constant:

$$\varepsilon_{\text{H.A.}}(\boldsymbol{k}, \omega) = 1 + \frac{4\,\pi\, e^2}{k^2}\, Q_0(\boldsymbol{k}, \omega) \cdot \left(1 - \frac{1}{2}\, \frac{4\,\pi\, e^2}{k^2 + k_F^2}\, Q_0(\boldsymbol{k}, \omega)\right)^{-1}. \tag{10}$$

Using (10) instead of (8), we can evaluate numerically $g(r)$ in the H.A. In the expression (10) $\dfrac{4\,\pi\, e^2}{k^2 + k_F^2}$ is a screened potential which reduces the correlation effects between electrons with parallel spin. The particular choice of the screening length k_F^{-1} comes from an approximation made by Hubbard to avoid a difficult integration. In practice one can show [6] that the range of the exchange hole is approximately equal to $(0.5\, k_F)^{-1}$. It is why we replace k_F by a parameter η to which we give the value k_F, $k_F/2$, and $k_F/4$. The pair functions obtained numerically using (10) and the parameter η are plotted in Fig. 1, 2, and 3.

In the R.P.A. as well as in the H.A. $g(r)$ shows for $r > 4$ an oscillatory character. The peaks

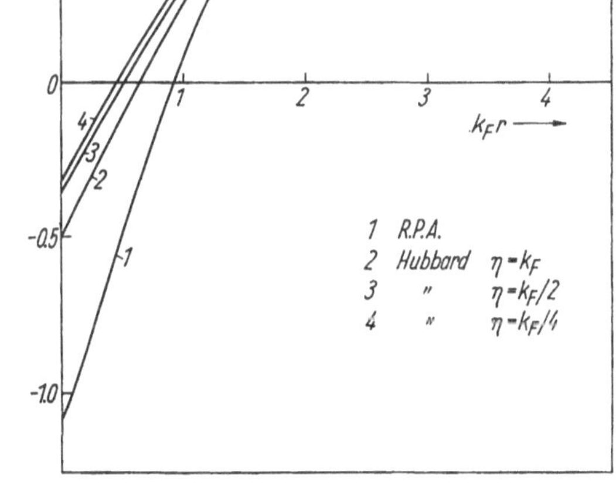

Fig. 2. Pair distribution function for $r_s = 3$ in the R.P.A. (1), in the H.A. (2), and in the modified Hubbard approximation (3, 4)

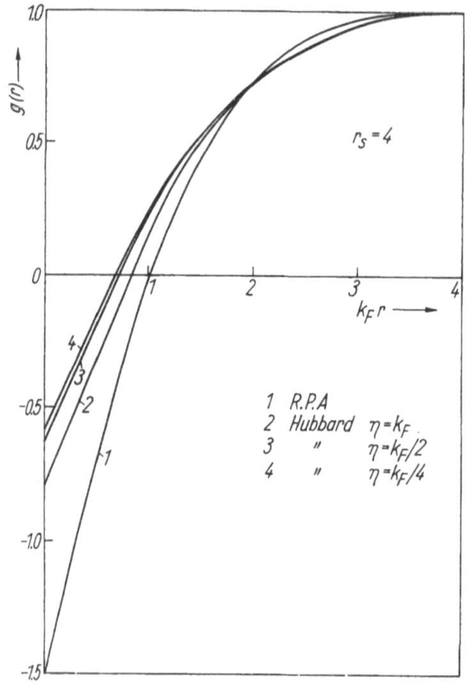

Fig. 3. Pair distribution function for $r_s = 4$ in the R.P.A. (1), in the H.A. (2), and in the modified Hubbard approximation (3, 4)

of the oscillations exceed unity, but too slightly to be seen in the figures. The same behaviour has been quoted by Gaskell.

One can see that the introduction of exchange terms included in the H.A. meets somewhat the difficulties although $g(r)$ remains negative at the origin for electron densities of real metals. Nevertheless, it appears that the inclusion of exchange terms omitted within the R.P.A. seems essential, while one considers small values of r. The comparison of the curves in Fig. 1, 2, and 3 shows that it would be more correct to choose a screening length greater than k_F^{-1} in (10).

Finally, it is worth noticing that the exchange corrections introduced by Hubbard give rise to a modification of the cut-off wave vector $\boldsymbol{k}_c$ separating the collective and the individual behaviour of the electron gas.

The relation

$$\varepsilon_1\left(\boldsymbol{k}, k\, v_F + \frac{\hbar\, k^2}{2\, m}\right) = 0 \tag{11}$$

which gives k_c in the R.P.A. (ε_1 is here the well-known Lindhard expression for ε_1 in the R.P.A.) becomes in the H.A.

$$\varepsilon_1\left(\boldsymbol{k}, k\, v_F + \frac{\hbar\, k^2}{2\, m}\right) \cdot \left(1 - f(k)\right) + f(k) = 0\,, \tag{12}$$

where

$$f(k) = \frac{1}{2}\,\frac{k^2}{k^2 + \eta^2}\,.$$

The k_c computed from (12) for different r_s are compared in Table 1 ($\beta = k_c/k_F$) with those obtained by the R.P.A. and by the Bohm and Pines [7] variational method.

Table 1

r_s	$\beta^{\text{R.P.A.}}$	$\beta^{\text{H.A.}}(\eta = k_F)$	$\beta^{\text{H.A.}}(\eta = k_F/2)$	$\beta^{\text{H.A.}}(\eta = k_F/4$	$\beta^{\text{B.P.}}$
1	0.560	0.535	0.501	0.464	0.353
2	0.729	0.683	0.634	0.592	0.498
3	0.849	0.753	0.727	0.682	0.611
4	0.945	0.865	0.800	0.756	0.706
5	1.029	0.921	0.863	0.818	0.791

Acknowledgements

The author whishes to express his thanks to Professor J. Pirenne for helpful discussions.

References

[1] A. Glick and R. Ferrell, Ann. Phys. (U.S.A.) **11**, 359 (1960).
[2] S. Ueda, Progr. theor. Phys. **26**, 45 (1961).
[3] T. Gaskell, Proc. Phys. Soc. **80**, 1091 (1962).
[4] D. Pines, Elementary Excitations in Solids, Benjamin Inc., New York 1963 (p. 150).
[5] J. Hubbard, Proc. Roy. Soc. **A243**, 336 (1957).
[6] P. Nozières and D. Pines, Phys. Rev. **111**, 442 (1958).
[7] D. Pines, Solid State Phys. **1**, 367 (1955).

(Received November 25, 1966)

Erratum

In the paper by M. Hagino and T. Takahashi

Photoelectric Properties of Cs–Sb Films Having Various Compositions

phys. stat. sol. **17**, 609 (1966)

on page 612, line 11, from the bottom, read "the samples ⑦ and ⑧" instead of "the samples ⑤ and ⑧".

See Short Note by E. IGRAS and T. WARMIŃSKI

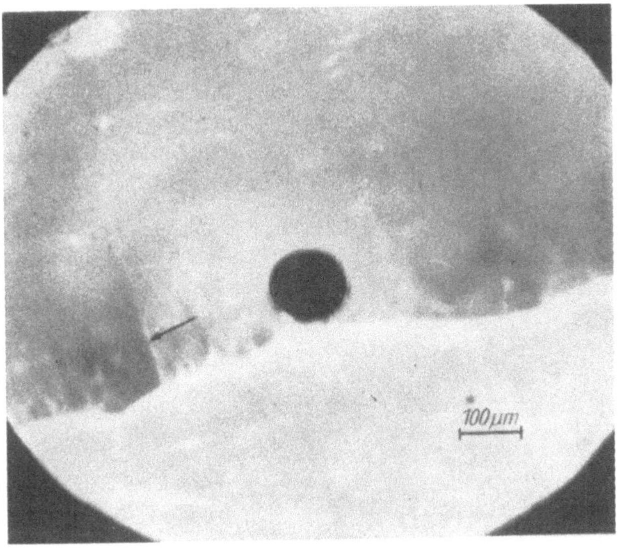

Fig. 1. Electron mirror micrograph of a silicon surface region doped with lithium, showing the appearance of regions with precipitated lithium. Temperature of the crystal: 120 °C

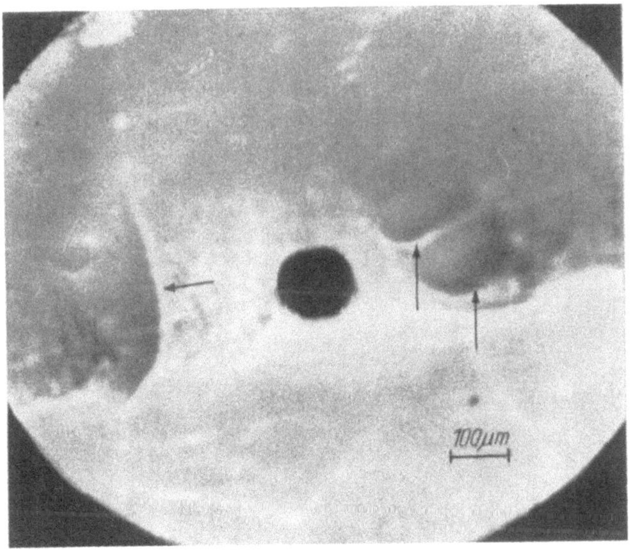

Fig. 2. The same as in Fig. 1. Temperature of the crystal: 90 °C

See Short Note by E. Igras and T. Warmiński

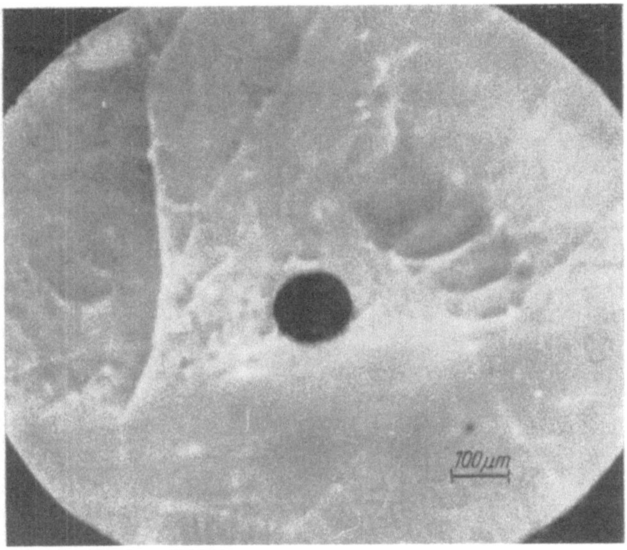

Fig. 3. The same as in Fig. 1. Temperature of the crystal: 60 °C

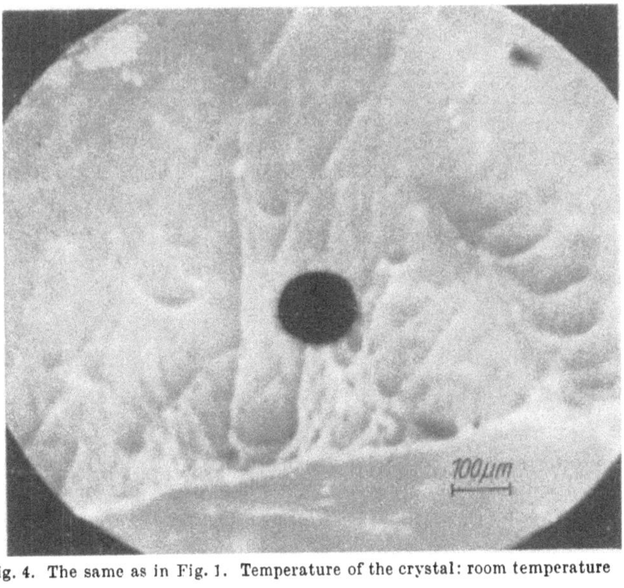

Fig. 4. The same as in Fig. 1. Temperature of the crystal: room temperature

See Short Note by K. WETZIG

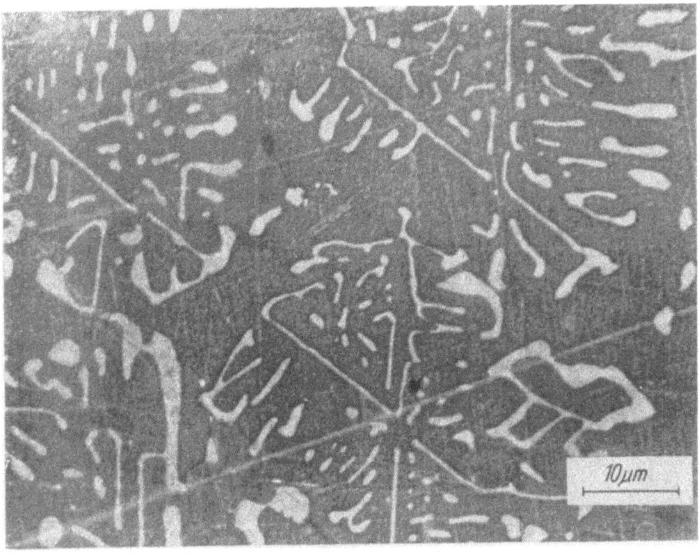

Fig. 1. $TaFe_2$ with inclusions of a second phase, identified as TaFe. Secondary electron emission image

a

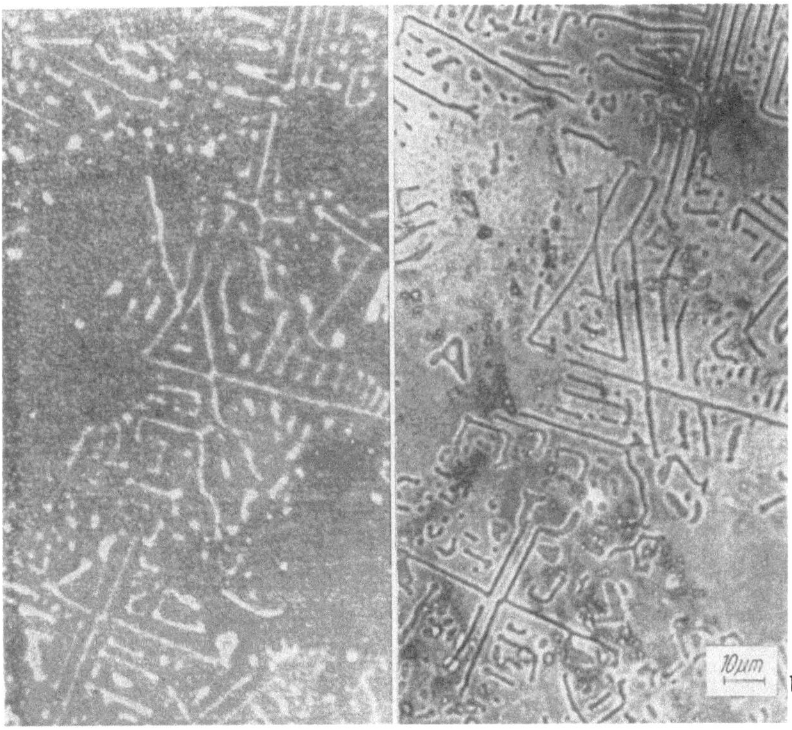

b

Fig. 2. This pair of figures shows the position of TaFe in the matrix. Image b was taken after grinding away a layer of 13 µm thickness

See Short Note by K. Wetzig

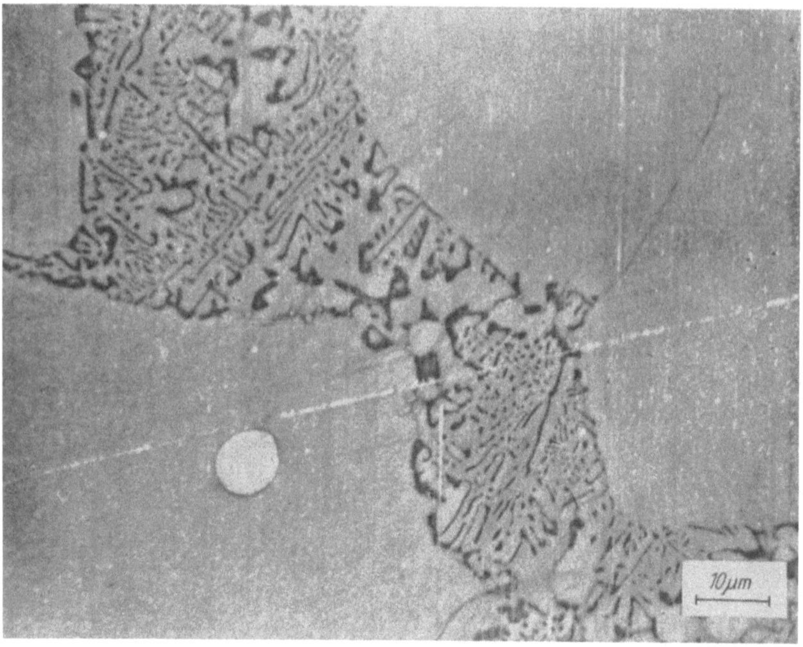

Fig. 3. Two-phase areas with a constant TaFe concentration enclose the grains of the matrix $TaFe_2$

phys. stat. sol. 19, K67 (1967)
Subject classification: 1.3; 4; 10; 22.1.2

Institute of Physics, Polish Academy of Sciences, Warsaw

Electron Mirror Observation of Lithium Precipitation at Scratch-Induced Dislocations in Silicon

By
E. IGRAS and T. WARMINSKI

It is well known that mechanical polishing of silicon introduces considerable damage into the surface layer (1 to 3). For example cracks may be produced with associated dislocations lying below the cracks. The precipitation process of impurities in semiconductors is highly structure sensitive. The precipitation of impurities on semiconductor surface layers makes it inhomogeneous with respect to electrical properties. Relatively strong inhomogeneities of surface conductivity or surface trap concentration can arise. The electron mirror microscope allows observations of the dynamic behaviour of electrical inhomogeneities in a given semiconductor micro-region (4, 5) and particularly of surface precipitation processes. The observations described here were made during measurements of surface diffusion of lithium in p-type silicon of resistivity 5 Ωcm. The electron mirror microscope was used to measure the positions of the p-n junctions after consecutive heat treatments in vacuum inside the microscope. The samples had the form of slices of dimensions of about 6x4x1.5 mm. They were polished with 1 μm aloxite on a woolen pad wetted with water, cleaned and degreased. Then metallic lithium was evaporated onto one half of the area of the hot (400 °C) silicon surface. The evaporation was carried out from a tantalum boat in vacuum. After about one minute from the beginning of the evaporation the crystal was cooled to room temperature. The alloyed lithium layer has been removed by slight mechanical polishing with 1 μm aloxite. Then the crystal was cleaned and mounted in the microscope. In this way half of the polished surface region was doped with lithium and became n-type.

Chemically polished crystals were also investigated. In the latter case about 100 μm of mechanically polished silicon was removed by chemical polishing. Then half of the crystal surface was doped with lithium in the way described above and cleaned in hydrofluoric acid. After a diffusion heat treatment inside the microscope (in the temperature range 100 to 200 °C) regions were found on the crystal surfaces which showed a very rapid and highly inhomogeneous shift of the position of the p-n junction. The latter may be explained by relatively high diffusivity of lithium along occasional cracks. When the temperature of the crystal was decreased then precipitation of lithium was observed around the scratch regions.

Figs. 1 to 4 show an example of the surface region which is visible after a diffusion heat treatment during three hours at about 180 °C. Fig. 1 corresponds to a temperature of the crystal of 120 °C, Fig. 2 to 90 °C, Fig. 3 to 60 °C, and Fig. 4 to room temperature. The regions with relatively strong lithium precipitation have the form of irregular narrow bands. The dimensions and the shape of these bands changed with decreasing temperature. The region were lithium began to precipitate already at about 120 °C is denoted by an arrow (Fig. 1). A white broad line on the same figure shows the boundary between p- (darker part of the figure) and n-type regions of the silicon surface. Fig. 2 shows that the region of precipitation of Fig. 1 became larger and simultaneously new regions with precipitated lithium appeared. These regions are marked by arrows. At still lower temperatures more defects with precipitated lithium can be seen (Figs. 3 and 4). The delay time between the appearance of new regions of precipitation with decreasing temperature can be explained by changes of the lithium concentration, i.e. by the degree of supersaturation along the surface of the crystal. The lithium concentration decreases in the direction from n-type surface layers to p-type ones. The figures are clear illustrations of lithium precipitation at dislocations formed below the cracks. The dislocations introduce a relatively high vacancy concentration. On the other hand, it is known that vacancies play a role of nuclei for precipitation (6). The nuclei for precipitation are formed by introducing lithium from an interstitial position to a vacancy position. In this way a complex Li □ is formed:

$$Li^{+} + e^{-} + \square = Li\,\square$$

The precipitation process decreases the concentration of electrically active lithium atoms. As a consequence the regions with high degree of precipitation differ in resistivity from the regions free of precipitates or with low degree of precipitation. The resistivity differences may also arise due to local strains which can be caused by precipitation. The investigation of chemically polished crystals did not reveal the inhomogeneous precipitation of lithium in our silicon samples. During the electron mirror microscopic observations a small electric current (of the order of 1 mA) was passed through the crystals. The change of the potential drop on surface regions with or without precipitates is responsible for the contrast formation.

The authors wish to thank Prof. Dr. L. Sosnowski for helpful discussions.

References

(1) E.M. PUGH and L.E. SAMUELS, J. Electrochem. Soc. 108, 1043 (1961); 109, 409 (1962).
(2) R. STICKLER and G.R. BOOKER, Phil. Mag. 8, 859 (1963).
(3) D.J.D. THOMAS, phys. stat. sol. 3, 2261 (1963).
(4) E. IGRAS and T. WARMIŃSKI, phys. stat. sol. 9, 79 (1965).
(5) E. IGRAS and T. WARMIŃSKI, phys. stat. sol. 13, 169 (1966).
(6) F.J. MORIN and H. REISS, J. Phys. Chem. Solids 3, 196 (1957).

(Received December 10, 1966)

phys. stat. sol. 19, K71 (1967)
Subject classification: 1.3; 4; 21

I. Institut für Experimentalphysik der Technischen Universität Dresden

Investigations of Surfaces of the Binary System Ta-Fe by Means of an Emission Electron Microscope

By
K. WETZIG

Surface investigations of polycrystals with the matrix $TaFe_2$ were carried out in a secondary electron emission microscope. The samples were prepared by melting at the stoichiometric ratio of $TaFe_2$. Unetched polished surfaces were imaged in a reflection light microscope without any contrast. In contrary to this the contrast of emission electron images was significant. A regular lamellar structure appears (Fig. 1). From the image contrast it was concluded that the observed structure proceeds from a second phase. This was proved by investigations with an X-ray microanalyzer. The phase has the composition of TaFe and has already been dealt with (1 to 3).

Polishing some µm from the surface between two imagings shows the TaFe-lamellas to go plate-like into the volume (Figs. 2a and 2b). To decide the position of these plates in the matrix it was tried to determine the surface orientation of small areas of the matrix by means of reflected electrons or X-ray diffraction. The lattice however was too deformed. Thereupon the following method of investigation was used. From a surface previously imaged in the emission microscope a layer of known thickness is ground away before new imaging. The variation of the lamella position to each other and with reference to marked points at the surface is measured. In connection with angle measurements the position of the second phase in the matrix may be calculated. The interpretation is simple in the case that lamellas form triangles on the surface (c.f. Fig. 2). The TaFe plates are supposed to precipitate along low indicated planes of the matrix. The variations Δh of altitude h and Δb of base b of the triangles appearing by grinding away a layer of known thickness d are calculated for several suitable planes. Those planes of the matrix where precipitation occurs are ascertained by comparison with the experimental results. In all cases analyzed precipitations of TaFe occured along pyramidal planes $\{10\bar{1}1\}$ of the matrix $TaFe_2$. Table 1 shows the results for a surface where TaFe lamellas form triangles of two different shapes (I resp. II).

Table 1

Determination of the position of TaFe plates in the matrix; $d = 3\ \mu m$

Planes	Shape of triangle I		Shape of triangle II	
	$\Delta h_I(\mu m)$	$\Delta b_I(\mu m)$	$\Delta h_{II}(\mu m)$	$\Delta b_{II}(\mu m)$
$(0001)\ (1\bar{1}00)\ (\bar{1}101)$	-7.4	-4.0	-4.0	-2.1
$(0001)\ (1\bar{1}01)\ (\bar{1}101)$	+5.6	+3.0	-4.0	-2.1
$(10\bar{1}0)\ (1\bar{1}00)\ (\bar{1}101)$	-0.1	0	+3.3	+1.7
$(10\bar{1}0)\ (1\bar{1}01)\ (\bar{1}101)$	+12.7	+6.6	+3.3	+1.7
$(\bar{1}011)\ (1\bar{1}00)\ (\bar{1}101)$	-3.7	-2.0	-0.4	-0.2
$(\bar{1}011)\ (1\bar{1}01)\ (\bar{1}101)$	+9.2	+5.0	-0.4	-0.2
Experimental data	+9.5	+4.9	0	0

The second phase is supposed to precipitate not exclusely along $\{10\bar{1}1\}$ planes of the matrix. In many cases the lamellas form exact right angles at the surface. This indicates a second phase precipitation along the basal plane and planes with a normal perpendicular to that of the basal plane – probably prism planes. In this case there are many surface orientations where right angles exist, e.g. $(11\bar{2}u)$ and (vwf(v,w)0), (u,v,w assuming any value).

The plates of the second phase exist in wide bands enclosing the grains of the matrix (Fig. 3). The volume concentration of TaFe in these bands was counted planimetrically for several melts. Within the limits of experimental error all numerical values agree (see Table 2).

Table 2

Concentration of TaFe in the two-phase areas

Melt No.	Part of surface	TaFe concentration in the bands
I	a	16.3 vol%
I	b	16.3 vol%
I	c	15.4 vol%
II		16.9 vol%
V		16.6 vol%
Average value		16.3 vol%

This result leads us to the assumption that we have to deal with an eutectic of TaFe and $TaFe_2$ with about 16 vol% of TaFe. Yet in the phase diagram of the system Ta-Fe given by Burlakov and Kogan (1) the growth of TaFe is peritectic.

Investigations of the discussed samples were made about contrast and phase analysis in the emission electron microscope. Results will soon be reported.

The author wishes to thank Prof. Dr. A. Recknagel and Prof. Dr. G.E.R. Schulze for valuable help in this work.

References

(1) V.D. BURLAKOV and V.S. KOGAN, Fiz. Metallov i Metallovedenie 7, 708 (1959).

(2) G. LEITNER, Diplomarbeit, Technische Universität Dresden 1964.

(3) E.I. GLADYSHEVSKII, P.I. KRIPYAKEVICH, and YA.P. YARMOLYUK, Izv. Akad. Nauk SSSR, Ser. neorg. Materialy 1, 1086 (1965).

(Received December 16, 1966)

phys. stat. sol. 19, K75 (1967)
Subject classification: 18.3

Physikalisches Institut der Hochschule für Verkehrswesen
"Friedrich List", Dresden

Crystal Anisotropy and Saturation Magnetisation of Hard Ferrites with a Composition near the M-Compound Prepared by Means of the Verneuil Method

By
L. JAHN

Single crystals of the permanent magnetic materials bariumferrite (after Rosenberg (1)) and strontiumferrite were produced by means of the Verneuil technique. It was tried to obtain the chemical composition BaM (= $BaO \cdot 6Fe_2O_3$) and SrM (= $SrO \cdot 6Fe_2O_3$).

Basing on a chemical analysis and the phase diagram of BaM (van Hook (2)), it seems impossible to obtain pure BaM single crystals by the flame fusion method. However, we have produced strontiumferrite single crystals with chemical compositions lying near the M-compound and with a content of Fe^{2+} ions of 5%. The vapour pressure which is essentially higher for BaO than for SrO at the same temperature (3) can explain this different behaviour. The homogeneity of the SrM single crystals was tested by means of chemical analysis and magnetic measurements. It was found to be sufficient over a length of some cm in axial and 1 cm in radial direction.

From the grown crystals of 10 to 30 g weight, surrounded by a crystalline zone 0.1 to 2 mm thick, we could easily cleave single crystal line plates with the following dimensions: 5 mm < a < 15 mm; 5 mm < b < 20 mm; 2 mm < c < 6 mm. The c-axis of the hexagonal lattice was parallel to the normal direction of the plates.

The saturation magnetisation J_s was measured in the easy direction on plates of 20 to 100 mg weight by the ballistic method[1] at 5 kOe and with a magnetic balance[1] at 20 kOe.

---

1) The author is indebted to the Institut für magnetische Werkstoffe der DAW Berlin, Jena, especially to Dr. G. Elbinger, for carrying out these measurements.

The constant K of the uniaxial crystalline anisotropy was determined from the torque maximum at 26 kOe.

Since it was very difficult to prepare exact spheres of these easily cleavable and mechanical anisotropic single crystals, the samples for K-measurements were approximately ellipsoids of revolution with a c'/a ratio of 0.6 to 0.95 and c' parallel to the hexagonal c-axis. Owing to this form anisot the K-value to be measured was about 1 to 10% lower. The results are shown in Table 1.

Table 1

Crystal No.	Chemical analysis		J_s (G)	ρ (gcm^{-3})	K ($ergcm^{-3}$)	
		Fe^{2+}(%)				c'/a
15	$BaO \cdot 7.4Fe_2O_3$	8	398	5.22	$3.2x10^6$	0.6
16	$BaO \cdot 9\ Fe_2O_3$	8	400	5.22	$3.2x10^6$	0.88
17	$BaO \cdot 6.5Fe_2O_3$ (technical)	6	379	5.26	$3.0x10^6$	0.93
28	$SrO \cdot 6.3Fe_2O_3$	5	378	5.11	$3.5x10^6$	0.7
29	$SrO \cdot 5.8Fe_2O_3$	4	386	5.11	$3.4x10^6$	0.94
30	$SrO \cdot 5.8Fe_2O_3$	4	382	5.10	$3.4x10^6$	0.55
33	$SrO \cdot 6.3Fe_2O_3$	5	380	5.11	$3.4x10^6$	0.95
34	$SrO \cdot 7.4Fe_2O_3$	5.5	374	5.10	$3.5x10^6$	0.50

Conclusions

Bariumferrite single crystals tend to the X- and W-composition (2). In comparison to this strontiumferrite single crystals have nearly the M-composition. In both cases the magnetic values do not strongly depend on the chemical composition in the investigated range.

The average measured K-value ($\approx 3.45x10^6$ $ergcm^{-3}$) of strontiumferrite is higher than the measured value of bariumferrite. This result lies beyond the limits of error.

This increase of the constant of anisotropy by substitution of Ba by Sr complies with the results of Bitetto (4) and Kojima and Goto (5), obtained from resonance measurements and observation of the remanent domain structure respectively. The measured J_s-values of strontiumferrite are in agreement with (4). With reference to (6) the J_s-values of the bariumferrite crystals (No. 15 and 16) are increased by 5% and the K-values decreased by 3%. A detailed paper is in preparation.

References

(1) M. ROSENBERG et al., Studii Cercetări Fizică (București) 3, 249 (1963).

(2) H.J. VAN HOOK, J. Amer. Ceram. Soc. 47, 579 (1964).

(3) M. VON ARDENNE, Tabellen der Elektronenphysik, Vol. 2, Berlin 1956.

(4) D.J. DE BITETTO, J. appl. Phys. 35, 3482 (1964).

(5) H. KOJIMA and K. GOTO, Proc. Internat. Conf. Magnetism, London 1965 (p. 727).

(6) J. SMIT and H.P.J. WIJN, Ferrite, Phil. Techn. Bibliothek 1962.

(Received December 10, 1966)

phys. stat. sol. 19, K79 (1967)
Subject classification: 10; 22.1.2

II. Physikalisches Institut der Humboldt-Universität zu Berlin

Density Changes due to Dislocations in Silicon Single Crystals

By
M. ARNDT and A.F. RUDOLPH

From density measurements details on the perfection of the crystal lattice as well as on the distribution of imperfections and impurities can be deduced. Till now informations about point defects from heat-treated, doped (1 to 4), neutron irradiated (5), and plastically deformed silicon crystals have been obtained. In this paper we study the influence of dislocations on the density of silicon single crystals.

For these studies a "compression-floating-method" (5) was used with which the density of the liquid is changed by hydrostatic pressure at constant temperature. The measuring set-up developed by Brogren and Stebler (5) was used because of its high temperature stability which was better than ±0.0001 °C. We used a dish with interchangeable vertical tubes. Modifications of the samples originating from sealing the sample cell are now avoidable and several specimens can be simultaneously measured. An optimum proportioning of the samples and tubes was guaranteed, so that they neither tilted nor adhered, and responded to pressure changes smaller than 1 Torr. We successfully used a solution of 78 ml bromoform (high-purity) and 25 ml n-hexan (high-purity) for these investigations. This solution has a lower evaporation rate than that used in (7) and didn't segregate itself.

The samples for measurements belonged to one dislocation-free crystal (Fig. 1) which was grown by the method of Dash (8) (in the [111] direction in a vacuum of 10^{-5} Torr). In these dislocation-free crystals dislocations can be generated at constant growth conditions by mechanical influence (9). For this reason, a little quartz crystal was put onto the surface of the melt which adheres to the growing crystal. The point of contact is the source of new dislocations which multiply by thermal stress. Now crystal regions of different dislocation

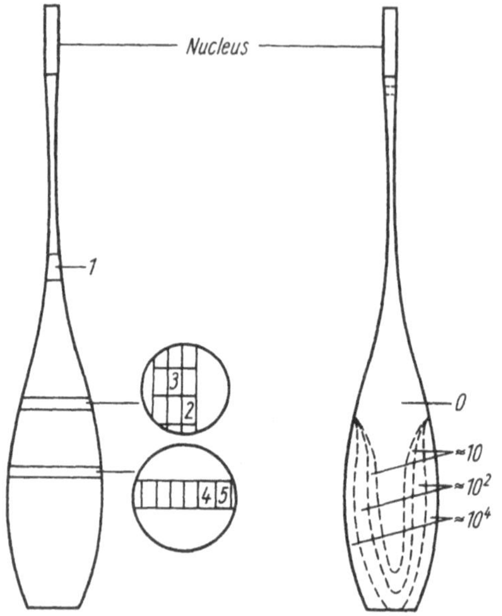

Fig. 1. Sample position and dislocation density of the crystal

densities and equal parameters are available for these investigations.

The specimens were mechanically polished, etched in one part of concentrated pure HF and three parts of concentrated HNO_3 and rinsed in HF and deionized water. The size of the specimens was approximately 4x7x0.7 mm^3. The dislocation density was determined after the measurement by etching (10).

We measured the "floating-pressure" of the specimens relative to each other at constant temperature. The resulting density difference was calculated from the compressibility of the liquid determined from the temperature-pressure-function and the specific thermal volume expansion.
We found

$$\frac{\Delta \varrho}{\Delta T} = 1.03\times10^{4} \text{ Torr deg}^{-1} .$$

The average expansion coefficient in the investigated temperature range was

$$\bar{\gamma} = 0.934\times10^{-3} \text{ deg}^{-1},$$

from which a relative density change per unit pressure was calculated to

$$\frac{\Delta \varrho}{\varrho \Delta \varrho} = 0.906\times10^{-7} \text{ Torr}^{-1}.$$

Table 1 lists the dislocation density and the relative density change of the samples of one crystal.

Table 1

specimen	1	2	3	4	5
dislocation density	0	0	0	1×10^3	2×10^5
$\frac{\Delta\varrho}{\varrho}$ relative to specimen 1	0	0	0	-4.5×10^{-7}	-2.8×10^{-6}

The absolute error of the measurement was under most unfavourable conditions

$$\Delta\left(\frac{\Delta\varrho}{\varrho}\right) = 4\times10^{-7}.$$

The investigated crystals could have only point defects in the dislocation-free area. Because the elimination of dislocations takes place by supersaturation with vacancies (8, 17), it is possible that the concentration of vacancies in specimen 1 is greater than the thermal equilibrium concentration (5×10^{12} cm^{-3} at room temperature) (12). Samples 2 and 3 belonged to a dislocation-free crystal region which was grown at normal growth velocity (10 cm/h). However, specimen 2 and 3 didn't differ from sample 1 within the experimental accuracy. Only in crystal regions of appropriate high dislocation density the crystal density was measurably changed (Table 1). This is to be expected because the dislocations cause a decrease of the density (11). The shearing of the crystal lattice caused by the generation of dislocations increased the distance of neighbouring lattice planes (13). Screw dislocations dilated the crystal structure in radial symmetry (14). Dislocations observed in the diamond-type lattice turned out to be predominantly mixed (60°-dislocations) and screw dislocations (15).

The density changes of the specimens could be influenced by the interaction of dislocations and point defects (8, 16 to 21). Doping with electrically active elements cannot explain the magnitude of the effect, because they are sufficiently homogeneously distributed. The content of carbon which decreases also the density was not investigated. On account of the growing parameters no distribution corresponding to the dislocation density was expected.

Surface effects were prevented by careful sample preparation (6).

References

(1) M.E. STRAUMANIS and E.Z. AKA, J. appl. Phys. 23, 330 (1952).

(2) A. SMAKULA and V. SILS, Phys. Rev. 99, 1747 (1955).

(3) F.H. HORN, Phys. Rev. 97, 1521 (1955).

(4) J.S. PRENER and F.E. WILLIAMS, J. chem. Phys. 35, 1803 (1961).

(5) G. BROGREN and B. STEBLER, Z. Naturf. 17a, 86 (1962).

(6) M.E. STRAUMANIS, P. BORGEAUD, and W.J. JAMES, J. appl. Phys. 32, 1382 (1961).

(7) H.L. JOHNSTON and C.A. HUTCHINSON, J. chem. Phys. 8, 869 (1940).

(8) W.C. DASH, J. appl. Phys. 30, 459 (1959).

(9) G. GREGER, Z. angew. Phys. 13, 47 (1961).

(10) E. SIRTL and A. ADLER, Z. Metallk. 52, 529 (1961).

(11) H. STEHLE and A. SEEGER, Z. Phys. 146, 217 (1956).

(12) S. MAYBURG, Acta metall. 4, 52 (1956).

(13) A. SEEGER, Hdb. Phys., Vol. VII/1, Springer-Verlag, 1955 (p. 618).

(14) H. STEHLE and A. SEEGER, Z. Phys. 146, 217 (1956).

(15) J. HORNSTRA, J. Phys. Chem. Solids 5, 129 (1958).

(16) B. OKKERSE, Philips tech. Rev. No. 11, 335 (1959/60).

(17) W.C. DASH, Metall. Soc. Conf. 5, 195 (1959).

(18) F. SEITZ, Rev. mod. Phys. 23, 328 (1951).

(19) E. TEGHTSOONIAN and B. CHALMERS, Canad. J. Phys. 29, 370 (1951).

(20) C. ELBAUM, Phil. Mag. 5, 669 (1960).

(21) A.G. TWEET, J. appl. Phys. 29, 1520 (1958).

(Received December 10, 1966)

phys. stat. sol. 19, K83 (1967)
Subject classification: 14.1; 21.6

Department of Solid State Physics, Metal Physics Section,
Charles University, Prague

Resistivity Changes in Air-Quenched Gold Wires

By

B. SPRUŠIL and P. VOSTRÝ

The behaviour of point defects in a crystal lattice is requently studied by observations of resistivity changes. By his method many data were obtained on the behaviour of vaancies, especially in gold. The problem is none the less far rom being solved, owing to various experimental difficulties ncountered (e.g. influence of heat treatment before quenching, mbient atmosphere etc. - see e.g. (1)). In this letter we ish to give a preliminary report on data obtained on slowly uenched gold wires.

The specimens (diameter 0.012 cm, purity before quenching 9.5 at%[1], average length 10 cm) were spirally wound on a ica frame in a glass tube. This tube was held in liquid air, nd was heated by a dc current and quenched by switching off he current. The initial quenching velocity was estimated to e $\approx 1.6 \times 10^3$ deg/s. Resistivity changes were measured by a tandard potentiometric method using a dummy specimen, as escribed elsewhere (2, 3).

The values of the resistance R_T (at a temperature of T °C) nd the tables of the function $f(T) = R_T/R_0$ (R_0 is the reistance at 0 °C) were used for the determination of the pecimen temperature (4, 5). It was found that this function epends strongly on the previous heat treatment. From an X-ray nalysis of heat-treated wires it follows that grain size and riginal fiber texture change during the treatment.

The change of the resistance of the sample is more ronounced for R_0 than for the high-temperature values R_T. lthough this change is the greatest for the first minutes f previous annealing, it does not attain a constant value

) Supplied by "Safina", Vestec.

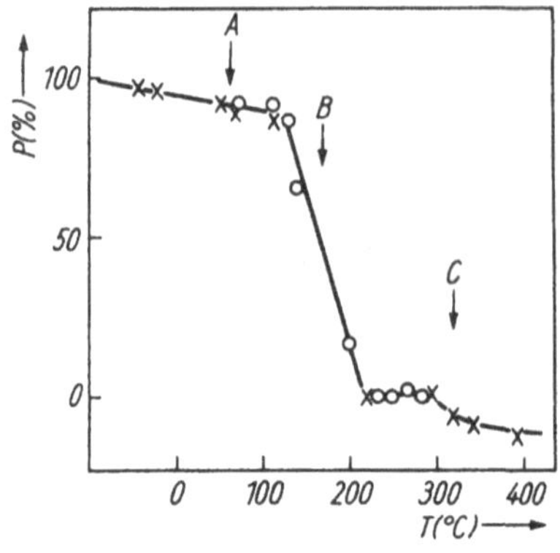

Fig. 1. Characteristic annealing spectrum (annealed for 8 min at each temperature). The quenching temperature was 838 °C (o) and 730 °C (x)

even after several hours of annealing. On the other hand, it was found that when taking for R_0 the value for the untreated sample, the function $R_T/R_0 = f(T)$ remains almost constant. The function f(T) experimentally determined in this way was used for the calculation of the specimen temperature.

After quenching from various temperatures the specimens were isochronally annealed to reveal the annealing stages. A typical curve is shown in Fig. 1. In the temperature range 0 to 80 °C a slight decrease of the quenched-in resistance was observed, this decrease being in some cases more pronounced in the range 80 to 110 °C (A). The amount of the extra resistivity in the interval 0 to 110 °C varies for different samples, and we were not able to correlate it with any of the quenching characteristics. The most prominent feature of the annealing curve is a sharp decrease in the interval

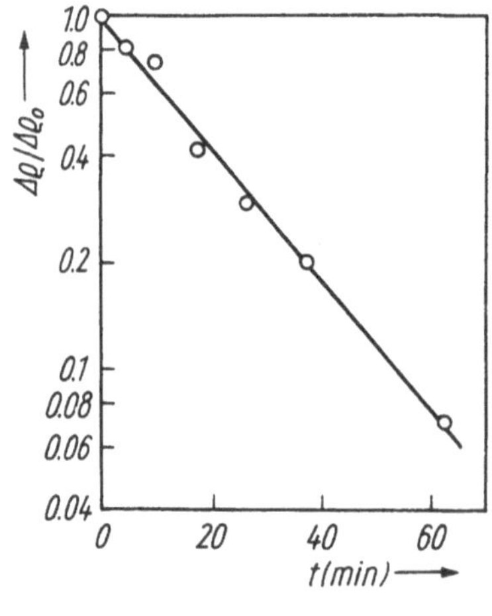

Fig. 2. Experimental values (o), compared with a theoretical curve for a first order annealing kinetics (solid line) for a specimen quenched from 802 °C and annealed at 150 °C

Fig. 3. Quenched-in resistivity as a function of quenching temperature

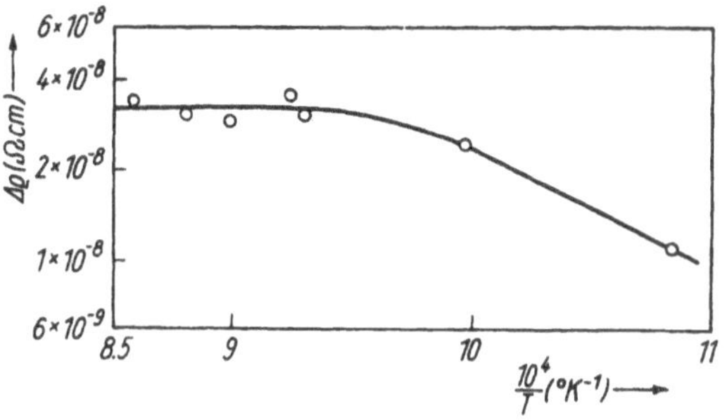

110 to 200 °C (B). One of the samples which was annealed at temperatures higher than 250 °C exhibits a further decrease in the range 300 to 400 °C (C).

In order to determine the annealing kinetics in stage (B) two samples were isothermally annealed in the appropriate temperature region. It was found that the annealing kinetics is of first order (see Fig. 2). The corresponding activation energy, determined for one specimen by the slope intersection method, was $E = (0.80 \pm 0.06)$ eV which lies well within the range of activation energies ascribed to vacancy migration in gold (see for example (6, 7)).

We suppose therefore that stage (B) of Fig. 1 corresponds to van Bueren's stage IV and is due to the annealing of single vacancies.

In Fig. 3 the resistivity increments $\Delta\rho = \rho_{0^oC} - \rho_{250^oC}$ are plotted against $1/T_q \times 10^4$, T_q being the quenching temperature. A tentative determination of a "formation" energy from the decreasing part of the curve gives $E \geq 0.8$ eV.

The results of these experiments can be briefly summarized as follows:

a) The function R_T/R_0 strongly depends on the state of the specimen, e.g. on its degree of recrystallization. It was not possible to reach a steady value even after an annealing during 6 hours at ≈ 800 °C in nitrogen atmosphere.

b) It was not possible to find any relationship between the observed annealing spectrum and the heat treatment before quenching.

c) The values of activation energies for vacancy formation and migration obtained are in agreement with previous results. The annealing of vacancies was observed at temperatures approximately 100 °C higher than published elsewhere (see for example (6)).

References

(1) J.A. YTTERHUS and R.W. BALLUFFI, Phil. Mag. 11, 707 (1965).

(2) K. MÍŠEK, Čes. Čas. Fys. 12, 332 (1962).

(3) B. SPRUŠIL, Czech. J. Phys. B15, 287 (1965).

(4) A.N. GERRITSEN, Hdb. Phys., Vol. XIX/1, Springer-Verlag, 1956 (p. 197).

(5) K. MÍŠEK and J. POLÁK, J. Phys. Soc. Japan 18, 180 (1963).

(6) J.E. BAUERLE and J.S. KOEHLER, Phys. Rev. 107, 1493 (1957).

(7) A.H. COTTRELL, Symposium on Vacancies and Other Point Defects in Metals and Alloys, AERE Harwell, 1958.

(Received December 27, 1966)

phys. stat. sol. 19, K87 (1967)
Subject classifications: 13; 20; 18; 22

Materials Research Laboratory, Pennsylvania State University, University Park, Pennsylvania (a), and
Electrical Engineering Department, United States Naval Academy, Annapolis, Maryland (b)

Magnetic and Optical Properties of Dioptase

By
R.E. NEWNHAM (a) and R.P. SANTORO (b)

Dioptase, $CuSiO_3 \cdot H_2O$, is a comparatively rare mineral, found in a few localities associated with other copper ores. Magnetic and optical measurements on a well-crystallized specimen from Reneville, Congo Francaise, are reported in this note. Identification of the mineral was confirmed by X-ray diffractometer patterns.

The crystal structure of dioptase has been refined by Heide and Boll-Dornberger (1). The space group is $R\overline{3} = C_{3i}^2$ with eighteen formula-units in the hexagonal cell, a = 14.61 and c = 7.80 Å. Dioptase is a cyclosilicate containing Si_6O_{18} rings similar to beryl. The silicate rings are inter-connected by copper atoms surrounded by a deformed octahedron of four oxygens and two water molecules. The octahedron is distorted to such an extent that it may alternatively be considered a square planar configuration with four Cu-O bonds of about 1.96 Å, and an average O-Cu-O angle of 91°. H_2O molecules lie at 2.61 and 2.75 Å; since one water is appreciably closer than the other, the Cu site symmetry is approximately C_{4v}. As illustrated schematically in Fig. 1, the Cu^{2+} ions share oxygen neighbors forming a three-dimensional network; each Cu shares a square planar edge with another Cu and corners with two more

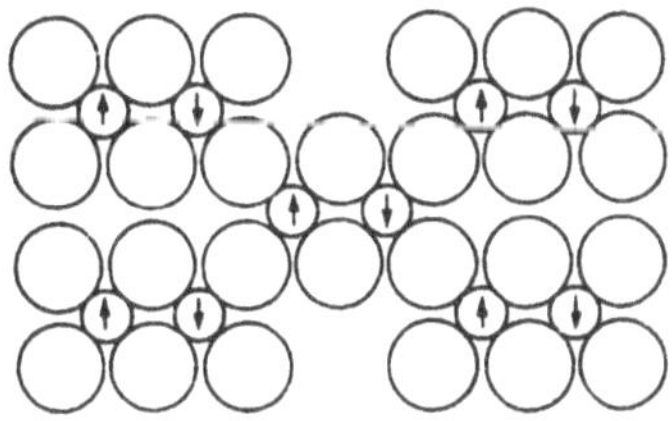

Fig. 1. Idealization of the copper (small circles) - oxygen (large circles) network in dioptase. Arrows indicate relative spin directions for the proposed magnetic structure

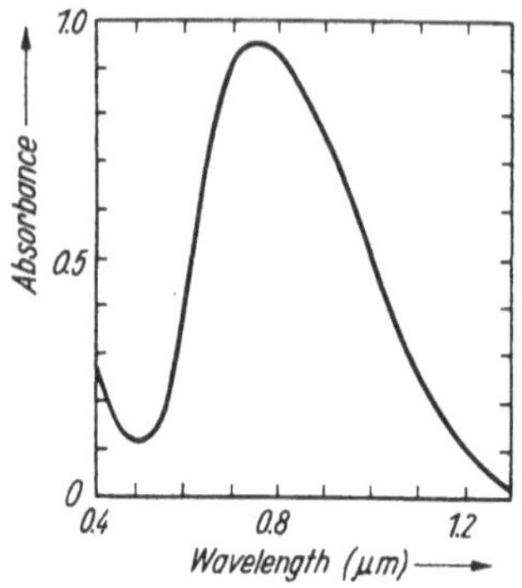

Fig. 2. Diffuse reflectance spectrum of dioptase

The diffuse reflectance spectrum of powdered dioptase (Fig. 2) was obtained using a Beckman DK-2A spectrophotometer with MgO as a reference material. Dominating the spectrum is an intense absorption peak at 13300 cm^{-1} with a band width at half maximum of 6500 cm^{-1}. The characteristic greenish-blue color of dioptase results from transmission near 5000 Å, between the broad peak and the ultraviolet absorption edge. Petrographic examination of small single crystals revealed little or no dichroism, so that the visible absorption spectrum does not depend markedly on polarization.

In a pure octahedral field the broad peak would be attributed to the $E_g \rightarrow T_{2g}$ transition of the Cu^{2+} $3d^9$ electron configuration. As previously mentioned, however, the near neighbor environment more nearly approximates C_{4v} symmetry, resulting in further splitting as shown in Fig. 3. The doubly degenerate E_g level divides into a B_1 ground state and an A_1 exited state, while triply degenerate T_{2g} separates into B_2 and doubly degenerate E. Since the Cu^{2+} site symmetry does not contain inversion, electric-dipole transitions are possible. Combining the representations of B_1 with those of the three exited states shows that only the transition to E contains an electric-dipole operator, suggesting $B_1 \rightarrow E$ as the origin of the intense absorption peak. The vacancy in the d shell shifts from $d_{x^2-y^2}$ to d_{yz} or d_{zx} during this

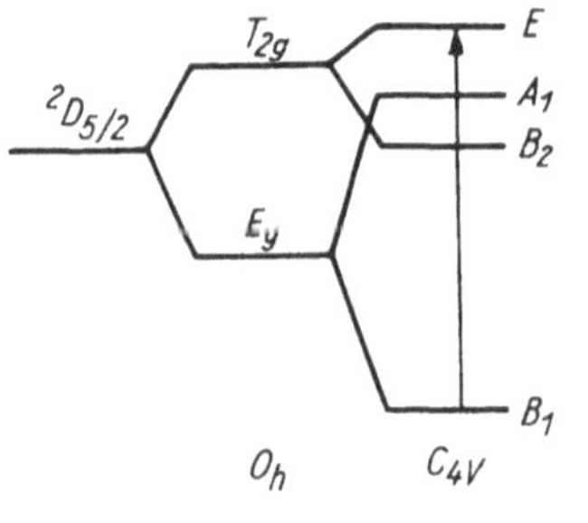

Fig. 3. Energy level diagram for a d^9 ion in an octahedral field modified by a square planar distortion with $c > a$

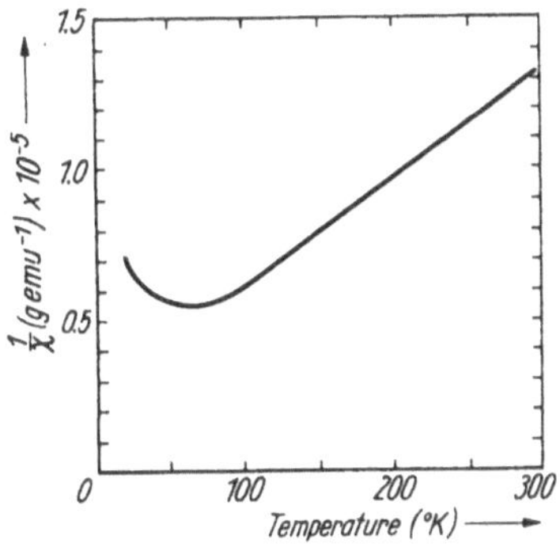

Fig. 4. Reciprocal susceptibility of dioptase plotted as a function of temperature

transition.

Magnetic susceptibility measurements were made with a vibrating-sample magnetometer. The field strength normally used was 10^4 Oe, but higher and lower fields were used to test the proportionality between field and magnetization. Results for polycrystalline $CuSiO_3 \cdot H_2O$ are shown in Fig. 4. The material becomes antiferromagnetic near $T_N = 70$ °K. Above the Néel point, the data obey a Curie-Weiss law, $\chi = C/(T + \theta)$, with $\theta = 70$ °K. The effective Bohr magneton number is $\mu_{eff} = 1.86$, comparable to other salts of divalent copper.

Molecular field calculations were carried out assuming a two-sublattice model, with α and β representing Weiss constants for the shared edge and shared corner interactions, respectively. Allowing α and β to be either ferromagnetic (+) or antiferromagnetic (-) leads to four possible situations. Both interactions positive results in ferromagnetism, obviously not the case with dioptase. Three different antiferromagnetic spin arrays are obtained when one or both signs are negative:

$$\text{I.} \quad +\alpha, \quad -\beta \quad T_N/\theta = (\alpha + 2\beta)/(2\beta - \alpha),$$

$$\text{II.} \quad -\alpha, \quad +\beta \quad T_N/\theta = (\alpha + 2\beta)/(\alpha - 2\beta),$$

$$\text{III.} \quad -\alpha, \quad -\beta \quad T_N/\theta = 1.$$

Type III seems to give the best agreement with experiment ($T_N \approx \theta \approx 70$ °K); relative spin orientations are illustrated in Fig. 1. The predicted magnetic structure is twice the size of the chemical unit cell with a = 14.61, c = 15.60 Å.

The authors are indebted to Prof. W.B. White for his advice and assistance. This work was supported by Advanced Research Projects Agency, Contract DA49-083, OSA-3140, and by the U.S. Air Force Materials Laboratory, Contract

AF33(615)-2199.

References

(1) H.G. HEIDE and K. BOLL-DORNBERGER, Acta cryst. 8, 425 (1955).

(Received January 4, 1967)

phys. stat. sol. 19, K91 (1967)
Subject classification: 14.4.1; 22.6

Laboratoire Central des Industries Électriques,
Fontenay-aux-Roses

On Charge Storage in Anodic Tantal Oxide Layers

By

C. CHERKI and R. COELHO

1. Introduction

The processes of charge storage and release in most insulators, and particularly in anodic oxide films (1) are still rather obscure, although permanent internal polarization (p.i.p.) is known to contribute (2). In this note, the importance of p.i.p. is confirmed by describing some temperature effects on isothermal and non-isothermal transient currents through anodic Ta_2O_5 films.

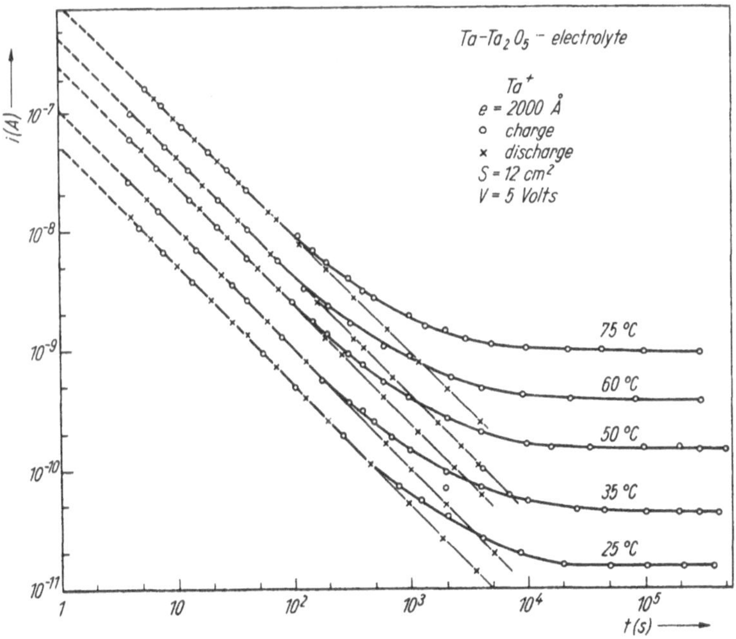

Fig. 1. Polarization and depolarization currents at various temperatures ($E \approx 2.5x10^5$ Vcm^{-1})

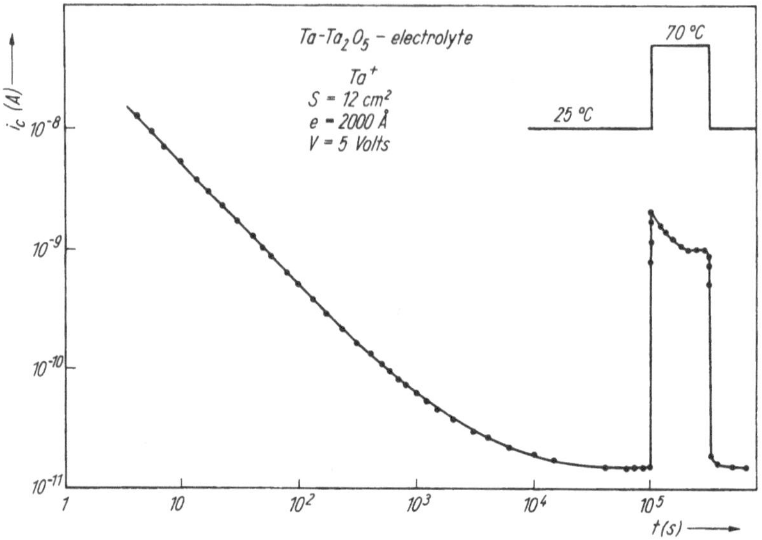

Fig. 2. Result of a heating step on the conduction current $i_c(T_1)$

2. Experimental conditions

The films were prepared by anodic oxidation in sulfuric solution of 12 cm^2 tantalum platelets, as described in (1) and (3). The measurements were performed in weak sulfuric solution. The voltage was applied in the anodizing direction but was smaller by one order of magnitude. In view of the small thickness of the Ta plates, (0.2 mm), the sample temperature was assumed equal to that of the electrolyte.

3. Experimental results

Fig. 1 represents the decay of isothermal polarization and depolarization currents under 5 V applied voltage (approximately 2.5×10^5 Vcm^{-1}). As shown previously (1), the decay of the polarization and depolarization are identical up to a few minutes. They obey a law of the form

$$i(t,T) = A(T)\, t^{-1}. \qquad (1)$$

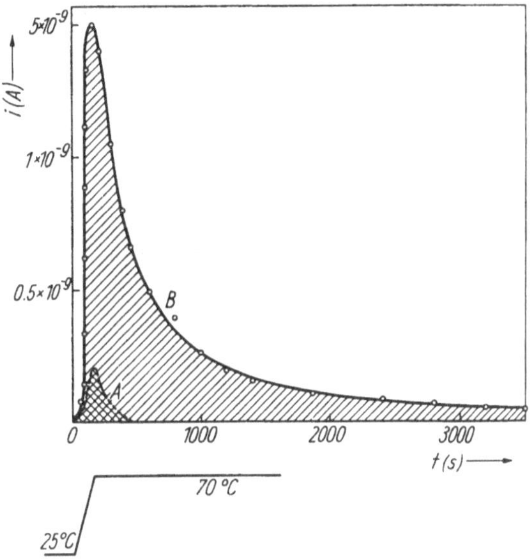

Fig. 3. Charge released in a short-circuited sample under the action of a heating step.
A without preheating during polarization,
B with preheating during polarization

As long as the temperature of polarization is equal or higher than the temperature of depolarization, i(t,T) is independent of the polarization temperature.

After a few minutes, the polarization current i(t,T) decays towards a steady value $i_c(T)$, while the depolarization current goes on according to (1). Within the investigated temperature range, A(T) and $i_c(T)$ obey Arrhenius laws, with activation energies of 0.5 and 0.72 eV, respectively.

Fig. 2 refers to experiments similar to those of Gross and Denard on electrets (4). First, we observe the decay of the polarization current at T_1 = 25 °C. Then, after 10^5 s, the sample is heated to T_2 = 70 °C. This heating results in a sudden rise of the current followed by a slow decay towards $i_c(T_2)$. The initial value of the transient depends on both the rate of temperature rise and on the difference $(T_2 - T_1)$. Furthermore, the charge $Q = \int i dt$ represented by the transient increases with T_2 and happens to be equal to the difference between $Q_2 = \int i(T_2)dt$ and $Q_1 = \int i(T_1)dt$ of the isothermal depolarization charges at T_2 and T_1, respectively.

After keeping $T = T_2$ until a steady current flows, T is lowered from T_2 to T_1 (either continuously or by small steps) in a time varying between 30 and 240 min. Under these conditions, the current recorded at a temperature T is equal to $i_c(T)$, in agreement with the observation of Gross on electrets. The same result is obtained even if the sample is heated during the polarization.

Fig. 3 represents the current pulses obtained when a short-circuited sample is heated from T_1 to T_2. Curves A and B represent the current pulses recorded, respectively, without and with previous heating during polarization. The amount of charge released, represented by the area under the curves, is considerably increased in the latter case. The difference between the two charges remains frozen in the dielectric until it is released by heating (or, eventually, irradiation).

Finally, we have investigated the effect of a temperature rise during depolarization and observed a striking analogy between a heat pulse and a voltage pulse.

4. Conclusions

a) Quick measurements of the conduction currents $i_c(T)$ can be performed by first heating the oxide film, then slowly decreasing the temperature.

b) As long as the temperature during polarization is equal or higher than that of depolarization, the depolarization current depends only on the latter temperature.

c) The charge absorbed during a heating step is not released during depolarization unless the sample is heated in short-circuit, or short-circuited at room temperature for an exceedingly long time.

These observations are typical of electret behaviour.

References

(1) C. CHERKI, R. COELHO, and J.L. MARIANI, Solid State Commun. 4, 411 (1966).

(2) J.R. FREEMAN, H.P. KALLMANN, and M. SILVER, Rev. mod. Phys. 33, 553 (1961).

(3) C. CHERKI and J.L. MARIANI, Rev. gén. Electr. 75, 769 (1966)

(4) B. GROSS and L.F. DENARD, Phys. Rev. 67, 253 (1945).

(Received January 4, 1967)

phys. stat. sol. 19, K95 (1967)
Subject classification: 10.1; 21.1.1

Inorganic Materials Research Division, Lawrence Radiation Laboratory, and Department of Mineral Technology, College of Engineering, University of California, Berkeley, California

Camel-Humps Peierls Hills in Iron[1)]

By
P. GUYOT

Recently, Chang and Graham (1) calculated the Peierls hills for an edge [111] dislocation lying in the (110) plane in Fe. They used relaxation techniques with a two atom anharmonic interaction potential so adjusted as to give the known second and third order elastic properties of Fe. E, the energy per identity distance along such straight dislocation, is plotted in Fig. 1 versus the distance moved in the y-direction; the hills have two equal peaks and primary and secondary minima. If $\Gamma(y)$ is the line tension of the dislocation, periodic in the y-direction with a period equal to the Burgers vector $b = 1/2\,[111]$, the plot can be analytically fitted by the equation

$$\Gamma(y) = \frac{\Gamma_c + \Gamma_o}{2} + \frac{\Gamma_c - \Gamma_o}{2}\left(\frac{\alpha}{4} + \cos\frac{2\pi y}{b} - \frac{\alpha}{4}\cos\frac{4\pi y}{b}\right) \qquad (1)$$

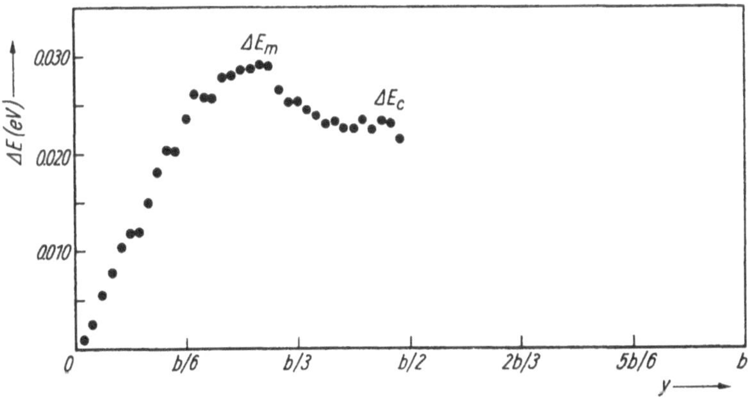

Fig. 1. ΔE versus distance moved in the [111] direction

1) AEC Contract No. W-7405-eng-48.

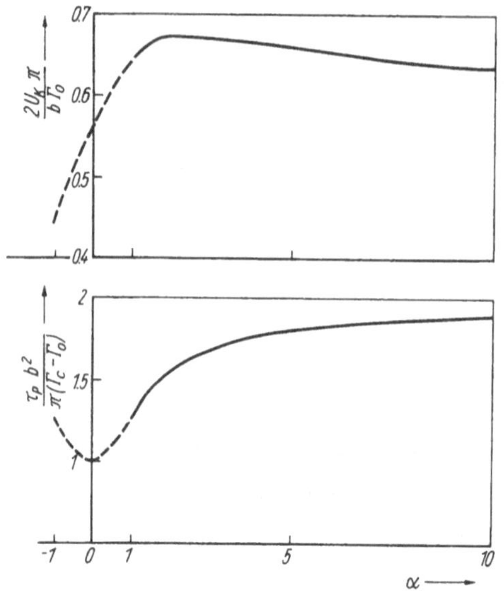

Fig. 2. Kink energy and Peierls stress. R = 1.01

with $\Delta E = (\Gamma(y) - \Gamma_o) 2\sqrt{2}\, b$, and $\alpha = 3.865$, $\Gamma(y) = \Gamma_o$ at $y = 0$, $\Gamma(y) = \Gamma_c$ at $y = b/2$. More generally, for $\alpha > 1$, such an equation represents camel-hump types of hills. Γ_m, the maximum value of $\Gamma(y)$, is related to Γ_o and Γ_c by the relation

$$\Gamma_c = \frac{2\Gamma_m - \Gamma_o\left(1 - \frac{1}{2\alpha} - \frac{\alpha}{2}\right)}{1 + \frac{1}{2\alpha} + \frac{\alpha}{2}} \;. \quad (2)$$

On the basis of the Dorn-Rajnak theory (2), the kink energy under zero stress, U_K, and the energy to nucleate a pair of kinks on a perfect dislocation moving beyond the hills under an applied stress τ^*, U_n, are given by

$$U_K = \Gamma_o \int_{-b/2}^{+b/2} \left\{\frac{\Gamma(y)}{\Gamma_o} - 1\right\}^{1/2} dy \;, \quad (3)$$

$$U_n = 2 \int_{y_o}^{y_c} \left\{\Gamma(y)^2 - \left[\tau^* b(y - y_o) + \Gamma(y_o)\right]^2\right\}^{1/2} dy . \quad (4)$$

The Peierls stress[2)], τ_p, is given by

$$\tau_p = \frac{(\Gamma_c - \Gamma_o)}{16 b^2 \alpha} \left(3 + \sqrt{1 + 8\alpha^2}\right) \sqrt{8\alpha^2 - 2 + 2\sqrt{1 + 8\alpha^2}} . \quad (5)$$

In the computations, $R = \Gamma_m/\Gamma_o$ and α were used as independent variables.

---

2) From Fig. 1a Peierls stress of 53.6 kg/mm^2 is deduced for iron, in good agreement with experimental values, Wynblatt and Dorn (4).

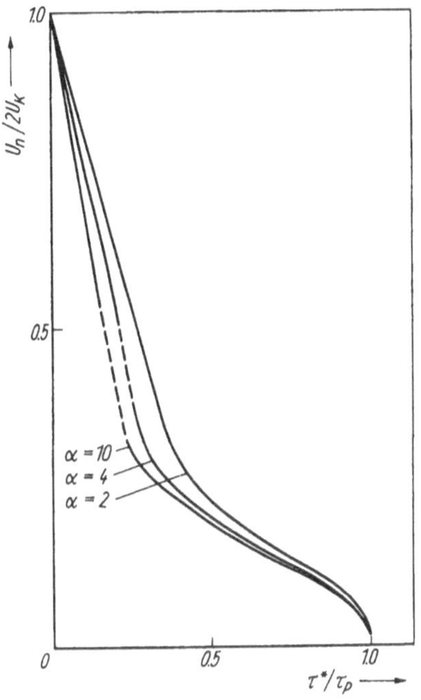

Fig. 3. Energy to nucleate a pair of kinks

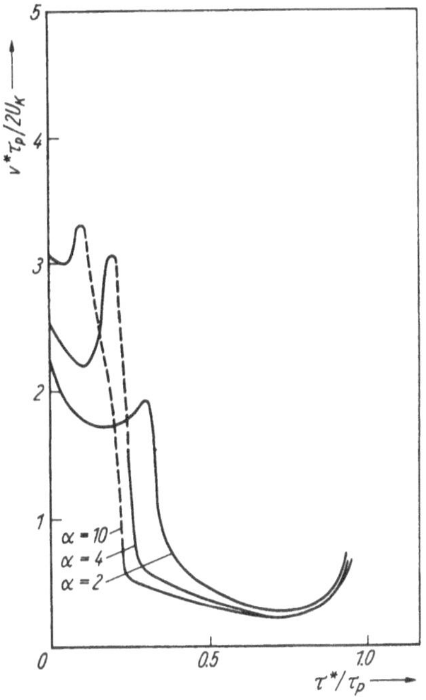

Fig. 4. Activation volume

Fig. 2 shows the variations of $\tau_p b^2/\pi(\Gamma_c - \Gamma_o)$ with α, and the variations of $2U_K\pi/b\Gamma_o$ with α for R = 1.01. The dotted lines for $-1 \leq \alpha \leq 1$ correspond to sinusoidal hills type, Dorn and Rajnak (2). $2U_K\pi/b\Gamma_o$ is closely proportional to $\sqrt{R-1}$.

Figs. 3 and 4, respectively, show the variations of $U_n/2U_K$ and $v^*\tau_p/2U_K$ (v^* being the activation volume defined by $v^* = -\partial U_n/\partial\tau^*$) as a function of the applied stress τ^*, for some values of α, and R = 1.01. For a given α and ratio τ^*/τ_p, $U_n/2U_K$ and $v^*\tau_p/2U_K$ were found insensitive to physically acceptable ranges in R (i.e. $1.01 \geq R \geq 1.004$).

Curves of Figs. 3 and 4 are broken for $0.1 < \tau^*/\tau_p < 0.3$. In this range of stresses the dislocation can lie in the intermediate valley. The present data refer to the case of a single activation for surmounting the double hump; the plotted activation energy is obtained by adding the two partial activation energies necessary to overcome each hump. Nevertheless the strain-rate, in fact, consists of three terms (one for each partial jump, and one for the previously considered single jump). Because the densities of dislocation lying in the deep and intermediate valleys are different, and likewise the areas swept out and the double kink widths for each partial jump, the formulation of the strain-rate no longer follows the simple form of a single thermally activated process. For these reasons the plots of Figs. 3 and 4 in the range $0.1 < \tau^*/\tau_p < 0.3$ are probably approximate.

As outlined by Guyot and Dorn (3) the curves of Figs. 3 and 4 differ appreciably from the corresponding curves for sinusoidal hills at high stresses. For a Peierls mechanism U_n is proportional to the absolute temperature, and therefore, as shown in Fig. 3, a zero slope stress-temperature is predicted at the absolute zero.

Similar analyses will be done in the future from the knowledge of the potential barriers for screw dislocations and edge dislocations of other orientations in body-centered cubic metals.

Acknowledgement

This research was conducted as part of the Inorganic Materials Research Division of the Lawrence Radiation Laboratory, Berkeley. The author wishes to express his appreciation to the United States Atomic Energy Commission for their support of this effort. He also thanks Prof. J.E. Dorn for stimulating discussion.

References

(1) R. CHANG and L.J. GRAHAM, phys. stat. sol. 18, 99 (1966).
(2) J.E. DORN and S. RAJNAK, Trans. AIME 230, 1052 (1964).
(3) P. GUYOT and J.E. DORN, Conf. on Deformation of Crystalline Solids, Ottawa 1966, to be published.
(4) P. WYNBLATT and J.E. DORN, UCRL-16601 (1965).

(Received January 5, 1967)

phys. stat. sol. 19, K99 (1967)
Subject classification: 1; 22.1

Materials Research Laboratory, Martin Company, Orlando, Florida (a), Metallurgy and Ceramics Research Laboratory, ARL Wright-Patterson AFB, Ohio (b), and Lockheed Missiles and Space Company, Palo Alto, California (c)

Further Comments on the Paper "On the Interpretation of Electron Diffraction Patterns from 'Amorphous' Boron"

By

H.M. OTTE (a), H.A. LIPSITT (b), P.F. LINDQUIST (c), M.L. HAMMOND (c), and R.H. BRAGG (c)

In a recent paper, hereinafter referred to as (1), Otte and Lipsitt presented a preliminary analysis of the structure of vapor-deposited boron. Comments on (1) were presented in this journal by Lindquist, Hammond, and Bragg, hereinafter referred to as (2).

It was clear from (2) that some misunderstanding of (1) existed; it was also stated that (1) had made some errors in their analysis. Fortunately, it was possible for the five principal authors of (1) and (2) to meet in September 1966 at the Air Force Materials Laboratory, Wright-Patterson Air Force Base, Ohio and to hold a day long discussion of the current status of the understanding of the structure of vapor-deposited boron. As part of that discussion the comments of (2) were also discussed and clarified.

The authors of (2) have recently completed a full investigation of the structure of vapor-deposited boron under the sponsorship of the U.S. Air Force Materials Laboratory, the results of which will shortly be submitted for publication. A discussion of those results prior to their publication would clearly be premature; however, in view of the serious nature of the comments of (2) it was thought advisable to clarify the issues already at hand.

It is now understood that (1) interpreted the boron halos in terms of an f.c.c. lattice, giving also certain indices in terms of a hexagonal axis system, but that (1) did not conclude that the B_{12} groups were necessarily centered on the lattice points, since they did not take intensity data into consideration. Therefore, a complete structure determination was not made. Likewise, the discussion of the relationship between faulted f.c.c. and h.c.p. was mistaken by (2) to imply that (1) believed their boron to consist of equal volume fractions of h.c.p. and f.c.c. material. It is clear that the data of (1) were sufficient only to conclude, as was actually done, that heavy faulting existed, that many of the characteristics of layered structures had been observed and that the best partial indexing of the X-ray and electron diffraction data that could then be accomplished indicated that the structure could best be treated as an f.c.c. one. In discussing the effects of beam heating, (1) "suggests" in the text that the effect would be to anneal out the faults, whereas in the abstract this was stated positively, and as such was misleading.

(2) actually enumerated ten weaknesses in the paper of (1) at first very briefly and then again in further detail. These points have been resolved as follows:

1. (2) agrees that (1) merely indexed the few apparent lines as f.c.c. but do not postulate that as a new structure.
2. (2) understands that (1) only doubled the "c" parameter for the purpose of an error computation - a valid procedure - and did not mean to associate this with an improper division of the β-rhombohedral structure.
3. (2) is correct. (1) only states but does not demonstrate why small crystallite size is rejected as the sole source of line broadening. (2) has made clear in more recent work that small particle size diffraction effects are present to some degree. If a very small particle size (range 10 to 100 Å, average 30 Å) can be accepted, then small crystallite size diffraction effects can possibly account for all the line broadening.
4. (2) is in error. (1) states in section 3.2 that "samples obtained from the crushed filaments <u>generally</u> showed three to four diffuse rings in the electron diffraction pattern ... Spot patterns could also be observed."

5. (2) is correct, except that (1) was not actually postulating a final structure for boron. (1) and (2) still regard this point to be a vital issue.
6. This comment is moot; the point cannot be decided in the absence of new and unequivocal data.
7. (2) is in error. It was pointed out in Item 2 above that the mathematical manipulation described is satisfactorily used in an error computation.
8. (2) is in error. Both extended relrod effects and diffuse diffraction lines are presented in (1) as evidence of faulting in the unheated material. No visual image evidence was presented, nor is any available at this time.
9. If the premise developed above is acceptable, then (1) does present evidence that the fault density is decreased by increased beam heating. The absence of striations in an EM image does not necessarily allow one to conclude that stacking faults must be absent, as was concluded by (2). However, further observations are required to establish the validity of the suggestion made by (1) that faults are annealed out on heating. Additional observervations in this regard have recently been presented by Lipsitt and Otte (3).
10. (2) is correct. The partial indexing of Fig. 3b would have been satisfactory in real space but not in reciprocal space.

References

(1) H.M. OTTE and H.A. LIPSITT, phys. stat. sol. 13, 439 (1966).
(2) P.F. LINDQUIST, M.L. HAMMOND, and R.H. BRAGG, phys. stat. sol. 17, K25 (1966).
(3) H.A. LIPSITT and H.M. OTTE, Proc. VI. Internat. Congr. Electron Microscopy, Vol. I, 1966 (p. 463).

(Received January 9, 1967)

phys. stat. sol. 19, K103 (1967)
Subject classification: 20

II. Physikalisch-Technisches Institut der Deutschen Akademie der Wissenschaften zu Berlin, Berlin-Adlershof

Ein Beispiel für die Existenz eines "Brewster-Winkels" in der nichtlinearen Optik anisotroper Kristalle

Von
R. FISCHER

Die Existenz eines "Brewster-Winkels" für die von der Oberfläche eines optisch isotropen, quadratischen Kristalls reflektierte 2. Harmonische wurde in (1) theoretisch vorhergesagt und in (2) experimentell nachgewiesen. Durchgeführt wurden die Experimente mit GaAs, einem Kristall der kubischen Klasse $\bar{4}3m$.

Hier wird die tetragonale Klasse $\bar{4}2m$ betrachtet, und es werden Bedingungen angegeben, unter denen für Kristalle dieser Klasse ein "Brewster-Winkel" einfach beobachtbar ist. Die Klasse $\bar{4}2m$ wurde gewählt, weil einige der in der nichtlinearen Optik häufig verwendeten Kristalle wie KDP und ADP zu dieser Klasse zählen und außerdem die Tensoren der quadratischen Suszeptibilitäten der Klassen $\bar{4}3m$ und $\bar{4}2m$ bekanntlich ((3), S. 121) nahezu identisch sind. Der wesentliche Unterschied besteht darin, daß Kristalle des tetragonalen Systems optisch einachsig sind. Beziehungen für die Vektoramplitude der von der ebenen Oberfläche derartiger Kristalle reflektierten 2. Harmonischen wurden in (4) angegeben.

Wir untersuchen nun folgenden Spezialfall: In der Einfallsebene der aus dem Vakuum einfallenden Grundwelle liege die optische Achse; weiterhin ist es zweckmäßig ((3), S. 157), den Kristall so zu orientieren, daß die Einfallsebene den Winkel zwischen den Richtungen [100] und [010] halbiert. Die Grundwelle sei senkrecht zur Einfallsebene polarisiert, im Kristall wird dann nur die ordentliche Grundwelle angeregt. Da im kristallographischen System nur die Komponenten des Tensors χ der quadratischen Suszeptibilität nicht verschwinden, deren drei Indizes verschieden sind, hat in dem betrachteten Spezialfall die Amplitude der nichtlinearen (quadratischen)

Polarisation das vom Einfallswinkel Θ_e der Grundwelle unabhängige Aussehen

$$\mathfrak{p}_1^{NL}(2\omega) = -\frac{b}{2} E_0^2(\omega)\mathfrak{t} \quad , \tag{1}$$

worin mit b die entsprechende Tensorkomponente bezeichnet wurde; ansonsten übernehmen wir die Bezeichnungen aus (4).

Aus den Gleichungen (11) bis (18) in (4) errechnet man für die Amplitude der speziellen Lösung der inhomogenen Wellengleichung den Ausdruck (2).

$$\mathfrak{g}_1^{NL}(2\omega) = \frac{\varepsilon_0(2\omega)\mathfrak{t} - \mathfrak{n}_0(\omega)\mathfrak{n}_0(\omega)\cdot\mathfrak{t}}{\varepsilon(2\omega):\mathfrak{n}_0(\omega)\mathfrak{n}_0(\omega) - \varepsilon_0(2\omega)\varepsilon_{ao}(2\omega)} \left|\mathfrak{P}_1^{NL}(2\omega)\right| . \tag{2}$$

Setzt man (2) in die in (4) angegebenen Gleichungen (40) bis (44) ein, sieht man, daß im Kristall nur die außerordentliche Harmonische angeregt wird und demzufolge auch die reflektierte Harmonische in der Einfallsebene der Grundwelle polarisiert ist; die Vektoramplitude der reflektierten Harmonischen kann aus (3) mit (4) und (5) berechnet werden.

$$\mathfrak{E}_r(2\omega) = \frac{v_3\left[\mathfrak{n}_r(2\omega) \times \left[\breve{\mathfrak{n}}_3 \times \mathfrak{n}_r(2\omega)\right]\right]}{\left[\mathfrak{n}_r(2\omega) \times \breve{\mathfrak{n}}_3\right]^2} , \tag{3}$$

$$\begin{aligned}\breve{\mathfrak{n}}_3 = \{&[\mathfrak{n}_0(2\omega) \times \mathfrak{q}][\mathfrak{n}_{ao}(2\omega) \times \mathfrak{n}_{ao}(2\omega)] - \\ &- [\mathfrak{n}_0(2\omega) \times \mathfrak{n}_r(2\omega)][\mathfrak{q} \times \mathfrak{n}_{ao}(2\omega)]\} \cdot \mathfrak{n}_0(2\omega) ,\end{aligned} \tag{4}$$

$$\begin{aligned}v_3 = \{&[\mathfrak{n}_0(\omega) \times \mathfrak{g}_1^{NL}(2\omega)][\mathfrak{q} \times \mathfrak{n}_{ao}(2\omega)] - \\ &- [\mathfrak{q} \times \mathfrak{g}_1^{NL}(2\omega)][\mathfrak{n}_{ao}(2\omega) \times \mathfrak{n}_{ao}(2\omega)]\} : \mathfrak{n}_0(2\omega)\mathfrak{n}_0(2\omega) .\end{aligned} \tag{5}$$

Als verallgemeinerter Brewster-Winkel wird der Einfallswinkel Θ_e^B der Grundwelle bezeichnet, bei dem die (in der Einfallsebene liegende) reflektierte 2. Harmonische verschwindet (1). Aus Gleichung (5) folgt, daß $\mathfrak{E}_r(2\omega)$ Null ist, wenn die Grundwelle unter dem durch

$$\sin \Theta_e^B = n_0(2\omega)\sin\vartheta \tag{6}$$

festgelegten Winkel von der Seite der Oberflächennormalen $\mathfrak{q}$ her, auf der auch die optische Achse $\mathfrak{t}$ liegt, auf die Grenzfläche fällt. In Gleichung (6) ist ϑ der Winkel zwischen der Oberflächennormalen und der optischen Achse. (6) ist wegen (1) ohne weiteres verständlich, denn man kann sich die nichtlineare Polarisation (1) als Anordnung von Dipolen vorstellen, die mit der Frequenz 2ω strahlen. Ein Dipol strahlt nicht in seiner Längsrichtung, dieser Längsrichtung entspricht im Vakuum die durch (6) bestimmte Richtung.

Für KDP muß bei der Wellenlänge des Rubinlasers wegen des großen Brechungsindex $n_o(2\omega)$ (5) der Winkel $\vartheta_{KDP} < 12^o$ sein, um den Brewster-Winkel für diesen Kristall experimentell nachweisen zu können.

Literatur

(1) N. BLOEMBERGEN und P.S. PERSHAN, Phys. Rev. 128, 606 (1962).
(2) R.K. CHANG und N. BLOEMBERGEN, Phys. Rev. 144, 775 (1966).
(3) S.A. AKHMANOV und R.V. KHOKHLOV, Probleme der nichtlinearen Optik, Moskau 1964 .
(4) R. FISCHER, phys. stat. sol. 19, 757 (1967).
(5) F. ZERNIKE Jr., J. Opt. Soc. Amer. 54, 1215 (1964).

(Received January 11, 1967)

phys. stat. sol. 19, K107 (1967)
Subject classification: 14.3; 10; 22.1.1

Natuurkundig Laboratorium der Vrije Universiteit, Amsterdam

Test of the Read Model Concerning the Electron Mobility in Plastically Deformed n-Type Ge

By
J.H.P. VAN WEEREN, R. STRUIKMANS, and J. BLOK

Dislocations with an edge component, introduced in n-type semiconductors, can have an influence on the electron mobility due to the space charge surrounding them (1). The model of Read (2) (elastic collision of conduction electrons with a smooth boundary of the space charge cylinder) implies that there is no effect on the mobility for current flow parallel to the dislocations. This is confirmed by experiments of Logan (3) on germanium bent plastically about the [211] direction at 800 °C. Broudy (4) who bent germanium about the [110] direction at 550 °C found a decreased "parallel" mobility.

In order to clear up this discrepancy, and more general to find the most favourable conditions to study the electrical effects of dislocations, we have investigated the effect of the direction of the bending axis and of deformation temperature.

We measured the Hall mobility between 100 and 280 °K on n-type Ge (10 Ωcm) plastically deformed at different temperatures in an N_2 atmosphere by four-point loading. Bars ($30x10x3\ mm^3$) were deformed in two ways:
[211] direction as bending axis, $(\bar{1}11)$ plane under 45° with upper face of the bar (single slip);
[110] direction as bending axis, $(\bar{1}11)$ and $(1\bar{1}1)$ planes under 55° with upper face (double slip).
The samples were plated with tin to prevent contamination during the heating. In order to minimize the effect of the known (5) variation in dislocation density in the direction normal to the neutral plane the bars were cut into three regions: one at the neutral plane (B-region), one above and one below the neutral plane (A-regions). The final dimensions of the crystals were about $7x1x0.5\ mm^3$.

Measurements were made with the current parallel and perpendicular to the bending axis. The results are mentioned in the Table 1 where the value of $(\mu^{100}/\mu^{280})/(\mu_o^{100}/\mu_o^{280})$ is given for [110]-type crystals. (The superscript gives the temperature, o refers to undeformed material.)
The load was the same for all crystals.

Table 1

T_{def} (°C)		560	665	760	760 + annealing for 60 min at 565 °C	830
Deformation time (min)		60	5	2	2.5	1
Radius of curvature (cm)		13	10	4	2.5	3
A	∥	0.67	0.68	0.33	0.61	0.27
	⊥	0.58	0.34	0.09	0.09	0.03
B	∥	1.0	0.95	0.78	0.84	0.92
	⊥	1.0	0.89	0.50	0.63	0.45

The values of $\mu_{\parallel}$ for [211] as the bending axis are nearly the same; those for $\mu_{\perp}$ are larger for deformation at $T > 700$ °C. In Fig. 1 we show the mobility after deformation at 830 °C for both bending directions.

From the table we see that in the A-regions for all deformation temperatures $\mu_{\parallel} < \mu_o$. This can be understood qualitatively as follows. The dislocations will be generated in the A-region as the shear stress is a maximum there. Simultaneously divacancies (6) are generated by the frequent intersection of the dislocations (parts of expanding dislocation rings). These point defects will reduce the mobility. Annealing at 565 °C during 60 min of a sample deformed at 760 °C gives evidence that there is only a small contribution of vacancies to the reduction of the mobility. Further there may be in the A-region dislocations not parallel to the bending axis which also results in a reduction in $\mu_{\parallel}$.

After being generated in the A-region the dislocations move under the applied stress inwards towards the neutral

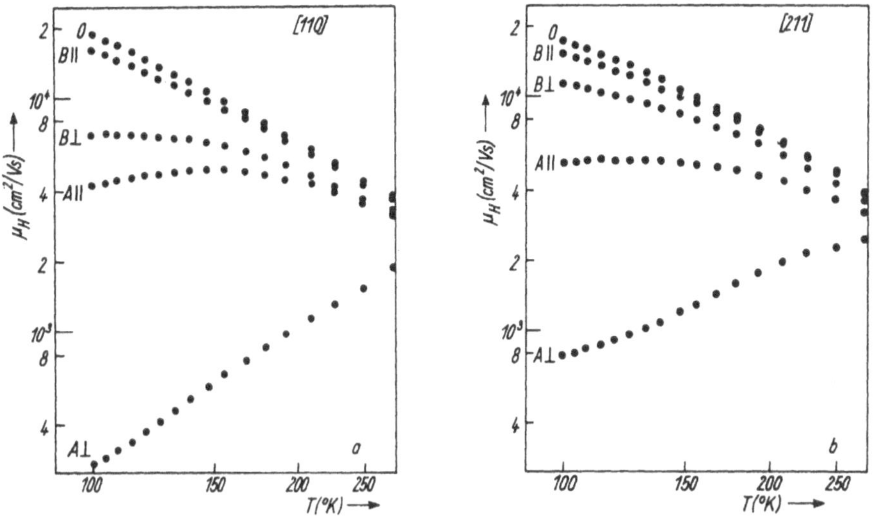

Fig. 1. Hall mobility vs. temperature in undeformed crystals and in crystals bent at 830 °C about the [110] (a) and [211] (b) direction

plane (5). We see from the table that at 560 °C the mobility of the dislocations is too small to reach the B-region. At 660 °C there is a small effect; at 760 °C the anisotropy indicates the presence of a sufficient number of dislocations in the B-region; still in this case $\mu_{\parallel} < \mu_0$. At 830 °C there is a large anisotropy, and $\mu_{\parallel} = 0.92\ \mu_0$. Thus the temperature is high enough to anneal out the divacancies to a reasonable amount in the short time of deformation. If the very small decrease in mobility can be ascribed to divacancies we may conclude that the model of Read is confirmed.

From Fig. 1 it is clearly seen that the anisotropy in [110]-type crystals is larger than in [211]-type crystals. We suppose that this fact can be ascribed to the distribution of dislocations. Preliminary etch pit studies on samples deformed at 830 °C (time of deformation 60 s) show that for [211]-type crystals polygonisation over short distances (about 15 μm) has occurred in the B-region. So, on [211]-type crystals it is not possible to measure the effect of the individual dislocations. In the A-region the dislocations are grouped also in slip lines. This grouping will decrease the anisotropy.

In [110]-type crystals no polygonisation is observed, the dislocations are distributed at random; a few slip lines were visible. We conclude that to study the electrical effects of dislocations it is favourable to bend about the [110] direction at a deformation temperature $\geq 830\ ^{\circ}C$.

We wish to express our acknowledgement to Dr. A. Lodder for valuable discussions.

References

(1) G.L. PEARSON, W.T. READ, and F.J. MORIN, Phys. Rev. 93, 666 (1954).

(2) W.T. READ, Phil. Mag. 46, 111 (1955).

(3) R.A. LOGAN, G.L. PEARSON, and D.A. KLEINMAN, J. appl. Phys. 30, 885 (1959).

(4) R.M. BROUDY, Adv. Phys. 12, 135 (1963).

(5) F.L. VOGEL, J. Metals 8, 946 (1956).

(6) P. HAASEN and A. SEEGER, Halbleiterprobleme IV, Ed. W. SCHOTTKY, F. Vieweg, Braunschweig 1958.

(Received January 11, 1967)

phys. stat. sol. 19, K111 (1967)
Subject classification: 20.1; 13.1; 22

Institute of Physics, Academy of Sciences of the Azerbaidzhan SSR, Baku

The Reflection Spectrum of GaSe and GaS Single Crystals near the Fundamental Absorption Edge

By
M.A. NIZAMETDINOVA

This note presents the results of reflection measurements on GaSe and GaS single crystals in the range 350 to 650 nm for the directions of polarization of the incident radiation $E \parallel c$ and $E \perp c$, where c is the hexagonal axis. Previously similar investigations of reflection have been performed in the range 220 to 400 nm (1). The reflection spectrum was obtained by using a spectrometer with quartz prism. The polarizer was Iceland spar. The measurements were carried out at near-normal incidence. The reflecting face was the surface polished parallel to the c-axis. The two-fold reflection was measured at room temperature.

Figs. 1 and 2 represent the reflection spectrum of GaSe and GaS. The peaks at 620 nm for GaSe and 438 nm for GaS are distinctly seen. The position of the peaks does not depend on the direction of the polarization. The shape of the peaks are slightly different, but nevertheless the peak for $E \parallel c$ is still somewhat sharper than for $E \perp c$.

For the identification of the reflection peaks in GaSe and GaS the analogy with the reflection spectrum of $A^{III}B^{V}$ and $A^{II}B^{VI}$ compounds was used in (2). In particular, the position of the reflection peaks in GaSe was determined as the average between the position of the peaks in ZnSe and GaAs, and in GaS between those of ZnS and GaP. Using this analogy, we also have found that the observed peaks correspond to the transition E_0, i.e. to direct transitions in the centre of the Brillouin zone. Thus, we obtain $E_0 = 2.0$ eV for GaSe and $E_0 = 2.83$ eV for GaS.

The value of E_0 in GaSe coincides with $E_g = 2.0$ eV found from the edge of the fundamental absorption, and the value of

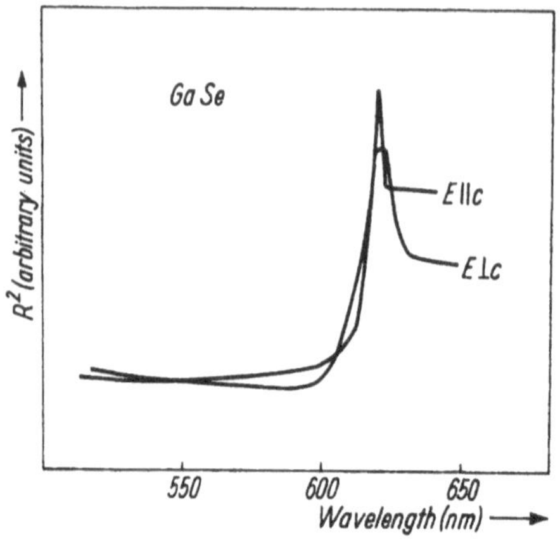

Fig. 1

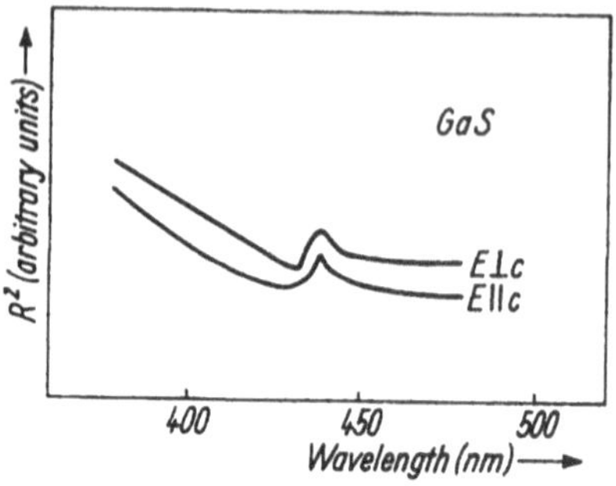

Fig. 2

E_0 = 2.83 eV for GaS is also related to the direct energy gap.

The peak at 2.83 eV seems to correspond to the exciton peak observed in (3).

These peaks can be connected with transitions in the centre of the Brillouin zone from the highest two-fold degenerate valence band (near which the non-degenerate band is located) into the lowest non-degenerate conduction band.

The author thanks Prof. G.B. Abdullaev for constant attention to the work.

References

(1) N.A. GASANOVA, G.A. AKHUNDOV, and M.A. NIZAMETDINOVA, phys. stat. sol. 17, K115 (1966).

(2) G.A. AKHUNDOV, N.A. GASANOVA, and M.A. NIZAMETDINOVA, phys. stat. sol. 15, K109 (1966).

(3) F. BASSANI, D. GREENWAY, and G. FISCHER, Proc. Internat. Conf. Phys. Semicond., Paris 1964 (p. 52).

(Received January 12, 1967)

Erratum

In the paper by YU.A. LOGATCHOV

Model Potential Approximation for Z_1-Centres in Ionic Crystals

phys. stat. sol. 19, K35 (1967),

in Table 1 the experimental values for $h\nu_o(Z_1)$ and $h\nu_o(Z_1)/h\nu_o(F)$ should read correctly 0.090 and 0.89, respectively.

Pre-printed Titles and Abstracts

Papers to be published in "physica status solidi"
Vol. 20, No. 1

Review Article

Y.P. VARSHNI, Department of Physics, University of Ottawa, Band-to-Band and Radiative Recombination in Group IV, VI, and III-V Semiconductors (II)

Original Papers

B. REPPICH, Institut für Metallphysik der Universität Göttingen, Plastische Verformung von Eisen-II-Oxid Einkristallines und polykristallines Eisen-II-Oxid, $Fe_{1-x}O$, mit einem Kationenleerstellengehalt von $0{,}05 \leq x \leq 0{,}11$ wurde im Stauchversuch bei Temperaturen $650\ ^{\circ}C \leq T \leq 1300\ ^{\circ}C$ statisch und dynamisch verformt. Stationäres Kriechen und dynamisches Fließen lassen sich beschreiben durch die Zustandsgleichung

$$\dot{\varepsilon} = C(1 + bx)\sigma^{m}\exp(-\frac{U}{kT})$$

mit $m = 4$ und $U = (3{,}54 \pm 0{,}21)$ eV $\hat{=}$ $(81{,}5 \pm 4{,}8)$ kcal/mol. Einkristalle sind weicher als Polykristalle. $Fe_{1-x}O$ wird mit steigendem Leerstellengehalt x duktiler. Für die beobachtete Streckgrenze werden die von Fe_3O_4-Ausscheidungen herrührenden Versetzungsringe verantwortlich gemacht. Die Ergebnisse lassen sich durch ein von der Spannung unterstütztes Klettern von Stufenversetzungen deuten.

B. ECKSTEIN, Mineralogisches Institut der Technischen Hochschule Aachen, A Disorder Model of Melting and Melts Verschiedene Materialeigenschaften von Metallschmelzen werden berechnet unter der Voraussetzung einer engen Verwandtschaft zwischen Kristall und Schmelze. Die Eigenschaften der Schmelzen setzen sich danach zusammen aus einem "Gitteranteil", der im wesentlichen dem Wert der Materialkonstante im Kristall am

Schmelzpunkt entspricht, und einem "defektbedingten" Anteil, der nach der Fehlordnungstheorie zu berechnen ist. Die Defektkonzentration wird dabei aus dem Volumensprung beim Schmelzen bestimmt. - Die so berechneten Werte zeigen zufriedenstellende Übereinstimmung mit den experimentellen Daten.

H. TRÄUBLE und U. ESSMANN, Institut für Physik am Max-Planck-Institut für Metallforschung, Stuttgart, und Institut für theoretische und angewandte Physik der Technischen Hochschule Stuttgart, Die Beobachtung magnetischer Strukturen von Supraleitern zweiter Art

Durch Aufdampfen kleiner ferromagnetischer Partikel mit Durchmessern von etwa 100 Å auf Blei-Indium-Kristalle ($\varkappa$ = 1,2 bei 1,1 °K) wurde die magnetische Struktur im Mischzustand bei 1,1 °K abgebildet und mit Hilfe von Abdruckverfahren elektronenmikroskopisch beobachtet. Unter günstigen Bedingungen liefert diese Methode ein Auflösungsvermögen von etwa 100 Å. Nach unseren bisherigen Beobachtungen dringt der magnetische Fluß in diskreten Punkten mit einem Durchmesser von etwa 1000 Å durch die Probenoberfläche. Die Mehrzahl der Punkte ist in unregelmäßiger Weise zu Bündeln von etwa 20 Punkten vereinigt. Mit steigendem Feld nimmt die Zahl der Bündel zu, und ihr Durchmesser wird kleiner. Identifiziert man die einzelnen Punkte mit Austrittsstellen einzelner Flußlinien, die jeweils ein Flußquantum tragen, so ergibt sich im Rahmen der zu erwartenden Genauigkeit befriedigende Übereinstimmung zwischen elektronenmikroskopischen Beobachtungen und gemessenen Magnetisierungskurven. Beobachtungen an dünnen Blei-Indium-Scheiben von 0,5 mm Dicke weisen auf die Existenz eines Zwischenzustands bei Supraleitern zweiter Art mit kleinem $\varkappa$ hin.

R.R. GAŁĄZKA, Institute of Physics, Polish Academy of Sciences, Warsaw, and L. SOSNOWSKI, Institute of Physics, Polish Academy of Sciences, Warsaw, and Institute of Experimental Physics, Warsaw University, Conduction Band Structure of $Cd_{0.1}Hg_{0.9}Te$

The Hall coefficient R_H and the conductivity σ are measured in the temperature range 4.2 to 300 °K. The thermoelectric power α as a function of the magnetic field and temperature are measured in the range 100 to 300 °K. The value of the effective mass m is calculated from these measurements for electron concentrations from 6.25×10^{16} cm^{-3} to 2.71×10^{18} cm^{-3}. The dependence of the effective mass m on k is interpreted according to Kane's model of the conduction band structure. The best agreement is obtained for values of parameters $\varepsilon_0 = \varepsilon_{\Gamma_6} - \varepsilon_{\Gamma_8} = 0.00$ eV and the coupling constant Q = 12 eV, that is for a linear ε-k dependence. The mobility of the electrons as a function of concentration and temperature is also analysed. For temperatures below 77 °K impurity scattering is dominant. Within the temperature range 77 to 300 °K the best agreement between theory and experiment is obtained for mixed ionic and optical phonon scattering.

R. BÖTTCHER, W. WINDSCH, Institute of Physics of the Karl-Marx-University, Leipzig, Division Structure of Matter, and W. LÜDKE, Jena, Carl-Zeiß-Straße 1, EPR Investigation on Mn^{2+}-Doped LiF Single Crystals Es werden die EPR-Spektren des Mn^{2+} von in der Schmelze und durch Diffusion mit MnF_2 dotierten LiF-Einkristallen diskutiert. Es treten zwei Komplexe auf. Der eine von ihnen hat orthorhombische und der andere axiale Symmetrie. Die Superhyperfeinstruktur wird erklärt. Es wird gezeigt, daß eine Diffusion von Mn^{2+}-Ionen in LiF-Kristallen möglich ist und daß die eindiffundierten Ionen im wesentlichen das gleiche Verhalten wie die bei der Dotierung in der Schmelze in den Kristall eingebauten Ionen zeigen.

P. SCHMELING, Institut für Radiochemie, Kernforschungszentrum Karlsruhe, Diffusion of Xenon in Barium Fluoride Die Diffusion von Xenon in Bariumfluorid nach Neutronenbestrahlung wurde untersucht. Bei Temperaturen oberhalb 1000 °C wird der Diffusionsprozeß mit einer Aktivierungsenergie von 3,7 eV beschrieben. Dieser ist von der Bestrahlungsdosis unabhängig.

D.J.D. THOMAS, Standard Telecommunication Laboratories, Harlow, The Effect of Heat Treatment on Silicon Nitride Layers on Silicon On heat treatment, structural changes may occur in silicon nitride layers deposited on silicon substrates by the glow discharge technique. The causes of the cracks and holes produced in the layer are shown to be thermal stresses and lack of adhesion between the layer and the substrate. Layers produced at above 600 °C, and particularly from silicon tetrachloride, remain perfect on heating up to 1200 °C. Dislocations are introduced into the silicon substrate at defects in the nitride layer. Deformation is produced by slip in the (111) planes parallel to the silicon surface. No defects are introduced into the silicon if the nitride layer remains perfect.

A.S. DAVYDOV, Institute of Physics, Academy of Sciences of the Ukrainian SSR, Kiev, On the Operator of Exciton-Phonon Interaction A general expression for the operator for the interaction of molecular excitons with acoustic and optical phonons is obtained in a linear approximation. An explicit expression is found for the dependence of exciton-phonon interactions on their wave vectors for a one dimensional crystal and for crystals of cubic, rhombic and tetragonal systems. The results obtained are compared with the operator for the interaction of free electrons and Wannier excitons with acoustic and optical vibrations in ionic crystals.

A.S. DAVYDOV and E.N. MYASNIKOV, Institute of Physics, Academy of Sciences of the Ukrainian SSR, Kiev, Absorption and Dispersion of Light by Molecular Exciton A calculation is made of the real and imaginary parts of the dielectric constant of molecular crystals due to exciton states, using retarded Green's functions. The interaction of excitons with acoustic and optical phonons is taken into account at different temperatures of the crystal. The dependence of the optical refractive index and the damping coefficient on the frequency of the incident light is obtained in the region of the exciton bands.

T. GESZTI, Research Institute for Technical Physics of the Hungarian Academy of Sciences, Budapest, On the Theory of Thermally Activated Processes The range of validity of "rate theory" is investigated and a clear distinction is made between "directional" and complete thermalization. It is suggested that the first process determines the hopping character of the motion, while the second determines whether the usual quasi-equilibrium description is applicable. Macroscopic diffusion is shown always to obey "rate theory", while frequency-dependent dielectric or inelastic loss may show non-equilibrium effects. In the case in which complete thermalization is slow (although directional thermalization may be fast) a succession of Lorentzian peaks is obtained instead of the single peak predicted by "rate theory". Qualitative dynamic conditions are presented for slow thermalization behaviour.

G. KUNZE, Kristallographische Abteilung des Mineralogischen Instituts der Technischen Hochschule Hannover, Refraktionsmodifizierte Braggsche Beugung (I) Da das normale Braggsche Gesetz wegen Vernachlässigung der Refraktion vielen festkörperphysikalischen Präzisionserfordernissen nicht mehr gewachsen ist, wurde auf der Basis der kinematischen Theorie der Röntgeninterferenzen für den Allgemeinfall der asymmetrischen Rückstrahl-Reflexion von ebenen Objekten ein allgemein gültiges Beugungsgesetz aufgestellt. Es gilt uneingeschränkt für Kristalle mit reellen Brechungsindizes und damit auch für lichtoptische Beugungserscheinungen an oder in nicht (oder höchstens schwach) absorbierenden Kristallen mit (quasi-) superperiodischen Inhomogenitäten (Antiphasenordnungsdomänen, Entmischungszonen, Ausscheidungslamellen, usw.).

M. BEVIS and N. SWINDELLS, Department of Metallury, Liverpool University, The Determination of the Orientation of Micro-Crystals Using a Back-Reflection Kossel Technique and an Electron Probe Microanalyser The determination of the orientation of small single crystals from back-reflection Kossel patterns obtained by using an electron probe microanalyser is considered. A procedure for determining the indices of directions in the crystal represented by the intersection of Kossel lines is presented, and the way in which this information may be used for determining the orientation of crystals without the aid of the stereographic projection is described. Complementary procedures for determining the full orientation of small crystals and the orientation of interface planes in these crystals are also described. To illustrate these procedures the orientation relationship between annealing twins and the matrix in an f.c.c. crystal has been determined together with the orientation of the twin matrix interface planes.

B.YA. PINES, I.P. GREBENNIK, R.I. BAZYURA, and G.V. ZOZULYA, State University, Kharkov, Surface and Volume Diffusion in Thin Films of the System Ag-Se The course of volume and surface diffusion in thin (10^{-5} to 10^{-6} cm) films of the system Ag-Se was investigated by means of electron diffraction and microscopic studies. The surface diffusion determines the growing speed of the two component region in specimens that have partly superimposed layers of components. Volume diffusion equalizes the concentration in the thickness direction of the film. The kinetics of the surface diffusion of Ag over Se could be stated by observing the displacement of the crystallization front of amorphous selenium. The surface diffusion of Se over Ag was controlled by the appearing α-Ag_2Se diffraction lines near the edge of the two component region.

J.R. PILBROW and J.M. SPAETH, Physics Department, Monash University, Clayton, Victoria, ESR Studies of Cu^{2+} in NH_4Cl Single Crystals between 4.2 and 453 °K (I) ESR spectra from two different Cu^{2+} centres with [100] tetragonal symmetry were observed in NH_4Cl single crystals. The spin-Hamiltonian parameters for both centres were determined in the temperature range 4.2 to 453 °K where NH_4Cl has the CsCl structure. The g-values and hyperfine constants for both centres show a marked anomaly in the vicinity of the λ-point and significant variations over the whole temperature range. Both centres have essentially a $|3z^2-r^2\rangle$ ground state, but with evidence of vibrational admixtures of the $|x^2-y^2\rangle$ state. In view of possible future calculations of the Cu^{2+} spin polarization and radial wave function for the ground state, the lattice parameter of NH_4Cl and NH_4Cl:Cu was measured in the temperature range 105 to 453 °K.

J.R. PILBROW and J.M. SPAETH, Physics Department, Monash University, Clayton, Victoria, ESR Studies of Cu^{2+} in NH_4Cl Single Crystals between 4.2 and 453 °K (II) The temperature dependence of the spin-Hamiltonian parameters for two Cu^{2+} centres in NH_4Cl single crystals, which were reported in the previous paper (I), is analysed in detail. The results show that, apart from zero point vibrations, higher lattice modes also contribute to the admixture of the $|x^2-y^2\rangle$ state into the $|3z^2-r^2\rangle$ ground state for these centres. For one of the centres there is also evidence that vibrations and reorientations of a single nearest neighbour NH_4^+ ion contribute to the admixture. From the analysis, the temperature dependence of $\langle r^{-3}\rangle_{3z^2-r^2}$, a crystal field splitting parameter and the contact contribution to the hyperfine field at the Cu^{2+} nucleus is obtained. All these quantities vary strongly with temperature and show sudden changes at the λ-point (-30.5 °C) for NH_4Cl. The temperature dependence of the contact term in the hyperfine interaction can be qualitatively understood in terms of core spin polarization by taking into account the observed changes in the radial distribution of the 3d ground state. The analysis gives further support for the models proposed for the two Cu^{2+} centres.

J. ŠEDIVÝ und H. ŠÍCHOVA, Lehrstuhl für Festkörperphysik der Mathematisch-Physikalischen Fakultät, Karlsuniversität, Prag, Röntgenographische Messung der charakteristischen Debye-Temperatur bei polykristallinem Silber mit Anwendung der Korrektur auf primäre Extinktion The Debye temperature Θ of polycrystalline silver at room temperature is determined from X-ray diffractometer measurements at room temperature only. The method (which is known as the "single temperature method") is based on a relatively precise correction for the effect of the primary extinction. Six polycrystalline silver specimens prepared in various ways are investigated. The average of these measurements gives a final value for Θ of (220 ± 8) °K. The difference between the values obtained for Θ for particular specimens is discussed taking into account the fact that each sample was studied 10 times.

H. SIGMUND and K. BERCHTOLD, Institut für Technische Elektronik der Technischen Hochschule München, Electrical and Photovoltaic Properties of PbS-Si Heterodiodes Es wird über

PbS-Si-Heterodioden berichtet, bei denen die PbS-Schicht aus wässriger Lösung auf Si-Kristalle bei Zimmertemperatur niedergeschlagen wurde. Die elektrischen und photoelektrischen Eigenschaften dieser Dioden werden gezeigt. Das Durchlaß- und Sperrverhalten ist mit anderen Heteroübergängen vergleichbar (Rekombinations-Tunnelmechanismus). Die spektrale Empfindlichkeit erstreckt sich von 1 bis 3 μm bei Zimmertemperatur, mit einer Empfindlichkeit im PbS-Maximum von ca. 6×10^3 V/W.

S.U. DZHALILOV and K.I. RZAEV, Institute of Physics, Academy of Sciences of the Azerbaidzhan SSR, Baku, On the Phenomenon of Selenium Vitrification On the basis of volumetric measurements the temperature region of selenium vitrification (18 to 36 °C) has been determined which correlates with that obtained from data on temperature dependence of viscosity (the function a_T), specific heat, and propagation velocity of ultra-sonic waves in selenium. It is shown that the vitrification temperature T_g of selenium depends on the experimental time base. A drop of the cooling rate by two orders (from 0.5 to 0.005 deg/min) causes the T_g value to decrease by 9 °C. Increasing concentration of Te in selenium lowers the vitrification temperature of the Se-Te copolymer linearly. The vitrification temperature of the specimen grows linearly with Sb concentration in selenium. This is explained by the binding action of Sb, consisting in the formation of chemical cross-links between the molecules of selenium.

R. STRAUBEL, Institut für Maschinelle Rechentechnik der Deutschen Akademie der Wissenschaften zu Berlin, Dresden, Außenstelle für magnetische Signalspeicher, Berlin, Anhysteretic Magnetization of a Collective of Magnetic Single-Domain Particles Für eine Ansammlung von einachsig anisotropen, vollständig ausgerichteten Einbereichsteilchen, bei der die Teilchen in linearen Ketten mit positiver oder negativer Wechselwirkung angordnet sind, wird gezeigt, daß sich die Ansammlung nach einem anhysteretischen Magnetisierungsvorgang mit kleinen Gleichfeldern in einem echten energetischen Gleichgewicht befindet. Damit entspricht die Ansammlung einem Ising-Modell entsprechender Spinanordnung. Darauf aufbauend wird die anhysteretische Remanenzkurve der Ansammlung in ihrem Anfangsteil und die anhysteretische Anfangssuszeptibilität für beide Typen von Wechselwirkungen berechnet.

H. BROSS and G. BOHN, Sektion Physik der Universität München, On the Theory of Electrical Resistivity of Multivalent Metals Im Falle mehrwertiger Metalle lassen sich die Pole in der Übergangswahrscheinlichkeit dadurch vermeiden, daß bei der Berechnung des Bardeenschen Ausdrucks (1) die Raumintegrale über die exakte Gestalt des Atompolyeders erstreckt werden. Der auf diese Weise erhaltene Ausdruck wird benützt, um die Temperaturabhängigkeit des elektrischen Widerstands von Aluminium zu ermitteln. Anisotrope Effekte infolge des Schwingungsspektrums werden durch ein Variationsverfahren berücksichtigt. Die Ergebnisse sind in befriedigender Übereinstimmung mit den experimentellen Werten.

(1) J. BARDEEN, Phys. Rev. **52**, 688 (1937).

Y.T. CHOU, Edgar C. Bain Laboratory for Fundamental Research, United States Steel Corporation, Research Center, Monroeville, Pennsylvania, On Stacked Screw Dislocation Arrays in an Anisotropic Medium A direct method is presented for finding the distribution of dislocations in uniformly stacked arrays in an anisotropic medium. A simple procedure for evaluating the stress field is given. The analytic technique is applied to a special case of an infinite sequence of identical double-ended pile-ups stacked one above the other. In this case, the number of dislocations in each of the pile-ups is found to be less than that in an isolated pile-up and decreases continuously as the separation between pile-ups decreases. The stress concentration around the tips of each array behaves in a similar manner. It is shown that the system is elastically equivalent to a vertical array of parallel shear cracks.

R.P. GUPTA, Department of Physics, Banaras Hindu University, Varanasi, Lattice Specific Heats of Thallium and Yttrium The lattice specific heats of thallium and yttrium are calculated using the electron gas model of Gupta and Dayal (1) for hexagonal metals. The interactions are restricted to fourth neighbours. The force constants are determined from experimental values of the elastic constants. The Θ-T curves calculated from the computed specific heats show a fairly satisfactory agreement with the experimental values, i.e. the difference is not greater than 3% for thallium and 6% for yttrium.

(1) R.P. GUPTA and B. DAYAL, phys. stat. sol. 8, 115 (1965).

G.B. ABDULLAEV, E.S. GUSEINOVA, and B.G. TAGIEV, Institute of Physics, Academy of Sciences of the Azerbaidzhan SSR, Baku, Electric Conductivity Studies of p-GaTe Polycrystals and Single Crystals in Strong Electric Fields It is shown that in the temperature range 77 to 220 °K and under fields increasing to 3×10^4 V/cm, conduction both of polycrystalline p-GaTe and of the single crystal increases according to Frenkel's formula ($\sigma = \sigma_0 e^{\beta\sqrt{E}}$). From the formula $\beta = \sqrt{e^3}/kT\sqrt{\varepsilon}$ and by determining the slope of the isotherms $\lg\sigma = f(\sqrt{E})$, the dielectric constants of p-GaTe, pertaining to the electric part of polarization, are found to be equal to 6 and 4 both along and across the layers, respectively. By extrapolating with respect to the cross-over point of isopotentials $\lg\sigma = f(1/T)$, the metallization temperature T_M = 20000 °K is determined, from which the value of kT_M = 1.74 eV is found. This agrees with the value for the gap-width of p-GaTe at absolute zero. By extrapolating the isotherms and finding a consistent solution to their equation for the electric strength, a value is obtained which is approximately equal to 6.5×10^5 V/cm.

P. DEVAUX and M. SCHOTT, Groupe de Physique du Solide, associé au CNRS, Laboratoire de Physique, École Normale Supérieure, Paris, Thermally Stimulated Currents without Optical Excitation, Application to Copper Phtalocyanine Thermally stimulated current peaks can occur in the absence of previous optical excitation, if at least one of the electrodes on the sample is non-ohmic. A variety of cases is considered. Most of these cases involve the study of deep-lying centres which give only very small current peaks. Their experimental characteristics are presented. They are particularly useful in the case of organic crystals, where optical filling of traps at low temperatures is inefficient and lengthy. This method, applied to copper phtalocyanine, enables centers lying at depths of 0.41 and 0.65 eV, which cannot be observed using optical excitation to be studied.

N.A. VLASENKO and S.A. ZYNO, Institute of Semiconductors, Academy of Sciences of the Ukrainian SSR, Kiev, Role of Carrier Polarization in the Mechanism of the Electroluminescence of ZnS-Mn Films A polarization, steady in time and caused by the accumulation of the charge carriers on the deep surface traps, are found in the electroluminescence of ZnS-Mn films. The polarization characteristics (the number of polarization charges, time dependence of its formation and destruction, ionization energy of traps, etc.) are determined in order to ascertain what a part the carrier polarization plays in mechanism of electroluminescence. It is established that the polarization mainly affects the change of internal field and therefore of impact ionization probability. In addition the polarization charge ionization in the moment of voltage-polarity reversal leads to an increase of the number of charge carriers which can accelerate. All electroluminescence properties and features of ZnS-Mn films may be explained from the view-point of impact-ionization mechanism by taking into account the polarization effects.

N.P. GUPTA and B. DAYAL, Department of Physics, Banaras Hindu University, Varanasi, Equation of State of Solid Argon and Krypton Equation of state, thermal expansion, and isothermal compressibilities of solidified argon and krypton are studied in the harmonic approximation. An exponential interatomic potential is used. Zero-point quantum effects are included. Good agreement between calculated and observed values is found.

L.L. HENCH and D.A. JENKINS, Department of Metallurgical and Materials Engineering, College of Engineering, University of Florida, Gainesville, AC Conductivity of a Glass Semiconductor The ac conductivity of a 70 mol% V_2O_5 - 30 mol% P_4O_{10} glass semiconductor has been measured over a frequency range of 20 to $2x10^6$ Hz and a temperature range of -23 to

200 °C. An activation energy of 0.34 eV was measured over the entire temperature range for frequencies up to 5×10^4 Hz. Below a transition temperature of 70 °C the activation energy for high frequency conduction decreased with increasing frequency.

S.Y. CHUANG and B.G. HOGG, Department of Physics, University of Manitoba, Winnipeg, Positron Annihilation in Alnico and $(La_{0.7}Pb_{0.3})MnO_3$ A study of the angular correlation of annihilation photons from positrons annihilating in alnico and single crystals of the perovskite $(La_{0.7}Pb_{0.3})MnO_3$ has been made. It is possible to account for the observed angular distributions by a simple-minded consideration of the physical properties of the two materials.

W. DREYBRODT and D. SILBER, Physikalisches Institut der Universität Frankfurt/Main, Electron Spin Resonance of Tl^{++} Centres in KCl Crystals Thallium-dotierte KCl-Kristalle werden bei 77 °K mit Röntgenlicht durchstrahlt, dann auf etwa 300 °K aufgewärmt und wieder auf 77 °K abgekühlt. Es werden ESR-Signale beobachtet, die einem Tl^{++}-Ion auf Kationen-Platz zugeordnet werden und eine aufgelöste HF-Struktur von den sechs nächsten Chlor-Ionen zeigen. Die Konstanten des Spin-Hamilton-Operators werden bestimmt. Nach einer Theorie von Watanabe (1) werden aus den Hyperfeinstruktur-Konstanten die Koeffizienten der LCAO-Funktion des Grundzustandes berechnet. Aus diesen und optischen Daten des Tl^{++}-Spektrums (2) wird die g-Verschiebung berechnet und mit dem experimentellen Wert verglichen.

(1) H. WATANABE, Phys. Rev. 149, 402 (1966).
(2) C.I. DELBECQ, A.K. GOSH, and P.H. YUSTER, Phys. Rev. 151, 599 (1966).

G.S. PAWLEY, Department of Natural Philosophy, University of Edinburgh, A Model for the Lattice Dynamics of Naphthalene and Anthracene The lattice dynamics of a model for naphthalene and anthracene is presented. Intermolecular forces of the 6-exp interatomic type are assumed, and dispersion curves are calculated. It is stressed that the Raman modes do not involve rotations about the principal inertia axes. Frequency distributions, Debye temperatures and Debye-Waller factors are calculated, and the validity of using the last to estimate Raman frequencies is assessed. The agreement with experiment is good, except perhaps for the elastic constants. Finally the thermal diffuse X-ray scattering is calculated and compared with experiment. The low resolution results agree both with experiment and with the "difference Fourier transform" theory. The high resolution calculation, however, explains the discrepancies between experiment and the earlier theory.

G. COSTE, Laboratoire Central des Industries Électriques, Fontenay-aux-Roses, Oscillatory Thermomagnetic Effects in Lead Telluride Les oscillations quantiques des coefficients de pouvoir thermoélectrique et de conductivité thermique ont été observées sur PbTe type p ($p = 3.6 \times 10^{18}$ cm^{-3}) à 4.2 °K et jusqu'à 50 kG.

N.A. PANGAROV, Institute of Physical Chemistry, Bulgarian Academy of Sciences, Sofia, Calculation of the Specific Surface Energies of Crystal Planes by Investigating the Orientation of Crystal Nuclei on Inert Substrates On the basis of the Stranski-Kaishev (1) method of the average works of separation, a method of calculation of the specific surface energies of crystal planes is elaborated, by investigating the axes of preferred orientation during the crystallization on inert substrates. This method is applicable both to the crystallization from the gaseous phase and to metal electrodeposition. The applicability of the method is shown for the electrodeposition of silver. The values of the specific surface energies for the (111), (100), and (110) planes are calculated.

(1) I.N. STRANSKI and R. KAISHEV, Z. phys. Chem. B26, 100, 114, 312 (1934); Ann. Phys. 23, 330 (1935); Z. Phys. 36, 393 (1935).

N.A. PANGAROV, Institute of Physical Chemistry, Bulgarian Academy of Sciences, Sofia, Twinning Processes in the Electrocrystallization of Face-Centered Cubic Metals A theory about the twinning process in crystals with a face-centered cubic lattice is developed. The Stranski-Kaishev (1) method of the average works of separation is used for calculating the work of formation of normal two-dimensional nuclei on the (111) plane and of nuclei in twinning position. The percentage probability of formation of twinning nuclei is calculated as a function of supersaturation. The theory is applicable both to the crystallization from vapours and to metal electrodeposition.

(1) I.N. STRANSKI and R. KAISHEV, Z. phys. Chem. B26, 100, 114, 312 (1934); Ann. Phys. 23, 330 (1935); Z. Phys. 36, 139 (1935).

F.T. GOLDSTEIN, Laboratory of Atomic and Solid State Physics, Cornell University, Ithaca, New York, F-Center Formation by Fundamental Absorption in KI Irradiadiation of KI in the fundamental edge results in a temperature dependent rate of F-center growth predominantly proportional to the inverse of the number of F-centers already produced. This is related to a new experimental result - the quenching of low temperature luminescence in irradiatively coloured crystals. In addition, the production of F-centers by absorption in perturbed exciton bands and the intrinsic production of α-centers are discussed.

K.-H. PFEFFER, Institut für Physik am Max-Planck-Institut für Metallforschung, Stuttgart, und Institut für theoretische und angewandte Physik der Technischen Hochschule Stuttgart, Mikromagnetische Behandlung der Wechselwirkung zwischen Versetzungen und ebenen Blochwänden (I) Infolge der inneren Spannungen, die von den Versetzungen hervorgerufen werden, wird der Magnetisierungsverlauf innerhalb und außerhalb einer Blochwand modifiziert. Für das Gesamtsystem Versetzung-Blochwand werden durch Minimalisierung der Gesamtenergie die mikromagnetischen Gleichungen in linearisierter Form abgeleitet, wobei die magnetoelastische Kopplung in Voigtscher Näherung (1) behandelt wird. Die linearen Differentialgleichungen besitzen im Unterschied zu den Gleichungen, die in der Theorie der ferromagnetischen Einmündung auftreten, Koeffizienten, die von dem Magnetisierungsverlauf der ungestörten Blochwand abhängig sind. Zur Lösung derselben wird das Differentialgleichungssystem entsprechend dem nicht-lokalen Charakter des magnetischen Streufeldes in ein System von zwei linearen, inhomogenen Integralgleichungen übergeführt. Die Energieänderung, die mit der

Änderung der Magnetisierungsverteilung verknüpft ist, ergibt sich in der zugrunde gelegten linearen Näherung als die Hälfte der hiermit verbundenen Änderung der magnetoelastischen Kopplungsenergie. Aus der Struktur der magnetoelastischen Kopplungsenergie werden qualitative Aussagen über die Magnetisierungsverteilung und das hieraus resultierende Wechselwirkungspotential im System Blochwand-Versetzung abgeleitet.

(1) W. VOIGT, Lehrbuch der Kristallphysik, Teubner, Leipzig-Berlin 1928.

R. ENDERLEIN, Department of Physics, State University, Moscow, The Influence of Collisions on the Franz-Keldysh-Effect An investigation is made of the effect of the electron-phonon and impurity scattering on the field-induced change in the optical interband absorption for photon energies near the direct band gap $\hbar\omega$. It is found that in the presence of collisions the exponential tail of the absorption coefficient at $\omega < \omega_0$ oscillates and that the oscillations at $\omega > \omega_0$ are damped. This damping seems to be in quantitative agreement with the experimental results of Hamakawa, Germano, and Handler (1)

(1) Y. HAMAKAWA, F. GERMANO, and P. HANDLER, Proc. 8th Internat. Conf. Phys. Semicond., Tokyo 1966 (to be published).

Original Papers published in "Fizika Tverdogo Tela"
Vol. 8, No. 12

V.I. IVERONOVA, A.N. TIKHONOV, P.N. ZAIKIN, and A.P. ZVYAGINA, Moscow, Determination of the Phonon Spectrum of Crystals from the Heat Capacity The phonon spectrum of crystals may be calculated approximately from the heat capacity and other thermodynamic functions. The approximate distribution function is obtained from the integral equation taking the phonon spectrum as a continuous partly smooth function with a finite number of transition points in the derivative. The present approach yields a smooth function with a minimum which gives characteristic information regarding the fine structure and satisfies the equation adequately. The frequency distribution function of Al is calculated from experimental heat capacity data.

YA.I. DUTCHAK, V.YA. PROKHORENKO, and N.M. KLYM, Lvov, Peculiarities in the Structure of Liquid Sn-Sb Alloys An X-ray study is made of the structure of liquid Sn-Sb alloys (8.8 at% Sb). An average statistical distribution of the different types of atoms is observed at 255 and 415 °C. The absolute thermo e.m.f. is measured over a wide range of concentrations and temperatures. A maximum is observed in the isotherms at concentrations corresponding to the high-temperature boundary of the inter-metallic β-phase. This fact is connected with the partial conservation of directed bonds.

V.V. VASKIN, V.S. METRIKIN, V.A. USKOV, and M.YA. SHIROBOKOV, Gorkii, Effect of an Internal Electric Field on the Diffusion of Impurities in Semiconductors A theoretical study is made of the effect of an internal electric field on the simultaneous diffusion of donors and acceptors in semiconductors. The theoretical concentration curves are compared with experimental curves obtained from the investigation of simultaneous diffusion of radioactive In^{114} and Sb^{124} in germanium at 795 °C.

I.V. GREKHOV, N.N. KRYUKOVA, and V.E. CHELNOKOV, Leningrad, Microplasma Effects in Silicon An experimental investigation is made of microplasma effects in Si during the avalanche breakdown of deep p-n junctions with large area. It is found that numerous microplasmas having different breakdown voltages are distributed homogeneously over the p-n junction plane. The current-voltage characteristic for breakdown of a large-area p-n junction has the form $I = I_b \exp \frac{U - U_b}{A}$ (I_b, U_b being the current and voltage at the beginning of the breakdown, and A being a constant for a given p-n junction). The calculated geometric dimensions of the microplasma agree satisfactorily with experiment.

M.I. KAGANOV and I.L. SKLOVSKAYA, Kharkov, Surface Wave in Piezoelectrics An explanation is given for the existence of an additional surface wave in a piezoelectric. The dispersion equation for this wave is investigated. It is shown that the velocity of the additional wave v lies in the range $\frac{\sqrt{3}}{2} c_t < v < c_t$ (c_t transverse velocity of sound). The electric field distribution in the wave is estimated.

R. KATILYUS, Leningrad and Vilnyus, Hot Electron Theory for Anisotropic Semiconductors The theory of an anisotropic semiconductor in a strong electric field developed in (1) and which does not adopt the relaxation time approximation is generalized to the case of an arbitrary quasi-elastic lattice scattering mechanism and inter-electronic scattering. The corresponding current-voltage characteristics are calculated.

(1) V.L. GUREVICH and R. KATILYUS, Zh. eksper. teor. Fiz. 49, 1145 (1965).

YU.E. PERLIN, YU.B. ROZENFELD, and B.S. TSUKERBLAT, Kishinev, On the Origin of Extrinsic Absorption and Luminescence Bands of Rare-Earth Activated Crystals A theoretical study is made of the origin of optical bands in doped crystals. The interaction between the external electrons of a localized centre of small radius with optical and acoustic phonons is investigated. It is shown that the existence of localized vibrations can under special conditions lead to a "periodic" intensity distribution in the optical spectrum. The $4f \rightleftarrows 5d$ transitions in Ce^{3+}-doped CaF_2, BeF_2, and SrF_2 crystals are studied, as examples.

B.V. SELYUK, Smolensk, Charge Localization in a Ferroelectric Condenser It is shown that a considerable proportion of the charge in a ferroelectric condenser is not localized on the electrodes but in the region near the surface or in the surface levels of the crystal. This charge localization occurs even under comparatively fast switching conditions. For a low density of surface levels a charged layer with anomalous properties is formed near the crystal surface.

V.P. DOBREGO, Leningrad, Calculation of Illumination Characteristics of Hopping Photoconductivity On the basis of the model of inter-impurity recombination the illumination characteristics of hopping photoconductivity are calculated for weakly compensated samples. Comparison with experimental data supports the validity of the model used.

ZH.I. ALFEROV, V.I. KOROLKOV, and M.K. TRUKAN, Leningrad, Electrical Properties of GaP-GaAs p-n Heterojunctions Measurements are made of the current-voltage and capacity-voltage characteristics of epitaxial p-n heterojunctions of the system GaP-GaAs in the temperature range 77 to 400 °K. The samples are prepared by the gas transport method in a closed system. The frequency dependence of the capacity is studied. The results are discussed on the basis of Anderson's energy band model of a p-n heterojunction (1) assuming the existence of a layer with increased concentration of deep donors near the boundary section in GaAs. Agreement is found between the experimentally determined values and the calculated values of the fundamental parameters of the band model of a step-like heterojunction and the energy separations of the conduction bands (ΔE_c) and the valence bands (ΔE_v).

(1) R.L. ANDERSON, Solid State Electronics 5, 341 (1962).

L.M. SOROKIN, Leningrad, Electron-Microscopic Investigation of the Initial Stage of the Decomposition of the Supersaturated Solid Solution Cu - 2% Be An electron-microscopic study is made of the decomposition of the solid solution Cu - 2% by weight Be in the initial stage (annealing at 200 °C: 5 h, at 250 °C: 1 h). It is found that a particle of the initial precipitate, because of its coherence to the matrix, and because of the elastic anisotropy of the matrix, acts both as stress distortion in a direction perpendicular to the plane of initial precipitation <100> and as <110> shear deformation. The contrast of the pictures observed can be explained on the basis of the kinetic theory of electron scattering. The results are compared with X-ray data for the alloys investigated.

A.M. KOSEVICH, Kharkov, Dispersion Law in the Extrinsic Lattice Vibration Band A theoretical investigation is made of the concentration broadening of localized impuritiy frequencies near the edge of the optical band of the spectrum of an ideal crystal. For particular impurity centres and properties of the optical frequency band, a broad impurity band of frequencies occurs even for low impurity concentrations. The width of the impurity band may be comparable with the value of the frequency gap separating the isolated localized frequency from the edge of the optical band. A dispersion law for impurity vibrations is obtained and analysed.

V.I. KONOVALOV and S.M. RYABCHENKO, Kiev, Some Aspects of the Broadening of EPR Lines in Weak Fields The zero and second moment of the satellites of the EPR absorption curve (obtained using first-order perturbation theory), and the second moment of the envelope curve (which takes into account all transitions to a first approximation) are calculated using a high-temperature approach subject to the experimental condition ν_0 = const. Dipole and exchange interactions are taken into account. The additional contribution to the transition to the fundamental frequency is calculated from the non-secular part of the perturbation. The ratio of the second moment of the envelope curve $\langle\overline{\Delta H^2}\rangle_\Sigma$ to the second moment of the fundamental line $\langle\overline{\Delta H^2}\rangle_1^{(o)}$ for a polycrystalline sample is found to be 4/3 for the experimental condition ν_0 = const. The first moment of the envelope curve with respect to H_0 arising from the contribution of the satellites is calculated. The limits of applicability of the results obtained are discussed and the results are compared with previous relevant data.

B.G. ZHURKIN, N.A. PENIN, and PREM SVARUN, Moscow, Effect of Electron Hopping on the EPR Spectrum of Phosphorus in

Strongly Doped n-Type Silicon An investigation is made of the dependences of the width and form of the EPR lines in strongly doped n-type silicon on the phosphorus concentration, the degree of compensation (with respect to boron), and the temperature. Increasing the phosphorus concentration from 4×10^{17} to 1×10^{18} cm^{-3} and the temperature from 2 to 20 °K causes a narrowing of the lines which have a Lorentz form in the temperature and concentration range considered. A high degree of compensation ($K \approx 0.8$ to 0.9) broadens the line and causes its form at the edge to change from Lorentz-like to Gaussian. The Bohr radius is estimated to be (18 ± 2) Å which agrees with the theoretical value.

A.V. KOGAN, V.D. KULKOV, and L.P. NIKITIN, Leningrad, Hyperfine Interaction Fields of Heavy Elements in Iron The internal field H_{eff} at lutetium nuclei in iron is determined from the anisotropy of radioactive Lu^{177}; its value does not exceed 70 kOe. The internal fields for other elements in the series from Lu to Au in the iron matrix are listed. The observable dependence of H_{eff} on the atomic number cannot be explained on the basis of existing theories.

A.G. ZHILIN and B.S. MONOZON, Leningrad, Quasi-Classical Treatment of the Spectrum of a Hydrogen-Like System in a Strong Magnetic Field The adiabatic method is proposed for the quasi-classical treatment of the spectrum of a hydrogen-like system in a strong magnetic field. The criteria for the applicability of the adiabatic approach are deduced. A qualitative comparison is made of the results obtained with data for magneto-optical absorption in Cu_2O crystals.

V.G. LYAPIN, Kiev, The Origin of the Long-Wavelength "Tail" in the X-Ray Emission Spectra of Diamond-Type Crystals The energy gaps ΔE of the valence bands in Si and Ge are determined from the multi-electron theory of the band structure of diamond-like semiconductors (1). This treatment yields some variation in the position of the $E(\mathbf{k})$ branches of the valence band. For the valence band level with the highest value Γ_{15}, the X-ray emission intensity near the long-wavelength boundary of the fundamental band has the form $I(\xi) \sim \nu^4(\xi - \xi_0)^{1/2}$ for K-spectra as well as for $L_{2,3}$-spectra (I emission intensity, ξ energy of the emitted photon with frequency ν, ξ_0 the long-wavelength boundary of the fundamental band). A relation is deduced which connects the difference between the hole energy levels $E(\Gamma_{15}) - E(\Gamma_{25'})$ with the lattice periodicity and the cyclotron constants. It yields lower boundaries for ΔE in Si and Ge, which are greater than the values for ΔE determined in (2 to 4). Hence it appears that large parts of the emission spectra in Si and Ge which have been related in (2 to 4) to the long-wavelength "tail" belong to the fundamental band and are caused by transitions between X-ray and deep "additional" levels of the valence band.

(1) V.G. LYAPIN and K.B. TOLPYGO, Fiz. tverd. Tela 6, 1158 (1964); 8, 3156 (1966).
(2) B. KERN, Z. Physik 159, 178 (1960).
(3) R.S. CRISP and S.E. WILLIAMS, Phil. Mag. 6, 365 (1961).
(4) D.H. TOMBOULIAN and D.E. BEDO, Phys. Rev. 104, 590 (1956).

V.G. VESELAGO, Moscow, Properties of Substances with Simultaneously Negative Dielectric and Magnetic Susceptibility The fundamental properties of substances with negative dielectric and magnetic susceptibility ((ε) and (μ) respectively) are considered theoretically, treating conducting ferromagnetics as examples. It is shown that for $\varepsilon < 0$ and $\mu < 0$ the refractive index (relative to vacuo) is negative.

M.A. SAVCHENKO and K.N. STEPANOV, Kharkov, Theory of Coupled Electromagnetic and Spin Waves in Helicoidal Magnetic Structures A theoretical investigation is made of the coupling of spin and electromagnetic waves in helicoidal magnetic structures in the ferromagnetic resonance region in the absence of an external magnetic field. It is shown that the plasma waves and one of the electromagnetic waves do not interact with the spin waves. For the electromagnetic wave which does interact with the spin wave, and for the spin wave itself the frequency corrections far from the point of intersection of the non-interacting branches are of the order of ς, but near the point of intersection are of the order of $\sqrt{\varsigma}$ (where $\varsigma = \frac{g\mu^2}{Sa^3\sqrt{KK'}}$

g Landé factor, μ Bohr magneton, S is the atomic spin, a the lattice constant, and K,K' the anisotropy constants of the crystal; for known substances $\varsigma \approx 10^{-2}$). Expressions for the frequencies of the interacting waves are also obtained in the tight binding range when $\varsigma \gtrsim 1$.

V.V. TIKHONOV, A.V. GOLUBKOV, and I.A. SMIRNOV, Leningrad, Heat Capacity of NdS, LaSe, and LaTe The heat capacity of NdS, LaSe, and LaTe is measured in the temperature range 90 to 390 °K. The Debye temperatures Θ_D of these compounds are determined. From the linear dependence of Θ_D on the average atomic weight observed experimentally, approximate values are calculated for Θ_D and the melting points for the complete series of rare-earth monochalcogenides.

E.A. EGOROV and V.V. ZHIZHENKOV, Leningrad, NMR Study of the Molecular Mobility in Elastically Deformed Oriented Polymers NMR is used to observe changes in the flexibility of macromolecules during the process of elastically reversible tension in the oriented polymers capron, lavsan, polyvinyl alcohol, polyethylene, and polytetrafluorethylene. The experiments show that the micro-Brownian mobility of the polymer chains is retarded under load. This effect is reversible. The dependence of the number of the "mobile" segments on the magnitude of the reversible deformation and temperature is deduced.

A.N. NAUMOV and V.B. PTASHNIK, Leningrad, Methods for Determining the Isotope Effect of the Diffusion in Electrotransport in Solids A discussion is given of graphical and approximate analytical methods for determining the isotope effect of diffusion in electrotransport in solids. Formulae are derived relating the distribution factor of the isotopes determined experimentally with the ratio of the diffusion coefficients of the isotopes.

N.V. VOLKOVA, V.A. LIKHACHEV, V.M. SALMANOV, and I.D. YAROSHETSKII, Leningrad, Kinetics of the Formation and Recovery of Laser-Induced Damage in LiF Single Crystals A systematic investigation is made of the character and microstructure of the damage region, and also of the kinetics of the formation and subsequent recovery of the damage centres which are formed under the action of intensive laser radiation in LiF single crystals.

B.P. ZACHARCHENYA, I.B. RUSANOV, and I.I. TAKHISTOVA, Leningrad, Magneto-Optics of "Tetragonal Centres" in CaF_2-Eu^{3+} Crystals The Zeeman splitting of emission and absorption lines in the spectrum of CaF_2-Eu^{3+} is connected with centres of different symmetry (cubic, tetragonal, orthorhombic) which are formed by the heterovalent substitution of the cation by the ion Eu^{3+}. The regularities in the Zeeman splitting of "tetragonal" lines observed experimentally are described. A theoretical analysis of these regularities is presented which relates these lines to the magnetic-dipole transitions $^1\Gamma_1 \longleftrightarrow {}^2\Gamma_5$ in the field C_{4v}. From the concentration dependence of the intensity of the lines and their connection with cubic and orthorhombic centres found by a magneto-optical experiment, it is concluded that the excess charges at the centres of different symmetry can be compensated by the same doubly negatively charged lattice defect.

M.I. IGLITSYN, E.G. PEL, L.YA. PERVOVA, and V.I. FISTUL, Moscow, Instability of the Electron-Hole Plasma of a Semiconductor due to Non-Linear Current-Voltage Characteristics It is shown that conductivity oscillations can be excited in a crystal in which the electron and hole components of the conductivity are non-linear (for example, as result of field dependence of the recombination cross section). This instability has a resonant character. The conditions for the existence of these oscillations are formulated, and expressions are obtained for the frequency of the oscillations and for the critical field. The period of oscillation and the value of the critical field observed experimentally in Au-doped p-type germanium agree with the calculated values.

N.V. FOMIN, Leningrad, Theory of the Electronic Effect in Thermal Expansion Calculations are made of the Grüneisen coefficients and the non-linear correction of Hooke's law via the anharmonicity parameters of third order. The increase of the Grüneisen coefficient observed in metals with decreasing temperature in the range $T \lesssim \Theta_D$ can be explained on the basis of electron-phonon coupling (Θ_D Debye temperature).

YA.I. DUTCHAK, O.P. STETSKIV, and V.P. OSYPENKO, Lvov, Thermo E.M.F. of Some Liquid Metals and Alloys A study is made of the thermo e.m.f. of six metals during the transition through the crystallization point. Some liquid alloys of the systems In-Sn and In-Pb are also investigated.

V.F. ELESIN and E.A. MANYKIN, Moscow, Instabilities in Semiconductors with Absolute Negative Conductivity A theoretical investigation is made of the high-frequency properties of the absolute negative conductivity previously assumed in (1). The conditions for electrical instability of the semiconductor in the state with negative conductivity are obtained. It is shown that electrical domains may be formed in the absence of an external electric field in a homogeneous semiconductor.

(1) V.F. ELESIN and E.A. MANYKIN, Pisma Zh. eksper. teor. Fiz. 3, 26 (1966).

A.I. PILSHCHIKOV and V.I. DUDKIN, Moscow, Ferromagnetic Resonance in the Case of Domain Structures A theoretical and experimental study is made of the effect of domain structures on the ferromagnetic resonance in ferrite single crystals with cubic structure and negative anisotropy constant. A constant external magnetic field H is applied in the [011] direction. The domains are considered theoretically as plane layers perpendicular to the (100) plane and inclined at an arbitrary angle α to the direction of H. For the case $\alpha \neq 0$ and $\alpha \neq \frac{\pi}{2}$ coupled vibrations excited both by longitudinal and transverse high-frequency fields can occur. The experimental dependence of the resonance frequencies on the constant external magnetic field in Mg-Mn ferrite single crystals agrees with the theoretical dependence for a specific range of constant magnetic fields. Also there is experimental confirmation for the occurrence of coupled vibrations excited by longitudinal as well as by transverse fields.

V.P. SAKUN, Moscow, EPR Line Shape in the Case of Interactions which are Quadratic in Spin Using the equations for the spin density matrix, the shape of the EPR curves for spin 3/2 is calculated for a magnetically diluted solid. A small constant term quadratic in spin is included in the lattice interaction. The case is considered in which the resonance curve consists of unresolved lines.

Short Notes published in "Fizika Tverdogo Tela"
Vol. 8, No. 12

E.A. MAKAROV, Novosibirsk
Piezoresistance Constant π_{44} in n-Type Silicon

K.S. ALEKSANDROV, L.M. RESHCHIKOVA, and B.V. BEZNOSIKOV, Krasnoyarsk
Anomaly of Elastic Properties in $KMnF_3$

B.YA. YAZYTSKII and YU.M. POPLAVKO, Kiev
High-Frequency Dispersion of $BaTiO_3$ below the Curie Point

L.A. GOLUBEVA and T.I. NIKITINSKAYA, Leningrad
Effect of X-Ray Irradiation on the Electrical Properties of Fluorite

I.KH. AKOPYAN, S.S. GRIGORYAN, and A.S. YAKOVLEV, Leningrad
Luminescence of $ZnSiP_2$ Crystals

L.G. ELISEENKO, V.N. SHCHEMELEV, and M.A. RUMSH, Leningrad
Relation between the Directional and Diffusion Part of the Path of Kilovolt Electrons in a Solid

L.G. ELISEENKO, V.N. SHCHEMELEV, and M.A. RUMSH, Leningrad
Absorption of Kilovolt Electrons by Solids

V.S. POSTNIKOV, YU.M. ELKIN, and S.I. MESHKOV, Voronezh
Internal Friction in F.C.C. Metals during Tension

I.V. IOFFE, Leningrad
Method for Determining Piezoelectric Characteristics of Crystals

M.V. KURIK, Kiev
Correction to the Paper: M.S. BRODIN, M.V. KURIK, and S.P. YURTSENYUK, Optical Absorption and Energy Band Structure of CdS-CdSe Crystals

M.L. MEILMAN,
EPR of Gd^{3+} in Synthetic $PbWO_4$

I. N. KURKIN and L.YA. SHEKUN, Kazan
EPR of Nd^{3+} in $CdMoO_4$ Scheelites

V.N. BOGOMOLOV, Leningrad
Effect of Hydrostatic Pressure on the Electrical Conductivity and Hall Effect in Rutile

V.A. IGNATYUK and F.T. NOVIK, Leningrad
High-Voltage Photo E.M.F. in Epitaxial ZnTe Films

V.M. GENKIN, G.M. GENKIN, and V.M. FAIN, Gorkii
Theory of Non-Linear Properties of Ferromagnetics

N.N. KRAINIK, S.N. POPOV, and I.E. MYLNIKOVA, Leningrad
Observation of Nuclear Quadrupole Resonance in Ferroelectric SbSJ

G.B. STUCHINSKII,
Temperature Dependence of the Secondary Electron Emission in MgO

V.A. SAAKYAN and I.A. SMIRNOV, Leningrad
Thermal Conductivity in Semiconductors with Two Types of Carrier

V.S. BOIKO and E.P. FELDMAN, Kharkov
Elastic Twinning near a Grain Boundary

A.A. GALKIN, A.D. PROKHOROV, and G.A. TSINTSADZE, Donetsk
Hyperfine Structure of the EPR of Co^{2+} in $ZnWO_4$

I.A. KARPOVICH and B.N. ZVONKOV, Gorkii
Saturation of the Photocurrent in Strong Fields in CdSe Layers

V.A. IGNATCHENKO and YU.A. KUDENKO, Krasnoyarsk
Inversion of the Nuclear Magnetization during Domain Boundary Motion

M.M. ZARIPOV, V.S. KROPOTOV, L.D. LIVANOVA, A.L. STOLOV, and ZH.S. YAKOVLEVA, Kazan
EPR and Optical Spectrum of Cr^{3+} in MgF_2

A.A. ANDREEV and A.P. REGEL, Leningrad
Hall Coefficient in the Liquid Metals Hg, Ga, In, Sn, and in the Alloys In_2Bi and Hg-Sn

L.V. VOLODKO, M.T. LAPPO, V.M. LOMAKO, and V.D. TKACHEV, Minsk
Modulation of Reflected Light from Fast-Neutron Irradiated Silicon p-n Junctions

A.A. BUGAI, P.T. LEVKOVSKII, V.M. MAKSIMENKO, L.I. POTKIN, and A.B. ROITSIN, Kiev
Effect of an External Constant Electric Field on the EPR Spectrum of Cr^{3+} in $MgWO_4$

G.M. ZVERER, L.V. MAKARENKO, and A.I. SMIRNOV
EPR of Ce^{3+} and Nd^{3+} in $SrMoO_4$ Single Crystals

A.S. VOLKOV, V.V. GALAVANOV, B.V. IZVOZCHIKOV, and I.A. TAKSAMI, Leningrad
Effect of Hydrostatic Pressure on Current Carrier Lifetime in p-Type InSb

G.M. NEDLIN and R.KH. SHAPIRO, Leningrad
Excitation of Magneto-Elastic Vibrations in a Ferromagnetic by an Electromagnetic Pulse

B.T. KOLOMIETS, V.G. ROMANOV, and P.K. KHODOSEVICH, Leningrad
Photoconductivity of Selenium at Low Temperatures under the Action of Ultraviolet and X-Rays

A.R. SHULMAN and N.P. BAZHANOVA, Leningrad
Energy Spectrum of Secondary Electrons from Tungsten Covered with BaO for Low Energies of the Primary Electrons

I.S. GORBAN and V.M. KOSAREV, Kiev
Radiative and Non-Radiative Recombination of Non-Equilibrium Carriers in GaP Crystals

G.B. RAIS, M.I. BROMBERG, and E.M. GUGEL, Kharkov
Determination of the Stored Energy of Plastic Deformation in Cu from the Difference in the Electrolytic Potentials

A. NASYROV, B.M. LEBED, and G.A. SMOLENSKII, Leningrad
Peculiarities in the Ferro-Acoustic Resonance in the Solid Solution Yttrium Ferrite-Gallate

I.A. ABROYAN and L.A. TSEKHNOVICHER, Leningrad
Electron-Excitation in CdS Crystals by Hydrogen Ions

B.A. VISHNYAKOV and K.A. OSIPOV, Moscow
Investigation of the Precipitation of MoC Films from $Mo(CO)_6$ under the Action of an Electron Beam

L.N. PELIKH and V.V. EREMENKO, Kharkov
Quantum Oscillations of the Chemical Potential of Bismuth in Crossed Electric and Magnetic Fields

YU.P. KHUKHRYANSKII, M.I. SHCHEVELEV, and B.N. POPOVICH, Voronezh
Observation of Dislocations in Silicon

A.S. GERSHUN, L.A. SYSOEV, and B.L. TIMAN, Kharkov
Investigation of the Charge Formed in CdS Crystals under the Action of an Electric Field

L.S. PALATNIK, YU.V. ZOLOTNITSKII, and A.G. RAVLIK, Kharkov
Coercive Force and Anisotropy Field of Multi-Layer Permalloy Films

V.I. BAIBAKOV, G.S. PADO, and A.A. KARTUSHINA, Mendeleevo
Continuous Amplification Region of Ultrasonics in CdSe

V.M. KORSUN and A.M. NEMCHENKO, Dnepropetrovsk
Electroluminescence during the Diffusion of Cu in ZnS Crystals

B.G. LURE, A.N. MURIN, and I.V. MURIN, Leningrad
Self-Diffusion of Ag in Mixed Crystals of AgCl + $CdCl_2$

V.V. MAKAROV and N.N. PETROV, Leningrad
Penetration of Light Atomic and Molecular Ions in SiC Single Crystals

M.I. BARNIK, I.M. TARASOVA, S.G. YUDIN, N.I. ASTAFEV, and B.I. BEGLOV
Photon Cross Section of Zn and Hg Atoms in Ge

V.M. MUZHDABA and S.S. SHALYT, Leningrad
Carrier Scattering in PbTe (Analysis of the Wiedemann-Franz Ratio)

A.G. ARONOV and G.E. PIKUS, Leningrad
Theory of Tunnel Electroluminescence in Semiconductors

19. Magnetic Resonance in Solids

20. Optical Properties of Solids
    - 20.1 Spectra, Optical Constants (Colour Centres see 10.2; X-Ray Spectra see 20)
    - 20.2 Lasers
    - 20.3 Luminescence of Solids (X-Ray Spectra see 20)

21. Metals. Alloys
    - 21.1 Transition Metals (Group Sc to Zn) and their Alloys
        - 21.1.1 Fe and Fe Alloys
    - 21.2 Alkali Metals
    - 21.3 Alkali-Earth Metals
    - 21.4 Rare-Earth Metals
    - 21.5 Actinides
    - 21.6 Noble Metals
    - 21.7 Semi-Metals (Semiconducting Alloys see 22.7)

22. Semiconductors and Ionic Crystals
    - 22.1 Elements
        - 22.1.1 Germanium
        - 22.1.2 Silicon
        - 22.1.3 Group VI Elements
    - 22.2 III—V Compounds
        - 22.2.1 Arsenides
        - 22.2.2 Phosphides
        - 22.2.3 Antimonides
    - 22.3 Carbides
    - 22.4 II—VI Compounds
        - 22.4.1 Sulphides
        - 22.4.2 Selenides
        - 22.4.3 Tellurides
    - 22.5 Halides
        - 22.5.1 Silver Halides
        - 22.5.2 Alkali Halides
        - 22.5.3 Alkali-Earth Halides
    - 22.6 Oxides (CdO, ZnO see 22.4; Spinels see 18.3 or 14.4.2)
    - 22.7 Semiconducting Intermetallic Compounds
    - 22.8 Tertiary and Higher Compounds (Carbonates, Phosphates, Silicates and Mica see 22)
    - 22.9 Organic Semiconductors

23. Solidified Gases

Attention

The preceding classification scheme is closely akin to the scheme used by Physics Abstracts to classify the various fields of solid state physics. A paper will either be classified under one ore more of the sub-sections or, if none of these are relevant, it will just appear under the general section heading (i.e. sections and subsections are equivalent in this sense).

The *subject classification* given at the head of each paper begins with the main subject matter of the paper, and this is followed, if necessary, by figures referring to subsidiary subject matter and the type of solid investigated.

The papers published in this issue are listed in the order of their subject classification at the end of the table of contents.

Manuscripts and letters should be sent to

physica status solidi, 102 Berlin, Neue Schönhauser Str. 20
or Internationaler Buchversand GmbH, 1 Berlin 19, Eschenallee 21.

The object of this journal is to publish rapidly important and topical results of fundamental research in the field of solid state physics. Each issue will be published on the first day of the month. Items published will fall into the following categories:

1. Review articles written by arrangement with the editors.
2. Original papers which have not been submitted for publication elsewhere, after being reviewed. (The latest date for acceptance of contributions is the 8th day of each month, 50 days before the delivery of the journal.)
3. Short notes published in the off-set part of the journal. (The latest date for acceptance of manuscripts for this part is the 12th day of each month, 16 days before the delivery of the journal.)
4. Pre-printed titles and abstracts of papers to be published in this or in the Soviet journal „Физика Твердого Тела" (Fizika Tverdogo Tela), the latter translated into English.

Editorial Information

Manuscripts should be submitted preferentially in English. Besides, papers in German, French, and Russian are admitted. Authors are asked to observe the following rules:

1. Manuscripts have to be written as concise as possible. As the extreme limit, 20 and 3 pages (28 lines per page; abstracts, tables, references, captions of figures included) are admitted for original papers and short notes, respectively. All manuscripts should be typewritten on one side olny. Two copies are required.
2. The author's name, initials, and institution should be given.
3. An abstract of about ten lines in English and a second one in German, French, or Russian are required for original papers.
4. Drawings should be enclosed on separate sheets of paper; photographs should be submitted, on white paper, only for review articles and original papers. Duplicates are required.
5. Captions for the figures should be submitted on separate sheets of paper, and special instructions with regard to the size of illustrations should be given on these.
6. References should be numbered in order of appearance and should be given, on a separate sheet, in the form used in this journal.
7. Units, symbols, etc. should be used in accordance with the recommendations given by the IUPAP.
8. Each paper should contain a proposal for the subject classification according to the classification scheme appearing on the second and third cover of each issue.

The authors will be sent two galley proofs of review articles and original papers. Proof reading of short notes and pre-printed titles and abstracts will be done by the editorial staff.

For each original paper, and short note, a total of 75 reprints is available free of charge. The number of reprints free of charge per review article is 25. In addition, review articles will be remunerated.

FSC
www.fsc.org
MIX
Papier | Fördert
gute Waldnutzung
FSC® C083411

Kolibri 360 GmbH
Ferdinand-Jühlke-Straße 7
99095 Erfurt, Deutschland
produktsicherheit@kolibri360.de